Lecture Notes in Computer Science 2755

Edited by G. Goos, J. Hartmanis, and J. van Leeuwen

Springer
Berlin
Heidelberg
New York
Hong Kong
London
Milan
Paris
Tokyo

Martin Wirsing   Dirk Pattinson
Rolf Hennicker (Eds.)

Recent Trends in Algebraic Development Techniques

16th International Workshop, WADT 2002
Frauenchiemsee, Germany, September 24-27, 2002
Revised Selected Papers

Springer

Series Editors

Gerhard Goos, Karlsruhe University, Germany
Juris Hartmanis, Cornell University, NY, USA
Jan van Leeuwen, Utrecht University, The Netherlands

Volume Editors

Martin Wirsing
Dirk Pattinson
Rolf Hennicker
Ludwig-Maximilians-Universität München
Institut für Informatik
Oettingenstr. 67, 80538 München, Germany
E-mail: {wirsing,pattinso, hennicke}@informatik.uni-muenchen.de

Cataloging-in-Publication Data applied for

A catalog record for this book is available from the Library of Congress.

Bibliographic information published by Die Deutsche Bibliothek
Die Deutsche Bibliothek lists this publication in the Deutsche Nationalbibliografie;
detailed bibliographic data is available in the Internet at <http://dnb.ddb.de>.

CR Subject Classification (1998): F.3.1, F.4, D.2.1, I.1

ISSN 0302-9743
ISBN 3-540-20537-3 Springer-Verlag Berlin Heidelberg New York

Springer-Verlag is a part of Springer Science+Business Media

springeronline.com

© Springer-Verlag Berlin Heidelberg 2003
Printed in Germany

Typesetting: Camera-ready by author, data conversion by Olgun Computergrafik
Printed on acid-free paper   SPIN: 10971710   06/3142   5 4 3 2 1 0

Preface

This volume contains selected papers from WADT 2002, the 16th International Workshop on Algebraic Development Techniques. Like its predecessors, WADT 2002 focussed on the algebraic approach to the specification and development of systems, an area that was born around the algebraic specification of abstract data types and encompasses today the formal design of software systems, new specification frameworks and a wide range of application areas.

WADT 2002 took place at the convent of Frauenchiemsee, Germany, on September 24–27, 2002, and was organized by Rolf Hennicker, Dirk Pattinson and Martin Wirsing. The workshop also included a special track on Formalism, Logic, Institution – Relating, Translating and Structuring (FLIRTS), and three satellite events: a meeting of the IFIP Working Group 1.3 on Foundations of System Specification, a Workshop on Global Computing organized by the AGILE project, and a Workshop on Multimedia Instruction in Safe and Secure Systems (MMISS).

The program consisted of invited talks by Egidio Astesiano (Genoa, Italy), Andrew Gordon (Cambridge, UK), and Jan Rutten (Amsterdam, The Netherlands), and 44 presentations describing ongoing research on main topics of the workshop: formal methods for system development, specification languages and methods, systems and techniques for reasoning about specifications, specification development systems, methods and techniques for concurrent, distributed and mobile systems, and algebraic and co-algebraic foundations.

The steering committee of WADT consisted of Michel Bidoit, Hans-Jöerg Kreowski, Peter Mosses, Fernando Orejas, Francesco Parisi-Presicce, Donald Sannella, and Andrzej Tarlecki. With the help of Till Mossakowski for the FLIRTS track, and the local organizers, several papers were selected, and the authors were invited to submit a full paper for possible inclusion in the workshop proceedings. All submissions underwent a careful refereeing process. The selection committee then made the final decisions.

This volume contains the final versions of the 20 contributions that were accepted. It contains also two invited papers, those of Egidio Astesiano and Jan Rutten, and two invited presentations of the AGILE project on Software Architecture for Global Computing and of the MMISS project on Multimedia Instruction in Safe and Secure Systems. We are also proud to present in this volume the cantata "Zero, Connected, Empty" which was specially written for the banquet of the workshop. This cantata was composed by Ryoko Goguen with words by Joseph Goguen; an essay by these authors explains some of the thoughts behind the cantata.

We are extremely grateful to all the referees who helped the selection committee in reviewing the submissions: M. Cerioli, R. Diaconescu, J.A. Goguen, B. Hoffmann, M. Hofmann, S. Katsumata, R. Klempien-Hinrichs, B. Konikowska,

A. Kurz, D. Llywelyn, C. Lüth, F. Nickl, P. Ölveczky, C. Oriat, W. Pawlowski, G. Reggio, H. Reichel, M. Roggenbach, G. Rosu, and L. Schröder.

The workshop was jointly organized by IFIP WG 1.3 (Foundations of System Specification), GI Fachgruppe 0.1.7 (Specification and Semantics), and the Graduiertenkolleg "Logic in Computer Science" of LMU Munich and TU Munich. WADT 2002 received generous sponsorship from the following organizations:

DFG (Deutsche Forschungsgemeinschaft),
Institut für Informatik, Ludwig-Maximilians-Universität München
Münchner Universitätsgesellschaft

As organizers of the workshop, we would like to express our deepest thanks to Sister Scholastica of the convent of Frauenchiemsee for hosting the workshop with dedication and care, and for greatly facilitating the innumerable local organization tasks. Hubert Baumeister, Marianne Diem, Anton Fasching, Florian Hacklinger, Piotr Kosiuczenko, Philipp Meier and several other members of the department provided invaluable help throughout the preparation and organization of the workshop. We are grateful to Springer-Verlag for their helpful collaboration and quick publication.

Finally, we thank all workshop participants both for lively discussions and for creating a friendly and warm atmosphere in spite of all the rain!

July 2003   Martin Wirsing, Dirk Pattinson, Rolf Hennicker

Table of Contents

Invited Technical Papers

Invited Non-technical Papers

Contributed Papers

AGILE: Software Architecture for Mobility*

L. Andrade[6], P. Baldan[8], H. Baumeister[1], R. Bruni[2], A. Corradini[2],
R. De Nicola[3], J. L. Fiadeiro[6], F. Gadducci[2], S. Gnesi[4], P. Hoffman[7],
N. Koch[1], P. Kosiuczenko[1], A. Lapadula[3], D. Latella[4], A. Lopes[5],
M. Loreti[3], M. Massink[4], F. Mazzanti[4], U. Montanari[2], C. Oliveira[5],
R. Pugliese[3], A. Tarlecki[7], M. Wermelinger[5], M. Wirsing[1], and A. Zawłocki[7]

[1] Institut für Informatik, Ludwig-Maximilians-Universität München
[2] Dipartimento di Informatica, Università di Pisa
[3] Dipartimento di Sistemi e Informatica, Università di Firenze
[4] Istituto di Scienze e Tecnologie dell'Informazione "A. Faedo" CNR Pisa
[5] Faculdade de Ciências da Universidade de Lisboa
[6] ATX Software SA
[7] Institute of Informatics, Warsaw University
[8] Dipartimento di Informatica, Università di Venezia

Abstract. Architecture-based approaches have been promoted as a means of controlling the complexity of system construction and evolution, in particular for providing systems with the agility required to operate in turbulent environments and to adapt very quickly to changes in the enterprise world. Recent technological advances in communication and distribution have made mobility an additional factor of complexity, one for which current architectural concepts and techniques can be hardly used. The AGILE project is developing an architectural approach in which mobility aspects can be modelled explicitly and mapped on the distribution and communication topology made available at physical levels. The whole approach is developed over a uniform mathematical framework based on graph-oriented techniques that support sound methodological principles, formal analysis, and refinement. This paper describes the AGILE project and some of the results gained during the first project year.

1 Introduction

Architecture-based approaches have been promoted as a means of controlling the complexity of system construction and evolution, in particular for providing systems with the agility that is required to operate in turbulent environments and to adapt very quickly to new business requirements, new design technologies, or even to changes in the enterprise world which, like mergers and acquisitions, require new levels of openness and interoperability. However, the architectural approach offers only a "logical" view of change; it does not take into account

* This research has been partially sponsored by the EC 5th Framework project AGILE (IST-2001-32747) (www.pst.informatik.uni-muenchen.de/projekte/agile).

M. Wirsing, D. Pattinson, and R. Hennicker (Eds.): WADT 2002, LNCS 2755, pp. 1–33, 2003.
© Springer-Verlag Berlin Heidelberg 2003

the properties of the "physical" distribution topology of locations and communication links. It relies on the fact that the individual components can perform the computations that are required to ensure the functionalities specified for their services at the locations in which they are placed, and that the coordination mechanisms put in place through connectors can be made effective across the "wires" that link components in the underlying communication network. Whereas the mobility of computations is a problem that we are becoming to know how to address in the field of "Global Computation", the effects of mobility on coordination are only now being recognised as an additional factor of complexity, one for which current architectural concepts and techniques are not prepared for. As components move across a network, the properties of the wires through which their coordination has to take place change as well, which might make the connectors in place ineffective and require that they be replaced with ones that are compatible with the new topology of distribution. In addition, updates on the communication infrastructure will lead, quite likely, to revisions of the coordination mechanisms in place, for instance to optimise performance.

The AGILE project aims to contribute to the engineering of Global Computation and Coordination Systems. It is funded by EU initiative on "Global Computing". The partners of the AGILE project are Ludwig-Maximilians-Universität München, Università di Pisa, Università di Firenze, Istituto di Scienze e Tecnologie dell'Informazione "A. Faedo" CNR Pisa, ATX Software SA, Faculdade de Ciências da Universidade de Lisboa, and more recently the University of Warsaw and the University of Leicester.

The objective of AGILE is to develop an architectural approach in which mobility aspects can be modelled explicitly as part of the application domain and mapped to the distribution and communication topology made available at physical levels. The whole approach is developed over a uniform mathematical framework based on graph-oriented techniques that support sound methodological principles, formal analysis, and refinement across levels of development. Application areas of AGILE include E-Business, Telecommunications, Wireless Applications, Traffic Control Systems and decision support systems which need to collect global information.

More precisely, AGILE pursues the following three main research topics:

– The development of primitives for explicitly addressing mobility within architectural models. This work is based on CommUnity and its categorical framework [17, 16] supporting software architectures on the basis of the separation between "computation" and "coordination" with an additional dimension for "distribution". Consequently, primitives for the third dimension of "mobility", are developed with which the distribution topology can be explicitly modelled and refined across different levels of abstraction.
– The definition of algebraic models for the underlying evolution processes, relating the reconfiguration of the coordination structure and the mobility of components across the distribution topology. This work is based on graph transformation techniques [7] and Tile Logic [18], and is also the basis for logical analysis of evolution properties as well as for tools for animation and early prototyping.

- The development of an extension of UML for mobility that makes the architectural primitives available to practitioners, together with tools for supporting animation and early prototyping.

The following main aspects are pursued in all three research topics:

- analysis techniques for supporting compositional verification of properties addressing evolution of computation, coordination and distribution, and
- refinement techniques for relating logical modelling levels with the distribution and communication topology available at physical levels

In this paper we give an introduction to the approach of the AGILE project and present some of the results gained during the first project year. In particular, we present extensions of three well-known formalisms to mobility: extensions of CommUnity, of UML, and of Graph Transformation Systems. We also introduce an extension of KLAIM, an experimental kernel programming language specifically designed to model and to program distributed concurrent applications with code mobility.

In the developed extension of CommUnity, primitives were added to CommUnity that support the design of components that can perform computations in different locations and be interconnected to other components over a distributed and mobile network. Patterns of distribution and mobility of components (or groups of components) can be explicitly represented in architectures through a primitive called distribution connector. These patterns include coordination patterns that are location-dependent or even involve the management of the location of the coordinated parties. The semantics of the architectural aspects of this extension were developed over the categorical formalisation already adopted for CommUnity.

The extensions of UML cover class diagrams, sequence diagrams, and activity diagrams. The idea for all of these extensions is similar to the idea of ambients or Maude, in that a mobile object can migrate from one host to another and it can be a host for other mobile objects. It may interact with other objects. Like a place, a mobile object can host other mobile objects, and it can locally communicate and receive messages from objects at other places. Objects can be arbitrarily nested, generalising the limited place-agent nesting of most agent and place languages.

Graph Transformations Systems are used to give an operational semantics to the UML extensions. Object diagrams and the actions of activity diagrams are represented using Typed Hyperedge Replacement Systems. Each action of an activity diagram is modelled by a unique graph transformation rule. We can show that under suitable assumptions a set of graph transformation rules implements correctly the dependencies in an activity diagram. In case stronger synchronisation is necessary, we use a specialisation of the tile model: Synchronised Typed Hyperedge Replacement Systems.

As for KLAIM, although designed for dealing with mobility of processes located over different sites of local area networks, it lacked specific primitives for properly handling *open systems*, namely systems with dynamically evolving

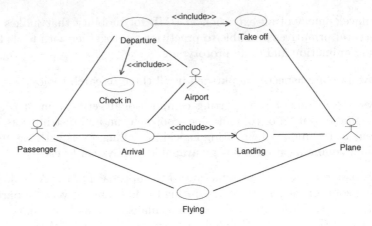

Fig. 1. Use Case diagram.

structures where new nodes can get connected or existing nodes can disconnect. In the paper we present OPENKLAIM, an extension of KLAIM with constructs for explicitly modeling connectivity between network nodes and for handling changes in the network topology.

The paper is organised as follows. The next section introduces the airport case study, which serves as a running example for the material in the following sections. Section 3 presents our UML extensions for modelling mobile systems. Next, in Sect. 4 we present our research on the structural aspects of the architectural approach. The starting point for this research are CommUnity (Sect. 4.1) and KLAIM (Sect. 4.2). Section 5 presents the way graph transformation and its synchronised version (which is a specific instance of the Tile Mode) can be used for the specification of a fragment of the airport case study. Finally, in Sect. 6 a conclusion and an outlook to future work is presented.

2   The Airport Case Study

As an example of mobile objects we consider planes landing and taking off from airports. These planes transport other mobile objects: passengers. In a simplified version of this scenario, departing passengers check in and board the plane. After the plane has arrived at the destination airport, passengers deplane, and claim their luggage. We consider also actions performed by the passengers during the flight, like the consumption of a meal, or making and publishing pictures.

Figure 1 shows these requirements as a UML use case diagram. A use case diagram consists of use cases and actors. The identified actors are the Airport, the Passenger and the Plane. The actor Airport starts the use cases Departure and Arrival, which allow passengers to check in and to deplane, respectively, and allow planes to take off and land, respectively (included use cases TakingOff and Landing). Planes control the use case Flying.

The flow of events of a use case can either be detailed textually or graphically using UML activity diagrams or UML sequence diagrams. The objects involved and their classes are described by class diagrams.

Parts of this case study serve as running example for the modelling techniques presented in the following sections.

3   UML for Global Computation

UML is extended using the extension mechanisms provided by the UML itself, i.e. stereotypes, tagged values and OCL constrains as well as by improvements in the visual representation used in activity and sequence diagrams.

The objective of this research is to develop an extension of the UML to support mobile and distributed system design. This includes linguistic extensions of the UML diagrammatic notations using the extension mechanisms provided by the UML itself, i.e. stereotypes, tagged values, and OCL constrains, as well as introducing new visual representations. Further, the objective includes extensions of the Unified Process and a prototype for simulating and analysing the dynamic behaviour of designs of mobile and distributed systems.

In this section, we give a UML (Unified Modeling Language) [34] specification of the flight. The specification consists of use case diagrams, class diagrams, activity diagrams and sequence diagrams. In the following we show only a part of the solution; a more comprehensive solution can be found in [3, 27].

3.1   Class Diagrams

We model the simplified airport problem domain using UML class diagrams. We identify the following classes: Airport, Plane, Flight, Passenger and Country, where:

- Airport is an origin or destination location.
- Flight is the trip that happens along a particular route on a particular day.
- Plane is the machine that operates a flight.
- Passenger is a person who is waiting for boarding a plane at an airport, is on a plane or has just arrived at the airport.
- Country is a place where an airport is located.

In our extension of class diagrams for mobility, we distinguish between objects and locations which are movable and which are not (cf. [3]). Movable objects are denoted with the UML stereotype «mobile», and objects which can serve as locations are indicated with the stereotype «location». Movable objects and locations are required to have a unique attribute atLoc whose value is the location they are at. We require that the relation given by the atLoc attribute is acyclic. Note that this implies that locations form a hierarchy.

We can only move objects and locations that are movable. In our example, Passenger and Plane are mobile objects. In addition, Plane has a location role, the same as Airport. The problem domain is visualised as a UML class diagram as it is shown in Fig. 2. Note that OCL constraints [34] can be attached to modelling elements to express semantic conditions.

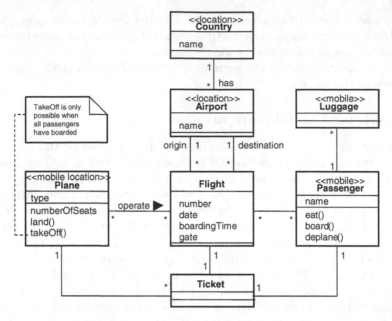

Fig. 2. Class diagram of the airport example.

3.2 Activity Diagrams

In this section we introduce two variants of activity diagrams for modelling mobility. These diagrams were introduced in [3]. The first variant is responsibility centered and uses swimlanes to model who is responsible for an action. The second is location centered and uses the notation of composite objects to visualise the hierarchy of locations.

A typical activity of mobile objects is the change of location, i.e., the change of the atLoc attribute. An object can move from one location to another. In our UML extension, we distinguish these activities by representing them as a stereotyped activity that we call «move». (cf. 4). Not included in the example, but not less important is the stereotyped activity called «clone» that first clones the object to be moved afterwards.

Figure 3 shows the activity diagram corresponding to the Departure use case. Note that the fact that the plane can only take off if the passenger boarded and the luggage is loaded is expressed by the use of joins in the UML activity diagram. Compare this with the CommUnity approach presented in Sec. 4.1. Note that partitions marked with actor's names are defined to organise responsibilities for these activities. Such an activity diagram with partitions gives a responsibility-centred view of the flow of events.

Once the objects are identified, the activity diagrams can be enhanced with object flows, showing relationships among successive loci of control within the interaction. Figure 4 shows such an enhanced activity diagram. The objects are attached to the diagram at the point in the computation at which they are suit-

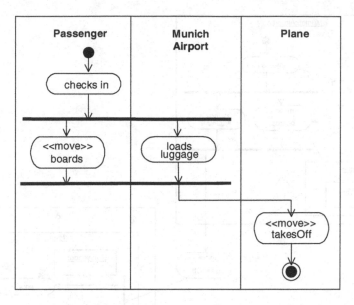

Fig. 3. Activity diagram of the departure scenario.

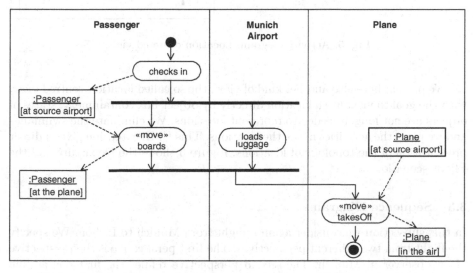

Fig. 4. Activity diagram departure scenario: Responsibility centered view.

able as an input or produced as an output. In our example, the in-going objects
to an activity are very often the same as the outgoing. The corresponding state
of the objects is specified in the square brackets (cf. objects Plane and Passenger
in Fig. 4). This is what we call responsibility-centred view; the responsibilities
are given by swimlanes and the locations are represented indirectly by object
states.

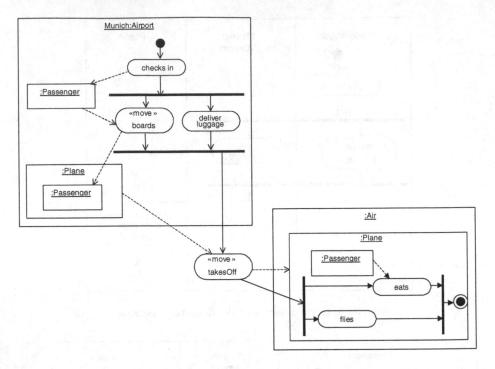

Fig. 5. Activity diagram: Location centered view.

We present here also another kind of view, the so called location-centred view, with the goal to visualise a location directly by object box containment: states of objects are not longer needed to represent locations. We eliminate the swimlanes and we place the activities inside the locations. This UML extension gives a direct presentation of the topology of locations. Figure 5 shows the Departure and the Flight scenario.

3.3 Sequence Diagrams

In this subsection we consider again a flight from Munich to Lisbon. We specify the flight from two different perspectives. The first perspective is the perspective of an observer in Munich. The second perspective refines the first one adding several details. We use here the extension of UML sequence diagrams proposed in [27] for modelling mobile objects. The behaviour of mobile objects is modelled by a generalised version of lifelines which allows us to represent directly the topology of locations within a sequence diagram. For different kinds of actions like creating, entering or leaving a mobile object stereotyped messages are used. This notation provides also a zoom-out, zoom-in facility allowing us to abstract from specification details.

Figure 6 shows a simple story of a passenger x who boards an airplane in Munich airport, flies to Lisbon and publishes a picture in a WAN. The domain

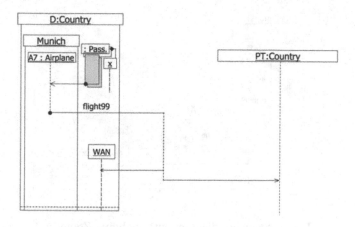

Fig. 6. Sequence diagram with mobility.

model of this sequence diagram is based on the class diagram of Fig. 2 and uses additionally the class Network. The story is described from the perspective of an observer on the German side. The person x together with other passengers enters the airport and then boards the airplane A7. The airplane flies to Lisbon (the flight number is 99), but the only thing the observer can see is that the airplane is airborne, but not what happens inside the airplane nor further details of this flight. The next event which the observer is able to notice is the appearance of a picture in the WAN. To model several passengers (i.e. objects of class Passenger), we use the UML multi-object notation, which allows us to present in a compact way several passengers playing the same role. Person x is distinguished using the composition relationship. The observer does not care about the order in which the passengers board or leave the plain and what they do during the flight. We abstracted here also from the architecture of WAN and the person's position. This simple view shows some of the barriers person x has to cross while flying from Munich to Lisbon.

In the view presented in Fig 7, we show much more details. We show political boundaries which regulate the movement of people and devices, like airplanes, computers and so on. Within those boundaries, there are other boundaries like those protecting airports and single airplanes against intruders. Only people with appropriate passports and tickets may cross those boundaries. Therefore, in our model we make explicit those boundaries and moving across them. The airplane A7 is a very active mobile computing environment, full of people who are talking, working with their laptops, calling their families, making pictures or connecting to Web via phones/modems provided in the airplane. We can see here, what happens inside the airplane during the flight; the jump arrow contains the action box of the airplane A7. Passenger x makes pictures with his digital camera, the pictures are send then to the WAN. As usual, a digital camera does not allow him to send pictures directly to WAN. It is also forbidden to use mobile phones during the flight. Therefore the passenger safes the pictures to

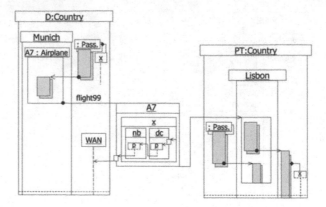

Fig. 7. Sequence diagram with mobility: Zoom in.

his notebook nb, logs into the onboard network and then transmits the pictures to WAN via the onboard network. We abstract here from the structure of the WAN network (indicated by dashed line). Let us point out that the sending of the picture by passenger x is not temporally related to crossing any border like those of Germany, or Munich and so on. The only thing we can say is that it happens between the start of the airplane and its landing. Finally, all the passengers leave the airplane and the airport. The passenger can see that the airplane is boarded by new passengers.

3.4 Statechart Diagrams

In this section we show the use UML statecharts for the design and the specification of the dynamic behaviour of the airport system. A statechart diagram is defined for each class of the model, providing a complete operational description of the behaviour of all the objects of the class. The full system is then represented by a set of class objects. The UML semantics [34, 37, 31, 38] associates to each active object a state machine, and the possible system behaviours are defined by the possible evolutions of these communicating state machines. All the possible system evolutions can be formally represented as a bi-labelled transition system in which the states represent the various system configurations and the edges the possible evolutions of a system configuration. The topology of the system is modelled by an atLoc attribute, associated to each class, which represents its locality. Mobility is realized by all the operations which update the atLoc attribute of an object (the «move» operations).

The verification of the system is done with a prototypal "on-the-fly" model checker (UMC) (cf. [20]) for UML statecharts. On-the-fly verification means intuitively that, when the model checker has to verify the validity of a certain temporal logic formula on one state, it tries to give an answer by observing the internal state properties (e.g. the values of its attributes) and than by checking recursively the validity of the necessary subformulas on the necessary next states.

In this way (depending on the formula) only a fragment of the overall state space might need to be generated and analysed to be able to produce the correct result (cf. [5, 14]). The logic supported by UMC, $\mu ACTL^+$ (cf. [20]) is an extension of the temporal logic $\mu ACTL$, (cf. [13]) which has the full power of μ-calculus (cf. [28]). This logic allows both to specify the basic properties that a state should satisfy, and to combine these basic predicates with advanced logic or temporal operators. More precisely the syntax of $\mu ACTL^+$ is given by the following syntax where the χ formulae represent evolution predicates over, signal events sent to a target object (here square parenthesis are used to denote optional parts):

$$\chi ::= true \mid [target.]event[(args)] \mid \neg\chi \mid \chi \wedge \chi$$

$$\phi ::= true \mid \phi \wedge \phi \mid \neg\phi \mid assert(VAR = value) \mid EX\tau\phi \mid EX\chi\phi \mid EF\phi \mid \mu Y\phi(Y) \mid Y^{*}$$

where Y ranges over a set of variables, *state formulae* are ranged over by ϕ, $EX\chi$ is the *indexed existential next* operator and EF is the *eventually* operator.

Several useful derived modalities can be defined, starting from the basic ones. In particular, we will write $AG\phi$ for $\neg EF\neg\phi$, and $\nu Y.\phi(Y)$ for $\neg\mu Y.\neg\phi(\neg Y)$; ν is called the maximal fixpoint operator.

The formal semantic of $\mu ACTL^+$ is given over bi-labelled transition systems. Informally, a formula is true on an LTS, if the sequence of actions of the LTS verifies what the formula states. We hence say that the basic predicate $assert(VAR = value)$ is true if and only if in the current configuration the attribute VAR has value equal to $value$. The formula $EX\chi\phi$ holds if there is a successor of the current configuration which is reachable with an action satisfying χ and in which the formula ϕ holds. The formula $AG\phi$, illustrates the use of the "forall" temporal operator and holds if and only if the formula ϕ holds in all the configurations reachable from the current state.

Following the above syntax we will write using $\mu ACTL^+$ formulae such as:

$$EX \ \{Chart.my\_event\} \ true$$

that means: in the current configuration the system can perform an evolution in which a state machine sends the signal *myevent* to the state machine Chart. Or the formula:

$$AG \ ((EX \ \{my\_event\} \ true) \rightarrow assert(object.attribute = v))$$

meaning that the signal *myevent* can be sent, only when the object attribute has value v.

Coming back to the airport example, let us consider an extremely simplified version of the system composed of two airports, two passengers (one at each airport), and one plane. The plane is supposed to carry exactly one passenger and flies (if it has passengers) between the two airports. Departing passengers try to check in at the airport and than board the plane. We contemplate only one observable action performed by the passengers during the flight, namely the consumption of a meal. The complete dynamic behaviour of the objects of classes Passenger, Airport and Plane, is shown in Fig. 8 and Fig. 9, in the form of statecharts diagrams.

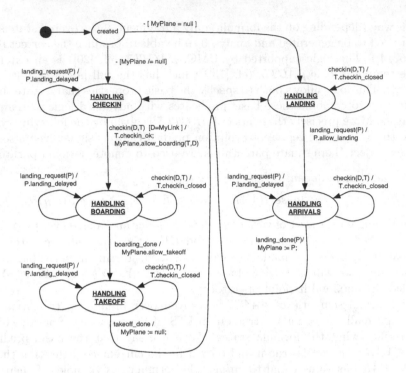

Fig. 8. Airport statemachine.

The initial deployment of the system is defined by the following declarations

object	class	initial values for attributes
Airport1	Airport	MyLink $\Rightarrow$ Airport2, MyPlane $\Rightarrow$ Plane1
Airport2	Airport	MyLink $\Rightarrow$ Airport1
Traveler1	Passenger	atLoc $\Rightarrow$ Airport1, Destination $\Rightarrow$ Airport2
Traveler2	Passenger	atLoc $\Rightarrow$ Airport2, Destination $\Rightarrow$ Airport1
Plane1	Plane	atLoc $\Rightarrow$ Airport1

An example of property which can be verified over this system is the following: It is always true that Traveler1 can eat only while he/she is flying on Plane1. This property can be written in $\mu ACTL^+$ as:

$AG~((EX~\{eating(Traveler1)\}~true)~\rightarrow$

$\qquad (assert~(Traveler1.atLoc = Plane1)~\&~assert(Plane1.atLoc = null)))$

We wish to point out that that the development activity of UMC is still in progress and we have reported here some preliminary results on its application to the airport case study. Indeed several aspects of UML statcharts are not currently supported (e.g. the execution of "synchronous call" operations, the use of "deferred events", the use of "history states"), and the logic itself needs

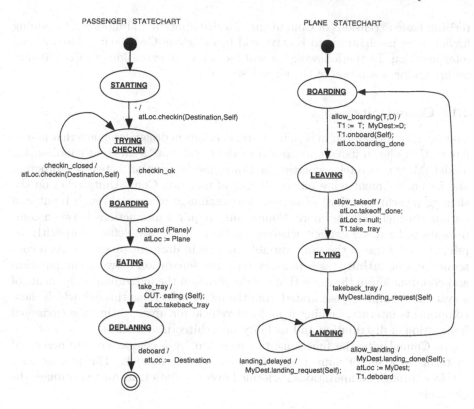

Fig. 9. Plane and passenger statemachines.

to be better investigated, (e.g. its relation with localities). Work in this direction is planned in the next future.

4   The Structural Aspects of the Architectural Approach

The goal of the research on architectures is to develop the structural aspects of the architectural approach to mobility, including semantic primitives, categorical semantics, refinement mechanisms, and a toolbox of connectors and operations, as well as modularisation and structuring facilities for the systems considered and their specifications. The starting point for this research are three complementary formalisms: the parallel program design language CommUnity [17] as a platform in which the separation between "computation" and "coordination" has been achieved; the language KLAIM [11] as a programming language with appropriate coordination mechanisms that permit negotiating the cooperation activities of mobile components, services and resources; the specification language CASL [1] as a means for providing architectural specification and verification mechanisms. During the first year of the project, the grounds for the integration of distribution and mobility in architectures were set by designing and mathematically characterising basic extensions of CommUnity for distribution and mobility, by adding higher-order mechanisms to KLAIM, and by enriching CASL with observational interpretation. In the following we will focus on the extensions to CommUnity by presenting a solution of the airport scenario.

4.1   CommUnity

CommUnity, introduced in [17], is a parallel program design language that is similar to Unity [6] in its computational model but adopts a different coordination model. More concretely, whereas, in Unity, the interaction between a program and its environment relies on the sharing of memory, CommUnity relies on the sharing (synchronisation) of actions and exchange of data through input and output channels. Furthermore, CommUnity requires interactions between components to be made explicit whereas, in Unity, these are defined implicitly by relying on the use of the same variables names in different programs. As a consequence, CommUnity takes to an extreme the separation between computation and coordination in the sense that the definition of the individual components of a system is completely separated from the interconnections through which these components interact, making it an ideal vehicle for investigating the envisaged integration of distribution and mobility in architectural models.

In CommUnity the functionalities provided by a component are described in terms of a set of named actions and a set of channels. The actions offer services through computations performed over the data transmitted through the channels.

Channels. In a component design channels can be declared as input, output or private. Private channels model internal communication. Input channels are used for reading data from the environment of the component. The component has no control on the values that are made available in input channels. Moreover, reading a value from an input channel does not consume it: the value remains available until the environment decides to replace it. Output and private channels are controlled locally by the component, i.e. the values that, at any given moment, are available on these channels cannot be modified by the environment. Output channels allow the environment to read data produced by the component. Private channels support internal activity that does not involve the environment in any way. Each channel is typed with the sort of values that it can transmit.

Actions. The named actions can be declared either as private or shared. Private actions represent internal computations in the sense that their execution is uniquely under the control of the component. Shared actions offer services to other components and represent possible interactions between the component and the environment, meaning that their execution is also under the control of the environment. The significance of naming actions will become obvious below; the idea is to provide points of rendezvous at which components can synchronise.

Space of Mobility. We adopt an explicit representation of the space within which mobility takes place, but we do not assume any fixed notion of space. This is achieved by considering that space is constituted by the set of possible values of a special data type Loc included in the fixed data type specification over which components are designed.

The data sort Loc models the positions of the space in a way that is considered to be adequate for the particular application domain in which the system is or will be embedded. The only requirement that we make is for a special location ⊥ to be distinguished whose role will be discussed further below. In this way, CommUnity can remain independent of any specific notion of space and, hence, be used for designing systems with different kinds of mobility. For instance, in physical mobility, the space is, typically, the surface of the earth, represented through a set of GPS coordinates. In some kinds of logical mobility, space is formed by IP addresses. Other notions of space can be modelled, namely multi-dimensional spaces, allowing us to accommodate richer perspectives on mobility such as the ones that result from combinations of logical and physical mobility, or logical mobility with security concerns.

Unit of Mobility. In components that are location-aware, we make explicit how their constituents are mapped to the positions of the fixed space. Mobility is then associated with the change of positions. By constituents we mean channels, actions, or any group of these. This means that the unit of mobility — the smallest constituent of a system that is allowed to move — is fine-grained and different from the unit of execution.

The constituents of a component are mapped to the positions of the space through location variables. These variables (locations, for short) can be regarded as references to the position of a group of constituents of a component that are permanently colocated. In a component design, locations can be declared as input or output in the same way as channels but are all typed with sort Loc. Input locations are read from the environment and cannot be modified by the component. Hence, if l is an input location, the movement of any constituent located at l is under the control of the environment. Output locations can only be modified locally but can be read by the environment. Hence, if l is an output location, the movement of any constituent located at l is under the control of the component.

Each local channel x of a design is associated with a location l. We make this assignment explicit by writing $x@l$. At every given state, the value of l indicates the position of the space where the values of x are made available. A modification in the value of l entails the movement of x as well as of the other channels and actions located there. Input channels are located at a special output location whose value is invariant and given by ⊥. The intuition is that this location variable is a non-commitment to any particular location. The idea is that input channels will be assigned a location when connected with a specific output channel of some other component of the system.

Each action name g is associated with a set of locations including λ, meaning that the execution of action g is distributed over those locations. In other words, the execution of g consists of the synchronous execution of a guarded command in each of these locations.

Airport Example. We consider a system that is required to control the check-in and boarding of passengers as well as the take-off of planes at airports. In the case of flights with stops, the system should also control the boarding and deplane of passengers during the intermediary stops. The design solution we shall adopt distributes the system over hosts at airports and planes and comprises mobile agents moving from host to host. Moreover, some of the hosts are themselves mobile. Flights, seats, airports and planes identifiers are modelled by data types.

```
ArpId              % airports identifiers
PlId               % plane identifiers
Flight             % flight info
   src: Flight->ArpId            %source of flights
   dest: Flight->ArpId           %destination of flights
   next: ArpId*Flight->ArpId     % next stop relationship of flights
```

We need a bi-dimensional space in order to model (1) the physical movement of planes and, consequently, the movement of the hosts they hold, and (2) code mobility. We define the data types Phy and Host to model these two dimensions. Locations consist of a physical location followed by a logical one — Phy.Host. For simplicity, we consider that airports and planes are associated with single hosts.

```
Host               % logical dimension
   ahost:ArpId->Host
   phost:PlId-> Host
Phy                % physical dimension
   = ArpId+{air}
Loc                % locations
   = Phy.Host
   ph:Loc->Phy       % 1ˢᵗ projection
   host:Loc->Host    % 2ⁿᵈ projection
```

In the envisaged airport system, we may easily identify two component types — passenger and plane; both have a dynamic set of instances in the running system. Passengers have a seat in a given flight and are involved in activities such as check-in, boarding and exiting the plane. Planes operate flights, transporting luggage and passengers. They take off and land, possibly more than once.

```
design  passenger is
inloc     1
prv    s@l: [0..2], seat@l: StId, fl@l: Flight
do     checkin@l: [ s=0 → s:=1 ]
[]     boards@l: [ s=1 → s:=2 ]
[]     leaves@l: [ s=2 → s:=1 ]

design  plane is
outloc    1
out    fl@l: Flight,
prv    s@l: [0..3], id@l: PlId, a@l:AirId
do     load_lug@l: [ s=0→ s:=1 ]
[]     takeoff@l: [ s=1 ∧ ph(l)≠dest(fl) → s:=2 || l:=air.host(l) || a:=next(a,fl)]
[]     lands@l: [ s=2 → s:=1 ||l:=a.host(l) ]
[]     unload_lug@l: [ s=1 ∧ ph(l)=dest(fl) → s:=3 ]
```

Planes offer one output channel so that their behaviour can be coordinated according to the flights they operate. Whereas planes have output locations, passengers have input locations: this is because planes control their own mobility whereas passenger movement is determined by the environment, namely the planes that they board. It remains to define the coordination of the activities of planes and passengers at departure, namely the fact that a plane can take off only when all passengers that checked-in are on board. In CommUnity, the mechanisms through which coordination between system components is achieved can be completely externalised from the component programs and modelled explicitly as first-class citizens. The global property of the airport system just described can be achieved by interconnecting a plane and each of its passengers through a connector that ensures that the action takesoff of plane cannot precede the action boards of passenger. This coordination activity can be established by interconnecting plane and passenger to a scheduler: a program seq with two actions ac1 and ac2 that have to be executed in order. The required interconnection is expressed through the following diagram

connector departure(passenger,plane) is

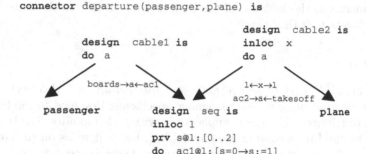

This connector type ensures, for the instances of planes and passengers to which it is applied, that the plane takes off only when the passenger is on board (cf. 3).

The physical presence of a passenger in a check-in counter has to give rise to the creation of an instance of passenger somewhere. Recall that the location of passenger was defined to be controlled by the environment but, so far, we have not specified by who and how. We opt for a solution where the instances of passenger are mobile agents. They are initially placed on the host of the source airport but boarding triggers their migration to the host of the corresponding plane. The required pattern of distribution and mobility of passenger can be regarded as part of the necessary coordination between the passenger and the corresponding plane in the system. In fact, it can be completely externalised from the component design and modelled explicitly as a first-class citizen through a binary distribution connector. The passenger and the corresponding plane have to be interconnected through a program driver as shown in the following diagram.

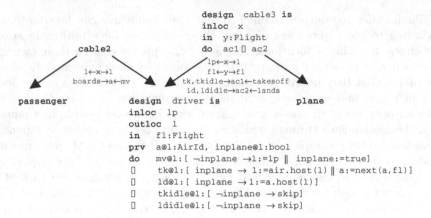

```
                              design  cable3 is
                              inloc  x
                              in  y:Flight
            cable2            do  ac1[] ac2
                                  1p←x→1
        1←x→1                     f1←y→f1
       boards→a←mv         tk,tkidle→ac1←takesoff
                             1d,1didle→ac2←lands
    passenger          design  driver is           plane
                       inloc  1p
                       outloc  1
                       in   f1:Flight
                       prv  a@1:AirId, inplane@1:bool
                       do   mv@1:[ ¬inplane →1:=1p ‖ inplane:=true]
                       []     tk@1:[ inplane → 1:=air.host(1) ‖ a:=next(a,f1)]
                       []     1d@1:[ inplane →1:=a.host(1)]
                       []     tkidle@1:[ ¬inplane → skip]
                       []     1didle@1:[ ¬inplane → skip]
```

This diagram defines that the program driver controls the location of the
passenger. The boarding is defined to be the trigger for the migration and the
new position is provided by the plane — its own current location. Moreover, from
that moment on, the location of the passenger is subject to the same changes
that the location of that plane.

4.2 KLAIM

KLAIM [11, 4] (*Kernel Language for Agent Interaction and Mobility*) is an experimental kernel programming language specifically designed to model and to
program distributed concurrent applications with code mobility. The language is
inspired by the Linda coordination model [19], hence it relies on the concept of
tuple space. A tuple space is a multiset of *tuples*; these are containers of information items (called *fields*). There can be *actual fields* (i.e. expressions, processes,
localities, constants, identifiers) and *formal fields* (i.e. variables). Syntactically,
a formal field is denoted with !*ide*, where *ide* is an identifier. For instance, the
sequence (*"foo"*, *"bar"*, !*Price*) is a tuple with three fields: the first two fields are
string values while the third one is a formal field.

Tuples are anonymous and content-addressable. *Pattern-matching* is used to
select tuples in a tuple space. Two tuples match if they have the same number
of fields and corresponding fields match: a formal field matches any value of
the same type, and two actual fields match only if they are identical (but two
formals never match). For instance, tuple (*"foo"*, *"bar"*, 100 + 200) matches with
(*"foo"*, *"bar"*, !*Val*). After matching, the variable of a formal field gets the value
of the matched field: in the previous example, after matching, *Val* (an integer
variable) will contain the integer value 300.

Tuple spaces are placed on *nodes* that are part of a *net*. Each node contains
a single tuple space and processes in execution; a node can be accessed through
its *address*. There are two kinds of addresses: *Sites* are the identifiers through
which nodes can be uniquely identified within a net; *Localities* are symbolic
names for nodes. A reserved locality, **self**, can be used by processes to refer
to their execution node. Sites have an absolute meaning and can be thought of

as IP addresses, while localities have a relative meaning depending on the node where they are interpreted and can be thought of as aliases for network resources. Localities are associated to sites through *allocation environments*, represented as partial functions. Each node has its own environment that, in particular, associates self to the site of the node.

KLAIM processes may run concurrently, both at the same node or at different nodes, and can perform five basic operations over nodes. $\mathbf{in}(t)@\ell$ evaluates the tuple t and looks for a matching tuple t' in the tuple space located at ℓ. Whenever the matching tuple t' is found, it is removed from the tuple space. The corresponding values of t' are then assigned to the formal fields of t and the operation terminates. If no matching tuple is found, the operation is suspended until one is available. $\mathbf{read}(t)@\ell$ differs from $\mathbf{in}(t)@\ell$ only because the tuple t', selected by pattern-matching, is not removed from the tuple space located at ℓ. $\mathbf{out}(t)@\ell$ adds the tuple resulting from the evaluation of t to the tuple space located at ℓ. $\mathbf{eval}(P)@\ell$ spawns process P for execution at node ℓ. $\mathbf{newloc}(s)$ creates a new node in the net and binds its site to s. The node can be considered a "private" node that can be accessed by the other nodes only if the creator communicates the value of variable s, which is the only way to access the fresh node. Finally, KLAIM processes are built up from the special process **nil**, that does not perform any action, and from the basic operations by using standard operators borrowed from process algebras [33], namely *action prefixing*, *parallel composition* and *process definition*. In particular, recursive behaviours are modelled via process definitions. It is assumed that each process identifier A, parameterised w.r.t. $\widetilde{P}$, $\widetilde{\ell}$ and $\widetilde{e}$, has a *single* defining equation of the form $A(\widetilde{P}, \widetilde{\ell}, \widetilde{e}) \stackrel{def}{=} P$ (notation $\widetilde{\cdot}$ denotes a list of objects of a given kind).

A KLAIM Extension: OpenKlaim. OPENKLAIM, that has been first presented in [4], is an extension of KLAIM that was specifically designed for enabling users to give more realistic accounts of *open systems*. Indeed, open systems are dynamically evolving structures: new nodes can get connected or existing nodes can disconnect. Connections and disconnections can be temporary and unexpected. Thus, the KLAIM assumption that the underlying communication network will always be available is too strong. Moreover, since network routes may be affected by restrictions (such as temporary failures or firewall policies), *naming* may not suffice to establish connections or to perform remote operations. Therefore, to make KLAIM suitable for dealing with open systems, the need has arisen to extend the language with constructs for explicitly modelling connectivity between network nodes and for handling changes in the network topology.

OPENKLAIM is obtained by equipping KLAIM with mechanisms to dynamically update allocation environments and to handle node connectivity, and with a new category of processes, called *coordinators*, that, in addition to the standard KLAIM operations, can execute privileged operations that permit establishing new connections, accepting connection requests and removing connections. The new privileged operations can also be interpreted as movement operations: entering a new administrative domain, accepting incoming nodes and exiting from

Table 1. OPENKLAIM Syntax.

$f ::= e \mid P \mid \ell \mid *l \mid !x \mid !X \mid !\ell$		TUPLE FIELDS		
$t ::= f \mid f,t$	TUPLES		$P ::=$	PROCESSES
			nil	*null process*
$\ell ::= l \mid s$	LOCALITIES & SITES		$\mid a.P$	*action prefixing*
			$\mid P_1 \mid P_2$	*parallel composition*
$a ::=$	ACTIONS		$\mid A\langle \widetilde{P}, \widetilde{\ell}, \widetilde{e}\rangle$	*process invocation*
$\quad$ **out**(t)@ℓ	*output*			
$\mid$ **in**(t)@ℓ	*input*		$\mathbb{C} ::=$	COORDINATORS
$\mid$ **read**(t)@ℓ	*read*		P	*(standard) process*
$\mid$ **eval**(P)@ℓ	*migration*		$\mid pa.\mathbb{C}$	*action prefixing*
$\mid$ **bind**(l,s)	*bind*		$\mid \mathbb{C}_1 \mid \mathbb{C}_2$	*parallel composition*
			$\mid A\langle \widetilde{\mathbb{C}}, \widetilde{\ell}, \widetilde{e}\rangle$	*coordinator invocation*
$pa ::=$	PRIVILEGED ACTIONS			
$\quad a$	*(standard) action*		$C ::= \langle et\rangle \mid \mathbb{C} \mid C \mid C$	NODE COMPONENTS
$\mid$ **newloc**$(s,\mathbb{C})$	*creation*			
$\mid$ **login**(ℓ)	*login*		$N ::=$	NETS
$\mid$ **logout**(ℓ)	*logout*		$\mid s ::_\rho^S C$	*single node*
$\mid$ **accept**(s)	*accept*		$\mid N_1 \parallel N_2$	*net composition*

an administrative domain, respectively. The syntax of OPENKLAIM processes is presented in Table 1.

OPENKLAIM processes can be thought of as user programs and differs from KLAIM processes in the following three respects.

- When tuples are evaluated, locality names resolution does not take place automatically anymore. Instead, it has to be explicitly required by putting the operator $*$ in front of the locality that has to be evaluated. For instance, $(3,l)$ and $(s, \textbf{out}(s_1)@s_2.\textbf{nil})$ are fully-evaluated while $(3, *l)$ and $(*l, \textbf{out}(l)@\textbf{self}.\textbf{nil})$ are not.
- Operation **newloc** cannot be performed by user processes anymore. It is now part of the syntax of coordinator processes because, when a new node is created, it is necessary to install one such process at it and, for security reasons, user processes cannot be allowed to do this.
- Operation **bind** has been added in order to enable user processes to enhance local allocation environments with new aliases for sites. For instance, $\textbf{bind}(l,s)$ enhances the local allocation environment with the new alias l for s.

Coordinators can be thought of as processes written by node managers, a sort of superusers. Thus, in addition to the standard KLAIM operations, such processes can execute local (namely they are not indexed with a locality) coordination operations to establish new connections (viz. $\textbf{login}(\ell)$), to accept connection requests (viz. **accept**(s)), and to remove connections (viz. **logout**(ℓ)). Coordinators are stationary processes (namely, they cannot occur as arguments of **eval**) and cannot be used as tuple fields. They are installed at a node either when the node is initially configured or when the node is dynamically created, e.g. when a coordinator performs **newloc**($s, \mathbb{C}$) (where $\mathbb{C}$ is a coordinator).

An OPENKLAIM network node is a 4-tuple of the form $s ::^S_\rho C$, where s is the site of the node (i.e. its physical address in the net), ρ is the local allocation environment, S gives the set of nodes connected to s and C are the components located at the node, i.e. is the parallel composition of evaluated tuples (represented as $\langle et \rangle$) and of (user and) coordinator processes. A net can be either a single node or the parallel composition of two nets N_1 and N_2 with disjoint sets of node sites.

If $s ::^S_\rho C$ is a node in the net, then we will say that the nodes in S are *logged in s* and that s is a *gateway* for those nodes. A node can be logged in more than one node, that is it can have more than one gateway. Moreover, if s_1 is logged in s_2 and s_2 is logged in s_3 then s_3 is a gateway for s_1 too. Gateways are essential for communication: two nodes can interact only if there exists a node that acts as gateway for both. Moreover, to evaluate locality names, whenever s_1 is logged in s_2, if a locality cannot be resolved by just using the allocation environment of s_1, then the allocation environment of s_2 (and possibly that of nodes to which s_2 is logged in) is also inspected.

The OPENKLAIM approach puts forward a clean separation between the coordinator level (made up by coordinator processes) and the user level (made up by standard processes). This separation makes a considerable impact. From an abstract point of view, the coordinator level may represent the network operating system running on a specific computer and the user level may represent the processes running on that computer. The new privileged operations are then system calls supplied by the network operating system. From a more implementation oriented point of view, the coordinator level may represent the part of a distributed application that takes care of the connections to a remote server (if the application is a client) or that manages the connected clients (if the application is a server). The user level then represents the remaining parts of the application that can interact with the coordinator by means of some specific protocols.

To save space, here we do not show OPENKLAIM operational semantics (we refer the interested reader to [4]). Informally, the meaning of the coordination primitives is the following. Operation **newloc**($s, \mathbb{C}$) creates a new node in the net, binds the site of the new node to s and installs the coordinator $\mathbb{C}$ at the new node. Notice that a **newloc** does not automatically log the new node in the generating one. This can be done by installing a coordinator in the new node that performs a **login**. Differently from the standard KLAIM **newloc** operation, the environment is not explicitly inherited by the created node, instead it is subsumed by using the "logged in" relationships among nodes. Operation **login**(ℓ) logs the executing node, say s, in ℓ but only if at ℓ there is a coordinator willing to accept a connection, namely a coordinator of the form **accept**(s').$\mathbb{C}$. As a consequence of this synchronisation, s is added to the set S of nodes logged in ℓ and s' is replaced with s within $\mathbb{C}$. Operation **logout**(ℓ) disconnects the executing node, say s, from ℓ. As a consequence, s is removed from the set S of nodes logged in ℓ and any alias for s is removed from the allocation environment of ℓ.

An OpenKlaim Implementation of the Airport Scenario. As an example of the use of OPENKLAIM, we consider in this section the simplified airport scenario, with planes landing and taking off and passengers arriving and departing. The scenario we want to implement has the following specification:

- a passenger has to check in before board a plane;
- a plane is ready to take off when all passengers have boarded and the luggage has been loaded.

Passengers already have a boarding card, thus each passenger knows plane and seat assigned to him/her and, moreover, the number of passengers that must board on a plane is known. For simplicity sake, we only model one airport, one plane, and the passengers that must board on that plane.

We can identify two participants – *passenger* and *plane*; both have a dynamic set of instances in the running system. The implementation we present models each instance as an OPENKLAIM node. Methods (i.e. *checkin*, *loadLug* and so on) are implemented by OPENKLAIM processes. We model passenger and plane mobility via the OPENKLAIM primitives for reconfiguring open nets (i.e. **login**, **accept** and **logout**).

The model of the physical space is represented by an OPENKLAIM net. For each airport there exists a node in the net where passengers and planes can be host (i.e. connected). Airport nodes represent the immobile nodes of the system, passenger and plane are mobile nodes. In the system we present, each mobile node (plane and passengers) is initially connected with an immobile node (airport).

Passengers already have assigned a seat in a given flight and are involved in activities such as check-in and boarding the plane. Planes make flights by transporting luggage and passengers. A passenger first checks-in, then he can board the plane. A plane has to load all the luggage and all passengers having a boarding card for that flight before it can take off.

Node passenger hosts processes *checkin* and *boards* defined as follows:

$$checkin(airport, flight, seat) \stackrel{def}{=} \mathbf{out}(\text{``}checkin\text{''}, *\mathtt{self}, flight, seat)@airport.$$
$$\mathbf{in}(\text{``}checkinOk\text{''})@\mathtt{self}.$$
$$\mathbf{out}(\text{``}boardOk\text{''})@\mathtt{self}$$

$$boards(airport, plane) \stackrel{def}{=} \mathbf{in}(\text{``}boardOk\text{''})@\mathtt{self}.$$
$$\mathbf{login}(plane).$$
$$\mathbf{logout}(airport).$$
$$\mathbf{out}(\text{``}boards\text{''}, *\mathtt{self})@plane$$

Process *checkin*, parameterised w.r.t. airport, flight and seat, merely sends a "checkinOk" request to the airport and waits for a reply. Process *boards*, parameterised w.r.t. airport and plane, after checkin has been completed, allows passengers to log in the plane and to log out the airport (this implements the physical mobility of passengers).

Node plane hosts processes $loadLug$ and $takesoff$ defined as follows:

$$loadLug(airport) \overset{def}{=} \textbf{out}(\text{``}loadLug\text{''}, *\texttt{self})@airport.$$
$$\textbf{in}(\text{``}loadLugOk\text{''})@\texttt{self}.$$
$$\textbf{out}(\text{``}takeoffOk\text{''})@\texttt{self}$$

$$takesoff(airport) \overset{def}{=} \textbf{accept}(s_1).$$
$$\textbf{in}(\text{``}boards\text{''}, s_1)@\texttt{self}.$$
$$\dots$$
$$\textbf{accept}(s_n).$$
$$\textbf{in}(\text{``}boards\text{''}, s_n)@\texttt{self}.$$
$$\textbf{in}(\text{``}takeoffOk\text{''})@\texttt{self}.$$
$$\textbf{logout}(airport).$$
$$\textbf{out}(\text{``}On\ air!\text{''})@\texttt{self}$$

Process $loadLug$, parameterised w.r.t. airport, simply represents the load of luggage. Process $takesoff$, parameterised w.r.t. airport, allows the plane to log out the airport only when all the passengers are on board (i.e. have been accepted by the plane).

Node airport hosts processes $handleCheckin$ and $handleLoadLug$, used to handle check-in requests from passengers and loading-luggage messages from planes respectively, defined as follows:

$$handleCheckin \overset{def}{=} \textbf{in}(\text{``}checkin\text{''}, !l, !flight, !seat)@\texttt{self}.$$
$$\textbf{out}(\text{``}checkinOk\text{''})@l.$$
$$handleCheckin$$

$$handleLoadLug \overset{def}{=} \textbf{in}(\text{``}loadLug\text{''}, !l)@\texttt{self}.$$
$$\textbf{out}(\text{``}loadLugOk\text{''})@l.$$
$$handleLoadLug$$

Finally, the overall system is defined by a net with a node for each instance of airport, plane and passenger:

$$system \overset{def}{=} airport \parallel plane \parallel passenger_1 \parallel \dots \parallel passenger_n$$

where each node is defined as follows:

$$airport \overset{def}{=} s_{arp} ::^{\{s_1, s_2, \dots, s_n, s_{pln}\}}_{\{s_{arp}/\texttt{self}\}} handleCheckin \mid handleLoadLug$$

$$plane \overset{def}{=} s_{pln} ::_{\{s_{pln}/\texttt{self}, s_{arp}/\texttt{airport}\}} loadLug(airport) \mid takesoff(airport)$$

$$passenger_i \overset{def}{=} s_i ::_{\{s_i/\texttt{self}, s_{arp}/\texttt{airport}, s_{pln}/\texttt{plane}\}} checkin(airport, flight, seat_i) \mid$$
$$boards(airport, plane)$$

5   Specification Framework for Evolution

The objective of this research is to design a framework for the specification and analysis of the system's evolution arising from reconfiguration and mobility. This includes extensions to Graph Transformation and Tile Logic to include key features for representing distribution and mobility, the application of such extensions to model the evolution arising from the reconfiguration of the distributions connectors introduced in the research on architectures, and the development of analysis techniques for the verification of security and behavioural properties of mobile distributed systems, including the design of topological modalities. The logical techniques developed will be generalised to allow for combination with other formalisms.

During the first year of the project the grounds were set for the integrated specification framework by extending graph transformations and tile logic by encoding of Single-Pushout graph rewriting into Tiles, defining transactions in the Tile Model, by adding higher order features for graph rewriting, and by defining an appropriate graph transformation framework for the operational semantics of UML. To obtain analysis techniques for security and behavioural properties, two ambient-like calculi were developed [35, 32] and a technique for the analysis of graph transformation systems was proposed [2].

In the following we will focus on the operational semantics of UML object and activity diagrams by Graph Transformation Systems. Starting from the UML specification, we first show how to encode instance diagrams as graphs of a suitable kind, in order to define rule-based transformations on them. Next we represent behavioural diagrams as graph transformation systems: we consider a simple *activity diagram*, and we present one graph transformation rule for each activity in it. Each rule will describe the local evolution of the system resulting from the corresponding activity. Most importantly, by resorting to the theory of graph transformation we are able to show that the proposed rules implement correctly the dependencies among the various activities, as described in the activity diagram. Finally, we show that a generalisation of the example by allowing *a list* of passengers boarding to a plane (instead of a single passenger), can be modelled conveniently by an extension of graph transformation with *synchronisation*, which is a specific Tile Model.

5.1   Modelling the Airport Scenario with Graph Transformation

The various kinds of diagrams used in a UML specification essentially are graphs annotated in various ways. Therefore it comes as no surprise that many contributions in the literature use techniques based on the theory of graph transformation to provide an operational semantics for UML behavioural diagrams (see, among others, [12, 22, 30, 29, 15, 21]). Clearly, a pre-requisite for any such graph transformation based semantics is the formal definition of the structure of the graphs which represent the states of the system, namely the *instance graphs*. However, there is no common agreement about this: we shall present a novel formalisation, which shares some features with the one proposed in [23].

An *instance graph* includes a set of *nodes*, which represent all data belonging to the state of an execution. Some of them represent the elements of primitive data types, while others denote instances of classes. Every node may have at most one outgoing *hyperedge*, i.e., an edge connecting it to zero or more nodes[1].

Conceptually, the node can be interpreted as the "identity" of a data element, while the associated hyperedge, if there is one, contains the relevant information about its state. A node without outgoing hyperedges is either a *constant* or a *variable*.

Typically, an instance of a class C is represented by a node n and by an hyperedge labelled with the pair $\langle instanceName : C \rangle$. This hyperedge has node n as its only *source*, and for each attribute of the class C it has a link (a *target tentacle*) labeled by the name of the attribute and pointing to the node representing the attribute value. Every instance graph also includes, as nodes, all constant elements of primitive data types, like integers (0, 1, -1, ...) and booleans (true and false), as well as one node null:C for each relevant class C.

Figure 10 (a) shows an instance diagram which represents the initial state of the airport scenario. As usual, the attributes of an instance may be represented as directed edges labeled by the attribute name, and pointing to the attribute value. The edge is unlabeled if the attribute name coincides with the class of the value (e.g., lh123 is the value of the plane attribute of tck). An undirected edge represents two directed edges between its extremes. The diagram conforms to a class diagram that is not depicted here.

Figure 10 (b) shows the instance graph (according to the above definitions) encoding the instance diagram. Up to a certain extent (disregarding OCL formulas and cardinality constraints), a class diagram can be encoded in a corresponding *class graph* as well; then the existence of a graph morphism (i.e., a structure preserving mapping) from the instance graph to the class graph formalizes the relation of conformance.

In the following we shall depict the states of the system as instance diagrams, which are easier to draw and to understand, but they are intended to represent the corresponding instance graphs.

Figure 3 shows the activity diagram of the Use Case Departure of the airport case study. This behavioural diagram ignores the structure of the states and the information about which instances are involved in each activity, but stresses the causal dependencies among activities and the possible parallelism among them. More precisely, from the diagram one infers the requirement that board and load_luggage can happen in any order, after check_in and before take_off.

By making explicit the roles of the various instances in the activities, we shall implement each activity as a graph transformation rule. Such rules describe local modifications of the instance graphs resulting from the corresponding activities. We will show that they provide a correct implementation of the activity diagram, since the above requirement is met.

Let us first consider the activity board. Conceptually, in the simplified model we are considering, its effect is just to change the location of the passenger (i.e.,

---

[1] Formally, the graphs are *term graphs* [10, 36].

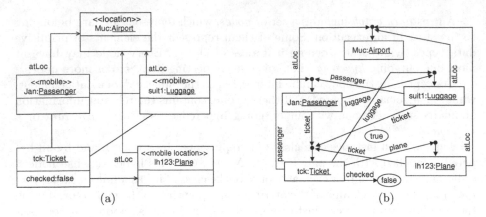

Fig. 10. An instance diagram (a) and the corresponding instance graph (b).

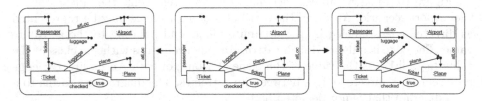

Fig. 11. The graph transformation rule for boarding.

its atLoc attribute) from the airport to the plane. In the rule which implements the activity, we make explicit the preconditions for its application: 1) the passenger must have a ticket for the flight using that plane; 2) the value of the checked attribute of the ticket must be true; 3) the plane and the passenger must be at the *same* location, which is an airport.

All this is represented in the graph transformation rule implementing the activity board, shown in Fig. 11. Formally, this is a *double-pushout graph transformation rule* [9], having the form $L \xleftarrow{l} K \xrightarrow{r} R$, where L, K and R are instance graphs, and the l and r are graph morphisms (inclusions, in this case; they are determined implicitly by the position of nodes and edges).

Intuitively, a rule states that whenever we find an occurrence of the *left-hand side* L in a graph G we may replace it with the *right-hand side* R. The *interface* graph K and the two morphisms l and r provide the *embedding information*, that is, they specify where R should be glued with the *context* graph obtained from G by removing L. More precisely, an *occurrence* of L in G is a graph morphism $g : L \to G$. The context graph D is obtained by deleting from G all the nodes and edges in $g(L - l(K))$ (thus all the items in the interface K are preserved by the transformation). The embedding of R in D is obtained by taking their disjoint union, and then by identifying for each node or edge x in K its images $g(x)$ in G and $r(x)$ in R: formally, this operation is a *pushout* in a suitable category.

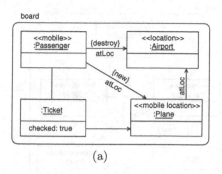

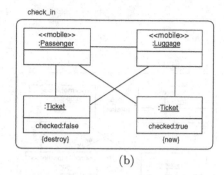

(a)   (b)

Fig. 12. The rules for boarding (a) and for checking in (b) as collaboration diagrams.

Comparing the three graphs in the rule, one can see that in order to change the value of the attribute atLoc of the Passenger, the whole hyperedge is deleted and created again: one cannot delete a single attribute, as the resulting structure would not be a legal hypergraph. Instead, the node representing the identity of the passenger is preserved by the rule. Also, all the other items present in the left-hand side (needed to enforce the preconditions for the application of the rule) are not changed by the rule.

It is possible to use a much more concise representation of a rule of this kind, by depicting it as a single graph (the union of L and R), and annotating which items are removed and which are created by the rule. Figure 12 (a) shows an alternative but equivalent graphical representation of the rule of Fig. 11 as a degenerate kind of *collaboration diagram* (without sequence numbers, guard conditions, etc.) according to [8].

Here the state of the system is represented as an instance diagram, and the items which are deleted by the rule (resp. created) are marked by {destroy} (resp. {new}: beware that these constraints refer to the whole Passenger instance, and not only to the atLoc tentacle. For graph transformation rules with injective right-hand side (like all those considered here), this representation is equivalent to the one above, and for the sake of simplicity we will stick to it.

Figure 12 (b) and Fig. 13 (a, b) show the rules implementing the remaining three activities of Fig. 3, namely check_in, load_luggage and take_off: the corresponding full graphical representation can be recovered easily. Notice that the effect of the take_off rule is to change the value of the atLoc attribute of the plane: we set it to null, indicating that the location is not meaningful after taking off; as a different choice we could have used a generic location like Air or Universe.

The next statement, by exploiting definitions and results from the theory of graph transformation, describes the causal relationships among the potential rule applications to the instance graph of Fig. 10 (b), showing that the dependencies among activities stated in the diagram of Fig. 3 are correctly realized by the proposed implementation.

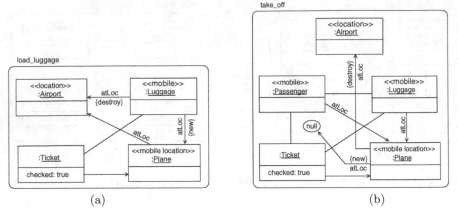

Fig. 13. The rules for loading the luggage (a) and for taking off (b).

Proposition 1 (Causal dependencies among rules implementing activities). *Given the start instance graph G_0 of Fig. 10 (b) and the four graph transformation rules of Fig. 12 and 13,*

- *the only rule applicable to G_0 is* check_in, *producing, say, the instance graph G_1;*
- *both* board *and* load_luggage *can be applied to graph G_1, in any order or even in parallel, resulting in all cases in the same graph (up to isomorphism), say G_2;*
- *rule* take_off *can be applied to G_2, but not to any other instance graph generated by the mentioned rules.*

5.2   Enriching the Model with Synchronised Graph Transformation

Quite obviously, the rule take_off presented in the previous subsection fits in the unrealistic assumption that the flight has only one passenger. Let us discuss how this assumption can be dropped by modeling the fact that the plane takes off only when ALL its passengers and ALL their luggages are boarded.

We shall exploit the expressive power of Synchronized Hypergraph Rewriting [24, 26, 25], an extension of hypergraph rewriting which uses some basic features inspired by the Tile Model [18], to model this situation in a very concise way. Intuitively, the plane has as attribute the collection of *all* the tickets for its flight, and when taking off it broadcasts a synchronization request to all the tickets in the collection. Each ticket can synchronize only if its passenger and its luggage are on the plane. If the synchronization fails, the take_off rule cannot be applied. This activity can be considered as an abstraction of the check performed by the hostess/steward before closing the gate.

Conceptually, a graph transformation rule *with synchronization* is a rule where one or more nodes of the right-hand side may be annotated with an

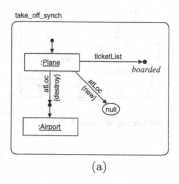
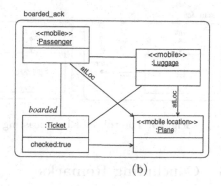

(a)   (b)

Fig. 14. The rules for taking off while checking that all passengers are on board (a), and for acknowledging the synchronization (b).

action. If the node is a variable, the action can be interpreted as a synchronization request issued to the instance which will be bound to the variable when applying the rule. If instead the annotated node is the source of an instance, the action can be interpreted as an acknowledgment issued by that instance. Given an instance graph, a bunch of such rules with synchronization can be applied simultaneously to it only if all synchronization requests are properly matched by a corresponding acknowledgment.

To use this mechanism in our case study, consider the association Plane $\overset{1}{\underset{*}{\rightleftharpoons}}$ Ticket with the obvious meaning: we call TicketList the corresponding attribute of a plane (cf. Fig. 2). Figure 14 (a) shows rule take_off_synch: the plane takes off, changing its location from the airport to null, only if its request for a synchronization with a *boarded* action is acknowledged by its collection of tickets. In this rule we depict the state as an instance graph, because we want to show explicitly that a node representing the value of the attribute ticketList of the plane is annotated by the *boarded* action. On the other side, according to rule boarded_ack, a ticket can acknowledge a *boarded* action only if its passenger and its luggage are both located on its plane. Here the state is depicted again as an instance *diagram*, and the *boarded* action is manifested on the node representing the identity of the ticket.

To complete the description of the system, we must explain how the tickets for the flight of concern are linked to the ticketList attribute of the plane. In order to obtain the desired synchronization between the plane and all its tickets, we need to assume that there is a subgraph which has, say, one "input node" (the ticketList attribute of the plane) and n "output nodes" (the tickets); furthermore, this subgraph should be able to "match" synchronization requests on its input to corresponding synchronization acknowledgments on its ouputs.

More concretely, this is easily obtained, for example, by assuming that the collection of tickets is a linked list, and by providing rules for propagating the synchronization along the list: this is shown in Fig. 15, where the rules should be intended to be parametric with respect to the action *act*.

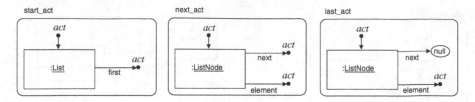

Fig. 15. The rules for broadcasting synchronizations along a linked list.

6   Concluding Remarks

The AGILE project is developing an architectural approach to software development for global computing in which mobility aspects can be modelled explicitly at several levels of abstraction. The whole approach is developed over a uniform mathematical framework based on graph-oriented techniques to support sound methodological principles, formal analysis, and refinement. In this paper we have presented some of the results gained during the first project year. AGILE has obtained many other results concerning specification, verification and analysis of global computation systems we could not present here because of lack of space.

Using the running example of the simple airport case study we have shown how several well-known modelling, coordination and programming languages can be extended or directly used to model mobility. In particular, we have presented

- an extension of the UML for modelling mobility,
- an extension of the program design language CommUnity to support mobility, and
- OpenKLAIM, a language for programming distributed open systems,
- and have shown how a graph transformations and tile logic can be used to give a mathematical basis to a kernel of UML with mobility.

Currently, we pursue our goal of developing a mathematically well-founded architectural approach to software engineering of global computing systems. We are working on a tighter integration of the different formalisms by analysing their relationships, by defining further translations between each of the formalisms, and by studying analysis, verification and refinement techniques where also institutions and categorical techniques will play a major role. We have also started to design and implement tools for supporting software development with our methods.

Acknowledgments

We would like to acknowledge the contribution of E. Tuosto, L. Bettini, D. Gorla, A. Knapp, S. Merz, J. Zappe, G. Marczynski, M. Bednarczyk, T. Borzyszkowski, A. Borzyszkowski, and J. Gouveia.

References

1. E. Astesiano, M. Bidoit, H. Kirchner, B. Krieg-Brückner, P. D. Mosses, D. Sannella, and A. Tarlecki. CASL: The common algebraic specification language. *Theoretical Computer Science*, 2003. To appear. See also the CASL Summary at http://www.brics.dk/Projects/CoFI/Documents/CASL/Summary/.
2. P. Baldan, A. Corradini, and B. König. Static analysis of distributed systems with mobility specified by graph grammars – A case study. In H. Ehrig, B. Krämer, and A. Ertas, editors, *Integrated design and Process Technology*. Society for Design and Process Science, 2002.
3. Hubert Baumeister, Nora Koch, Piotr Kosiuczenko, and Martin Wirsing. Extending activity diagrams to model mobile systems. In M. Aksit, M. Mezini, and R. Unland, editors, *Objects, Components, Architectures, Services, and Applications for a Networked World. International Conference NetObjectDays, NODe 2002, Erfurt, Germany, Oct. 7–10, 2002. Revised Papers*, volume 2591 of *LNCS*, pages 278–293. Springer, 2003.
4. L. Bettini, M. Loreti, and R. Pugliese. Structured nets in Klaim. In *Proc. of the 2000 ACM Symposium on Applied Computing (SAC'00), Special Track on Coordination Models, Languages and Applications*, pages 174–180, 2000.
5. G. Bhat, R. Cleaveland, and O. Grumberg. Efficient on-the-fly model checking for CTL*. In *Proceedings of Symposium on Logics in Computer Science*, pages 388–397. IEEE, 1995.
6. K. Chandy and J. Misra. *Parallel Program Design — A Foundation*. Addison-Wesley, 1988.
7. A. Corradini and R. Heckel. Graph transformation and visual modelling techniques. In J.D.P. Rolim, A.Z. Broder, A. Corradini, R. Gorrieri, R. Heckel, J. Hromkovic, U. Vaccaro, and J.B. Wells, editors, *ICALP Workshops 2000*, volume 8 of *Proceedings in Informatics*, pages 357–486. Carleton Scientific, 2000.
8. A. Corradini, R. Heckel, and U. Montanari. Graphical Operational Semantics. In Rolim et al. [7], pages 357–486.
9. A. Corradini, U. Montanari, F. Rossi, H. Ehrig, R. Heckel, and M. Löwe. Algebraic Approaches to Graph Transformation I: Basic Concepts and Double Pushout Approach. In Grzegorz Rozenberg, editor, *Handbook of Graph Grammars and Computing by Graph Transformation. Vol. 1: Foundations*. World Scientific, 1997.
10. A. Corradini and F. Rossi. Hyperedge replacement jungle rewriting for term rewriting systems and logic programming. *Theoretical Computer Science*, 109:7–48, 1993.
11. R. De Nicola, G.-L. Ferrari, and R. Pugliese. KLAIM: A kernel language for agents interaction and mobility. *IEEE Transactions on Software Engineering*, 5(24):315–330, 1998.
12. G. Engels, J.H. Hausmann, R. Heckel, and S. Sauer. Dynamic meta modeling: A graphical approach to the operational semantics of behavioural diagrams in UML. In *Proceedings of the Third International Conference on the Unified Modeling Language (UML'2000)*, volume 1939 of *LNCS*, pages 323–337. Springer, 2000.
13. A. Fantechil, S. Gnesi, F. Mazzanti, R. Pugliese, and E. Tronci. A symbolic model checker for ACTL. In *International Workshop on Current Trends in Applied Formal Methods*, number 1641 in LNCS. Springer - Verlag, 1999.
14. J.-C. Fernandez, D. Jard, T. Jéron, and C. Viho. Using on-the-fly verification techniques for the generation of test suites. In *Proceedings of Conference on Computer-Aided Verification (CAV '96)*, number 1102 in LNCS, pages 348–359. Springer-Verlag, 1996.

15. G. L. Ferrari, U. Montanari, and E. Tuosto. Graph-based models of internetworking systems. In A. Haeberer, editor, *Formal Methods at the Crossroads: from Panaces to Foundational Support*, LNCS. Springer, 2003. To appear.
16. J. L. Fiadeiro, A. Lopes, and M. Wermelinger. A mathematical semantics for architectural connectors. In R.Backhouse and J.Gibbons, editors, *Generic Programming*, number 2793 in LNCS. Springer-Verlag, 2003.
17. J. L. Fiadeiro and T. Maibaum. Categorical semantics of parallel program design. *Science of Computer Programming*, 28:111–138, 1997.
18. F. Gadducci and U. Montanari. The Tile Model. In G. Plotkin, C. Stirling, and M. Tofte, editors, *Proof, Language and Interaction: Essays in Honour of Robin Milner*. MIT Press, 2000.
19. D. Gelernter. Generative communication in Linda. *ACM Transactions on Programming Languages and Systems*, pages 80–112, 1985.
20. S. Gnesi and F. Mazzanti. On the fly model checking of communicating UML State Machines. Technical report, ISTI, 2003. submitted for publication.
21. M. Gogolla, P. Ziemann, and S. Kuske. Towards an integrated graph based semantics for UML. In Paolo Bottoni and Mark Minas, editors, *Electronic Notes in Theoretical Computer Science*, volume 72. Elsevier Science, 2003.
22. Martin Gogolla. Graph transformations on the UML metamodel. In *ICALP Workshop on Graph Transformations and Visual Modeling Techniques*, pages 359–371. Carleton Scientific, 2000.
23. R. Heckel, J.M. Küster, and G. Taentzer. Confluence of typed attributed graph transformation systems. In *Proceedings of the First International Conference on Graph Transformation (ICGT'2002)*, volume 2505 of *LNCS*, pages 161–176. Springer, 2002.
24. D. Hirsch and U. Montanari. Synchronized hyperedge replacement with name mobility. In K.G. Larsen and M. Nielsen, editors, *Proceedings of CONCUR 2001*, volume 2154 of *LNCS*, pages 121–136. Springer, 2001.
25. Dan Hirsch. *Graph Transformation Models for Software Architecture Styles*. PhD thesis, Universidad de Buenos Aires, Argentina, 2003. To appear.
26. B. König and U. Montanari. Observational Equivalence for Synchronized Graph Rewriting. In *Proceedings TACS'01*, volume 2215 of *LNCS*, pages 145–164. Springer, 2001.
27. P. Kosiuczenko. Sequence diagrams for mobility. In J. Krogstie, editor, *Proc. of MobIMod workshop, Tampere, Finland*, LNCS. Springer, Berlin, 2002. to appear.
28. D. Kozen. Results on the propositional $\mu-calculus$. *Theoretical Computer Science*, 27, 1983.
29. S. Kuske, M. Gogolla, R. Kollmann, and H.J. Kreowski. An integrated semantics for UML class, object and state diagrams based on graph transformation. In *Integrated Formal Methods, Third International Conference, IFM 2002*, volume 2335 of *LNCS*, pages 11–28. Springer, 2002.
30. Sabine Kuske. A formal semantics of UML state machines based on structured graph transformation. In M. Gogolla and C. Kobryn, editors, *Proceedings of the Fourth International Conference on the Unified Modeling Language (UML'2001)*, number 2185 in LNCS, pages 241–256. Springer, 2001.
31. D. Latella, I. Majzik, and M. Massink. Towards a formal operational semantics of UML statechart diagrams. In *Proceedings of IFIP TC6/WG6.1 FMOODS '99*. Kluwer Academic Publishers, 1999.

32. S. Merz, M. Wirsing, and J. Zappe. A spatio-temporal logic for the specification and refinement of mobile systems. In M. Pezzè, editor, *Fundamental Approaches to Software Engineering (FASE 2003)*, Lecture Notes in Computer Science, Warsaw, Poland, April 2003. Springer-Verlag. To appear.
33. R. Milner. *Communication and Concurrency*. Prentice-Hall, 1989.
34. OMG. Unified Modeling Language (UML), version 1.5. www.omg.org, March 2003.
35. D. Pattinson and M. Wirsing. Making components move: A separation of concerns approach. In *Proc. 1st International Workshop on Formal Methods for Components (FMCO'02)*, Lecture Notes in Computer Science. Springer-Verlag, 2002. To appear.
36. Detlef Plump. Term graph rewriting. In H. Ehrig, G. Engels, H.-J. Kreowski, and G. Rozenberg, editors, *Handbook of Graph Grammars and Computing by Graph Transformation. Vol. 2: Applications, Languages, and Tools*. World Scientific, 1999.
37. M. von der Beeck. Formalization of UML-statecharts. In M. Gogolla and C. Kobryn, editors, *Proceedings of the Fourth International Conference*, number 2185 in LNCS, pages 406–421. Springer-Verlag, 2001.
38. T. Wieringa and J. Broersen. A minimal transition system semantics for lightweight class and behavioral diagrams. In *ICSE'98 Workshop on Precise Semantics for Software Modeling Techniques*, 1998.

A Coinductive Calculus
of Component Connectors

F. Arbab[1] and J.J.M.M. Rutten[1,2]

[1] CWI, P.O. Box 94079, 1090 GB Amsterdam, The Netherlands
{farhad,janr}@cwi.nl,
http://www.cwi.nl/{~farhad,~janr}
[2] Division of Mathematics and Computer Science,
Faculty of Sciences, Vrije Universiteit, De Boelelaan 1081a,
1081 HV Amsterdam, The Netherlands

Abstract. Reo is a recently introduced channel-based model for coordination, wherein complex coordinators, called connectors, are compositionally built out of simpler ones. Using a more liberal notion of a channel, Reo generalises existing dataflow networks. In this paper, we present a simple and transparent semantical model for Reo, in which connectors are relations on timed data streams. Timed data streams constitute a characteristic of our model and consist of twin pairs of separate data and time streams. Furthermore, coinduction is our main reasoning principle and we use it to prove properties such as connector equivalence.

1 Introduction

Reo (from the Greek word $\rho\epsilon\omega$ which means *"[I] flow"*) is a recently introduced [Arb02,AM02] channel-based coordination model, wherein complex coordinators, called connectors, are compositionally built out of simpler ones. Reo is intended as a "glue language" for construction of connectors that orchestrate component instances in a component-based system. The emphasis in Reo is on connectors and their composition only, not on the components that are being connected. In this paper, we present a simple and transparent semantical model of connectors and connector composition, which can be used as a compositional *calculus*, in which properties such as connector equivalence, optimization, and realization can be expressed and proved.

The basic connectors are channels, each of which is a point-to-point communication medium with two distinct ends. Channels can be used as the only communication constructs in communication models for concurrent systems, because the primitives of other communication models (e.g., message passing or remote procedure calls) can be easily defined using channels. In contrast to other channel-based models, Reo uses a generalised concept of channel. In addition to the common channel types of synchronous and asynchronous, with bounded or unbounded buffers, and with fifo and other ordering schemes, Reo allows an open ended set of channels, each with its own, sometimes exotic, behaviour. For instance, a channel in Reo need not have both an input and an output end; it can

M. Wirsing, D. Pattinson, and R. Hennicker (Eds.): WADT 2002, LNCS 2755, pp. 34–55, 2003.

have two input ends or two output ends instead. In addition to channels, Reo
has one more basic connector, called the merge operator. More complex connectors can then be constructed from the basic connectors (channels and merge)
through an operation of connector composition.

Because Reo is not concerned with the internal activity of the components
that it connects, we represent components by their interfaces only. Therefore,
we model the input ends and output ends of connectors as *streams* (infinite
sequences) of abstract (uninterpreted) data items. Moreover, we associate with
every such data stream an infinite sequence of (natural or non-negative real)
numbers. These numbers stand for the respective moments in time at which
their corresponding data items are being input or output. This allows us to
describe and reason about the precise timing constraints of connectors (such as
synchronous versus asynchronous, and bounded versus unbounded delay). Thus,
we model the potential behaviour of connector ends as *timed data streams*, which
are pairs consisting of a data stream and a time stream. Note that we use pairs of
streams rather than streams of pairs (of timed data elements), since this enables
us to reason about time explicitly, which turns out to be particularly useful.

The main mathematical ingredients of our model are sets A^ω of streams
over some set A (of data items or time moments). These sets A^ω carry a so-
called *final coalgebra* structure, consisting of the well-known operations of head
and tail (here called initial value and derivative). As a consequence, we can
benefit from some basic but very general facts from the discipline of coalgebra,
which over the last decade has been developed as a general behavioral theory
for dynamical systems (see [JR97,Rut00] for an overview). In particular, the
final coalgebra A^ω satisfies principles of *coinduction*, both for definitions and
for proofs. The latter are formulated in terms of so-called *stream bisimulations*,
an elementary variation on Park's and Milner's original notion of bisimulation
for parallel processes [Mil80,Par81]. As we shall see, these coinduction principles
are surprisingly powerful. They will be applied to both data streams and time
streams.

Having modelled connector ends as timed data streams, we then model connectors as *relations* on timed data streams, expressing which combinations of
timed data streams are mutually consistent. This relational model is, in spite of
its simplicity, already sufficiently expressive to study a number of notions and
questions about component connectors, such as equivalence (when do two connectors have the same behavior?), expressiveness (which connectors can I build
out of a given set of basic connectors?), optimization (given a connector, can
I build an equivalent connector out of a smaller number of basic connectors?),
verification (given the specification of a certain connector behavior and given
a connector, does the connector meet the specification?), realization (given the
specification of a certain connector behavior, can I actually build a connector
with precisely that behavior out of a given set of basic connectors?), and the
like. In the present paper, we shall mainly focus on connector equivalence and,
to a lesser extent, the expressiveness of our connector calculus.

Reo is more general than dataflow models, Kahn-networks, and Petri-nets, which can be viewed as specialised channel-based models that incorporate certain basic constructs for primitive coordination. While Reo is designed to deal with the flow of data, it, more specifically, differs fundamentally from classical dataflow models in four important aspects:

1. Although not treated here, the topology of connections in 'full' Reo is inherently dynamic and accomodates mobility.
2. Reo supports a much more general notion of channels. Amongst others, it allows the combined use of synchronous and asynchronous channels.
3. The model of Reo is based on a clear separation of data and time.
4. Coinduction is the main reasoning principle.

Of all related work, Broy and Stølen's book [BS01] deserves special mention, since it is also based on (timed) data streams. However, the points mentioned above distinguish our model also from theirs. In particular, the separation of data and time, in our model, in combination with the use of coinduction, leads to simpler specifications (definitions) and proofs. See Section 9 for a concrete example. Finally, coalgebra and coinduction have been used in models of component-based systems in [Bar01] and [Dob02]. Also these models are distinguished from ours by the (first three) points above. Moreover, our model is far more concrete, and therefore allows actual equivalence proofs.

2   Streams and Coinduction

Let A be any set and let A^ω be the set of all streams (infinite sequences) over A:

$$A^\omega = \{\alpha \mid \alpha : \{0, 1, 2, \ldots\} \to A\}$$

We present some basic facts on A^ω, notably how to give definitions and proofs by coinduction. In this section, the set A is arbitrary but later, we shall look in particular at streams over some data set D and streams over the time domain $\mathbb{R}_+$.

Individual streams will be denoted as $\alpha = (\alpha(0), \alpha(1), \alpha(2), \ldots)$ (or $a = (a(0), a(1), a(2), \ldots)$). We call $\alpha(0)$ the *initial value* of α. The *(stream) derivative* α' of a stream α is defined as

$$\alpha' = (\alpha(1), \alpha(2), \alpha(3), \ldots)$$

Note that $\alpha'(n) = \alpha(n+1)$, for all $n \geq 0$. Later we shall also need 'higher-order' derivatives $\alpha^{(k)}$, for any $k \geq 0$, defined as $\alpha^{(0)} = \alpha$ and $\alpha^{(k+1)} = (\alpha^{(k)})'$. These satisfy $\alpha^{(k)}(n) = \alpha(n+k)$, for any $n \geq 0$.

Stream initial values and derivatives will be used both in definitions of (operations on) streams and in proofs of properties of streams. In this manner, a *calculus* of streams is obtained, in close analogy to classical analytical analysis. More specifically, we formulate definitions using the so-called *behavioural differential equations*, which specify both the initial value and the derivative of the stream being defined. Such definitions are also called *coinductive*. We illustrate this type of definition through a few basic examples. Let the operations *even*, *odd*, and *zip* be defined by the following system of equations (one for each $\alpha, \beta \in A^\omega$):

behavioural differential equation	initial value
$even(\alpha)' = even(\alpha'')$	$even(\alpha)(0) = \alpha(0)$
$odd(\alpha)' = odd(\alpha'')$	$odd(\alpha)(0) = \alpha'(0)$
$zip(\alpha, \beta)' = zip(\beta, \alpha')$	$zip(\alpha, \beta)(0) = \alpha(0)$

The reader should have no trouble convincing himself that these operations satisfy the following identities:

$$even(\alpha) = (\alpha(0), \alpha(2), \alpha(4), \ldots)$$
$$odd(\alpha) = (\alpha(1), \alpha(3), \alpha(5), \ldots)$$
$$zip(\alpha, \beta) = (\alpha(0), \beta(0), \alpha(1), \beta(1), \ldots)$$

These equalities could in fact have been taken as definitions, but we prefer the coinductive definitions instead, because they allow the use the coinduction proof principle, as we shall see shortly.

As in analysis, whether a differential equation has a (unique) solution or not depends in general on the shape of the equation. For the three elementary behavioural differential equations above (and in fact all other equations that we shall encounter in the present paper), the existence of a unique solution can be easily established by some elementary reasoning. (For the general case, see the remark at the end of this section.)

Proofs about streams will be given in terms of the following elementary notion. A *(stream) bisimulation* is a relation $R \subseteq A^\omega \times A^\omega$ such that, for all α and β in A^ω:

$$\text{if } \alpha \, R \, \beta \text{ then } \begin{cases} (1) & \alpha(0) = \beta(0) \text{ and} \\ (2) & \alpha' \, R \, \beta' \end{cases}$$

(The union of all bisimulation relations is itself a bisimulation, called *bisimilarity*.) Bisimulations are used in the formulation of the following *coinduction proof principle*. For all $\alpha, \beta \in A^\omega$:

$$\text{if } \alpha \, R \, \beta, \text{ for some bisimulation } R, \text{ then } \alpha = \beta \qquad (1)$$

In other words, in order to prove $\alpha = \beta$, it is sufficient to establish the existence of a bisimulation relation $R \subseteq A^\omega \times A^\omega$ such that $\alpha \, R \, \beta$.

Consider for instance the following three identities on streams, for all $\alpha, \beta \in A^\omega$,

1. $even(zip(\alpha, \beta)) = \alpha$
2. $odd(zip(\alpha, \beta)) = \beta$
3. $zip(even(\alpha), odd(\alpha)) = \alpha$

Since the following three relations on streams:

1. $\{\langle even(zip(\alpha, \beta)), \alpha\rangle \mid \alpha, \beta \in A^\omega\}$
2. $\{\langle odd(zip(\alpha, \beta)), \beta\rangle \mid \alpha, \beta \in A^\omega\}$
3. $\{\langle zip(even(\alpha), odd(\alpha)), \alpha\rangle \mid \alpha \in A^\omega\} \cup$
   $\{\langle zip(odd(\alpha), even(\alpha'')), \alpha'\rangle \mid \alpha \in A^\omega\}$

are bisimulations, the above three identities follow, respectively, by coinduction.

The validity of the proof principle itself can be easily established by proving $\alpha(n) = \beta(n)$ for all $n \geq 0$, by induction on n. More abstractly, both the coinduction proof and definition principle are ultimately based on the fact that the set A^ω carries a *final coalgebra* structure, which is given by the combination of the operations of initial value and stream derivative:

$$A^\omega \to A \times A^\omega, \quad \alpha \mapsto \langle\alpha(0), \alpha'\rangle$$

See [JR97,Rut00] for general references on coalgebra. For a detailed treatment of the final coalgebra of streams, see [Rut01]. The latter paper contains in particular detailed results about behavioural differential equations (for streams over the set $A = \mathbb{R}$ of real numbers).

3 Coinduction and Greatest Fixed Points

There is yet another, in fact more classical, way of understanding bisimulations and the coinduction proof principle. Consider the set

$$\mathcal{P}(A^\omega \times A^\omega) = \{R \mid R \subseteq A^\omega \times A^\omega\}$$

of binary relations on A^ω, and the function $\Phi : \mathcal{P}(A^\omega \times A^\omega) \to \mathcal{P}(A^\omega \times A^\omega)$ defined, for any $R \subseteq A^\omega \times A^\omega$, by

$$\Phi(R) = \{\langle\alpha, \beta\rangle \mid \alpha(0) = \beta(0) \ \wedge \ \langle\alpha', \beta'\rangle \in R\}$$

As an immediate consequence of the definition of bisimulation, we have

$$R \text{ is a bisimulation} \quad \Leftrightarrow \quad R \subseteq \Phi(R)$$

Bisimulation relations are, in other words, *post-fixed points* of Φ. (The characterisation of bisimulations as post-fixed points goes back, in the context of nondeterministic transition systems, to [Par81,Mil80].) Consequently, the coinduction proof principle (1) is equivalent to the following equality, where $id_{A^\omega} = \{\langle\alpha, \alpha\rangle \mid \alpha \in A^\omega\}$:

$$id_{A^\omega} = \bigcup\{R \mid R \subseteq \Phi(R)\}$$

Since id_{A^ω} is itself a (bisimulation and thus a) post-fixed point, it is in fact the greatest fixed point of Φ. Therefore the above equality is an instance of the following well-known greatest fixed point theorem [Tar55]. Let X be any set and let $\mathcal{P}(X) = \{V \mid V \subseteq X\}$ be the set of all its subsets. If $\Psi : \mathcal{P}(X) \to \mathcal{P}(X)$ is a monotone operator, that is, $R \subseteq S$ implies $\Psi(R) \subseteq \Psi(S)$ for all $R \subseteq X$ and $S \subseteq X$, then Ψ has a greatest fixed point $P = \Psi(P)$ satisfying

$$P = \bigcup \{R \mid R \subseteq \Psi(R)\} \tag{2}$$

This equality can be used as a proof principle in the same way as (1): in order to prove that $R \subseteq P$, for any $R \subseteq X$, it suffices to show that R is a post-fixed point of Ψ, that is, $R \subseteq \Psi(R)$. We shall see further instances of this theorem (and as many related proof principles) in Section 6.

4   Timed Data Streams

We model connectors as relations on timed data streams, which we introduce in this section.

For the remainder of this paper, let D be an (arbitrary) set, the elements of which will be called *data* elements. The set DS of *data streams* is defined as

$$DS = D^\omega$$

that is, the set of all streams $\alpha = (\alpha(0), \alpha(1), \alpha(2), \dots)$ over D. Let $\mathbb{R}_+$ be the set of non-negative real numbers, which play in the present context the role of *time* moments. Let $\mathbb{R}_+^\omega$ be the set of all streams $a = (a(0), a(1), a(2), \dots)$ over $\mathbb{R}_+$. Let $<$ and $\leq$ be the relations on $\mathbb{R}_+^\omega$ that are obtained as the pointwise extensions of the corresponding ('strictly smaller' and 'smaller than') relations on $\mathbb{R}_+$. That is, for all $a = (a(0), a(1), a(2), \dots)$ and $b = (b(0), b(1), b(2), \dots)$ in $\mathbb{R}_+^\omega$,

$$a < b \;\equiv\; \forall n \geq 0,\; a(n) < b(n), \quad a \leq b \;\equiv\; \forall n \geq 0,\; a(n) \leq b(n)$$

The set TS of *time streams* is defined by the following subset of $\mathbb{R}_+^\omega$:

$$TS = \{a \in \mathbb{R}_+^\omega \mid a < a'\}$$

Note that time streams $a \in TS$ satisfy, for all $n \geq 0$,

$$a(n) < a'(n) = a(n+1)$$

and thus consist of increasing time moments $a(0) < a(1) < a(2) < \cdots$.

Finally, the set TDS of *timed data streams* is defined by

$$TDS = DS \times TS$$

and contains pairs $\langle \alpha, a \rangle$ consisting of a data stream $\alpha = (\alpha(0), \alpha(1), \alpha(2), \dots)$ in DS, and a time stream $a = (a(0), a(1), a(2), \dots)$ in TS.

As we shall see shortly, connectors will be modelled as relations on timed data streams. Each of the arguments of such a relation will be viewed as an input or as an output end of the connector that is modelled by the relation. Thus the following operational interpretation of a timed data stream $\langle \alpha, a \rangle$ can be given: the time stream a specifies for each $n \geq 0$ the time moment $a(n)$ at which the nth data element $\alpha(n)$ is being input or output:

$\alpha:$	$\alpha(0)$	$\alpha(1)$	$\alpha(2)$	$\cdots$	$\alpha(n)$	$\cdots$
$a:$	$a(0)$	$a(1)$	$a(2)$	$\cdots$	$a(n)$	$\cdots$

Connectors being relations, there are typically many timings a possible at a specific connector's end, which together with a given data stream α form admissible timed data streams $\langle \alpha, a \rangle$, that is, satisfying the connector's relation. A timed data stream $\langle \alpha, a \rangle$ could therefore also be viewed as a *scenario*, one out of many, for the behaviour of a connector end. Connectors, then, relate various such scenarios that together are mutually consistent.

As we already observed in the introduction, one could have, alternatively and equivalently, defined timed data streams as (a subset of) $(D \times \mathbb{R}_+)^\omega$, because of the existence of an isomorphism

$$D^\omega \times \mathbb{R}_+^\omega \cong (D \times \mathbb{R}_+)^\omega, \quad \langle \alpha, a \rangle \mapsto (\langle \alpha(0), a(0) \rangle, \langle \alpha(1), a(1) \rangle, \langle \alpha(2), a(2) \rangle, \ldots)$$

We prefer to work with pairs of streams rather than streams of pairs, because this will allow us to reason about the data streams and time streams separately, which turns out to be of crucial importance for much of what follows.

We could also have used streams of *natural* numbers $0, 1, 2, \ldots$ for our timings, rather than (positive) real numbers. This difference would leave most of our model unaffected. Our model with 'continuous time', however, is more abstract than the model with 'discrete time' would be, in the sense that more connector equivalences can be proved. (In the world of temporal logic, this observation goes back to at least [BRP86].) An example is the equivalence of a *fifo$_2$* buffer (with capacity 2) with the composition of two *fifo$_1$* buffers in Section 7.

Finally, it is often useful to require time streams a to be not only increasing: $a < a'$, but also *progressive*: for every $N \geq 0$ there exists $n \geq 0$ with $a(n) > N$. This assumption prevents 'Zeno' paradoxes, where infinitely many actions take place in a bounded time interval. In most of what follows, the progressive time assumption is not used, but whenever it is, we shall mention it explicitly.

5   Basic Connectors: Channels

The most basic connectors are *channels*, which are formally defined as binary relations

$$R \subseteq TDS \times TDS$$

on timed data streams. For such relations, we distinguish between input and output argument positions, called *input ends* and *output ends*, respectively. This information will be relevant for the definition of connector composition in Section 7. In the pictures that we draw of channels and connectors, input and output ends are denoted by the following arrow shaft and head:

input: $\longmapsto$ $\cdots$ output: $\cdots$ $\longrightarrow$

Here are the (for our purposes) most important examples of channels:

1. The *synchronous channel* $\longmapsto$ is defined, for all timed data streams $\langle \alpha, a \rangle$ and $\langle \beta, b \rangle$, by

$$\langle \alpha, a \rangle \longmapsto \langle \beta, b \rangle \quad \equiv \quad \alpha = \beta \ \wedge \ a = b$$

This channel inputs the data (elements in the) stream α at times a, and outputs the data stream β at times b. All data elements that come in, come out again (in the same order): $\alpha = \beta$. Moreover, each element enters and exits the channel at the very same time moment: $a = b$.

2. The *synchronous drain* $\overset{syndr}{\longmapsto\!\!\dashv}$ is defined, for all timed data streams $\langle \alpha, a \rangle$ and $\langle \beta, b \rangle$, by

$$\langle \alpha, a \rangle \overset{syndr}{\longmapsto\!\!\dashv} \langle \beta, b \rangle \quad \equiv \quad a = b$$

The corresponding data elements in the streams α and β enter the two input ends of this channel simultaneously: $a = b$. No relation on the data streams is specified (the data elements enter and 'disappear').

3. The *fifo buffer* $\overset{fifo}{\longmapsto}$ is defined, for all timed data streams $\langle \alpha, a \rangle$ and $\langle \beta, b \rangle$, by

$$\langle \alpha, a \rangle \overset{fifo}{\longmapsto} \langle \beta, b \rangle \quad \equiv \quad \alpha = \beta \ \wedge \ a < b$$

This is an unbounded fifo (first-in-first-out) buffer. What comes in, comes out (in the same order): $\alpha = \beta$, but later: $a < b$ (which is equivalent to $a(n) < b(n)$, for all $n \geq 0$).

4. The *fifo$_1$ buffer* $\overset{fifo_1}{\longmapsto}$ is defined, for all timed data streams $\langle \alpha, a \rangle$ and $\langle \beta, b \rangle$, by

$$\langle \alpha, a \rangle \overset{fifo_1}{\longmapsto} \langle \beta, b \rangle \quad \equiv \quad \alpha = \beta \ \wedge \ a < b < a'$$

This models a 1-bounded fifo buffer. What comes in, comes out: $\alpha = \beta$, but later: $a < b$. Moreover, at any moment the next data item can be input only after the present data item has been output: $b < a'$, which is equivalent to $b(n) < a(n+1)$, for all $n \geq 0$.

5. The *fifo$_k$ buffer* $\overset{fifo_k}{\longmapsto}$ for any $k \geq 1$, is defined, for all timed data streams $\langle \alpha, a \rangle$ and $\langle \beta, b \rangle$, by

$$\langle \alpha, a \rangle \overset{fifo_k}{\longmapsto} \langle \beta, b \rangle \quad \equiv \quad \alpha = \beta \ \wedge \ a < b < a^{(k)}$$

(Recall from Section 2 that $a^{(k)}$ denotes the k-th derivative of the stream a.) This models a k-bounded fifo buffer, generalizing the fifo$_1$ buffer above. What comes in, comes out: $\alpha = \beta$, but later: $a < b$. Moreover, at any moment the kth-next data item can be input only after the present data item has been output: $b < a^{(k)}$ (which is equivalent to $b(n) < a(n + k)$, for all $n \geq 0$).

6. Let $x \in D$ be any fixed data element. The *fifo(x) buffer* $\overset{fifo(x)}{\longmapsto}$ is defined, for all timed data streams $\langle \alpha, a \rangle$ and $\langle \beta, b \rangle$, by

$$\langle \alpha, a \rangle \overset{fifo(x)}{\longmapsto} \langle \beta, b \rangle \quad \equiv \quad \beta(0) = x \; \wedge \; \alpha = \beta' \; \wedge \; a < b'$$

This channel behaves precisely as the unbounded fifo buffer above, but for the fact that, initially, it contains the data element x, which is the first element to come out: $\beta(0) = x$ (at time $b(0)$).

6   More Channels and the Merge Connector

The definitions of the basic (channel) connectors so far have been, mathematically speaking, fairly straightforward. Next, we introduce some further basic connectors, including the merge operator, using greatest fixed point definitions. As we saw in Section 3, each of these definitions will come together with its own proof principle, similar to the coinduction principle in Section 2.

1. The *asynchronous drain* $\overset{asyndr}{\longmapsto\!\!\dashv}$ inputs any two streams of data items at its two input ends, but never at the same time (in contrast to the synchronous drain of Section 5). It is defined, for all timed data streams $\langle \alpha, a \rangle$ and $\langle \beta, b \rangle$, by

$$\langle \alpha, a \rangle \overset{asyndr}{\longmapsto\!\!\dashv} \langle \beta, b \rangle \quad \equiv \quad a \bowtie b$$

where $\bowtie \subseteq TS \times TS$ is a relation on time streams, given by

$$a \bowtie b \equiv a(0) \neq b(0) \; \wedge \; \begin{cases} a' \bowtie b \text{ if } a(0) < b(0) \\ a \bowtie b' \text{ if } b(0) < a(0) \end{cases}$$

More precisely, $\bowtie$ is defined as the greatest fixed point of the following monotone operator, $\Phi_\bowtie : \mathcal{P}(TS \times TS) \to \mathcal{P}(TS \times TS)$, defined for $R \subseteq TS \times TS$, by

$$\Phi_\bowtie(R) = \{\langle a, b \rangle \mid a(0) \neq b(0) \; \wedge \; \begin{cases} \langle a', b \rangle \in R \text{ if } a(0) < b(0) \\ \langle a, b' \rangle \in R \text{ if } b(0) < a(0) \end{cases} \}$$

Thus $\bowtie = gfp(\Phi_\bowtie)$. A $\bowtie$-*bisimulation* is a relation $R \subseteq TS \times TS$ with $R \subseteq \Phi_\bowtie(R)$. There is, as an immediate consequence of (2) in Section 3, the following $\bowtie$-*coinduction* proof principle. For all time streams a and b:

$$\text{if } a\,R\,b, \text{ for some } \bowtie\text{-bisimulation } R, \text{ then } \alpha \bowtie \beta \tag{3}$$

For an example of a proof by $\bowtie$-coinduction, see the end of the present section.

2. The connector *merge* is a ternary relation M having two input ends and one output end, and is defined, for $\langle \alpha, a \rangle, \langle \beta, b \rangle, \langle \gamma, c \rangle$, by

$$\begin{array}{l} \langle \alpha, a \rangle \\ \qquad\qquad\searrow \\ \qquad\qquad\quad M \longrightarrow \langle \gamma, c \rangle \\ \langle \beta, b \rangle \nearrow \end{array}$$

$$\equiv M(\langle \alpha, a \rangle, \langle \beta, b \rangle, \langle \gamma, c \rangle)$$
$$\equiv a(0) \neq b(0) \ \wedge$$
$$\begin{cases} \alpha(0) = \gamma(0) \ \wedge \ a(0) = c(0) \ \wedge \ M(\langle \alpha', a' \rangle, \langle \beta, b \rangle, \langle \gamma', c' \rangle) \ \text{if } a(0) < b(0) \\ \beta(0) = \gamma(0) \ \wedge \ b(0) = c(0) \ \wedge \ M(\langle \alpha, a \rangle, \langle \beta', b' \rangle, \langle \gamma', c' \rangle) \ \text{if } b(0) < a(0) \end{cases}$$

This connector merges the two data streams α and β on its input ends into a stream γ on its output end, on a 'first come first served' basis. It inputs one data element at a time: $a(0) \neq b(0)$. The data element that is handled first, say $\alpha(0)$ at time $a(0) < b(0)$, is the first element to come out: $\alpha(0) = \gamma(0)$, at exactly the same moment: $a(0) = c(0)$. After that, the connector handles the remainder of the streams in the same manner again: $M(\langle \alpha', a' \rangle, \langle \beta, b \rangle, \langle \gamma', c' \rangle)$. (Similarly for the case that $\beta(0)$ is handled first, at time $b(0) < a(0)$.) The relation M can be formally defined as the greatest fixed point of a monotone operator Φ_M defined, for any $R \subseteq TDS \times TDS \times TDS$, by

$$\Phi_M(R)(\langle \alpha, a \rangle, \langle \beta, b \rangle, \langle \gamma, c \rangle)$$
$$\Leftrightarrow a(0) \neq b(0) \ \wedge$$
$$\begin{cases} \alpha(0) = \gamma(0) \ \wedge \ a(0) = c(0) \ \wedge \ R(\langle \alpha', a' \rangle, \langle \beta, b \rangle, \langle \gamma', c' \rangle) \ \text{if } a(0) < b(0) \\ \beta(0) = \gamma(0) \ \wedge \ b(0) = c(0) \ \wedge \ R(\langle \alpha, a \rangle, \langle \beta', b' \rangle, \langle \gamma', c' \rangle) \ \text{if } b(0) < a(0) \end{cases}$$

An M-*bisimulation* is a relation R with $R \subseteq \Phi_M(R)$ and we have a M-*coinduction* proof principle: if $R(\langle \alpha, a \rangle, \langle \beta, b \rangle, \langle \gamma, c \rangle)$, for some M-bisimulation R, then $M(\langle \alpha, a \rangle, \langle \beta, b \rangle, \langle \gamma, c \rangle)$.

Under the assumption that our time streams are progressive (defined at the end of Section 4), the merge operator is *fair*: from both input ends, infinitely many data elements will be input.

3. The *lossy synchronous channel* $\xrightarrow{lsyn}$ is defined, for all $\langle \alpha, a \rangle$ and $\langle \beta, b \rangle$, by

$$\langle \alpha, a \rangle \xmapsto{lsyn} \langle \beta, b \rangle \quad \equiv$$

$$a(0) \leq b(0) \ \wedge \ \begin{cases} \alpha(0) = \beta(0) \ \wedge \ \langle \alpha', a' \rangle \xmapsto{lsyn} \langle \beta', b' \rangle \ \text{if } a(0) = b(0) \\ \langle \alpha', a' \rangle \xmapsto{lsyn} \langle \beta, b \rangle \qquad\qquad\qquad \text{if } a(0) < b(0) \end{cases}$$

This channel passes an input data element instantaneously on as an output element: $\alpha(0) = \beta(0)$, in case $a(0) = b(0)$; after that, it continues with the remainder of the streams as before. If $a(0) < b(0)$, that is, if it is too early for the output end of the channel to be active, the input data element $\alpha(0)$ is simply discarded (lost), and the channel proceeds with $\langle \alpha', a' \rangle$ on its input and $\langle \beta, b \rangle$ on its output end. As before, this channel can be formally defined as the greatest fixed point of a monotone operator on the set of binary relations on timed data streams.

Next, we illustrate the use of the coinduction proof principles that were introduced above. A look at the definition of M and one moment's thought suffice to see that, for all timed data streams $\langle \alpha, a \rangle$, $\langle \beta, b \rangle$, $\langle \gamma, c \rangle$,

$$\text{if } M(\langle \alpha, a \rangle, \langle \beta, b \rangle, \langle \gamma, c \rangle) \text{ then } a \bowtie b \tag{4}$$

since the merge connector never inputs two data items at its two input ends at the same time. But how to prove it formally? The answer is provided by what we have called $\bowtie$-coinduction above. Consider the following relation:

$$R = \{\langle k, l \rangle \mid \exists \kappa,\ \lambda,\ \mu,\ m:\ M(\langle \kappa, k \rangle, \langle \lambda, l \rangle, \langle \mu, m \rangle)\}$$

Using the definition of M, it is straightforward to prove that this is a $\bowtie$-bisimulation. As a consequence of (3), $R \subseteq \bowtie$, which implies (4). For a second example, consider the following equivalence, for all timed data streams $\langle \alpha, a \rangle, \langle \beta, b \rangle, \langle \gamma, c \rangle$:

$$M(\langle \alpha, a \rangle, \langle \beta, b \rangle, \langle \gamma, c \rangle) \Leftrightarrow M(\langle \beta, b \rangle, \langle \alpha, a \rangle, \langle \gamma, c \rangle) \tag{5}$$

The implication from left to right (and thus the equivalence) follows by M-coinduction from the trivial observation that

$$S = \{(\langle \alpha, a \rangle, \langle \beta, b \rangle, \langle \gamma, c \rangle) \mid M(\langle \beta, b \rangle, \langle \alpha, a \rangle, \langle \gamma, c \rangle)\}$$

is an M-bisimulation, which implies $S \subseteq M$.

7   Composing Connectors

Connectors are relations and their composition can therefore be naturally modelled by relational composition. For instance, the composition of two copies of the synchronous channel yields the following binary relation, defined for all timed data streams $\langle \alpha, a \rangle$ and $\langle \beta, b \rangle$, by

$$\langle \alpha, a \rangle \longmapsto \circ \longmapsto \langle \beta, b \rangle$$

$$\equiv \exists \langle \gamma, c \rangle : \ \langle \alpha, a \rangle \longmapsto \langle \gamma, c \rangle \ \wedge \ \langle \gamma, c \rangle \longmapsto \langle \beta, b \rangle$$

$$\equiv \exists \langle \gamma, c \rangle : \ (\alpha = \gamma \ \wedge \ a = c) \ \wedge \ (\gamma = \beta \ \wedge \ c = b)$$

(which happens to be equivalent to $\langle \alpha, a \rangle \longmapsto \langle \beta, b \rangle \ \equiv \ \alpha = \beta \ \wedge \ a = b$). Composition essentially does two things at the same time: the output end (argument) of the first connector is identified with the input end of the second, and the resulting 'mixed' end is moreover hidden (encapsulated) by the existential quantification. We shall also use the following picture to denote connector composition:

$$\langle \alpha, a \rangle \longmapsto \exists \langle \gamma, c \rangle \longmapsto \langle \beta, b \rangle$$

Here $\exists \langle \gamma, c \rangle$ is used to indicate that this is an internal end of the connector, which is no longer accessible (for further compositions) from outside. The two aspects of connector composition: identification and hiding, could, and for the full version of Reo actually should, be separated. But for the basic examples we shall be dealing with, this type of composition is sufficient.

Note that the identification of connector ends, which are timed data streams, includes the identification of the respective time streams, thus synchronising the timings of the two connectors.

The general definition of the composition of an arbitrary n-ary connector R and an m-ary connector T, is essentially the same. One has to select a number of (distinct) output ends and input ends from R, and equal numbers of input ends and output ends from T, which then are connected in pairs in precisely the same manner as in the example above. To describe this in full generality, one would have to be slightly more formal and explicit about the (input and output) types of argument positions. Although not very difficult, such a formalisation would not be very interesting. Moreover, it will not be necessary for the instances of connector composition that will be presented here. In all cases, the relevant typing information will be contained in the pictorial representations with which connector compositions will be introduced.

We shall also allow an output end to be connected to *several* input ends at the same time, to each of which the output is copied. Here is an example, in which the output end of a synchronous channel is connected to the input ends of two other synchronous channels:

$$\langle \alpha, a \rangle \longmapsto \exists \langle \delta, d \rangle \longmapsto \langle \beta, b \rangle$$
$$\downarrow$$
$$\langle \gamma, c \rangle$$

$$\equiv \exists \langle \delta, d \rangle : (\alpha = \delta \ \wedge \ a = d) \ \wedge \ (\delta = \beta \ \wedge \ d = b) \ \wedge \ (\delta = \gamma \ \wedge \ d = c)$$
$$\equiv \alpha = \beta = \gamma \ \wedge \ a = b = c$$

The connection of several *output ends* to one and the same input end, can be modelled by means of the merge operator introduced in Section 6.

Finally, it is relevant to note that nothing in the above prevents us from connecting an output end to an input end of one and the same connector, simply by connecting them to the input end and the output end of a synchronous channel. In other words, we have in passing included the possibility of *feedback* loops into our calculus.

Here are various examples of composite connectors (including examples of feedback) built out of a number of basic connectors. As usual, $\langle \alpha, a \rangle$, $\langle \beta, b \rangle$, $\langle \gamma, c \rangle$ are arbitrary timed data streams.

1. The composition of two unbounded fifo buffers yields again an unbounded fifo buffer:

$$\xrightarrow{\;fifo\;} \circ \xrightarrow{\;fifo\;} = \xrightarrow{\;fifo\;}$$

because of the following equivalences:

$$\langle\alpha,a\rangle \xrightarrow{\;fifo\;} \circ \xrightarrow{\;fifo\;} \langle\beta,b\rangle$$

$$\equiv \langle\alpha,a\rangle \xrightarrow{\;fifo\;} \exists\langle\gamma,c\rangle \xrightarrow{\;fifo\;} \langle\beta,b\rangle$$

$$\equiv \exists\langle\gamma,c\rangle : \alpha=\gamma \;\wedge\; a<c \;\wedge\; \gamma=\beta \;\wedge\; c<b$$

$$\equiv \alpha=\beta \;\wedge\; a<b$$

$$\equiv \langle\alpha,a\rangle \xrightarrow{\;fifo\;} \langle\beta,b\rangle$$

2. The composition of two $fifo_1$ buffers yields a $fifo_2$ buffer:

$$\xrightarrow{\;fifo_1\;} \circ \xrightarrow{\;fifo_1\;} = \xrightarrow{\;fifo_2\;}$$

because

$$\langle\alpha,a\rangle \xrightarrow{\;fifo_1\;} \circ \xrightarrow{\;fifo_1\;} \langle\beta,b\rangle$$

$$\equiv \langle\alpha,a\rangle \xrightarrow{\;fifo_1\;} \exists\langle\gamma,c\rangle \xrightarrow{\;fifo_1\;} \langle\beta,b\rangle$$

$$\equiv \exists\langle\gamma,c\rangle : \alpha=\gamma \;\wedge\; a<c<a' \;\wedge\; \gamma=\beta \;\wedge\; c<b<c'$$

$$\Rightarrow \alpha=\beta \;\wedge\; a<b<a''=a^{(2)} \quad [\text{since } c<a' \text{ implies } c'<a'']$$

$$\equiv \langle\alpha,a\rangle \xrightarrow{\;fifo_2\;} \langle\beta,b\rangle$$

Given $\alpha=\beta$ and $a<b<a''$, the converse of the above implication can be proved by defining $\gamma=\alpha$ and

$$c(n) = 1/2 \times (\,\max\{a(n),b(n-1)\} + \min\{a(n+1),b(n)\}\,)$$

for all $n \geq 0$ (where $b(n-1)=0$ for $n=0$).

3. Consider the following composition of three synchronous channels:

$$\langle\alpha,a\rangle \xrightarrow{\hspace{2cm}} \circ \xrightarrow{\hspace{2cm}} \langle\beta,b\rangle \quad \equiv \quad \alpha=\beta=\gamma \;\wedge\; a=b=c$$

$$\downarrow$$

$$\langle\gamma,c\rangle$$

This connector can be viewed as a 'take-cue' regulator: any time a data item is taken from γ (by some future context), that same data item is allowed to flow from left to right. This constitutes one of the most basic examples of what could be called *exogenous coordination*, that is, coordination from outside.

4. The following connector is a variation on the previous one, in that the lower channel now is a synchronous drain:

$$\langle \alpha, a \rangle \longmapsto \circ \longmapsto \langle \beta, b \rangle \quad \equiv \quad \alpha = \beta \;\wedge\; a = b = c$$
$$\bigg\vert \; syndr$$
$$\langle \gamma, c \rangle$$

It is a 'write-cue' regulator that regulates the flow of data items from left to right by inputs or writes on the lower channel end. Note that what is being input there is irrelevant. What matters is that such inputs are synchronised, through the synchronous drain, with both the channels above: $a = b = c$.

5. With four synchronous channels and one synchronous drain, the following barrier synchroniser can be constructed:

$$\langle \alpha, a \rangle \longmapsto \circ \longmapsto \langle \beta, b \rangle \quad \equiv \quad \alpha = \beta \;\wedge\; \gamma = \delta \;\wedge\; a = b = c = d$$
$$\bigg\vert \; syndr$$
$$\langle \gamma, c \rangle \longmapsto \circ \longmapsto \langle \delta, d \rangle$$

The synchronous drain in the middle ensures that data items pass through the upper and lower channels simultaneously.

6. Here is a simple example of a feedback loop, consisting of one (unbounded) fifo buffer containing an initial data element $x \in D$, and two synchronous channels:

$$fifo(x)$$
$$\circ \longmapsto \circ \longmapsto \langle \alpha, a \rangle$$

$$fifo(x)$$
$$\equiv \exists \langle \beta, b \rangle \qquad \exists \langle \gamma, c \rangle \longmapsto \langle \alpha, a \rangle$$

$$\equiv \exists \langle \beta, b \rangle \, \exists \langle \gamma, c \rangle : \; (\gamma(0) = x \;\wedge\; \beta = \gamma' \;\wedge\; b < c') \;\wedge\; (\gamma = \beta \;\wedge\; c = b)$$
$$\wedge \; (\gamma = \alpha \;\wedge\; c = a)$$
$$\equiv \exists \langle \beta, b \rangle \, \exists \langle \gamma, c \rangle : \; \alpha = \beta = \gamma \;\wedge\; a = b = c \;\wedge\; \gamma(0) = x \;\wedge\; \gamma = \gamma' \;\wedge\; c < c'$$
$$\equiv \exists \langle \beta, b \rangle \, \exists \langle \gamma, c \rangle : \; \alpha = \beta = \gamma \;\wedge\; a = b = c \;\wedge\; \gamma = (x, x, x, \ldots)$$
$$\equiv \alpha = (x, x, x, \ldots)$$

For the last but one equivalence, note that $(\gamma(0) = x \wedge \gamma = \gamma')$ is equivalent to $\gamma = (x, x, x, \ldots)$; moreover, the inequality $c < c'$ is redundant, because c is by assumption a time sequence and hence satisfies this inequality by definition. The behaviour of this connector is thus pretty much what it should be. It outputs perpetuously the data element x. Note that there are no constraints on the time stream $a \; (= b = c)$. The only requirement is that it is indeed a time stream, that is, satisfies $a < a'$.

7. Let $x \in D$ be again some fixed data element. The following connector acts as a *sequencer* on its two connector ends:

$$\langle \alpha, a \rangle \xmapsto{\;syndr\;} \circ \overset{fifo}{\underset{fifo(x)}{\rightleftharpoons}} \circ \xmapsto{\;syndr\;} \langle \beta, b \rangle$$

$$\equiv \langle \alpha, a \rangle \xmapsto{\;syndr\;} \exists \langle \gamma, c \rangle \overset{fifo}{\underset{fifo(x)}{\rightleftharpoons}} \exists \langle \delta, d \rangle \xmapsto{\;syndr\;} \langle \beta, b \rangle$$

$$\equiv \exists \langle \gamma, c \rangle \, \exists \langle \delta, d \rangle : \; a = c \; \wedge \; d = b \; \wedge$$
$$(\gamma = \delta \; \wedge \; c < d) \; \wedge \; (\gamma(0) = x \; \wedge \; \gamma' = \delta \; \wedge \; d < c')$$
$$\equiv \exists \langle \gamma, c \rangle \, \exists \langle \delta, d \rangle : \; \gamma = \delta = (x, x, x, \ldots) \; \wedge \; c = a \; \wedge \; d = b \; \wedge \; (c < d < c')$$
$$\equiv a < b < a'$$

Thus, arbitrary data elements can be input alternatingly from the left and the right channel ends, at times

$$a(0) < b(0) < a(1) < b(1) < \cdots$$

For future reference, we introduce the following notation for the sequencer connector:

$$\langle \alpha, a \rangle \xmapsto{\;seq\;} \langle \beta, b \rangle \; \equiv a < b < a' \tag{6}$$

8. One can construct a connector that serialises any number k of channel ends by combining $k + 1$ sequencers. For instance,

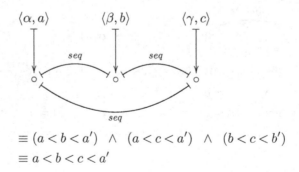

$$\equiv (a < b < a') \; \wedge \; (a < c < a') \; \wedge \; (b < c < b')$$
$$\equiv a < b < c < a'$$

8 Connector Equivalence

In Section 7, we already saw some elementary examples of connector equivalence, such as:

$$\xmapsto{\quad} \circ \xmapsto{\quad} = \xmapsto{\quad}$$

$$\xmapsto{\ fifo\ } \circ \xmapsto{\ fifo\ } = \xmapsto{\ fifo\ }$$

$$\xmapsto{\ fifo_1\ } \circ \xmapsto{\ fifo_1\ } = \xmapsto{\ fifo_2\ }$$

Below we present some further, slightly less elementary examples.

1. Recall the definition of the sequencer in Section 7:

$$\langle \alpha, a \rangle \xmapsto{\ seq\ } \langle \beta, b \rangle \equiv \langle \alpha, a \rangle \xmapsto{\ syndr\ } \circ \overset{fifo}{\underset{fifo(x)}{\rightleftarrows}} \circ \xmapsto{\ syndr\ } \langle \beta, b \rangle$$

$$\equiv a < b < a'$$

(where $x \in D$ is some fixed data item). Here is an alternative way of constructing the sequencer, now with a 1-bounded fifo buffer and a synchronous drain:

$$\langle \alpha, a \rangle \xmapsto{\ fifo_1\ } \circ \xmapsto{\ syndr\ } \langle \beta, b \rangle$$

$$\equiv \langle \alpha, a \rangle \xmapsto{\ fifo_1\ } \exists \langle \gamma, c \rangle \xmapsto{\ syndr\ } \langle \beta, b \rangle$$

$$\equiv \exists \langle \gamma, c \rangle : (\alpha = \gamma \ \wedge \ a < c < a') \ \wedge \ c = b$$

$$\equiv a < b < a'$$

$$\equiv \langle \alpha, a \rangle \xmapsto{\ seq\ } \langle \beta, b \rangle$$

2. Conversely, a 1-bounded fifo buffer can be constructed using two synchronous channels, a sequencer, and an unbounded fifo buffer:

$$\langle \alpha, a \rangle \xmapsto{\ fifo_1\ } \langle \beta, b \rangle \equiv \langle \alpha, a \rangle \longmapsto \circ \overset{fifo}{\underset{seq}{\rightleftarrows}} \circ \longmapsto \langle \beta, b \rangle$$

because we have the following equivalences:

$$\langle \alpha, a \rangle \longmapsto \circ \overset{fifo}{\underset{seq}{\rightleftarrows}} \circ \longmapsto \langle \beta, b \rangle$$

$$\equiv \langle \alpha, a \rangle \longmapsto \exists \langle \gamma, c \rangle \overset{fifo}{\underset{seq}{\rightleftarrows}} \exists \langle \delta, d \rangle \longmapsto \langle \beta, b \rangle$$

$$\equiv \exists \langle \gamma, c \rangle, \exists \langle \delta, d \rangle : \ \langle \alpha, a \rangle = \langle \gamma, c \rangle \ \wedge \ \langle \delta, d \rangle = \langle \beta, b \rangle \ \wedge$$
$$(\gamma = \delta \ \wedge \ c < d) \ \wedge \ c < d < c')$$

$$\equiv \alpha = \beta \ \wedge \ a < b < a'$$

$$\equiv \langle \alpha, a \rangle \xmapsto{\ fifo_1\ } \langle \beta, b \rangle$$

3. Recall the definition of asynchronous drain in Section 6:

$$\langle \alpha, a \rangle \xrightarrow{asyndr} \langle \beta, b \rangle \ \equiv \ a \bowtie b$$

Here is another way of constructing the asynchronous drain, using the merge connector and a synchronous drain:

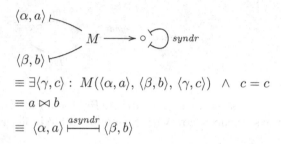

$$\equiv \exists \langle \gamma, c \rangle : \ M(\langle \alpha, a \rangle, \langle \beta, b \rangle, \langle \gamma, c \rangle) \ \wedge \ c = c$$
$$\equiv a \bowtie b$$
$$\equiv \ \langle \alpha, a \rangle \xrightarrow{asyndr} \langle \beta, b \rangle$$

For the middle equivalence, the implication from left to right follows from (3) in Section 6, which was proved by $\bowtie$-coinduction. The converse implication can be proved in a similar fashion, using M-coinduction.

4. Next we look at a connector that is built from two unbounded fifo buffers, the sequencer, and the merge operator:

Using the coinduction proof principle, we shall prove that this connector has the following behaviour (with the operations *zip*, *even* and *odd* as in Section 2):

$$zip(\alpha, \beta) = \gamma \ \wedge \ a < even(c) \ \wedge \ b < odd(c)$$

The proof consists of the following sequence of equivalences:

$$\equiv \exists \langle \delta, d \rangle \exists \langle \epsilon, e \rangle : \ (\alpha = \delta \ \wedge \ a < d) \ \wedge \ (\beta = \epsilon \ \wedge \ b < e) \ \wedge \ (d < e < d')$$

$$\wedge\ M(\langle\delta,d\rangle,\langle\epsilon,e\rangle,\langle\gamma,c\rangle)$$
$$\equiv \exists d,e: \ a<d\ \wedge\ b<e\ \wedge\ (d<e<d')\ \wedge\ M(\langle\alpha,d\rangle,\langle\beta,e\rangle,\langle\gamma,c\rangle)$$
$$\equiv \exists d,e: \ a<d\ \wedge\ b<e\ \wedge\ zip(\alpha,\beta)=\gamma\ \wedge\ zip(d,e)=c$$
[using (7) below]
$$\equiv \exists d,e: \ a<d\ \wedge\ b<e\ \wedge\ zip(\alpha,\beta)=\gamma\ \wedge\ d=even(c)\ \wedge\ e=odd(c)$$
$$\equiv zip(\alpha,\beta)=\gamma\ \wedge\ a<even(c)\ \wedge\ b<odd(c)$$

We have used the following equivalence, which will be proved by coinduction:

$$(d<e<d')\wedge M(\langle\alpha,d\rangle,\langle\beta,e\rangle,\langle\gamma,c\rangle) \tag{7}$$
$$\Leftrightarrow (zip(\alpha,\beta)=\gamma\wedge zip(d,e)=c)$$

For the implication from left to right, define the following relation on streams:

$$R=\{\langle zip(\kappa,\lambda),\ \mu\rangle\ |\ \exists k,l,m: \ (k<l<k')\ \wedge\ M(\langle\kappa,k\rangle,\langle\lambda,l\rangle,\langle\mu,m\rangle)\}$$

We show that R is a bisimulation. Consider a pair $\langle zip(\kappa,\lambda),\mu\rangle$ in R, with corresponding time streams k,l,m. Because $k<l$ it follows from

$$M(\langle\kappa,k\rangle,\langle\lambda,l\rangle,\langle\mu,m\rangle)$$

that $\mu(0)=\kappa(0)$, and since $\kappa(0)=zip(\kappa,\lambda)(0)$, this proves the first of the two bisimulation conditions. Next consider the pair of derivatives

$$\langle zip(\kappa,\lambda)',\mu'\rangle\ =\ \langle zip(\lambda,\kappa'),\mu'\rangle$$

It follows from the definition of M that the latter pair is again in R, since

$$(k<l<k')\ \wedge\ M(\langle\kappa,k\rangle,\langle\lambda,l\rangle,\langle\mu,m\rangle)$$
$$\Rightarrow (l<k'<l')\ \wedge\ M(\langle\kappa',k'\rangle,\langle\lambda,l\rangle,\langle\mu',m'\rangle)$$
$$\equiv (l<k'<l')\ \wedge\ M(\langle\lambda,l\rangle,\langle\kappa',k'\rangle,\langle\mu',m'\rangle)\ \ \text{[by equivalence (5)]}$$

This proves that R is a bisimulation. Assuming now $(d<e<d')$ and

$$M(\langle\alpha,d\rangle,\langle\beta,e\rangle,\langle\gamma,c\rangle)$$

$zip(\alpha,\beta)=\gamma$ follows by coinduction. In the same manner, one shows $zip(d,e)$ $=c$. This proves the implication from left to right of equivalence (7). The implication from right to left can be proved along similar lines, using M-coinduction.

9   Protocol Verification

Our calculus of component connectors also allows the formulation and formal verification of communication protocols. We present a simple example, taken from [BS01, pp. 29–36]. It consists of an (unbounded fifo) lossy buffer composed

with a driver that corrects the lossiness of the buffer. Below we specify both connectors, and prove that their composition is equivalent to an ordinary (correct)
unbounded fifo buffer.

Let $\langle \alpha, a \rangle$, $\langle \beta, b \rangle$, and $\langle \delta, d \rangle$ be timed data streams over an arbitrary data set
D and let $\langle \gamma, c \rangle$ be a timed data stream with $\gamma \in \{0, 1\}^\omega$. We define the lossy
buffer as a ternary relation L on timed data streams with one input end and two
output ends as follows:

$$
\langle \beta, b \rangle \longmapsto L \overset{\textstyle\longrightarrow \langle \delta, d \rangle}{\underset{\textstyle\longrightarrow \langle \gamma, c \rangle}{}}
$$

$$
\equiv L(\langle \beta, b \rangle, \langle \gamma, c \rangle, \langle \delta, d \rangle)
$$

$$
\equiv b < c \ \wedge \ b < d \ \wedge \ \begin{cases} \gamma(0) = 1 \ \wedge \ \delta(0) = \beta(0) \ \wedge \ L(\langle \beta', b' \rangle, \langle \gamma', c' \rangle, \langle \delta', d' \rangle) \\ \vee \\ \gamma(0) = 0 \ \wedge \ L(\langle \beta', b' \rangle, \langle \gamma', c' \rangle, \langle \delta, d \rangle) \end{cases}
$$

This connector inputs data items at the input end β. For every data item that
is input, there are two possible scenarios: (1) the data item is stored successfully and is output (at some later moment) at the upper output end δ together
(not necessarily simultaneously) with a success signal 1 along γ, after which the
connector proceeds as before with the remainder of all streams involved. *Or*:
(2) storage of the data item fails, no data item is output along the end β, a 0
signalling the failure is output along γ, and the connector proceeds as before,
now with $\langle \beta', b' \rangle$ and $\langle \gamma', c' \rangle$, but with $\langle \delta, d \rangle$ unchanged.

It follows from the definition of the lossy buffer that eventually some data
item gets successfully stored and output. In other words, there exists $n \geq 0$ with
$\gamma(n) = 1$. In order to prove this, assume $L(\langle \beta, b \rangle, \langle \gamma, c \rangle, \langle \delta, d \rangle)$ and suppose that
$\gamma(n) = 0$, for all $n \geq 0$. Then

$$
L(\langle \beta^{(n)}, b^{(n)} \rangle, \langle \gamma^{(n)}, c^{(n)} \rangle, \langle \delta, d \rangle)
$$

for all $n \geq 0$ (recall that the superscript (n) stands for the n-th derivative). As a
consequence, $b^{(n)}(0) = b(n) < d(0)$, for all n. Under the assumption that our time
streams are progressive (cf. the end of Section 4): for any $N \geq 0$ there exists
$n \geq 0$ with $b(n) > N$, this is a contradiction. Therefore, there exists $n \geq 0$ with
$\gamma(n) = 1$. We see that, somewhat surprisingly, the fact that all our streams are
infinite and the assumption that time streams are progressive, together imply
here the right type of fairness (or liveness) behaviour.

Next we turn to the driver, which has two input ends and one output end,
and is defined as the following ternary relation D:

$$
\begin{array}{c} \langle \alpha, a \rangle \longmapsto \\ \qquad\qquad D \longrightarrow \langle \beta, b \rangle \\ \langle \gamma, c \rangle \longmapsto \end{array}
$$

$$\equiv D(\langle \alpha, a \rangle, \langle \beta, b \rangle, \langle \gamma, c \rangle)$$

$$\equiv a = b \wedge a < c < a' \wedge \beta(0) = \alpha(0) \wedge$$

$$\begin{cases} \gamma(0) = 1 \wedge D(\langle \alpha', a' \rangle, \langle \beta', b' \rangle, \langle \gamma', c' \rangle) \\ \vee \\ \gamma(0) = 0 \wedge D(\langle \alpha, a' \rangle, \langle \beta', b' \rangle, \langle \gamma', c' \rangle) \end{cases}$$

The driver inputs data items at α and outputs them at β. Before proceeding with the next data item, it checks its input at γ. If $\gamma(0) = 1$ then the last data item that has been output is considered to have been handled correctly (by the lossy buffer in the composition below), and D proceeds as before with the remainder of all streams involved. If $\gamma(0) = 0$, however, something has gone wrong (the buffer has lost the data item), and D sends the data item again. This is modelled here by $D(\langle \alpha, a' \rangle, \langle \beta', b' \rangle, \langle \gamma', c' \rangle)$, in which all streams have progressed to their derivatives but for α, which remains unchanged. As a consequence, $\alpha(0)$ is (again) the next data item that D will output (but note that the time stream a *has* changed into a').

Composing the driver and the lossy buffer as below yields a connector that is equivalent with the (non-lossy) unbounded fifo buffer: for all timed data streams $\langle \alpha, a \rangle$ and $\langle \delta, d \rangle$,

$$\langle \alpha, a \rangle \longmapsto D \quad \overset{\exists \langle \beta, b \rangle}{\underset{\exists \langle \gamma, c \rangle}{\rightleftharpoons}} \quad L \longrightarrow \langle \delta, d \rangle \quad \equiv \quad \langle \alpha, a \rangle \overset{fifo}{\longmapsto} \langle \delta, d \rangle$$

For the implication from left to right, we have to show that $\alpha = \delta$ (and $a < d$). To this end, define the following relation on data streams:

$$R = \{ \langle \alpha, \delta \rangle \mid \exists a, d, \langle \beta, b \rangle, \langle \gamma, c \rangle :$$

$$D(\langle \alpha, a \rangle, \langle \beta, b \rangle, \langle \gamma, c \rangle) \wedge L(\langle \beta, b \rangle, \langle \gamma, c \rangle, \langle \delta, d \rangle) \}$$

In order to prove that R is a bisimulation relation, consider a pair $\langle \alpha, \delta \rangle$ in R with 'witnesses' $a, d, \langle \beta, b \rangle, \langle \gamma, c \rangle$ such that

$$D(\langle \alpha, a \rangle, \langle \beta, b \rangle, \langle \gamma, c \rangle) \text{ and } L(\langle \beta, b \rangle, \langle \gamma, c \rangle, \langle \delta, d \rangle)$$

Let n be the smallest natural number such that $\gamma(n) = 1$ (which exists by the remark above). It follows that

$$D(\langle \alpha, a^{(n)} \rangle, \langle \beta^{(n)}, b^{(n)} \rangle, \langle \gamma^{(n)}, c^{(n)} \rangle) \wedge L(\langle \beta^{(n)}, b^{(n)} \rangle, \langle \gamma^{(n)}, c^{(n)} \rangle, \langle \delta, d \rangle)$$

Together with $\gamma^{(n)}(0) = \gamma(n) = 1$, this implies $\alpha(0) = \beta^{(n)}(0) = \delta(0)$ and, moreover,

$$D(\langle \alpha', a^{(n+1)} \rangle, \langle \beta^{(n+1)}, b^{(n+1)} \rangle, \langle \gamma^{(n+1)}, c^{(n+1)} \rangle) \wedge$$

$$L(\langle \beta^{(n+1)}, b^{(n+1)} \rangle, \langle \gamma^{(n+1)}, c^{(n+1)} \rangle, \langle \delta', d' \rangle)$$

Thus, $\langle \alpha', \delta' \rangle \in R$, which concludes the proof that R is a bisimulation. It now follows by coinduction that $\alpha = \delta$. (A minor variation on this argument proves that $a < d$.)

For the implication from right to left, choose $\langle \beta, b \rangle = \langle \alpha, a \rangle$, $\gamma = (1, 1, 1, \ldots)$, and $c = 1/2 \times (a + a')$ (in a hopefully self explanatory notation). It is now a straightforward proof by (D- and L-)coinduction to show that

$$D(\langle \alpha, a \rangle, \langle \beta, b \rangle, \langle \gamma, c \rangle) \text{ and } L(\langle \beta, b \rangle, \langle \gamma, c \rangle, \langle \delta, d \rangle)$$

10 Conclusion

We have provided a simple and transparent semantical model for Reo, in which connectors are defined as relations on timed data streams. We use coinduction to reason about both time streams and data streams, leading to some initial formal results on expressiveness and connector equivalence in Reo. Our work on Reo and this model is on-going. One of the first questions to address is to decide what set(s) of basic channels and connectors to choose as the basis for a connector calculus (or calculi). Another plan is to look at more instances of connector protocol verification. On the basis of the example of Section 9 (and other examples not included in the present paper, such as the alternating bit protocol), we expect that the present model will be competitive with both traditional dataflow networks and with process algebra, by combining the best of those two worlds. Like data flow and unlike process algebra, Reo is channel-based and models the (communication) topology of connectors explicitly. Like process algebra and unlike data flow, (our model of) Reo is a calculus in which complex connectors are compositionally built out of simpler ones. Moreover, unlike data flow, the model for Reo that we presented is both simple and formal enough to allow actual verification. And unlike process algebra, there is no need to use nondeterministic transition systems and computationally complicated notions such as weak or branching bisimulation. Instead, streams and coinduction are all that is needed.

Acknowledgements

Many thanks to Jaco de Bakker, Falk Bartels, Frank de Boer, Clemens Kupke, Alexander Kurz, and the members of the ACG Colloquium, for many stimulating discussions, suggestions, and corrections.

References

[AM02] F. Arbab and F. Mavaddat. Coordination through channel composition. In *Proceedings of Coordination Languages and Models*, volume 2315 of *Lecture Notes in Computer Science*. Springer, 2002.

[Arb02] F. Arbab. A channel-based coordination model for component composition. Report SEN-R0203, CWI, 2002. Available at URL www.cwi.nl.

[Bar01]   L. Barbosa. *Components as Coalgebras*. PhD thesis, Universidade do Minho, Braga, Portugal, 2001.

[BRP86]   H. Barringer, R.Kuiper, and A. Pnueli. A really abstract concurrent model and its temporal logic. In *Proceedings of the 13th Annual ACM Symposium on Principles of Programming Languages*, pages 173–183. ACM, 1986.

[BS01]   M. Broy and K. Stølen. *Specification and development of interactive systems*, volume 62 of *Monographs in Computer Science*. Springer, 2001.

[Dob02]   E.-E. Doberkat. Pipes and filters: modelling a software architecture through relations. Report 123, Chair for software technology, University of Dortmund, 2002.

[JR97]   Bart Jacobs and Jan Rutten. A tutorial on (co)algebras and (co)induction. *Bulletin of the EATCS*, 62:222–259, 1997. Available at URL: www.cwi.nl/~janr.

[Mil80]   R. Milner. *A Calculus of Communicating Systems*, volume 92 of *Lecture Notes in Computer Science*. Springer-Verlag, Berlin, 1980.

[Par81]   D.M.R. Park. Concurrency and automata on infinite sequences. In P. Deussen, editor, *Proceedings 5th GI conference*, volume 104 of *Lecture Notes in Computer Science*, pages 167–183. Springer-Verlag, 1981.

[Rut00]   J.J.M.M. Rutten. Universal coalgebra: a theory of systems. *Theoretical Computer Science*, 249(1):3–80, 2000.

[Rut01]   J.J.M.M. Rutten. Elements of stream calculus (an extensive exercise in coinduction). In Stephen Brooks and Michael Mislove, editors, *Proceedings of MFPS 2001: Seventeenth Conference on the Mathematical Foundations of Programming Semantics*, volume 45 of *Electronic Notes in Theoretical Computer Science*, pages 1–66. Elsevier Science Publishers, 2001.

[Tar55]   A. Tarski. A lattice-theoretical fixpoint theorem and its applications. *Pacific Journal of Mathematics*, 5:285–309, 1955.

An Attempt at Analysing
the Consistency Problems in the UML
from a Classical Algebraic Viewpoint*

E. Astesiano and G. Reggio

DISI, Università di Genova, Italy

Abstract. In this paper, after introducing the problem and a brief survey of the current approaches, we look at the consistency problems in the UML in terms of the well-known machinery of classical algebraic specifications. Thus, first we review how the various kinds of consistency problems were formulated in that setting. Then, and this is the first contribution of our note, we try to reduce, as much as possible, the UML problems to that frame. That analysis, we believe, is rather clarifying in itself and allows us to better understand what is new and what instead could be treated in terms of that machinery. We conclude with some directions for handling those problems, basically with constrained modelling methods that reduce and help precisely individuate the sources of possible inconsistencies.

1 Introduction

Few years after its introduction, the UML has become, for good and bad, a "lingua franca" for an object-oriented support to software development. Among the many problems raised by its use, the consistency of UML artifacts/documents has emerged as crucial and challenging. Indeed, it has attracted in recent years, especially after the year 2000, a significant amount of work; that interest is also witnessed by the organization of a workshop within the conference ≪UML≫ 2002 on "Consistency Problems in UML-based Software Development" [1].

Within that context "consistency" bears roughly the same meaning of logical consistency; indeed the UML artifacts/documents are organized in "models" that correspond to logical specifications. However the nature of those artifacts is, at least apparently, so different from the traditional specifications used in the formal method community that even the definition of consistency is somewhat controversial. Among the sources of difficulty we can mention: the nature of the notation, that is visual and not directly defined adopting inductive techniques; the UML multiview approach, which describes a system, at some level of abstraction, as a collection of sub-descriptions dealing with possibly overlapping aspects; the use of UML artifacts throughout a software development process typically consisting of many phases, in which the same kind of document may have different meanings. With Engels et al. [18], we can say that *"Altogether the*

---

* Work supported by the Italian National Project SAHARA (Architetture software per infrastrutture di rete ad accesso eterogeneo).

M. Wirsing, D. Pattinson, and R. Hennicker (Eds.): WADT 2002, LNCS 2755, pp. 56–81, 2003.
© Springer-Verlag Berlin Heidelberg 2003

consistency conditions depend on the diagrams [constructs] involved, the development process employed, and the current stage of the development."

Confronted with the consistency problems in the UML, first during some work on UML semantics done within the CoFI Initiative [30] and now in the development of a UML-based method [5,6], because of our long acquaintance with algebraic development techniques, we have been tempted to look at them trying to borrow the classical frame of logical-algebraic specifications, also for understanding and isolating the possible novelties. Indeed, in that framework one could define concepts and problems in a rather rigorous way, though of course that does not mean to solve all consistency problems, many of them being of uncomputable nature. However, among the many related valuable research attempts at dealing with consistency in UML (consider, e.g., [1] also for further references) that view is not explicit; though it appears to be underlying some treatment, notably in [18], where some rather clean informal definitions have a clear logical origin. The analysis in our paper will help understand why most of the related papers departs from the logical-algebraic approach; indeed the relationship is not obvious at all.

The purpose of this paper is first of all informative and exploratory, to establish a link between the work and terminology in the UML world and in the logical-algebraic setting. To achieve that purpose we first present extensively the origin and the nature of the consistency problems in the UML in Sect. 3 and briefly review the current terminology and research directions in Sect. 4. Then in Sect. 5, avoiding many technicalities, in a simple language adopting some conceptual object-oriented notation (a subset of UML), we recall the basic setting of logical-algebraic specifications; and, in Sect. 6, we propose a view of the UML setting in terms of the classical concepts of the logical-algebraic setting. On the basis of the new setting, we then propose, in Sect. 7, a first attempt at exploiting and learning from the outlined correspondence to deal in practice with consistency issues in the UML framework. We will use in the paper a simple UML model as running example, presented in Sect. 2.

Here, by UML we intend UML 1.3 as presented by the official standard specification version 1.3, [37]; while [34] is the reference manual (also if for previous version 1.1) and [21] is a short introduction with examples. A warning for those acquainted with the current literature on the UML consistency: we do not intend here to pursue the so called translational approach, by which one converts the UML or parts of it to some semantically well-defined formalism and then handles consistency by translation. Our attempt is to stay at the UML level trying to individuate at that level the correspondence with a logical-algebraic setting. For a classical algebraic approach we mention, e.g., the one in [38], for a much more updated and comprehensive view see [3], and the CoFI initiative [30].

2   UML Model Example

The UML model presented in this section describes an abstract design of a system for handling the order invoicing in a company. This model includes a UML class diagram plus associated constraints introducing the elements used to model the

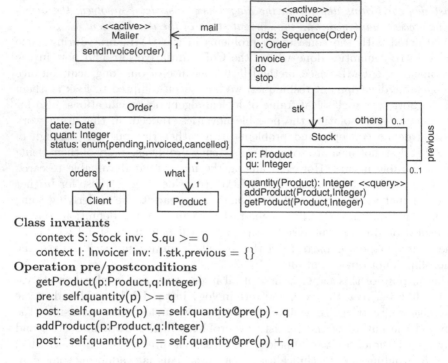

Class invariants
    context S: Stock inv: S.qu $>= 0$
    context I: Invoicer inv: I.stk.previous $= \{\}$
Operation pre/postconditions
    getProduct(p:Product,q:Integer)
    pre: self.quantity(p) $>= q$
    post: self.quantity(p) $=$ self.quantity@pre(p) - q
    addProduct(p:Product,q:Integer)
    post: self.quantity(p) $=$ self.quantity@pre(p) $+ q$

Fig. 1. UML Class Diagram with Constraints.

```
getProduct(p:Product,q:Integer) method:
    if (self.pr  = p) {self.qu  = self.qu - q}
    elseif (self.others  <> {}) {self.others. getProduct(p,q)}
    else {null}
quantity(p:Product):Integer method:
    if (self.pr  = p) {return self.qu}
    elseif (self.others  <> {}) {return self.others.quantity(p)}
    else {return 0}
```

Fig. 2. Methods Associated with Operations of the Passive Class Stock.

invoicing system (Fig. 1), some methods defining some operations of the passive class Stock (Fig. 2), a statechart defining the behaviour of the instances of the active class Invoicer (Fig. 3), and a UML sequence diagram, describing how three objects cooperate to successfully invoicing an order (Fig. 4).

3   Consistency in the UML: The Problem

Consistency is a heavily overloaded word in the computing field, and used in a particular way in the UML community, thus we first try to clarify what it means.

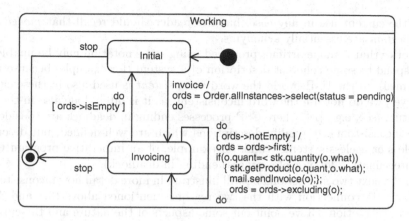

Fig. 3. Statechart Defining the Behaviour of the Active Class Invoicer.

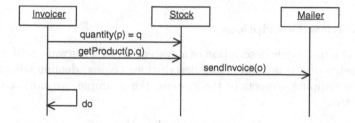

Fig. 4. Sequence Diagram Showing a Successful Order Invoice.

A first informal approximation is in the following statement from [18]:

"During the development process artifacts representing different aspects of the system are produced. The artifacts should be properly related to each other in order to form a consistent description of the developed system."

There are two main reasons to have many different artifacts describing the same system:

multiview description techniques: at some level of abstraction a system is described as a collection of sub-descriptions dealing with different, possibly overlapping, aspects;

phased development process: the system is developed throughout different phases and iterations, each one producing a new more refined description of the system.

In the words of Engels et al. [18] *"Altogether the consistency conditions depend on the diagrams [constructs] involved, the development process employed, and the current stage of the development."*

A terminological clarification: in the world of the object-oriented software engineering a system description is called a *model* (e.g., UML models), whereas in the world of the formal methods, a system description is called a *specification* (e.g., logical-algebraic specifications). We will use the terminology consistent with the current use in any case, but the reader should recall that *model* and *specification* are essentially synonyms.

Notice that a single artifact produced using some notation may be unable to correspond to some coherent description of a system (for example, because it is ill-formed); in the UML world the word *inconsistent* is used also in these cases. However, we do not use the word inconsistent, as it is often done also in the SE literature (see, e.g., [18] where CSP processes ending in deadlock are considered to be inconsistent), to qualify descriptions which are well-defined, but describe senseless or useless systems (think for example, of an imperative program with a nonterminating loop or of a process ending in deadlock).

In the next two subsections we will illustrate in more detail how inconsistency may arise in connection with the two reasons mentioned above (3.1 and 3.2); while in subsection 3.3 we point out some aspects of the nature and the current status of the visual notation used, which may complicate and influence also the treatment of the consistency of a single diagram.

3.1   Multiview Descriptions

We speak of a multiview description of a system (or of a software artifact) whenever it consists of a collection of sub-descriptions (*views*) dealing with different, possibly overlapping, aspects of the system. For example, we can have the following views:

static view: the types of the entities building the system;
behaviour view: the behaviour of the various entities building the system;
interaction view: how the entities building the system interact among them;
...

The UML model of Sect. 2 shows a simple concrete case of a multiview description. Indeed, Fig. 1 is the static view, Fig. 2 and 3 are two behaviour views, and Fig. 4 is an interaction view.

Notice that a description/model/specification split in many different views is not just any description/model/specification modularly decomposed or structured; indeed in the second case the structure/decomposition may follows the structure of the described system; think, e.g., of a description of a distributed system split into the description of the composing processes.

Multiview models have nice advantages but also some problematic aspects. First of all, splitting a model of a system in several views allows to decompose it in chunks having a sensible size, and this is quite relevant in the case of visual models. Moreover, each single view focuses on a different aspect, and this is useful to analyze and to understand the various features of the modelled system. The splitting of a model into views allows also to split the work of producing such model among different persons and/or along the time. The less nice aspect of a multiview model is that the consistency of the overlapping submodels, corresponding to the various views, has to be guaranteed. The example of Sect. 2 shows many cases where the views are overlapping, and so where possible inconsistencies may arise, e.g.,

- the description of the class Invoicer in the static view of Fig. 1 (operations, attributes, constraints) must be coherent with the behaviour of the same class as presented in Fig. 3;
- the behaviour of class Invoicer depicted in Fig. 3 must be coherent with the role played by its instances in the collaboration depicted in the interaction view of Fig. 4;
- the method associated with the operation getProduct in Fig. 2 must be coherent with the pre/postconditions associated with the same operation in Fig. 1.

Another problematic aspect is that when a multiview model is refined to a less abstract one, each submodel cannot be refined independently; think for example, of refining an object of class Invoicer into several cooperating objects, in such case three views of the model must be refined in a coherent way.

3.2 Phased Development Process

In the software engineering world it is now well established that the development process of a software artifact should be organized in phases and that each phase is organized in various activities, where some of them may be iterated several times. Furthermore, many models[1] of such process have been proposed, each one characterized by its own phases and tasks to be done in each of them. Among the most important, of very different nature, from general guidelines to formal standards, we have

- the basic old one waterfall, characterized by four phases: capture and specification of the requirements, design, coding and maintenance;
- the official standard of the German public administration V-Model;
- the most known one based on the UML RUP [31], quite heavy;
- the new Agile methods family (such as Extreme-Programming[2]).

Whatever software process development model we consider, the various phases and iterations require to produce different artifacts describing the system under development, which should be coherent among them; coherent means that they cannot present contradictory statements about an aspect of the system. In general such artifacts are related by particular relationships, such as realization, refinement, implementation, ...

Taking as example the trivial waterfall model, then the design must *realize* the requirements expressed by the requirement specification, and the code should be a correct *implementation* of the design specification.

We have also that in the UML world the models used in different phases may use different UML constructs (recall that UML has been defined by putting together many different notations). For example, we may have use case diagrams, sequence and collaboration diagrams for the requirements, and class diagrams plus statecharts and method definitions for the design.

---

[1] Here we have another occurrence of the word *model*, not to be confused with the UML models.

[2] http://www.ExtremeProgramming.org/

3.3   The UML Notation

In the UML community the word *consistency* is used also w.r.t. a single diagram, with the general idea of the diagram being statically well-formed and having one precise meaning. These aspects are now quite standard and non-problematic for usual programming and specification languages (see, e.g., Java or SDL), and also in the case of a quite refined syntax and sophisticated semantics. Consider, for example, CASL [2] a logical-algebraic specification language that offers a powerful combination of ISO-Latin characters, subtyping, mix-fix operation syntax, and overloading to write quite readable textual specifications, and a sophisticated formal semantics for partiality, subsorting and architectural specifications.

But, UML is a visual language, and so its syntax is not (and cannot be) given by the standard techniques, such as BNF and abstract syntax trees or terms built by combinators. Thus even syntactic correctness does pose new and relevant problems. The chosen technique is to use *metamodelling* to present the abstract syntax plus the associated visual notation (concrete syntax) given apart for each abstract construct. Presenting the abstract syntax (the chapter of the official UML specification concerning the abstract syntax is strangely titled *Semantics*) in a metamodelling style means to give a (meta) class diagram, where the classes (or better metaclasses) correspond to the various constructs, their attributes and the relationships among them (including aggregation/composition) to the elements characterizing such constructs; and specialization has the obvious meaning. The conditions corresponding to the properties guaranteeing the static correctness (well-formedness) are given as constraints attached to such (meta) class diagram. However, due to its dimension, the (meta) class diagram defining UML is split in several parts, each one concerning some particular construct (e.g., state machines) and presented in a separate chapter of the UML official specification [37] together with the relative constraints. Thus, the constraints consider always a unique construct (e.g., state machines: there is a unique initial state, there is no transition leaving a final state), and there are no constraints considering a whole model made by several constructs. So nothing is said about the mutual relationships among the constructs building a model (e.g., a call event appearing on a transition of a state machine must be built by an operation of the context class).

We present in Fig. 5 a simplified fragment of the UML metamodel showing some of the constructs needed to define a class diagram.

The concrete syntax of UML, which is in the official standard [37] called *notation*, correspond to give many different visual diagrams each one corresponding to a subset of a model. Notice that, however, the relationship between the abstract and the concrete syntax is quite weak; for example the class diagram does not appear as a metaclass in the abstract syntax but only in the notation, and a unique model element (collaboration) corresponds to two different visual diagrams: sequence diagram and collaboration diagram.

UML aims to being the *unified* notations to model all the different aspects of a software system during the various development phases, and thus it offers an

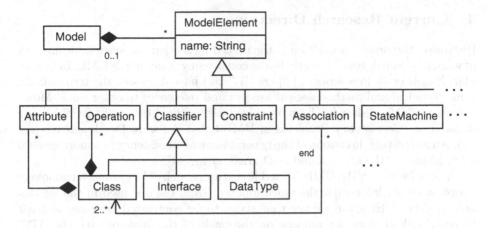

Fig. 5. UML Metamodel Simplified Fragment Concerning Class Diagram.

abundance of constructs originated in many different notations. Indeed, UML combines quite standard ways to present classes and associations, with more specific notations for the behavioural aspects (state machine, sequence diagram, collaboration diagram, activity diagram), with special notation for the use cases (use case diagram), with diagram for the physical/deployment view of the systems, and with constructs to structure the models (packages). We have, as nicely said in [18], a *"multitude of UML diagrams"*.

There is another aspect of UML that makes the treatment of its syntax problematic. Indeed, UML is not just a language to be used as it is in any case; instead to support its use in particular domains, or following particular development methods, or using particular applicative technologies UML offers already in itself mechanisms to define its own specializations (UML profiles). For example,

- stereotypes allow to define variants of the existing constructs,
- tagged values allow to add parts to existing constructs,
- additional constraints allow define restrictions on the subset of the UML which may be used (e.g., no visible attributes are allowed).

For what concerns the semantics (once called *dynamic* as opposed to *static*) of a UML model the situation is still worse. First of all the semantics is defined informally and in several points it is ambiguous, incomplete and even contradictory. Moreover, the existence of a variable semantics for some constructs is explicitly stated (*semantic variation points*); e.g., the policy for handling the event queues for the state machines is not fixed. Such feature allows to tune the notation to various possibly very different usages.

Finally, the same construct may be used for different purposes, even at different phases; for example, a state machine may describe the behaviour of an active or of a passive class, of an operation, of a use case, or of the whole system. In all these cases the semantics is slightly different.

4   Current Research Directions

Because of the problems outlined in the preceding section, a considerable amount of work has been devoted recently to the consistency issues in the UML. In particular, Engels et al. in a series of papers ([17–20]) have discussed the terminology, some technical and methodological aspects and references to other work. Moreover, a special workshop has been organized within the «UML» 2002 Conference devoted to "Consistency Problems in UML-based Software Development" (see [1]). An interesting discussion of the general issue of consistency in a more general setting than UML can be found in Derricik et al. [14].

As can be seen in the UML-related literature, usually the current terminology adopts a rather informal style, much different from the one used in the logical-algebraic field. Here too we use that style, to be contrasted later on, at least in part, with the one we propose on the basis of the analogy with the ADT viewpoint.

First we have to mention two orthogonal classifications of consistency, namely "Horizontal versus Vertical" and "Syntactic versus Semantic".

Horizontal consistency, also named intra-consistency or intra-model consistency, is *"related to consistency among diagrams within a given model"* [1], typically within a development phase.

Vertical consistency, also named inter-consistency or inter-model consistency, is *"concerning consistency between different models"*, typically at different development phases [1]; the qualification "vertical" is referred to the process of refining models and requires the refined model be consistent with the one it refines [19].

Syntactic consistency *"ensures that a specification conforms to the abstract syntax specified by the metamodel . . . this requires that the overall model be well-formed"* [19].

Semantic consistency

- *"with respect to horizontal consistency, requires models of different viewpoints to be semantically compatible with regards to the aspects of the system which are described in the submodels"* [18].
- *"with respect to vertical consistency, semantic consistency requires that a refined model be semantically consistent with the one it refines"* [18].

Finally and noteworthy, in [20] also *evolution consistency* is mentioned and addressed, namely consistency between different versions of the same submodel.

With reference to the aforementioned classification, let us briefly review some research directions.

As for static consistency, there are various attempts, but still a lack of methods for defining static semantics analogous/comparable to the methods centered around term structure and induction principles. As we have seen in the previous section, the static semantics is dealt with using metamodelling (class diagrams plus constraints). Thus the current work on static semantics is concerned with checking metamodelling constraints within a rule checking system, with various techniques. For example the xlinkit system, a generic tool for managing the consistency of distributed documents in XML format, is used and expanded to check consistency of XML documents in XMI format (Finkelstein et al. [23]). In [36] Surrouille and Caplat propose a transformation of OCL constraints into operational rules handled within a knowledge system. Sometimes, as it is typical of a good amount of UML-related work, there is not a a clear cut between syntax and semantics; in this line among other things the NEPTUNE project (Bodeveix et al. [10]) on one side extends OCL (also with temporal operators) and then provides tools for checking well-formedness of OCL rules at the application level and also satisfaction of OCL constraints defined at the metalevel.

As for semantic consistency, we can roughly distinguish three approaches.

In the transformational approach all viewpoints are translated to a common underlying semantic framework and deal with consistency there. The common underlying semantic framework is provided by

- an integration of Transition Systems, Algebraic Specifications and Transformation Rules (Grosse-Rhode [22]);
- Generalized Labelled Transition Systems (Reggio et al. [33], Bhaduri and Venkatesh [9]);
- RDS (Reactive System Design Support) (Lano et al. [26]);
- High Level Petri nets (Baresi and Pezzè [7]);
- Automata (Lilius [27]).

A related approach provides a common semantic model, but to derive consistency checks in the original specification; this is the case of the work of Davies and Crichton ([13]) that derives sets of allowable traces in an object diagram via the CSP semantic model and of Jurjens ([25]) that proposes a refinement for UML diagrams from a common ASM semantic model.

Another approach is by interpreting and testing; for example in [17] Engels et al. propose DMM, Dynamic Meta Modelling, a graphical version of SOS rules used to interpret UML models and to generate tests.

A most important point stressed in particular by Engels et al. [19, 17] is the following principle: in cases such as UML-based development, it is mandatory to provide a *methodological framework* for dealing with consistency issues. Thus we need development methods dealing with consistency, typically by restricting the use of notation to be consistent; for example this is included in the work of the Autofocus group (TUM [35]) and of Huzar et al. [24]. At the end of this paper we will show how consistency is addressed from that methodological viewpoint in our own work [5, 6]. In order to understand the complexity of the issue let us report the steps of a general frame for consistency management proposed by Engels at al. [19, 17]

1. Identification of conceptual model, aspects, notation
2. Identification of consistency problems
3. Choice of a (common) semantic domain
4. Partial mapping of aspects leading to consistency problems
5. Specification of consistency conditions
6. Location and analysis of potential inconsistencies.

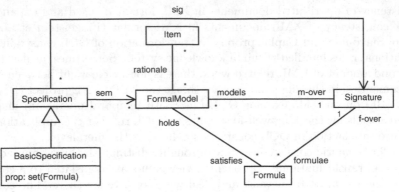

General
> context F: Formula inv: F.holds.m-over = F.f-over
> context FM: FormalModel inv: FM.satisfies.f-over = FM.m-over
> context S: Signature inv: S.models.includesAll(S.formulae.holds)

Well-Formedness Rules
> context BSP: BasicSpecification inv: BSP.prop.f-over = BSP.sig

Semantics
> context SP: Specification inv: SP.sig.models.includesAll(SP.sem)
> context BSP: BasicSpecification inv: BSP.sem = BSP.prop.holds

Fig. 6. Logical-Algebraic Specification Framework: Basic Specifications.

5   The Logical-Algebraic Specification Case

5.1   The Logical-Algebraic Framework

The logical-algebraic framework may be presented in a very refined way using
the categorical language and the concept of institution, see, e.g., [38, 3, 30]. Here,
to make smoother the transition to UML and also to be understandable by the
wider audience of UML specialists, we present it following the metamodelling
style typical of the UML and the concept of specification method of [4]. Of course
that means to remain at a more informal level and not to be concerned with more
sophisticated issues, like the consistency with the change of signatures. In Fig. 6
we summarize the essential structure with the comments below, and [*exemplify
them for the case of first-order specification of partial abstract data types, see
[3]*]. Notice below the different terminology from the one of the UML: the UML
models roughly correspond to the specifications here; while the (formal) models
here correspond to the basic semantic structures.

- The Item are the specified elements [*abstract data structures/types*].
- The FormalModel are formal structures corresponding to the specified items
  [*many-sorted partial algebras*].
- The association rationale describes how the formal models give an abstract
  and precise representation of the specified items [*it is easy to see how many-
  sorted algebras formally models data structure: carriers = sets of values,
  algebra functions = data operations and predicates*]; see [4] for more signi-
  ficative examples of rationales.

- The formal models are classified by their static structure, that in the logical-algebraic case is usually defined by a Signature; in general a signature is a list of the *constituent features* of the formal models [*many-sorted first-order signatures; in this case the constituent features are sorts, operations and predicates*].

  The association, whose association ends[3] are m-over and models links the formal models with the signatures describing their structure [*each many-sorted algebra is built over a many-sorted signature*].

- A Formula is a description of a property of interest about the formal models. More precisely, a formula describes a property that is sensible only for the formal models having a given structure represented by a signature; thus each formula is built over a signature (the association with ends f-over and formulae links signatures with the formulae built over them) [*first-order formulae over a many-sorted signature*].

- The association, whose ends are holds and satisfies, defines when a formula holds on a formal model/a formal model satisfies a formula. Clearly, this relationship is sensible only when the linked formal models and formulae are built over the same signature, see the constraints in Fig. 6 [*the usual interpretation of first-order formulae*].

- A Specification is characterized by a signature (sig) and by a set of formal models, its semantics (sem); such models are all over the signature of the specification (constraint SP.sig.models.includesAll(SP.sem)).

- A BasicSpecification is a specification that consists exactly of a signature and of a set of formulae over it, and determines the set of formal models satisfying all such formulae, constraint BSP.sem = BSP.prop.holds. The well-formedness constraint BSP.prop.f-over = BSP.sig requires that the formulae of a basic specification are built over its signature.

Typically, an algebraic specification language offers together with construct to present basic specifications several ways to structure complex specifications, each one given by a combinator which builds new specifications from existing ones. The most common combinators are

- sum or union ("SP1 + SP2" is the specification having all the constituent features and all the properties of SP1 and of SP2);
- reveal ("reveal SIG in SP" is the specification SP where only the constituent features present in the signature SIG are made visible/revealed);
- rename ("rename SP by ISOMORPH" is the specification SP where its constituent features, sorts, operations and predicates, are renamed as described by the signature isomorphism ISOMORPH).

In Fig. 7 we give a fragment of the extension of the class diagram of Fig. 6 to include structured specifications.

---

[3] In the UML it is possibly to name differently the two ends of an association, each association end will be used to navigate in the direction towards the class to which it is placed near.

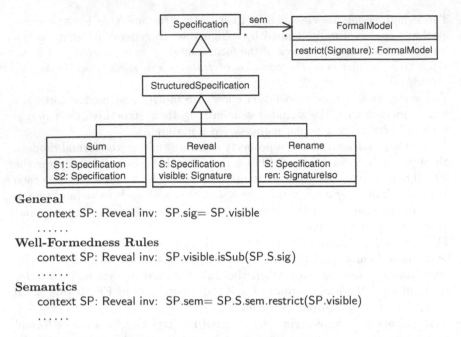

General
    context SP: Reveal inv: SP.sig= SP.visible
    ......

Well-Formedness Rules
    context SP: Reveal inv: SP.visible.isSub(SP.S.sig)
    ......

Semantics
    context SP: Reveal inv: SP.sem= SP.S.sem.restrict(SP.visible)
    ......

Fig. 7. Logical-Algebraic Specification Framework: Structured Specifications Fragment.

5.2   Consistency in Logical-Algebraic Specifications

In the logical-algebraic setting the consistency problems are defined along the following lines, where we use some current terminology found in the literature about UML.

First of all we define the so-called *syntactic consistency* that is called in the formal methods/programming language community *syntactic/static correctness* or *static semantics*.

– A basic logical-algebraic specification is *syntactically consistent* whenever it consists of a set of well-formed formulae over its signature, determine by the association f-over of Fig. 6 (see the constraint BSP.prop.f-over = BSP.sig in the same figure).

Usually in the algebraic setting the syntactically consistent basic specifications are defined in an inductive/constructive way, that is by defining directly by induction the set of all the correct formulae over a signature, and of all the correct (basic) specifications, instead of qualifying which elements in a larger set correspond to correct formulae, see, e.g., [38, 8] for inductive definition of first-order formulae. The syntactic consistency of structured specifications is handled in a similar way; here by means of some constraints on a class diagram, see, e.g., Fig. 7 for an example of such well-formedness rules; whereas in the algebraic

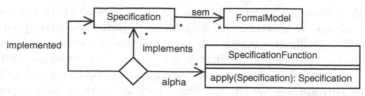

Semantics

context SP : Specification inv: SP.sem .includesAll(SP.alpha.apply(SP.implemented).sem)

Fig. 8. Logical-Algebraic Specification Framework: Implementation.

setting again the correct structured specifications are defined in an inductive/ constructive way [8].

For the semantic consistency, also in the logical-algebraic world we have the distinction between consistency of one specification (or intra-specification consistency) and vertical consistency (or inter-specifications consistency) of one specification w.r.t. another one.

- A logical-algebraic specification SP is *(horizontally) semantically consistent* (standard terminology *consistent*) iff its semantics, defined by the association sem of Fig. 6, is not empty (SP.sem <> {}).

Horizontal semantic inconsistencies in the logical-algebraic case are due to the fact that a specification includes some formula that implies the negation of another of its formulae (including the case of a formula that implies its negation).

Notice that a semantically consistent specification is not always a sensible specification of some data structure. Consider, for example, the case of a specification whose all models are isomorphic to the trivial algebra (the one whose carriers have exactly one element and the interpretation of operations and predicates is the obvious one); this is consistent, but *in most of the cases* it is not what the specifier intended. In the case of partial algebras, you can have also specifications where the unique model is the algebra whose carriers are empty sets. Unfortunately, it is not possible to define in general this kind of specifications, they depend on the particular setting, and cannot fully banned (sometimes they are really wanted, e.g., the specification of a token requires a unique sort with just one element). However, from the methodological point of view, it may be useful to define the class of such specifications, which we name *pseudo-inconsistent*, and perhaps to introduce techniques to detect them.

The problem of vertical consistency, concerning two algebraic specifications, one *implementing* the other, is handled in a very careful way. In this setting, we speak of vertical consistency relative to a given relationship between the structures of the two specifications due to Sannella and Wirsing (see [38]), given as a function mapping specifications into specifications. We summarize in Fig. 8 this notion of implementation, as a ternary association, where the end alpha shows the way a specification is implemented by the other one.

The presence of a semantics (association sem) allows to precisely define the notion of vertical consistency by the constraint in Fig. 8 stating that *"if SP is implemented by SP1, then the semantics of SP (a class of formal models) includes the semantics (another class of formal models) of $\alpha(SP1)$"."*

When the specification function is a composition of a renaming, an extension with derived operations and predicates, a reveal and an extension with axioms, we have the so called implementation by rename-extend-restrict-identify of [15, 16]; which corresponds, within the framework of abstract data types, to the Hoare's idea of implementation of concrete data types.

6   The UML Framework

Here we try to provide a framework for the UML corresponding to the previous one for algebraic specifications. After some preliminary considerations we give a schematic view of the possible correspondence. Our presentation is sketchy in the sense that we concentrate on the basic underlying ideas; indeed a too detailed treatment here would be senseless, since the UML is evolving; the version considered here 1.3 [37][4] will be replaced by a rather different one, UML 2.0.

Here we consider UML at the level of the abstract syntax, which is as it is presented by the metamodel in the *Semantic* chapter of the official standard [37]; reference to the corresponding visual diagrams (as presented in the *Notation* chapter of [37]) will be added to help relate what we present with the current view of UML.

Recall that in the following we use the terminology of Sect. 5.1, along the schema of Fig. 6.

The UML constructs to structure models (specifications) is the package, thus UML models without packages (and their variants as subsystems) may be considered as basic specifications, following the terminology of the previous section.

In this section we will explore the analogy with the algebraic specifications and argue that the role of basic specifications can be played in the UML by the notion of *UML basic model*.

UML Item. UML models are meant to describe real-world systems, as software systems, information systems, business organizations; and thus these are the elements of Item.

UML Signature. In the UML, neither in the metamodel, nor in the associated notation, there is an obvious construct that may play the role of the signature in the algebraic specification case. However, if we look carefully at the various (model) elements building a UML model we may discover that many of such model elements just state which *entities* will be used in the UML model to describe the system (giving their name and their kind/type), for example classes, operations and attributes. We call *structural* such model elements.

Now, a UML model made only of structural model elements may be considered a kind of signature, since it defines the structure of the modelled system; such models will be called *signature diagrams*.

---

[4] To be precise the last version is UML 1.5, but there are no big differences with 1.3.

– A *structural model element* may be

* a classifier of the following kinds: class (distinguished in active and passive), datatype, interface, use case, actor and signal; clearly it will be equipped with its own particular features (e.g., attributes, operations and signal receptions for classes), but without any form of semantic constraints (e.g., the isQuery annotation for an operation, or the specification part for a signal reception);
* an association, but without any semantic attribute (e.g., multiplicity and changeability)[5];
* a generalization relationship, but without any predefined constraint (only the subtype aspect matters at the structural level).

– A *signature diagram* is a UML model built only of structural model elements satisfying some well-formed constraints, as

* all classes have different names,
* all attributes of a class have different names,
* all operations of a class have different names,
* the type of an attribute is either an OCL type, or the name of a class appearing in the diagram,
* ...

Here for simplicity we have expressed the well-formedness constraints on signature diagrams by using the natural language, but they may be expressed precisely as OCL formulae.

We report in Fig. 9 the signature diagram for our running example of UML model of Sect. 2. To help make clear the difference with the class diagram of Fig. 1 we have shadowed the parts not belonging to the signature diagram.

Notice that our signature diagram may be visualized as a particular class diagram, and that any development method based on the UML requires to provide at least a static view of the modelled system, which is a class diagram. Thus our proposal is not peculiar, but it just makes explicit the underlying splitting between the static/structural part of a UML model and the dynamic/behavioural/ semantic part.

It is possible to define our signature diagrams in a very precise way at the metamodel level; we just need to introduce a new metaclass corresponding to the signature diagrams, to slightly modify the metaclasses corresponding to the structural model elements (e.g., by dropping the multiplicity from the associations) and to redefine the metaclass corresponding to a complete model (now it is an aggregate of one signature diagram and of several semantic model elements).

---

[5] We do not include aggregation/composition in the signature diagrams, because essentially they are just a normal associations plus some constraints concerning the creation/termination of the aggregated/composed objects and those of their subparts.

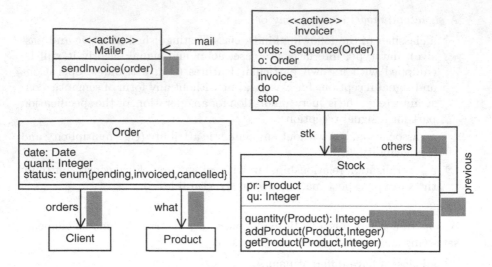

Fig. 9. The signature diagram of the example UML model.

UML FormalModel. There is not a standard choice for the formal models for the UML, because no official formal semantics is available. Many, mostly partial, proposals may be found in the literature; the important point is that the chosen formal models should be structures able to accommodate all the aspects supported by the UML models.

Just for having one at hand to be used as reference for explanatory purposes, we mention our proposal of [32, 33], where we advocated, according to a line of work developed within the CoFI initiative [30], the use of what we call UML-formal systems. They are sufficiently general to be used also for UML 2.0 we believe.

– A *generalized labelled transition system* is a 4-uple

$$(STATE, LABEL, INFO, \rightarrow),$$

where *STATE*, *LABEL* and *INFO* are sets and

$$\rightarrow \subseteq INFO \times STATE \times LABEL \times STATE.$$

A transition $(i, s, l, s') \in \rightarrow$, represented as $i : s \xrightarrow{l} s'$, describes a possibility/capability of the modelled system to pass from a situation represented by s into another one represented by s', where l describes the interaction with the external world during such transition, and i some additional information on the transition (e.g., moves of the subcomponents).

– The *UML-formal systems* are a particular class of generalized labelled transition system (see [33]).

We have then to match each signature diagram with the corresponding UML-formal systems. This can be done quite simply, because the UML-formal systems of [33] include a description of the static structure of the UML model to which they give semantics. For example, they define the names available in such model and classify them in class names, attribute names, operation names and so on.

UML Formula. Surprisingly or not, the UML constructs playing the role of formulae in our setting will be particular model elements (the non-structural ones) that state properties of the *entities* used in the modelled system. We name them *semantic model elements*, and list them below.

constraints including the implicit ones (i.e., those embedded in the definition of structural model elements, as association multiplicities and signal reception specifications);

methods in the UML official standard [37] they are considered as class features, but they just define the semantics of operations; thus they have no structural effect, but restrict the formal models to those where the interpretation of some operation matches the one described by the method itself;

state machines visually presented as statecharts, they fix the behaviour of the instances of classes, or of an operation, or of a use case; thus they restrict the formal models to those where the behaviour of such elements is the one described by the state machine.

collaborations visually presented as sequence or collaboration diagrams, impose restrictions on the possible interactions (exchange of messages, i.e., operation calling) among some objects used to model a system, and so they constrain the behavior of such objects.

activity graphs visually presented as activity diagrams, impose restrictions on the causal relationships among facts happening in the associated entity (an operation, the whole system, a part of the system, a use case, ...) and so they constrain the behavior of the mentioned entities.

Then, we need to define when a semantic model element is built (well-formed) over a signature diagram (association f-over of Fig. 6).

As we have pointed out repeatedly considering the current status of the UML, it is not worthwhile to present here in detail the well-formedness conditions for all constructs. Still it is worthwhile to illustrate that kind of statement w.r.t. some basic points and significant examples. Thus we single out the well-formedness of a state machine, as a paradigmatic case of UML semantic model element.

A state machine SM is well-formed over a signature diagram ΣD iff it fulfills the following rules

- Well-Formedness Rules from UML official specification (not depending on ΣD); e.g., *"A final state cannot have any outgoing transitions."*
- Well-Formedness Rules depending on ΣD, mainly about elements of the SM containing expressions/actions; e.g., for each transition of SM
  * the (OCL) expression of the guard is well-formed w.r.t. ΣD, the context class and the parameter list of the event, and has type Bool;

    * the event is well-formed w.r.t. ΣD and the context class, which depends
      on the kind of event:
      call event the operation is an operation of the context class,
      change event the (OCL) expression is well-formed w.r.t. ΣD and the
          context class, and has type Bool;
      ...
    * the action of the effect is well-formed w.r.t. ΣD, the context class and
      the parameter list of the event.

Notice that to check whether an OCL expression or an action is well-formed w.r.t. ΣD, a class and a possibly empty list of typed free variables, and to find the type of a correct expression are the basic ingredients to define not only when a state machine is well-formed, but also for the other semantic model element.

We want to point out that it is not advisable to follow the inductive/constructive style of the algebraic specifications to define the well-formed UML models. The reason is that UML is a visual notation whose constructs are in large part variants of graphs, and thus they cannot be naturally defined by means of combinators/constructors, which just support tree-like structuring[6].

For what concerns the association holds/satisfies, [33] defines when a UML-formal system is in agreement with an OCL constraint and a state machine. We have made also some feasibility studies for what concerns collaborations and activity graphs, whose results confirm that it is possible to define when a UML formal system is in accord with one of these UML constructs.

The *formulae* in our running UML model are obviously the diagrams of Fig. 3 and 4, the methods of Fig. 2, the explicit constraints in Fig. 1 and the implicit ones in the class diagram in the same figure (as the association multiplicity).

UML BasicModel. The basic specifications in the UML framework are the UML models without packages, which we name *UML basic model*; indeed, the package is the basic and unique UML construct to structure a model. A UML basic model may be rearranged to explicitly present a signature diagram and a set of semantic model elements. The semantics of a UML basic model (association sem) is, then, defined as for the algebraic specifications, that is the set of the UML formal systems over its signature diagram that *satisfies* all its semantic model elements.

Notice that a UML model without packages defined following the UML meta-model of [37] may be automatically transformed into the corresponding basic model, as already hinted before.

A structured UML model should be a UML model containing various packages related by many possible qualified associations, as import, access, and specialize. A careful investigation of the package construct is needed, to determine

---

[6] This aspect is common to any visual notation, not just to the UML. We followed the standard constructive style when presenting JTN [12] a Java targeted visual notation, but at the expense of quite hard labour and of very long and complex definitions.

which are precisely the underlying structuring mechanisms. Moreover, the actual package concept has been the subject of many criticisms, and new way to intend them has been proposed to be introduced in the forthcoming UML 2.0, see, e.g., [11]. For these reasons we have postponed the treatment of the UML structured models.

7 Dealing with Consistency in the UML Framework

7.1 Defining Consistency

Using the various ingredients of the UML framework defined in Sect. 6 we can now define the various kinds of consistency in the same way as for the logical-algebraic case of Sect. 5.

In the algebraic case we have that a basic specification is syntactically consistent (statically correct) whenever its signature is well-formed and all its formulae are well-formed over such signature.

For the UML case we can state

- A UML basic model UBM, essentially a pair $(\varSigma D, \{SME_1, \ldots, SME_k\})$ with $\varSigma D$ signature diagram and for $i = 1, \ldots, k$, SME_i semantic model element, is *syntactically consistent (well-formed/statically correct)* iff $\varSigma D$ is well-formed and for $i = 1, \ldots, k$ SME_i is well-formed over $\varSigma D$(defined by the association f-over, see Sect. 6).

Notice that in this way we do not need to define for all kinds of semantic model elements when they are pairwise syntactically consistent, just as in the logical-algebraic framework, we never had to define which pair of logical formulae are mutually syntactically consistent. Thus the technique that we have proposed is quite modular/scalable; indeed our definition of syntactic consistency can easily extended whenever the UML is extended; for each extension you have just to enlarge/restrict the set of the structural elements and of the semantic model elements, and if new kind of elements are added, just define the new well-formedness conditions w.r.t. the signature diagram.

The horizontal semantic consistency of the UML basic models is defined as in the algebraic case.

- A UML basic model UBM is *semantically consistent* iff its semantics, a set of UML-formal systems, is not empty ($UBM.\mathsf{sem} <> \{\}$).

Notice that this definition, obviously, requires to have at hand a UML formal, or at least quite precise, semantics.

We have thus given the general precise definition, but now we have to analyse the UML models to look for the possible causes of semantic inconsistency. In the algebraic case the only causes for inconsistency are the presence in the specification either of an unsatisfiable formula or of two mutually contradictory formulae, although to check it in the general case is an undecidable problem. Instead, we find that in a UML model there are many different causes of inconsistency and of very different nature. Here, we list some of the most relevant ones:

– a pre/postcondition is in contradiction with a method definition for the same operation;
– a pre/postcondition is in contradiction with an activity graph associated with the same operation;
– a pre/postcondition is in contradiction with a state machine associated with the same operation;
– a precondition on an operation is in contradiction with a state machine or an activity graph including a call of such operation;
– an invariant constraint is in contradiction with a state machine for the same class;
– a collaboration including a role for class C is in contradiction with a state machine for class C.
– ...

Notice, that several cases are quite subtle depending on which is the chosen semantics for the UML constructs; for example, it is quite hard to decide when two collaborations including a role for a class C are contradictory. If we assume that the semantics of a collaboration is to present a possible execution/life cycle/ ... of the modelled system, then two collaborations will never be in contradiction.

The semantics of the UML plays a fundamental role to discover the possible kinds of inconsistency. Moreover, such semantics should also help express in a precise (also if not formal way) the reason for the possible inconsistencies.

Quite surprisingly, the actual semantics of pre/postconditions does not produce inconsistent models, but just pseudo-inconsistent ones (see Sect. 5.2). Recall that the semantics of the pre/postconditions associated with an operation of [37] is precisely intended as follows[7]:

"postcondition: a constraint that must be true at the completion of an operation."

"precondition: a constraint that must be true when an operation is invoked."

Thus, with this semantics a pre/postcondition does not constrain the associated operation, but just its usage; for example, an operation that will be never called satisfies any pre/postcondition.

Some cases of pseudo-inconsistencies are:

– an unsatisfiable invariant constraint on class C (it holds on trivial UML-formal systems where there are no instances of class C);
– two invariant constraints are contradictory (as before);
– two preconditions (postconditions) for an operation are contradictory (it holds on trivial UML-formal systems where the operation will be never called);
– ...

For what concerns the vertical semantic consistency between two UML models we think that the approach chosen in the logical-algebraic framework could

---

[7] Notice that UML pre/postconditions are quite different from the Hoare's one (a postcondition is not related with any precondition).

be sensible also in this case. The problem, is that we have to define the UML correspondent of a *specification function* and to visually represent it, that is a convenient way to define transformations over UML models. Unfortunately, there is no a standard accepted proposal, but this is a hot topic in the UML community, also because of its importance within the MDA approach (Model Driven Architecture) [29].

7.2 Checking Semantic Consistency

The list of all possible causes for semantic inconsistencies in a basic UML model seems to be very long, and a very careful analysis is needed to complete it (see Sect. 7.1). Furthermore, each kind of inconsistency poses a different kind of technical problems, from classical satisfaction problems in the first-order logic or in the Hoare logic, to check whether a sequence is a possible path in a transition tree, or if a transition tree is in agreement with a partial order. As a consequence, a method for helping detect all possible inconsistencies is not feasible. Moreover, the semantics inconsistencies are based on the UML semantics that may change, e.g., in semantics variation point, or because we are using a UML profile for a particular development method, or for a particular phase of the development, or for a particular application domain. On the other side, a development method based on the UML in general uses only models having a particular form (for example, some construct can be never used, or used only in a particular context, or used only in particular form, e.g., state machines only associated with active classes).

Thus, to survive with (horizontal) semantic inconsistencies we propose to design development methods based on the UML with the following characteristics.

- First of all, the method should require to produce UML models with a precise syntactic structure. Such structure must also guarantee the syntactical consistency of those models.
  For example, the method may require that
  * for each operation of a class there is either a pre/postcondition or a method definition, but not both;
  * a use case is complemented by a set of sequence diagrams, and that any of them represents an alternative scenario;
  * at most one invariant is associated with each class and at most one pre/ postcondition is associated with an operation.
  The problem of checking whether a model has the required syntactic form should be computable, and thus it should be possible to develop tools to perform such check; for example, tools for evaluating OCL formulae or tools based on XML technologies, or for evaluating conditional rules.
- The intended (formal) semantics of the UML models produced following the method should be defined.
- The UML models having the particular form required by the method and the chosen semantics should be analysed w.r.t. consistency.
  Thus, it should be possible to factorize the checking of the consistency into a precise list of subproblems. If the possible causes of inconsistency are too

many or too subtle, perhaps the method needs to be revised, by making more stringent the structure of the produced models, or perhaps the troubles are due to the chosen semantics.

- For each inconsistency subproblem detected in the previous activity,
  * it should be described rigorously for the developers in terms of UML constructs without formalities [if you know it, then you can avoid it].
  * guidelines helping detect its occurrences should be developed, using the available techniques, such as automatic tools, inspection techniques, check lists, sufficient static conditions, ...

As an experiment, we have applied the approach proposed in this subsection to analyse a method for requirement specification based on UML, first introduced in [5], and presented in a more detailed way in [6]. Such method requires that the UML models presenting the requirements to have a precise form, and to be, obviously, statically consistent. The proposed approach seems to be quite effective in this case; indeed, the possible causes of semantical inconsistencies have been found explicitly, and they are not too many, because such method requires to produce UML models having a very tight structure, and makes precise the semantics of any used constructs. For what concerns the support to detect the various possible inconsistencies we are extending the tool ArgoUML[8]. with many new critiques, each one signalling either an inconsistency (for example, those concerning the static aspects) or a possible cause of inconsistency (e.g., a postcondition for an operation of a class with an invariant on the same class) with an explanation of the reason. One of the most problematic point, concerning semantic consistency, is to check whether a state machine is in contradiction with a sequence diagram; to help this check we are trying to use a prototyping tool [28].

The above experiment shown the fundamental importance of defining a precise explicit semantics of the used UML constructs. For example, in the method a state machine is used to define the behaviour of each use case, but its semantics is different from the original one. Indeed, following the new semantics a state machine describes A SET of the possible lives of the instances of the context class, whereas the original semantics states that a state machines describe ALL such lives. Thus, two state machines with the new semantics can never be in contradiction.

8   Conclusions

Consistency is a really big problem in practical software development, where scale, heterogeneity and methods do matter. The case of UML-based software development is a good example of the issues that may arise.

Here, contrary to the vast majority of the current literature on the subject, we have taken an unorthodox approach, starting from the experience we have gained in many years of involvement with the logical-algebraic techniques. As

---

[8] http://argouml.tigris.org/

an exploratory experiment, we have presented the problem and then tried to propose a framework for handling the consistency problems of the UML inspired by the framework of the logical-algebraic specifications. Admittedly at first sight the proposal may look naive; perhaps most people will be surprised by seeing a state machine playing the role of a formula; and indeed one may wonder what can be the benefit of that analogy. But, if we exploit that analogy to build a UML framework for consistency, then we can forget the terminology (formulae, formal models, signatures, etc.) and handle the consistency issues knowing exactly what should be done and thus also what we are not able to do; for example, because we do not have at hand a precise semantics. In particular we have a setting where to locate, with merits and limits, the proposed approaches.

Our analysis is preliminary and incomplete, especially for what concern vertical consistency. Still, we believe that it shows the feasibility of the following, even for the new UML versions to come:

- a precise and sensible way to define the various kinds of inconsistencies (static, semantic intra-model, semantic inter-models);
- a workable method for detecting the static inconsistency (in other communities just known under the name of static correctness), that is quite modular and scalable;
- a possible methodological approach to cope with semantic intra-model inconsistency, especially with approaches that are, in our own terminology, "well-founded" and use "tight structuring".

As aside remarks, we believe that some ADT concepts and frames are still useful to provide clarification and practical guidance; but also new problems are appearing, such as the syntax presentation of visual multiview notations, the aspect currently handled in the UML by metamodelling.

As already pointed out before, there are two aspects not treated at all in our analysis, namely structuring and vertical consistency. For the first it seems more sensible to wait for a new version of the UML. As for the second, a link has to be established with the so-called MDA, Model Driven Architecture approach [29] proposed by OMG to developing software, where the relevant relationships between different UML models will be singled out; for example, those from the so called Platform Independent Models and the corresponding Platform Specific Models.

References

1. Consistency Problems in UML-based Software Development: Workshop Materials. In L. Kuzniarz, G. Reggio, J.L. Sourrouille, and Z. Huzar, editors, *Consistency Problems in UML-based Software Development: Workshop Materials*, Research Report 2002-06, 2002.
   Available at http://www.ipd.bth.se/uml2002/RR-2002-06.pdf.
2. E. Astesiano, M. Bidoit, H. Kirchner, B. Krieg-Brückner, P. D. Mosses, D. Sannella, and A. Tarlecki. CASL : the Common Algebraic Specification Language. *T.C.S.*, 286(2), 2002.

3. E. Astesiano, B. Krieg-Brückner, and H.-J. Kreowski, editors. *IFIP WG 1.3 Book on Algebraic Foundations of System Specification.* Springer Verlag, 1999.
4. E. Astesiano and G. Reggio. Formalism and Method. *T.C.S.*, 236(1,2), 2000.
5. E. Astesiano and G. Reggio. Knowledge Structuring and Representation in Requirement Specification. In *Proc. SEKE 2002.* ACM Press, 2002. Available at ftp://ftp.disi.unige.it/person/ReggioG/AstesianoReggio02a.pdf.
6. E. Astesiano and G. Reggio. Tight Structuring for Precise UML-based Requirement Specifications: Complete Version. Technical Report DISI–TR–03–06, DISI, Università di Genova, Italy, 2003. Available at ftp://ftp.disi.unige.it/person/ReggioG/AstesianoReggio03c.pdf.
7. L. Baresi and M Pezzè. Improving UML with Petri Nets. *ENTCS*, 44(4), 2001.
8. H. Baumeister, M. Cerioli, A. Haxthausen, T. Mossakowski, P. Mosses, D. Sannella, and A. Tarlecki. Formal Methods '99 - CASL , The Common Algebraic Specification Language - Semantics, 1999. Available on compact disc published by Springer-Verlag.
9. P. Bhaduri and R. Venkatesh. Formal Consistency of Models in Multi-View Modelling. In [1].
10. J.-P. Bodeveix, T. Millan, C. Percebois, C. Le Camus, P. Bazes, and L. Ferraud. Extending OCL for Verifying UML Model Consistency. In [1].
11. T. Clark, A. Evans, and S. Kent. A Metamodel for Package Extension with Renaming. In J.-M. Jezequel, H. Hussmann, and S. Cook, editors, *Proc. UML'2002*, number 2460 in LNCS. Springer Verlag, Berlin, 2002.
12. E. Coscia and G. Reggio. JTN: A Java-targeted Graphic Formal Notation for Reactive and Concurrent Systems. In Finance J.-P., editor, *Proc. FASE 99*, number 1577 in LNCS. Springer Verlag, Berlin, 1999.
13. J. Davies and C. Crichton. Concurrency and Refinement in the UML. *ENTCS*, 70(3), 2002.
14. J. Derrick, D. Akehurst, and E. Boiten. A Framework for UML Consistency. In [1].
15. H.D. Ehrich. On the Realization and Implementation. In *Proc. MFCS'81*, number 118 in LNCS, pages 271–280. Springer Verlag, Berlin, 1981.
16. H. Ehrig, H.J. Kreowski, B. Mahr, and P. Padawitz. Algebraic Implementation of Abstract Data Types. *T.C.S.*, 20, 1982.
17. G. Engels, J.H. Hausmann, R. Heckel, and S. Sauer. Testing the Consistency of Dynamic UML Diagrams. In *Proceedings of IDPT 2002*, 2002.
18. G. Engels, J.M. Kuester, and L. Groenewegen. Consistent Interaction of Software Components. In *Proceedings of IDPT 2002*, 2002.
19. G. Engels, J. M. Kuster, L. Groenewegen, and R. Heckel. A Methodology for Specifying and Analyzing Consistency of Object-Oriented Behavioral Models. In V. Gruhn, editor, *Proceedings of the 8th European Software Engineering Conference (ESEC) and 9th ACM SIGSOFT Symposium on the Foundations of Software Engineering (FSE-9)*. ACM Press, 2001.
20. G. Engels, J. M. Kuster, and R. Heckel. Towards Consistency-Preserving Model Evolution. In *Proceedings ICSE Workshop on Model Evolution, Florida, USA, May 2002*, 2002.
21. M. Fowler and K. Scott. *UML Distilled: Second Edition.* Object Technology Series. Addison-Wesley, 2001.
22. M. Grosse-Rhode. Integrating Semantics for Object-Oriented System Models. In F. Orejas, P. G. Spirakis, and J. van Leeuwen, editors, *Proceeding of ICALP'01*, number 2076 in LNCS. Springer Verlag, Berlin, 2001.
23. C. Gryce, A. Finkelstein, and C. Nentwich. Lightweight Checking for UML Based Software Development. In [1].

24. B. Hnatkowska, Z. Huzar, L. Kuzniarz, and L. Tuzinkiewicz. A Systematic Approach to Consistency within UML Based Software Development Process. In [1].
25. J. Jurjens. Formal Semantics for Interacting UML subsystems. In B. Jacobs and A. Rensink, editors, *FMOODS 2002*, volume 209 of *IFIP Conference Proceedings*. Kluwer, 2002.
26. K. Lano, D. Clark, and K. Androutsopoulos. Formalising Inter-model Consistency of the UML. In [1].
27. X. Li and J. Lillius. Timing Analysis of UML Sequence Diagram. In R. France and B. Rumpe, editors, *Proc. UML'99*, number 1723 in LNCS. Springer Verlag, Berlin, 1999.
28. V. Mascardi, G. Reggio, E. Astesiano, and M. Martelli. From Requirement Specification to Prototype Execution: a Combination of Multiview Use-Case Driven Methods and Agent-Oriented Techniques. In *Proc. SEKE 2003*. ACM Press, 2003. To appear.
29. OMG Architecture Board MDA Drafting Team. Model Driven Architecture (MDA). Available at http://cgi.omg.org/docs/ormsc/01-07-01.pdf, 2001.
30. P.D. Mosses. CoFI: The Common Framework Initiative for Algebraic Specification and Development. In M. Bidoit and M. Dauchet, editors, *Proc. TAPSOFT '97*, number 1214 in LNCS. Springer Verlag, Berlin, 1997.
31. Rational. Rational Unified Process© for System Engineering SE 1.0. 2001.
32. G. Reggio, E. Astesiano, C. Choppy, and H. Hussmann. Analysing UML Active Classes and Associated State Machines – A Lightweight Formal Approach. In T. Maibaum, editor, *Proc. FASE 2000*, number 1783 in LNCS. Springer Verlag, Berlin, 2000.
33. G. Reggio, M. Cerioli, and E. Astesiano. Towards a Rigorous Semantics of UML Supporting its Multiview Approach. In H. Hussmann, editor, *Proc. FASE 2001*, number 2029 in LNCS. Springer Verlag, Berlin, 2001.
34. J. Rumbaugh, I. Jacobson, and G. Booch. *The Unified Modeling Language Reference Manual*. Object Technology Series. Addison-Wesley, 1999.
35. B. Schatz, P. Braun, F. Huber, and A. Wisspeintner. Consistency in Model-Based Development. In *Proc. of ECBS'03*. IEEE Computer Society, Los Alamitos, CA, 2003.
36. J. L. Sourroille and G. Caplat. Checking UML Model Consistency. In [1].
37. UML Revision Task Force. *OMG UML Specification 1.3*, 2000. Available at http://www.omg.org/docs/formal/00-03-01.pdf.
38. M. Wirsing. Algebraic Specifications. In J. van Leeuwen, editor, *Handbook of Theoret. Comput. Sci.*, volume B. Elsevier, 1990.

MultiMedia Instruction
in Safe and Secure Systems*

Bernd Krieg-Brückner[1], Dieter Hutter[2], Arne Lindow[1], Christoph Lüth[1],
Achim Mahnke[1], Erica Melis[3], Philipp Meier[6], Arnd Poetzsch-Heffter[4],
Markus Roggenbach[1], George Russell[1],
Jan-Georg Smaus[5], and Martin Wirsing[6]

[1] Bremen Institute for Safe and Secure Systems, Universität Bremen
[2] DFKI Saarbrücken
[3] Universität des Saarlandes
[4] Universität Kaiserslautern
[5] Universität Freiburg
[6] Ludwig-Maximilians-Universität München

Abstract. The aim of the MMiSS project is the construction of a multimedia Internet-based adaptive educational system. Its content will initially cover a curriculum in the area of Safe and Secure Systems. Traditional teaching materials (slides, handouts, annotated course material,
assignments, and so on) are to be converted into a new hypermedia
format, integrated with tool interactions for formally developing correct software; they will be suitable for learning on campus and distance
learning, as well as interactive, supervised, or co-operative self-study.
To ensure "sustainable development", i.e. continuous long-term usability of the contents, coherence and consistency are especially emphasised,
through extensive semantic linking of teaching elements and a particular
version and configuration management, based on experience in formal
software development and associated support tools.

1 Introduction and Overview

In the last few years the area of *safe and secure systems* has become more and
more important. Software is increasingly used to control safety-critical embedded systems, in aeroplanes, spaceships, trains and cars. Albeit its associated
security risks, electronic commerce over the internet is rapidly expanding. This
requires a better training of computer scientists in the foundations and practical application of formal methods used to develop these systems. The aim of
the MMiSS-project (*MultiMedia instruction in Safe and Secure Systems*) is to
set up a multimedia internet-based adaptive educational system, covering the
area of Safe and Secure Systems. With a consistent integration of hypermedia
course materials and formal programming tools, teaching in this area will at-
tain a level hitherto impossible in this form. The system will be as suitable for

---

* The MMiSS project has been supported by the German Ministry for Research and
Education, bmb+f, in its programme "New Media in Education".

M. Wirsing, D. Pattinson, and R. Hennicker (Eds.): WADT 2002, LNCS 2755, pp. 82–117, 2003.

learning on campus and for distance-learning with its associated management of assignments, as it is for interactive, supervised, or co-operative self-study.

At the core of the system is the hypermedial adaptation of a series of courses or lectures on the development of reliable systems. The teachers should be able to store various sorts of course material, such as overhead slides, annotations, lecture notes, exercises, animations, bibliographies, and so on, and retrieve them again for use in teaching, notably also re-using material of other authors. The system provides a formal framework for the integration of teaching materials based on a *semantic structure (ontology)* and enables fast directed access to individual teaching elements.

An initial collection of courses is already available and should be further hypermedially developed as part of the project in an Open Source Forum (cf. Sect. 8). It covers the use of *formal* methods in the development of (provably) *correct* software. Highlights include data modelling using algebraic specifications; modelling of distributed reactive systems; handling of real-time with discrete events; and the development of hybrid systems with continuous technical processes, so-called *safety-critical systems*. The curriculum also covers informal aspects of modelling, and introduces into the management of complex developments and into the basics of *security*.

The teaching material should, where possible, be available in several different *variants*. It should be left to the teachers, or the students, to choose between variants, according to the educational or application context. For example reactive systems could be modelled with either process algebras or Petri-nets; the material could be avaliable in English or other natural languages. The system also contains meta-data, representing ontological, methodological and pedagogical knowledge about the contents.

An important educational aspect is to teach about the possibilities and limits of formal tools. *Tools for formal software development* should be integrated in the system, to illustrate and intensify the contents to be taught. Thus students doing assignments can use the system to test their own solutions, while gathering experience with non-trivial formal tools. The integration of didactic aspects with formal methods constitutes a new quality of teaching. It will become possible both to present a variety of formal tools as a subject for teaching, and to use them as a new medium. Thus an algorithm can for instance be simultaneously developed, presented, and verified.

The goal of applying the MMiSS-system in as many universities and companies as possible, and the fact that the area of Safe and Secure Systems will further evolve in the future, requires the highest level of *flexibility, extensibility and reusability* of the content. It should be possible to incrementally extend or adapt content and meta-data, to suit the teacher's individual requirements, and to keep them up-to-date. We expect the system to be easily adapted and well usable in other subject domains.

As the individual parts of a curriculum rely on each other, there is a network of *semantic dependencies* that the system should be able to administer; at the least it has to offer a version- and configuration management. Additionally, an *ontology* allows a better support for orientation and navigation within the content. It forms the basis for adaptation to the user, for example by learning from exercises which concepts the students have understood, and by adapting future assignments accordingly.

The formalisation of semantic dependencies means that the system can help to *maintain the consistency (and completeness)* of the content. Definitions must be coordinated to suit each other; the removal or adaptation of some material may force the removal or adaptation of all dependent concepts. In formal software development, a similar problem has to be solved: there are also semantic dependencies between different parts of a development, for example between specification and implementation. Some of the project partners have already developed techniques for the administration of such dependencies as things change, and implemented them in development tools. Here we perceive an important *synergy* between expertise in formal software development – and support tools – and the demands of long-term *sustainable development for re-use* of consistent multimedia materials in an efficient and productive educational system.

Outline. Although the MMISS project is concerned with the development of a multi-media based eduction system for safe and secure systems, the techniques and tools developed in this project are not restricted to this particular area. In this sense we start with a description of general concepts to structure documents according to their semantics in Sect. 2 and briefly sketch an extension of LATEX, called MMiSSLATEX, that allows the specification of these additional structuring concepts. Sect. 3 describes the particular ontology used to structure the contents in this particular problem domain. Sect. 4 illustrates the tool support of MMISS to create or maintain such course material. MMISS provides various authoring tools to transfer LaTeX or Powerpoint based course material, enriched with additional semantic information, into the MMISS-Repository which maintains versions, configurations and the (user-defined) consistency of the material. Sect. 7 focusses on the presentation of the teaching material stored in the Repository. MMiSS supports two presentation mechanisms, one simply using the layout information as it is encoded in input documents written in MMiSSLATEX, and the ActiveMath environment that dynamically generates the presentations according to the skills and needs of the user. We conclude this presentation in Sect. 8 with a prelimary evaluation of teaching experiences using the MMiSS tools and with a view towards future developments.

2   Structuring Mechanisms for Documents

MMiSS aims at the support of the creation, the maintenance and the presentation of education material dedicated to various courses or lectures in a domain. As an author, one has to be aware of the various (mainly semantically oriented) structuring mechanisms hidden in these documents. Writing, for instance, a mathematical document in LATEX, there is an explicit syntactical structure of the document triggered by LATEX commands such as section, paragraph, etc.

Additionally, there are other more semantically oriented structuring mechanisms. Defining mathematical entities, we are likely to build up a hierarchy of definitions. In a conventional LATEX document, we do not represent these relations explicitly. However, we have to keep them in mind once we want to change documents in a consistent way. The overall design of MMiSS aims at an explicit representation of such relations, e.g.

- to navigate in the material along the semantical relations: during class, as a teacher; after class or during self-study, as a student;
- to support maintenance and update of course material.

2.1   An Ontology of Users

Ontologies provide the means for establishing a semantic structure. An ontology is a formal explicit description of concepts in a domain of discourse. Ontologies are becoming increasingly important because they provide also the critical semantic foundations for many rapidly expanding technologies such as software agents, e-commerce and knowledge management. Ontologies consist of concepts and relations between these concepts. Properties of a concept are specified by describing its various features and attributes. As instantiation we use a subset of the modeling language UML which is an actual de facto standard language for software development. As an ontology describes domain concepts abstractly by means of classes, subclasses and slots, UML seems to be particularly well-suited for the diagrammatic representation of the ontology [5].

Before we explore the variety of MMiSS Document Constructs for structuring in the sequel, let us consider a little example of an ontology in Fig. 1; for a more extensive treatment of the ontology subject see Sect. 2.6 and Sect. 3 below.

The example shows an *ontology* of potential Users of MMiSS and their Roles, resp. A *Professor*, as an *Academic User*, may assume the Role of a *Teacher*, thereby only having reading access to the material, or of an *Author* with a particular kind of writing access. This ontology is of course much simplified (there are also Developer and Administrator Roles, etc.). As notation we use a subset of UML class and object diagrams. It shows, however, the general principles:

- a taxonomic hierarchy of *classes* (the fat arrow with a triangular head denotes the "subclassOf" relation), e.g. a Professor is an Academic User, inheriting all its properties,
- individuals (*"objects"*) of these classes (not shown here, but cf. _deAttribute_ in Fig. 3 as an object of class LanguageAttribute, distinguished in the notation by underlining or leading and trailing underscores), and
- a (hierarchy of) *relations* (called associations in UML), declared between the classes and applied to objects, e.g. User assumesRole Role. Note that relations are inherited by the classes involved, e.g. Professor assumesRole Teacher.

We will use this notation to define and illustrate (parts of) the MMiSS ontology in the sequel. As an experiment in formatting, classes, objects and relations relating to the ontology used in this paper are highlighted by special fonts when

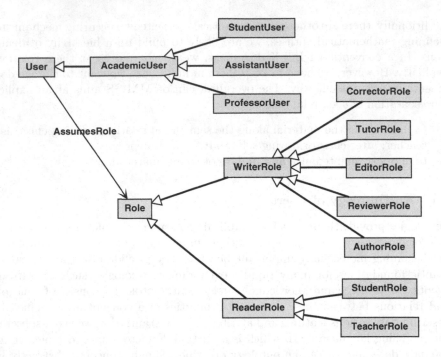

Fig. 1. Ontology of Users and Roles.

referred to in the text (as in ontology), and classes and objects pertaining to the MMiSS ontology are capitalised (as in LanguageAttribute).

2.2   Document Structure

Structural Entities. The primary purpose of structuring documents is conceptual. We are used to textually nesting paragraphs in sections, and sections in other sections, possibly classified into (sub)subsections, chapters, parts of documents or the like. The same is true here, cf. Fig. 2 that shows part of the ontology of MMiSS *Structural Entities*: *Sections* may be nested; they are not classified as chapters or the like to ease re-structuring without the need for renaming (section numbering etc. is, if desired, done automatically anyway during layout; the title of a Section will appear in a table of contents). A Section may contain smaller nested entities such as Units or Atoms, see below.

The largest Structural Entity is a *Package*. A Package is a document that corresponds to a whole course or book and contains all Structural Entities pertaining to it. A Package contains a Prelude that contains a kind of "global declarations" for it, e.g. a BibliographyPrelude, or an ImportPrelude for other Packages ("structuring in-the-large"), see Sect. 2.3.

Ontology Prelude. In particular, it may contain an OntologyPrelude, where the elements of an ontology may be declared (cf. Sect. 2.6) — it acts like a

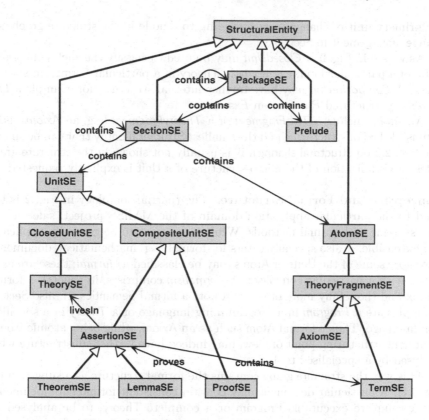

Fig. 2. (Partial) Ontology of Structural Entities.

signature of the Package for semantic interrelation, promising these elements to be defined in this Package, such that they become available when imported by another.

Sections, Units and Atoms. Each Section should contain three special sub-Sections or Units: The *Abstract* contains an overview of the Section; the *Introduction* gives a motivation for the content to come and sets a didactic goal ("what we are about to learn"); the *Summary* at the end recalls the highlights of the content ("what we have learned"). Note that there are no explicit "transition" Paragraphs between Sections since they would assume a given order; instead, the Introduction should refer to the upper context ("what we already know"), if necessary, and the Summary should provide forward references to the lower context ("what we will learn more about") in subsequent Sections.

A Section may contain *Units*, and Units may contain other Units or Atoms. A Unit is an entity one would like to be able to keep together and eventually present as a whole as far as possible, i.e. on a single slide in a lecture (or a single page in a book), possibly with a continuation slide with the same title. The Unit is the primary structuring facility; it is the minimal context for editing and the primary unit of change (corresponding to a node in the structure graph, see change management in Sect. 6.3).

As we see in Fig. 2, a *ClosedUnit* may be used to classify the enclosed content to be of a particular kind, e.g. a whole Theory in a particular Formalism such as CASL. A *CompositeUnit* may have further internal structure, for example a *List*, a *Table*, a structured *Proof*, or an *Example*.

An *Atom*, such as a *TextFragment* or a *TheoryFragment* (e.g. an *Axiom*), is an indivisible leaf of structuring and the smallest Structural Entity that can be shared (see Sec. 2.2 on structural sharing); it is usually not shown in the structure graph unless a visualisation of the micro-structure of a Unit is explicitly requested.

Conceptual and Formal Structure. The (partial) ontology in Fig. 2 is tailored to the particular application domain of the MMiSS project: safe and secure systems with formal methods. While it is meant to be generally applicable and extensible, it also specially caters for formal, e.g. mathematical, documents. Therefore some of the Units or Atoms may be classified as *formal*; these are associated with a particular Formalism. A *Formalism* comprises, in general, a formal Syntax and (hopefully more often than not) a formal Semantics, cf. also Sect. 3. Examples are a *Program* in a programming language or a *Theory* in a specification language. Thus a formal Atom such as an Axiom, while being atomic from a document structuring point of view, may indeed have further substructure when analysed by a specialised tool.

This way the structure graph contains the formal structure as a subgraph (cf. Fig. 14). A particular document may contain consistent formal sub-documents, e.g. a complete executable Program or a complete Theory, to be analysed together.

Sharing. Formal entities may be embedded piecewise (e.g. just an Axiom of a Theory), as they are being introduced and explained from a conceptual or pedagogical point of view. However, it is also a good idea to present them together, possibly in a separate part of the same document. Thus they are exhibited as a consistent whole, both from a conceptual point of view (e.g. a complete Theory with all TheoryFragments put together) and the technical consideration of having a complete formal document that can be treated by a tool (e.g. analysis of a complete Theory or compilation and execution of a Program with input data). Note also that it is often necessary for pedagogical purposes to be able to present alternatives and variations in a document, even incomplete or intentionally wrong ones that should not be subjected to formal analysis.

Units or Atoms of such a whole formal (sub)document may then be referred to repeatedly in other parts of the document; one would often wish them to be included as such instead of a mere Reference. Indeed, an entity will often appear in more than one place, e.g. as an Axiom in an explanatory Paragraph and as part of a consistent and complete Theory in an appendix. A copy will not do; common experience dictates that two copies of the "same" entity have a tendency to differ eventually. Thus *structural sharing* is needed, avoiding the danger of unintentional difference: an Axiom named by a Label in one part of a document (or a different document) may be included by an IncludeAtom operation, with a link to this Label, in another[1]. This operation will trigger a textual expansion in the presentation of the document such that both occurrences are indistinguishable in the presentation. In the source, the Axiom has a "home" where it can be edited, whereas it cannot be edited at the positions of the IncludeAtom operation. From a methodological point of view, it is preferable to maintain a complete Theory, which is, however, structured in such a way that links to a particular Axiom are possible from other places.

Sharing is not restricted to formal entities. Indeed, whole sub-documents can be shared when composing a new document from bits and pieces of existing ones.

Comprises and ReliesOn Relations. The textual nesting gives rise to *contains* relations and the include operations to *includes* relations corresponding to arrows in a directed acyclic graph, the structure graph, see Sect. 5.4 (cf. also Fig. 2 and Fig. 14). An Axiom, for instance, is contained in a Section, while Sections are themselves contained in Packages. MMiSS defines a hierarchy of Structural Entities to define the contains relation, e.g. Packages, Sections, Units, or Atoms. The contains and includes relations are special cases of the *comprises* relation; in the sequel we will make use of this comprises relation to define an appropriate change management for MMiSS-documents.

Besides the comprises relations, there is a family of *reliesOn* relations, reflecting various semantic dependencies between different parts of a document (cf. Fig. 2). For example, an Assertion (such as a Theorem) *livesIn* a Theory, a Proof *proves* an Assertion, an Example *illustrates* a Definition, and so on. In this case, we would usually like to insist on a linear order of appearance, i.e. the right-hand-side (target) of the relation should (textually) be presented before the left-hand-side.

2.3 Packages

Packages provide a means for modular document development by introducing *name spaces.* When writing a document, authors introduce identifiers as Labels for Structural Entities or as technical terms in an ontology. If these identifiers, subsumed as *names* in the sequel, are defined more than once, we say there is a *name clash.*

A Package encapsulates the name space of a document, such that names defined in a Package do not clash with names from other Packages. In order to use names from other Packages, these have to be imported explicitly (see below). In other words, Packages are very much like modules in programming languages such as Modula-2, Haskell, or Java.

Package Hierarchy. Packages are organised in a folder hierarchy, with the names given by *paths*. Because path names can get very long, paths can be renamed by *Path* declarations (*aliases*) of the form

---

[1] It is technically immaterial, whether the Axiom appears in the Paragraph and the IncludeAtom operation in the Theory, or vice versa.

$$a = p_1.p_2.\dots.p_n$$

where a is the new alias, and $p_1,\dots,p_n$ are either folder names or previously defined aliases, subject to the condition that each alias is defined exactly once, and Path declarations are acyclic.

There are three *special aliases*: Current refers to the folder of the Package it is used in; Parent refers to the parent folder; and Root refers to the root folder of the folder hierarchy (thus, the three special aliases correspond to '.', '..' and '/' in Posix systems). Users are discouraged to use the Root alias, since it makes reorganising the folder hierarchy difficult; it is mainly intended for usage by tools.

Export and Import. The *local names* are those defined in this package (as opposed to names imported into the package). By default, all local names are exported. Imported names may be re-exported. It is not possible to restrict the export; rather, name clashes and restrictions are resolved on import.

A Package specifies the imported packages in the *ImportPrelude*. The Import-Prelude contains a number of ImportPreludeDecls; each specifies a Package to be imported, plus a number of *import directives*. Import directives allow us to specify:

- Path aliases;
- Local or global import (when importing globally, the imported names are re-exported);
- Qualified or unqualified import (when an import is qualified, the imported Labels for Structural Entities are prefixed with the name of the Package from which they are imported);
- Hiding, revealing, or renaming of imported names (when we hide a name, it is not imported – when we reveal names, only these are imported);

2.4   Attributes

The possibility to define *Attributes* is a central feature for a Structural Entity, cf. Fig. 3. Standard *StructureAttributes* are e.g. the individual *Label* and *Title* of some Section or Unit.

Inheritance of Attributes. Most importantly, *attribute inheritance* to nested Structural Entities relieves the author from specifying Attributes over and over again and avoids cluttering; at the same time, an Attribute may be superseded for a nested "subtree" of Structural Entities.

Authors and Version Attributes. Each Structural Entity has an *Authors* and a *Version* Attribute (see also version control in Sect. 6.2). These Attributes record the author(s) of each fragment (inherited to nested Structural Entity). The System automatically keeps track of PriorAuthors and the authorship of individual Revisions.

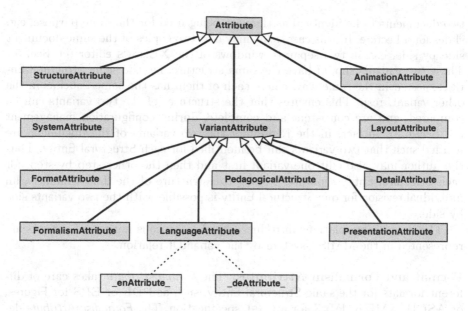

Fig. 3. (Partial) Attribute Ontology.

Layout and Animation Attributes. As will be discussed further in Sect. 7.1 and 7.2, presentation issues such as layout and animation should be separable from the "logical" content of a Structural Entity and should be confined to the necessary only. It is a relief that these can be specified independently as attributes and that attribute inheritance takes care of otherwise tedious repetition of logically irrelevant presentation detail. The specification, for example, that list items on a slide should be rolled out one after another could be specified at the root of a (sub)document and applies to it as a whole unless re-specified for a nested subdocument. Similarly, a revision of such a specification need only be made at its root.

2.5   Variants

Perhaps the most innovative feature of the MMISS project is the definition of *VariantAttributes* and the management of documents with several different *variants* in a consistent way.

Natural Languages. Let us take the *LanguageAttribute* as an example, cf. Fig. 3. A LanguageAttribute specifies the natural language in which a text is written, following the language codes of IETF RFC 1766 / ISO 639. The default is *en-GB* (British English), overriding the standard *ANY* attribute that is usually the default for the other VariantAttributes. Another example is *de* (German).

Let us now assume that an author wants to manage e.g. English and German documents in parallel. Most probably, the author would want the structure of the two documents to be identical as they are being used for the same purpose, e.g. slides for a Lecture. In this case, s/he may edit two copies of the same document side by side, e.g. in two separate windows of the XEMACS editor (cf. Sec. 5). These two variants should have the same structure, i.e. the same Structural Entities, nested in the same way, where each of them has the same Label as in the other variant, resp. This ensures that the structures of the two variants can be compared, and are consistent and complete, during configuration management (cf. Sec. 6.2). In fact, in the Repository the two variants of the document are merged such that two variants can be identified for each Structural Entity. Thus the author may also edit one variant first and then the other step by step, for each Structural Entity separately, along the structure of the first. Similarly, an individual revision for one Structural Entity is possible, with the two variants side by side.

The structuring relations introduced by these various notions of variants are represented in the MMISS-system by the variantOf relation.

Format and Formalism Attributes. The *FormatAttribute* takes care of different formats for the same Structural Entity, such as PDF or EPS for Figures, or ASCII, XML or LaTeX for a CASL specification. The *FormalismAttribute* defines the particular Formalism that a Structural Entity (and all its sub-entities) complies with, e.g. a specification language such as CASL; tools may then take advantage of this fact by checking for a special syntax in an ASCII source or generating a LaTeX variant from it for pretty formatting. The Formalism must be related to a particular ontology of FormalismAttributes, cf. also Sect. 3 and Sect. 2.2.

Detail and Presentation Attributes. A document or an individual Structural Entity may exist at several levels of detail during its development, and for different purposes (cf. the *DetailAttribute* in Fig. 3 and Table 1): a set of slides for a Lecture may be refined by adding annotations to LectureNotes, or further to a complete self-contained Course as a hyper-document for self-study. The Contents and Outline denote the underlying structure reflected in the table of contents, and this structure augmented by the various Summaries, resp. At the other end of the scale, conventional articles and books are located.

Table 1 contains another dimension — the *PresentationAttribute* specifies various kinds of presentation media: presentation on Paper, on a (black or white) Board, or as a Hyper document; further kinds specify presentation using an external tool by Replay of a previously conceived script, or by an Interactive presentation with the tool itself.

2.6 Semantic Interrelation

Declaration of an Ontology. Recall Fig. 1, the example of an ontology of Users and Roles. Such an ontology would be declared in MMISSLaTeX, the LaTeX extension of MMISS to represent the Document Constructs (cf. Sect. 5.1), in the OntologyPrelude as follows (partially shown here):

```
\DeclClass{User}{User}{}
 \DeclClass{AcademicUser}{Academic User}{User}
  \DeclClass{StudentUser}{Student}{AcademicUser}
\DeclClass{Role}{Role}{}
 \DeclClass{ReaderRole}{Reader}{Role}
  \DeclClass{TeacherRole}{Teacher}{ReaderRole}
  \DeclClass{StudentRole}{Student}{ReaderRole}
 \DeclClass{WriterRole}{Writer}{Role}
   \DeclClass{AuthorRole}{Author}{WriterRole}
\DeclRel{*-*}{assumesRole}{assumesRole}{}
\RelType{assumesRole}{User}{Role}
```

Consider e.g. \DeclClass{StudentRole}{Student}{ReaderRole}, the declaration of a class. The first parameter, StudentRole, denotes the particular technical term we use in the ontology; the second, Student, the textual phrase that should appear in the text as default (see below); the third, ReaderRole, the superclass of StudentRole, from which are inherited properties. Analogously, \DeclObject{...} and \DeclRel{...} declare objects and relations, resp., whereas \RelType{assumesRole}{User}{Role} declares the type of a relation, in this case from the class User to the class Role; such a declaration may appear several times for different (sub)classes to allow specific typing and "overloading". In the MMiSS ontology, such an *OntologyDecl* operation appears as a *Prelude Operation* in the OntologyPrelude, cf. Fig. 4.

Table 1. Detail and Presentation.

	Paper	Board	Hyper
	Text+Pictures	Manual	Hyper-Medium
Contents			
skeleton			
Outline			
abstracts			
Lecture			
presentation	handout	presentation on	laptop browsing
in class	before class	black/white board	during class
Lecture Notes			
annotated	handout	annotated	offline browsing
after presentation	after lecture	manuscript	personal annotation
Course			
self-contained	course script	integrated	personal navigation
for self-study		(manu)script	

Definition and References. An OntologyDecl is a kind of promise, that a corresponding *OntologyDef* will appear somewhere as an *Prelude Operation* in the source text of the document; e.g. \DefClass{StudentRole} is the defining

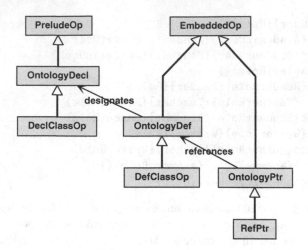

Fig. 4. PointsTo Relations.

occurrence for `StudentRole`, yielding *"Student"* in the formatted document, i.e. the default phrase declared above. Such an OntologyDef operation may appear as an *Embedded Operation* anywhere in the source.

Whenever a class, say, has been declared by an OntologyDecl, the technical term may be referred to simply as `\StudentRole{}` in the source text (or equivalently `\Ref{StudentRole}`), yielding "Student" in the formatted document. If an alternative phrase rather than the default phrase should appear, then e.g. `\Ref[my role as a student]{StudentRole}`, using an optional parameter for this phrase, will yield the desired "my role as a student".

An *OntologyPtr*, e.g. a *Ref*, may appear as an Embedded Operation anywhere in the source, before or after a corresponding OntologyDef. It will yield a hypertext link for the PresentationAttribute Hyper. A full reference may be obtained by `\Reference{StudentRole}`, yielding "StudentRole (see 2.6, on page 94)".

Note that a relation is predefined as a macro with two parameters, thus `\assumesRole{A }{ B}` yields "A assumesRole B", whereas `\Ref{assumesRole}` yields "assumesRole".

Resolution of Ambiguities. There are at least three reasons for having an extra technical term (the first parameter in an OntologyDecl):

- the default phrase (the second parameter) may be translated into a different language variant of the ontology, assuming that the technical term remains the same for uniformity of language variants,
- the technical term may be renamed upon Import from another package to avoid name clashes while the default phrase remains the same,
- apparent ambiguities may be resolved by having two different technical terms with the same default phrase.

To illustrate the ambiguity issue consider the following example and its source:

a Student assumes the role of a Student
a \StudentUser{} assumes the role of a \StudentRole{}

An apparent ambiguity (which is usually resolved by context in natural language) is resolved since there are two different technical terms in the example ontology. Note that a hyper-reference references the appropriate OntologyDef correctly.

PointsTo Relations. Consider Fig. 4: a Reference *references* an OntologyDef, an OntologyDef *designates* an OntologyDecl. Both relations belong to the family of *pointsTo* relations, quite similar to the relation family reliesOn.

Inheritance of Relation Properties. Consider an extract of the ontology of relations for Document Constructs:

```
\DeclRel{*-*}{comprises}{comprises}{relatesDocConstructs}
  \DeclRel{<-}{contains}{contains}{comprises}
  \DeclRel{*-*}{includes}{includes}{comprises}
\DeclRel{>}{reliesOn}{reliesOn}{relatesDocConstructs}
  \DeclRel{}{imports}{imports}{reliesOn}
  \DeclRel{}{livesIn}{livesIn}{reliesOn}
  \DeclRel{}{proves}{proves}{reliesOn}
  \DeclRel{}{after}{after}{reliesOn}
\DeclRel{->}{pointsTo}{pointsTo}{relatesDocConstructs}
  \DeclRel{}{designates}{designates}{pointsTo}
  \DeclRel{}{references}{references}{pointsTo}
\DeclRel{->}{variantOf}{variantOf}{relatesDocConstructs}
```

These form a hierarchy, where each relation inherits the properties for its super-relation. Formal properties are indicated by symbols whose semantics is only sketched here: > denotes a strict order, -> denotes an onto-relation, etc.

3   The Content Ontology for MMiSS Courses

In this section we present the variety of courses produced and presented in the MMiSS project (see Sect. 3.1). Moreover we describe the content ontology structure (see Sect. 3.2) and its development process (see Sect. 3.3) that we are using for the MMiSS courses. In general these content ontologies provide the means for establishing the semantic structure to relate different parts of the teaching material. In this sense an ontology is an explicit formal description of concepts in the domain of discourse.

3.1   MMiSS Courses

The MMiSS courses are divided into three areas for differently experienced audiences. Basic courses are provided for the subjects logic, data models and event models. We provide advanced courses for the subjects verification, data specification and reactive systems. Moreover, there exist specialised courses for subjects like formal software development, security and safety critical systems. For each subject shown in Fig. 5 several courses have been prepared and presented at the partner universities of MMiSS and also at other universities such as TU Berlin, TU Dresden, Univ. of Swansea, etc. For a detailed listing of all courses visit the MMiSS website (see [16]).

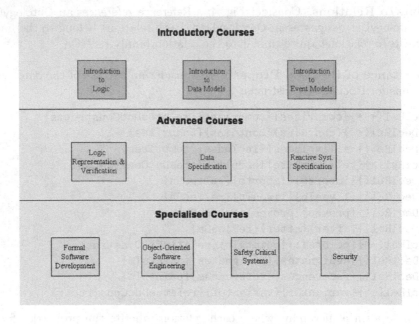

Fig. 5. Structure of MMiSS Courses.

3.2   The Ontology for Formal Methods

Although ontologies exist for many applications we are not aware of any ontology for formal methods. However, we base our ontology on several approaches for classifying and defining topics related to formal methods such as the ACM classification scheme [1], Astesiano and Reggio's work on defining a schema for formal development techniques [3], Clarke and Wing's survey on formal methods [8] and Steffen's framework for formal methods tools [26].

For describing the ontology of Formal Methods and its instances in UML we use class and object diagrams; cf. the introductory example in Sect. 2.1. The class diagrams serve as representation for the abstract notions such as *Domain, Engineering Method, Formal Method, Formalism, Language* and *Tool*. The object diagrams represent the instances of the abstract notions. Typically, particular concepts chosen in a course are represented by object diagrams (see Sect. 3.3).

The most general notion to describe a topic of research or teaching is the notion of Domain (see Fig. 6). A Domain is characterized by a number of *Concept*s and can have zero, one or more subdomains indicated by the association (relation) *isSubDomainOf*. Additionally, a Domain *uses* other Domains (the top level associations like isSubDomainOf and uses are not shown in Fig. 6).

Other classes are specializations of Domain and inherit its associations such as isSubDomainOf and uses. For example, since the class Engineering Method (see Fig. 6) is a specialization of Domain, it inherits the subdomain relation and the relation to Concept. Additionally, an Engineering Method *appliesTo* (zero,) one ore more Domains, it *isSupportedBy* Tools and its pragmatics are described by Processes (see [3]). Note that in our presentation the multiplicities of an association are indicated below the association name. Moreover, the multiplicities of the association ends are separated by a hyphen.

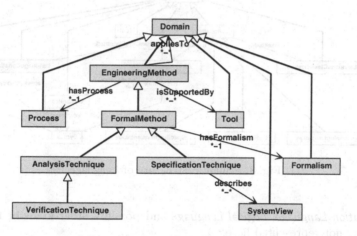

Fig. 6. Semantic Structure of Engineering Method and Formal Method.

The class Formal Method (see Fig. 6) is a specialization of Engineering Method with the particular feature that any instance of Formal Method is based on a Formalism. Formal Methods are classified into *Specification Technique* and *Analysis Techniques*; *Verification Techniques* is a subclass of Analysis Techniques (see [8]). Any Specification Technique serves to specify some *System Views* such as the data view, functional behaviour, concurrent behaviour, performance view etc. The class Formalism (see Fig. 7) is another specialization of the class Domain. A Formalism has one or more associated Languages and a Theory consisting of *Definitions* and *Theorems*. Any Language (see Fig. 8) has several *Language Constructs*, *Language Classifications* such as natural, functional, object-oriented or real time language (see [1]), and can be supported by some Tools. Moreover, any Language possesses a *Language Definition* consisting of a *Syntax* and possibly of one or more *Semantics*. We specialize languages into *Programming Language*,

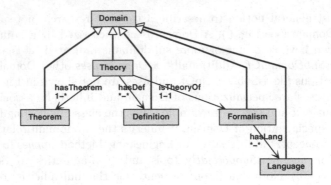

Fig. 7. Semantic Structure of Formalism.

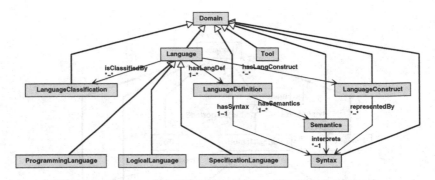

Fig. 8. Semantic Structure of Language.

Specification Language, *Logical Language* and possibly other kinds of Languages which are not represented here.

3.3   Systematic Construction of Ontologies

For constructing ontologies of particular courses in the area of Formal Methods we proceed as follows: We base the ontology of the course on the general model of Formal Methods as outlined above. In a first step, the general model is extended by new abstract Domains of the course that are not yet covered by the general model. In a second step, object diagrams of the ontology specific to the course are constructed according to the extended general model. We give an example of this procedure by describing (part of) the ontology of the course 'Foundations of System Specification' which is held regularly at LMU München. This course presents formal techniques for specifying and refining complex data structures, state-based systems and reactive systems. The underlying Formalisms are algebraic specifications based on the Language CASL for data structures; model-oriented specification techniques based on the Language Z for state-based systems, and Lamport's Temporal Logic of Actions for reactive systems. In the

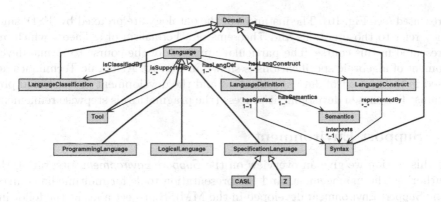

Fig. 9. Extension of the Model by the Languages CASL and Z of the Course.

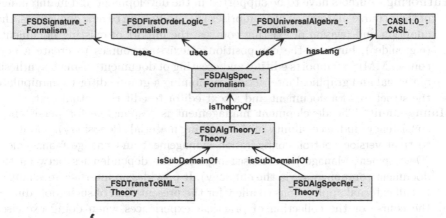

Fig. 10. Ontology for the Alg. Specification Formalism of the Course.

following we present the ontology for the specification of data structures and
state-based systems. In a first step the class diagram of Language is extended
by Z and CASL which form two new subclasses of Specification Language (see
Fig. 9).

The specific instance of CASL [2, 13, 6, 18] used in the course is the version
CASL 1.0. It is classified as Specification Language; its Language Constructs are
partitioned into *Basic CASL Specification*, *Structured CASL Specification*, *Architectural CASL Specification* and *Library CASL Specification*. CASL1.0 has formally
defined *CASL Syntax* and *CASL Semantics*; its *CASL Toolsuite* consists of parsers,
theorem provers and pretty printers (not detailed here; cf. [17]).

The specific instance of Formalism called *FSD Algebraic Specification* of the
course uses basic facts about *FSD Signatures*, *FSD First Order Logic* and *FSD
Universal Algebra* to explain the associated *FSD Algebraic Specification Theory*.
Different notions of refinement including their main properties and a translation
from executable specifications to the functional Programming Language SML are

presented (see Fig. 10). The instances of the ontology are prefixed by 'FSD' since they refer to those notions and Theorems of a Formalism of a Theory which are presented in this course. The particular approach of the course to formal development of algebraic specifications is shown in Fig. 10. Algebraic Techniques are used in the context of data specification and the development of functional programs. The chosen development process (the pragmatics) is stepwise refinement.

4 Support Environment

In this section we give an overview on the *Support Environment* integrating the authoring, the management, and the presentation tools for multimedia courses. The Support Environment developed in the MMiSS project aims at the following goals:

Authoring. Authors have to be supported in the development and maintenance of their course material. This comprises the production of new documents, the adaption and revision of existing courses, the import of existing documents (e.g. slides), but also the composition of existing courses to create a new course. MMiSS supports editing and creating of documents using a synthesis of textual and graphical interaction, combining a graph editor to manipulate the structure of a document and a text editor to edit the actual text.

Management. The development management is responsible for persistency, consistency, and accessibility of the teaching material (Repository), and it has to treat version control, configuration management and change management (Development Manager) for consistency and the dependencies between the document components (via the ontology). It provides an interface mechanism to call external applications to allow for the presentation of such tools during the course or the collection of practical experiences when doing exercises within these tools. Additional support is given by a flexible user management with administration support and the possibility for integrating typical tools for electronic communication.

Presentation The use of the learning material will be supported in different kinds of teaching scenarios: by a Teacher, Tutor or Student on various presentation platforms, or for individualised self-study by a Student using ACTIVE-MATH (see Sect. 7.3). Moreover, Students and Correctors use WEBASSIGN for assignments.

To achieve the above goals we have designed an open architecture that integrates subsystems developed by the different partners. Fig. 11 shows the major components of the Support Environment.

In the following sections we illustrate the individual components for authoring, maintenance and presentation in more detail.

5 Authoring Tools

The MMiSS tools allow users to produce, maintain and present course material. As mentioned before, one of the key design ideas is to provide means for making

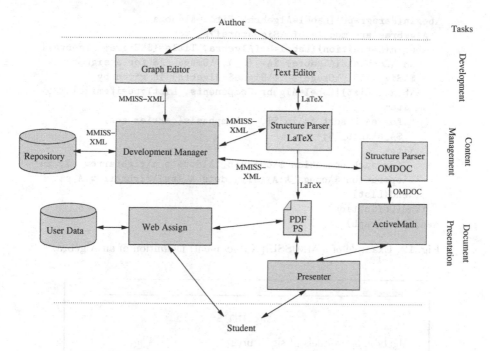

Fig. 11. MMiSS Support Environment System Architecture.

semantic relations in the course material explicit. However, standard languages that are normally used to represent course material, such as PowerPoint or LaTeX, do not support such semantic relations. Therefore, on the one hand MMiSS provides extensions of both languages to cope with such relations, and tools to import existing course material into MMiSS. On the other hand, MMiSS provides an authoring tool to create new course material in a structured way using graph and text editing facilities. Below we illustrate these features in more detail.

5.1 MMiSSLaTeX

LaTeX is a generally accepted language to produce technical documents and course material of high print quality. Although LaTeX comes with some structuring mechanisms it still lacks the expressiveness to formulate many of the structuring mechanisms presented in Sect. 2. In order to support all these mechanisms of MMiSS, we developed a LaTeX-style authoring language called MMiSS*LaTeX*, consisting essentially of a library of LaTeX class files. MMiSSLaTeX provides commands for each of the structuring operations presented in Sect. 2. Technically, most of them are defined as environments, e.g. for sections, definitions, or lists. The attributes are given as optional arguments to the environments or commands.

A document can be typeset using normal LaTeX with the help of the MMiSS-LaTeX class files to generate documents in PDF, Postscript or other formats. For

```
\begin{Paragraph}[Label=Algebra, Title=Algebra]
  Algebras are models of \Signature{}s.
  \begin{Definition}[Label=DefAlgebra, Title={$\Sigma$-Algebra}]
    An \Emphasis{Algebra} $A= (S_A, \Omega_A)$ for a signature
    $\Sigma=(S, \Omega)$ ($\Sigma$-Algebra) is given by
    \begin{List}[Label=AlgebraComponents, ListType=itemize]
    \item
      for each sort $s\in S$, a \Emphasis{carrier set}
      $A_s\in S_A$;
    \item
      for each operation $\omega:s_1\ldots s_n\rightarrow s$, an
      operation $\omega_A:A_{s1}\ldots A_{sn}\rightarrow A_s$.
    \end{List}
  \end{Definition}
\end{Paragraph}
```

Fig. 12. Example of a MMiSSLaTeX document: Definition of an Algebra.

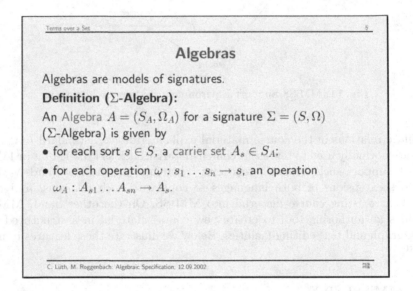

Fig. 13. The example from Fig. 12, rendered as a slide.

example, Fig. 12 shows a MMiSSLaTeX source text in which the concept of an algebra is defined; with LaTeX this is rendered as the slide shown in Fig. 13. Moreover, MMiSSLaTeX serves as an authoring and input language for the MMiSS repository.

5.2 PowerPoint

In order to migrate existing slides and to relieve the authors from the burden of writing content in an unfamiliar language, tools which allow the import of

POWERPOINT slides into the MMiSS-repository have been developed. The tool CPOINT, developed at the Carnegie Mellon University by Andrea Kohlhase, translates PowerPoint slides into OMDOC, which is another exchange language of the MMiSS-repository; CPOINT enriches POWERPOINT documents with additional semantical information, like the semantic structuring relations mentioned in Sect. 2, or with other metadata, like for instance authors or date, and finally translates these annotated slides into OMDOC. As OMDOC documents, the slides can be imported into the repository. CPOINT makes use of the QMATH tool to parse and translate mathematical formulas occurring inside the slides.

5.3   Interactive Course Creation

To edit material, MMiSS provides a graphical interface based on the graph visualisation system daVinci [9, 4] and the XEMACS editor [29]. The idea is to visualize and edit the various structuring relations contained in MMiSS documents in a graph editor while a text editor is used to deal with the basic text fragments (like Units, see Sect. 2.2). The predominant interaction paradigm is *direct manipulation* — authors do not have to learn cryptic command lines to interact with the system, they can just point at the entity of interest, or select from a menu of given choices.

5.4   Structure Graph

According to the structure of MMiSS documents in Sect. 2, a Structural Entity is an entity such as Section, Program, Exercise, TextFragment, etc. These entities are related, most obviously by textual nesting, but also by structural Links or References. This structure gives rise to the *structure graph*, which has the various entities as nodes, and the relations as labelled, directed edges. With regard to the comprises relation, the graph is directed and acyclic, but it is not a tree, since a Structural Entity may be included in more than one place (structural sharing).

As an example, consider a real-life lecture series introducing formal program development in the algebraic specification style to undergradute computer science students. One section of this lecture series, corresponding roughly to one lecture, introduces the basic concepts of algebraic specifications. The document is structured as follows:

- it has a short Introduction motivating what is going to come,
- an Abstract summarising the new concepts,
- the main part consisting of Paragraphs, introducing and defining the concepts of Signatures, Algebras, and Terms,
- followed by a short Example (the natural numbers in CASL),
- and closes with a Summary of the new concepts.

Introduction and Summary contain a list enumerating the concepts the user is about to learn, or has just learned. The main parts are Paragraphs, which are structured further: for example, Signatures, Algebra and Terms contain definitions of the corresponding concept. The resulting structure graph is shown in

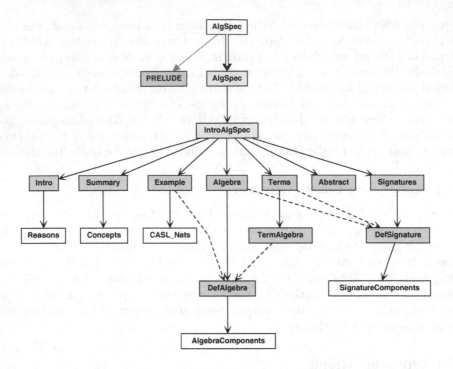

Fig. 14. Structure Graph of Example Document.

Fig. 14. Note how later Definitions and Examples refer back to earlier Definitions (indicated by dashed arrows), for example to define an Algebra we need to refer to Signatures.

Thus, each node represents a Section (shaded yellow), Unit (shaded green), or Atom containing the description of the corresponding notion in more detail. In order to edit this description the user can select a node, corresponding to a Repository object, and its code is loaded into the XEMACS editor (see Fig. 15). A special MMiSSLATEX mode gives the user additional editing assistance, e.g. to insert environments or commands. Documents can be edited in the MMiSSLATEX exchange format using a particular MMiSSLATEX mode; thus other editors may be used as well.

Since a Structural Entity can get quite large (for example, a whole Package), we only display one level in the XEMACS editor; nested Structural Entities are displayed by clicking buttons. For example, Fig. 15 shows the Paragraph labelled ALGEBRA being edited. It contains the definition labelled DEFALGEBRA, which has been opened, and the user is just about to open the list ALGEBRACOMPONENTS.

The Repository objects in the Repository are organised in folders, which allow the grouping of Repository objects much like directories in a file system. Folders may contain other folders, or MMiSS Structural Entities, i.e. Sections, Units or

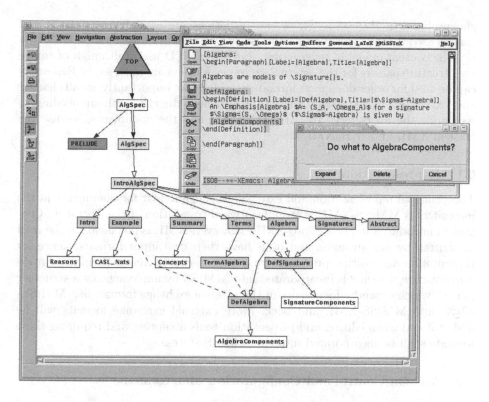

Fig. 15. The Structure Graph: Editing the Definition of an Algebra.

Atoms. The structure graph contains the hierarchy of folders and, at the leaves of this hierarchy, the structure of the Repository objects inside a Package, as nodes, with edges corresponding to the comprises relation (resulting from nesting and structural sharing).

6 Sustainable Development and Maintenance

The MMiSS *Repository* is the central database maintaining MMiSS documents. Sustainable development is supported by fine-grained version control, configuration management and a change management. The structured representation of documents as graphs allows operations to take the structuring into account (see e.g. change management described in Sect. 6.3). It is also the basis for a configuration management to control various versions of a document. We call such graphs together with the possible activities a *development graph*.

The Repository is implemented almost entirely in the functional programming language HASKELL [11] in about 60 Kloc. It uses the open source data base BerkeleyDB [25] to store documents. The graph visualisation system daVinci, the graphical user interface library Tcl/Tk [21] and the XEMACS editor are encapsulated in HASKELL. These encapsulations are available separately, and can be used independently, in particular the Tk encapsulation, called HTk [23].

The content model is generic over the XML DTD used (although of course the structure parsers for the external exchange formats are not), so the Repository can be used for other document formats as well. More importantly, small changes and extensions in the DTD can be implemented directly without needing to recompile the Repository. To parse the DTD and the documents, we use the Haskell XML library HaXML [28].

6.1 Representation

The principal representation and external exchange format for documents in the Repository is MMiSS-XML, a straightforward translation of the Document Constructs introduced in Sect. 2 into XML. However, XML is not meant to be read or written by human users, and tools have their own input formats, hence for presentation and editing purposes, we need external exchange formats. An *external exchange format* is incorporated into MMiSS by implementing a structure parser, which converts documents in the external exchange format, like MMiSS-LaTeX, into MMiSS-XML and back. More external exchange formats will be added if and when editing and presentation tools accepting and requiring these formats shall be incorporated into the MMiSS system.

6.2 Version Control and Configuration Management

The art of keeping track of the evolution of complex systems in general, and complex documents in this particular case, is called *configuration management*. Changes to a documents have to be organised and recorded, such that earlier configurations can always be retrieved. While usually configuration objects are source files, we follow here a *fine-grained* approach using Structural Entities of MMiSS, like Sections, Units or Atoms or associated Attributes, as configuration objects.

The *version graph* is the representation underlying *version control*. Nodes of the graph represent different *versions* of Repository objects while edges denote the *RevisionOf relation*. An author always starts interaction with the Repository by picking a version under development from the version graph. This version will be checked out into the user's local filespace and can then be edited. Fig. 16 (left) shows a typical version graph, as displayed by daVinci. Version 1.5 is the current *working version* (as visualised by the red shade); it is edited as shown in Fig. 15. When finished with editing, a user may *commit* changes back into the Repository (or may just dispose of them silently). New versions of the changed Repository object and of all Repository objects containing the changed Repository object are created. Thus, new versions are propagated upwards: a change in a constituting Repository object results in a new version of the parent Repository object, all the way up to the root folder.

When a Repository object has more than one Repository object which is a revision thereof, we say this is a *span* in the version graph. For example, in

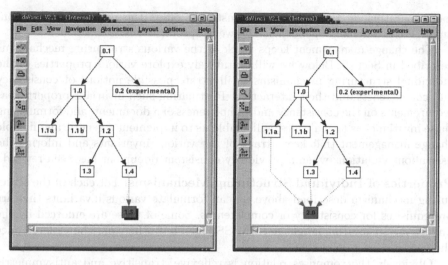

Fig. 16. Version Graph displayed by daVinci (left); Merging of Versions (right).

Fig. 16, there are three spans, starting at the versions 0.1, 1.0 and 1.2. A span corresponds to concurrent revisions of one object. It can be reconciled by a *merge*: users can pick the source versions from which they wish to incorporate all changes into one new version, which then becomes a revision of all the source versions. Fig. 16 (right) shows the merging of versions 1.1a, 1.3 and 1.4 to form a new version 2.0.

6.3   Change Management

The notion of *change management* is used for the maintenance and preservation of consistency and completeness of a development during its evolution. More precisely, we want to have a *consistent configuration* in which all constituents harmonise, versions are compatible, References and Links refer to the proper targets, etc. At the same time, it should be a *complete configuration*: e.g. the promises of forward References and Links should be fulfilled, i.e. they must not be dangling; if we have an English and a German variant of a whole document, then we expect to have a corresponding German variant for each English variant for all constituent Structural Entities, with the same overall structure and relations, and vice-versa.

Such notions are well-known for formal languages; in contrast, natural language used for writing teaching material does not usually possess a well-defined semantics; the notion of consistency is debatable. Different authors may postulate different requirements on the material in order to regard it as being consistent. The existence of an ontology already helps a great deal to check References.

It turns out that the notions of consistency and completeness are closely related to the Document Constructs and relatesDocConstructs relations. For special FormalismAttributes, additional structuring relations may be explored by special tools operating on these. CASL, for instance, offers the notions of extension, union, etc., to define dependencies between specifications.

The change management keeps track of the various structuring mechanisms described in Sect. 2. Below we will tentatively explore various properties of the individual structuring mechanisms to illustrate possible notions of consistency and completeness and their interaction. Postulating such invariant properties as requirements on the consistency and completeness of a document, and formulating these invariants as formal rules, will enable us to implement a generic and flexible change management that keeps track of the various invariants and informs the user about violations when a previously consistent document has been revised.

Properties of Individual Structuring Mechanisms. For each of the structuring mechanism described above we can formulate various invariants that are prerequisites for consistency or completeness. Some of these are enforced by the underlying structuring language (MMiSSLaTeX) but others may be violated once the user revises a document.

Obviously the comprises relation is reflexive, transitive and antisymmetric denoting an acyclic finite graph (which is actually a subgraph of the structure graph). These properties are trivially enforced by the Document Constructs. We may want to require additional invariants for consistency, e.g. that each major Structural Entity (such as Package or Section) contains at least one Unit or Atom, or that there is at most one Summary in a Section.

Each reliesOn relation or pointsTo relation is irreflexive and acyclic. We would also postulate as a consistency requirement that there is at most one target, i.e. the relations are in fact many-to-one; a completeness requirement is that that there is at least one target, e.g. References must not be dangling; both together require a unique target. Furthermore, for reliesOn relations, we require the target to be presented beforehand. However, the completeness requirement may be weaker for pointsTo relations as we tolerate forward pointers, even to other, future Packages (warnings should be given, though).

Regarding special FormalismAttributes, we adopt their reliesOn relations and corresponding properties. Axioms in CASL, for instance, depend on their global environment resulting from fragments of the Theory that specifies the signature of the symbols used in the Axioms.

The semantics of the variantOf relations depends on the various types of variants. Regarding variants in different languages (or on different levels of detail), we impose the completeness requirement that each variant in one language must have a corresponding variant in the other, for each constituent Structural Entity, with the same overall structure and relations (as an option, for each level of detail, and so on). Similarily one will be able to specify, as a consistency requirement, that all Programs should be in a particular FormalismAttribute, e.g. the programming language HASKELL. A corresponding completeness requirement would be that we have, for each Program, a variant in programming language C and JAVA, e.g. for different Teachers of a course.

Properties of Interactions between Structuring Mechanisms. While the properties mentioned above are specific to an individual structuring mechanism, we will explore possible interactions of different structuring mechanisms and how they can be used to refine consistency and completeness.

Relating the comprises and reliesOn relations (we subsume pointsTo relations here) allows us to formalize constraints regarding the closure of document parts with respect to the reliesOn relation. We may require, for example, that there is a Proof for each Theorem in a Package or that each Reference references an OntologyDef occurrence in this Package unless there is an explicit import. Furthermore, a reliesOn relation between two Structural Entities is propagated along the comprises relation towards the root of the hierarchy of nested Structural Entities. Consider, for example, a Proof in Section A that proves a Theorem in Section B, then Section A reliesOnSection B. Conversely, a reliesOn relation between two Structural Entities cannot be decomposed and propagated towards the leaves. Changing (parts of) one of them can affect the proposed reliesOn relation.

The interaction between the comprises and the variantOf relation is rather subtle and has not fully been investigated yet. For example, we expect the structure of a document with the DetailAttribute Lecture to be a homomorphic projection of the corresponding structure with the DetailAttribute Course.

Similarly, the interaction between the reliesOn relation (or pointsTo relation) and the variantOf relation merits further investigation. It is not clear what kind of relations across variants are desired, if any. In principle, each variant should be closed with respect to reliesOn relations, i.e. all targets should be provided in that variant. An exception might be an explicit pointer to material in a lecture from a course, but then this material should be included in the course anyway as a completeness requirement. The converse is more likely: one might want to make a pointer into a more detailed course or lecture notes document from slides in a lecture.

In any case, the more structure there is, the better are the chances for preserving consistency and completeness; any investment in introducing more reliesOn relations, for example, will pay off eventually. The change management will observe whether revisions by the user will affect these relations and, depending on the user's preferences, emit corresponding warnings. It is crucial to point out that, in contrast to formal developments such as in the MAYA-system [24], there is no rigorous requirement that a document should obey all the rules mentioned above. There may be good reasons, for instance, to present first a "light-weight" introduction to all notions introduced in a Section before giving the detailed definitions. In this particular case, one would want to introduce forward pointers to the definitions rather than making the definitions rely on the introduction; thus the rules are covered. The eventual aim of the MMiSS-design is to allow the user to specify her individual notion of consistency by formulating the rules the relations between the various structuring mechanisms have to obey.

6.4   Foreign Tools and Administration

A User Management component supports a simple user model with different Roles and handles the access rights of Authors, Teachers, Students, Tutors, Correctors, and also ToolDevelopers, SystemDevelopers, and Administrators (cf. Sect. 2.1).

WebAssign. The WEBASSIGN system developed at FernUniversität Hagen (see [7] or http://niobe.fernuni-hagen.de/WebAssign) supports web-based distribution, correction, and administration of course related assignments. Assignments may have interactive parts where system gives direct feedback to the student. WEBASSIGN also manages the integration of external tools (such as compilers) that check student answers or provide help in other ways. In addition, WEBASSIGN provides a flexible administrative support. The WEBASSIGN subsystem is presently being integrated.

7   Presentation

In this section, we concentrate on presentation issues such as layout and animation, and show how they can be realised using the authoring language MMiSSLaTeX.

In general, *presentation* issues should be separated from issues of *re*presentation in an abstract form (MMiSS-XML here), which can also serve as an external exchange format. In fact, the Author should be relieved from tedious formatting as much as possible. Therefore, work is under way to isolate layout and animation as attributes. Ideally, tools will generate different presentation forms automatically. The subsystem ACTIVEMATH, which is developed separately, is integrated via a mapping from MMiSS-XML to OMDOC and provides user-adaptive presentation based on pedagogic rules.

7.1   Layout in MMiSSLaTeX

Annotating Slides. At Universität Freiburg, experiments have been made to enhance slides for the course *Computer-Supported Modelling and Reasoning* (held regularly each year) towards a self-contained online course. We will report on our experiences below; they have led to new insights into the best ways of defining the layout (and animation) of the DetailAttribute LectureNotes and, to some extent, Course.

Usually, slides for a lecture are sketchy and rely on the oral presentation of the Teacher. So in order for the slides to be adequate for self-study, an apppendix has been added to each slide suite containing detailed explanations. There is a rich structure of pointers (hyperlinks resulting from References and Links), mainly going from some item in the lecture slides to an explanation of that item, but also many other pointers to even more detailed explanations, and forward and backward pointers within the slides. For example, whenever there is a sentence starting with "Recall that ...", there is a pointer to the corresponding previous item in the lecture, usually a Reference to the OntologyDef occurrence of an element declared in the ontology.

In fact, the slides of the lecture have been extended (as a refinement) to a level of detail that we would now regard as lecture notes for review by a Student after attending class; a self-contained online course without any tutoring would require yet a higher degree of verbosity. The tentative experience seems to indicate that different explicit levels of detail, depending on the Student's learning profile, are not so important. The level of detail develops dynamically during a learning session depending on the pointers the Student decides to follow.

However, in its current form the lecture notes is not very suitable for printing. At the least, the detailed explanations would have to be interleaved with the lecture slides so that an explanation immediately follows the slide referred to, and other pointers (References or Links), which are not hyperlinks anymore, have to be augmented by "(see Sect. ...)" or "(see page ...)". This emphasises the need for a more abstract representation format and tools for generating different output formats automatically, cf. Sect. 2.5. In fact, a different presentation for a Reference or Link is now being generated in MMiSSLATEX, depending on the PresentationAttribute: a hyperlink for Hyper and an extended text "(see Sect. ...)" for Paper.

To give a quantitative assessment of the material involved, one can say that extending lecture slides to a lecture notes at least doubles the size of the sources. A typical lecture notes slide will contain around five pointers. We will further extend the material as Student feedback reveals where more detail is needed.

Board Presentation. The Board PresentationAttribute of MMiSSLATEX (cf. Sect. 2.5) allows for preparing a 'shooting script' of courses. In addition to the slides to be presented, such a script may also include notes on

- what to write on the board,
- which interactions with the students shall take place,
- important oral remarks etc.

during a lecture. Thus, while the annotated slides for online learning provide help for the students, the Board PresentationAttribute is a means of support to the lecturer during the course presentation. Slides to be presented are included as pictures between text blocks to be written on the board. These text blocks are structured by the same environments as available for slides. During the lecture, this kind of presentation helps to keep overview on the course material: the lecturer sees more than the slide currently presented; personal notes of all kinds can be included; tedious but important things like a uniform numbering of chapters, sections, environments, etc. are done automatically.

While preparing a lecture in the Board style of MMiSSLATEX, text blocks or graphics can easily be shiftet between slide- and board-presentation thanks to the uniform naming of the structural entities. Technically, this is done by adding or removing the Board PresentationAttribute. This makes it possible to postpone the decision on how to present a certain item to the very last translation before presentation. In the electronic version of the resulting script, it is possible to run tool demonstrations included in the slides. Thus, one should consider the shooting script prepared in this way as the all inclusive document of a course's presentation.

Concerning the board content, one should be aware that it is of a different type than the material on slides: while slides are intended for presentation to students, only the lecturer will see the contents with the Board Presentation-Attribute. This allows board content to be less detailed, for instance the following might suffice: 'ex-tempore example: model an automaton with the signature provided by the above specification'. From a didactical point of view, such ex-tempore examples — maybe even suggested by the audience — are often better and far more impressive than examples, which are prepared in all details before the presentation. Of course, the same type of argument carries over to proof sketches instead of complete proofs. The Board PresentationAttribute allows also to include this kind of reminders in the course's shooting script.

7.2 Animation in MMiSSLaTeX

With respect to animation, we focus here on a presentation (e.g. of a slide in a lecture, but also of lecture notes in the Hyper variant) where parts are gradually appearing or disappearing in a sequence of displaying steps. The simplest and best-known case is that of an incremental buildup of a page: each step adds new text below the text already presented. These and more complex effects can be very useful in a lecture to illustrate how some complex object is built up step by step; the effect is similar to a presentation on the board. They are even more useful for lecture notes or a self-contained course as no Teacher is available.

So far, such steps have been realised by so-called pause levels in PDFLATEX using the PPOWER4 package [10].

Courses involving logic give rise to a particular application of animation effects, namely *animated derivation trees*. A derivation tree is shown in Fig. 17. The LATEX package PROOF for drawing such trees has been extended to support animation: the particular logical structure of

$$\cfrac{\cfrac{[A \vee B]^1 \quad \cfrac{[A]^2}{B \vee A}\text{ }\vee\text{-}IR \quad \cfrac{[B]^2}{B \vee A}\text{ }\vee\text{-}IL}{B \vee A}\text{ }\vee\text{-}E^2}{A \vee B \to B \vee A}\text{ }\to\text{-}I^1$$

Fig. 17. Derivation tree

derivation trees and the general input syntax for such trees is taken into account.

For each tree, one can specify at which pause levels it should be displayed. For each (sub)tree, one can again specify at which pause levels it should be displayed, *overriding* the specification for the surrounding tree. Derivation trees involve applications of *rules*, e.g. →-I. Each rule application can be associated with the *discharging* of an assumption, marked by brackets around the assumption and labelling both the rule and the brackets with a number. We have automated this process: the numbers are administrated using symbolic references (making it easy to compose trees). Moreover, the brackets and their label will by default inherit the pause level from the rule application. For example, one could specify that the whole tree (and hence its root step marked by rule →-I) appears from pause level 4 onwards, whereas the assumption $A \vee B$ at one of the leaves appears from level 2. Then, by default, the *brackets* around $A \vee B$ and the label (here: 1) will appear from level 4 onwards.

Derivation trees can be quite complex and the process of constructing them is very hard to understand based on static illustrations. We therefore found the new style package very useful.

7.3 User Adaptive Presentation in ActiveMath

The ACTIVEMATH project [14] was started independently from and before the MMiSS project and has provided a lot of valuable ideas.

Goals. In the previous sections it became clear that producing on-line learning material involves a lot of effort and that reusability in different contexts and for different presentations and presentation formats is a must in the development of future learning material. As one conclusion, a more abstract, semantical XML knowledge representation, OMDoc [12], has been developed and in addition, presentation tools and other functionalities of the learning environment are strictly separated from the knowledge representation of the learning content in ACTIVEMATH and can thus deliver different output formats, different hyperlinking, different presentations of symbols and formulas, personalized appearances etc.

Apart from these economically and technically-driven developments, a major goal of multimedia on-line learning is a better quality of learning. This objective calls for pedagogically and cognitively motivated features of a learning system which include personalization of content and appearance, the provision of feedback, and presentation according to the learning progress. For instance, a learner becomes bored and less motivated when the material and exercises are too easy for her and not challenging at all. Similarly, the learner's motivation will drop considerably when material and exercises are beyond her capabilities and knowledge mastery level. Therefore, a few advanced intelligent tutoring systems – including ACTIVEMATH – adapt the content and its sequencing to the learners goals, capabilities, and learning preferences/scenario.

Knowledge Representation. ACTIVEMATH was the first system that uses the knowledge representation OMDoc [20]. OMDoc is an extension of the OPENMATH XML-standard[2]. OPENMATH provides a grammar for the representation of mathematical objects and sets of standardized symbols (the content-dictionaries). OMDoc inherits the grammar for mathematical objects from OPENMATH and the existing content-dictionaries. In addition, OMDoc defines a framework for the definition of new symbols.

The objectives of OMDoc and MMiSS-XML are quite similar: OMDoc was originally more tailored towards mathematical content and is being extended now; MMiSS-XML has had more general objectives, is more tailored towards the document Document Constructs described above and the input language MMiSS-LaTeX; MMiSS-XML can be mapped to OMDoc and vice-versa — efforts are presently being made for further unification.

The metadata in core-OMDoc include the Dublin Core [27] metadata such as contributor and publisher. The ACTIVEMATH DTD extends OMDoc (see e.g. [15]) and contains additional – pedagogically motivated – metadata such as difficulty or field of an exercise and the prerequisite-of relation of instructional items for a concept that allow even more customization of the document delivery to the student and her learning situation.

[2] http://www.openmath.org

Adaptive Presentations. Thanks to a user model that stores and updates the learner's preferences, goals, activities, and mastery levels, ACTIVEMATH is able to present the learning material in a user-adaptive manner, content-wise and presentation-wise. In the table-of-contents a color-annotation informs the student about her mastery level for concepts to be learned.

The flexibility of the presentation process also chooses a low slide-verbosity or a high script-verbosity of the material according to the learner's needs. A slide presentation can automatically be (hyper-)linked to the more verbose explanations and other instructional items from the script sources.

Mathematical objects/symbols in the presentation have a semantic annotation that points to the meaning of the symbol in the content dictionary. This enables functionalities such as copy and paste of mathematical formulas to a service system's console.

The transformation from assembled XML content items to the actual output format is realized with a modular presentation process with style sheet application at its heart. Currently, ACTIVEMATH can realize LATEX PDF output formats that are well-suited for printing as well as HTML output format augmented by MATHML-presentation for mathematical symbols that is well-suited for browser presentation output.

Learning-Effective Features in ActiveMath. ACTIVEMATH offers several other features that are known to improve learning. In particular, it has a generic mechanism for integrating service systems/tools for active and exploratory learning, such as computer algebra systems or tools for formal software development.

A dictionary can be called from the material or by explicit search in the dictionary. It displays the definition of a concept and, if required, also the concepts and instructional items that are somehow related to that concept, e.g., examples illustrating the concept or exercises training the concept.

The learner can resume studying where she left off last time. She can manipulate (rename, delete) those (listed) materials she has studied previously. A notes facility enables the learner to take personal or group notes corresponding to items in the learning material. The user model is open and inspectable.

ACTIVEMATH is customisable to teacher's and learner's needs and easily configurable to pedagogical strategies and knowledge resouces.

8   Conclusion

In this paper we have presented the methodology, the techniques and tools of the MMiSS-project to support multimedia instructions in safe and secure systems. Summing up, the developed infrastructure allows a user

- to develop transparencies, lecture notes, complete courses
- to work on the board, with transparencies, interactively with tools
- to embed mathematical formulae, programs, etc.
- to manage e.g. English and German variants in parallel
- to publish complete and consistent packages

- to (partially) re-use the transparencies of a colleague
- to be made aware of the changes made by colleagues
- to develop a uniform terminology among various authors, and
- to have support for sustainable development.

Experiences. The system has been gradually introduced, over the duration of the project, into the normal teaching activities of the project partners. For example, the two semester course TECS (Techniques for the development of Correct Software) at Universität Bremen provides a gentle introduction to formal methods for software development. It deals with sequential as well as with reactive systems, using the algebraic specification language CASL [2, 13, 6, 18] and the process algebra CSP, e.g. [22], resp. On the tool's side, the theorem prover Isabelle and the model checker FDR play central roles. Besides simple exercises explaining single concepts, the TeCS problem sheets also include more complex tasks like specifying a family game (Nine Men's Morris) in CASL; verifying a simple interpreter within Isabelle/HOL; modelling a file system in CASL at both the requirements and the design level; proving the refinement relation between these two specifications in HOL/CASL.

Presenting TECS using the presentational part of MMiSSLaTeX has been quite successful. For the *author*, the overhead to produce course material within the MMiSSLaTeX format is negligable compared to other presentation systems. Besides the usual benefits of a computer based presentation like 'no slide confusion', the MMiSSLaTeX integration of tool demonstrations in the slides encourages the *teacher* to enliven the lectures by live demonstrations on the computer. The *students* are fond of the readability, the consistent markup, and the download-friendly PDF-filesize of the slides. It should be mentioned that these positive results also arise from a cautious usage of computer based presentations: about half of the course material has been taught in 'classical style' using a blackboard. A poll among the students of TECS gave the result that this is an optimal mixture.

State of the Project and Future Developments. The project has made good progress during its first two years. Many lectures have been converted to the initial LaTeX-oriented input format, with good quality output as slides in PDF-format. This material is now awaiting further coordination and refinement, as well as semantic interlinking via an ontology and using development graphs in the repository. The Development Manager, and other editing and authoring tools, have been made available in a first version.

While the project has achieved a satisfactory, consistent state, a lot still has to be done: the documentation has to be improved, various bits and pieces have to be completed (e.g. layout and animation attributes), etc. We hope that the planned extension mechanisms will facilitate future developments considerably.

MMiSSForum. As the open source model is used, teaching materials and tools are freely available to achieve a much wider national and international take-up.

To assist this, a MMiSSForum is has been set up with German, international, and industrial members, to evaluate the emerging curriculum and assist its development and distribution; you are welcome to join ([16]). The Advisory Board advises the project from a scientific as well as an industrial perspective, with a view to future applications. To go with the planned deployment at universities, a number of well-known German companies have already, through the various industrial contacts of the project partners, expressed an interest in measures for further in-house training.

Support and Partners. The MMiSS project is being supported by the German Ministry for Research and Education, bmb+f, in its programme *"New Media in Education"* from 2001 to 2004. The project partners are

- Universität Bremen (Krieg-Brückner, Drouineaud, Eckert (now at Darmstadt), Gogolla, Kreowski, Lindow, Lüth, Mahnke, Mossakowski, Peleska, Roggenbach (now at Swansea), Russell, Schlingloff (now at HU Berlin), Schröder, Shi)
- FernUniversität (Distance Education University) Hagen (Poetzsch-Heffter (now at Kaiserslautern), Bealu, Kraemer, Sun, Jelitto)
- Universität Freiburg (Basin (now at ETH Zürich), Klaedtke, Smaus,Wolff),
- Ludwig-Maximilians-Universität München (Wirsing, Kröger, Knapp, Hennicker, Meier, Zhang),
- Universität des Saarlandes (Hutter, Melis, Autexier, Siekmann, Stephan, Goguadze, Libbrecht, Ullrich).

Acknowledgement

We are grateful to the members of the Advisory Board, M. Kohlhase (Carnegie-Mellon University, Pittsburgh), V. Lotz (Siemens AG, München), H. Reichel (Technische Universität Dresden), W. Reisig (Humboldt Universität Berlin), D.T. Sannella (University of Edinburgh), and M. Ullmann (BSI [Federal Institute for Security in Information Technology], Bonn), for their advice.

References

1. Computing Classification System [1998 Version]. http://www.acm.org/class/.
2. E. Astesiano, M. Bidoit, B. Krieg-Brückner, H. Kirchner, P. D. Mosses, D. Sannella, and A. Tarlecki. CASL – the common algebraic specification language. *Theoretical Computer Science*, 286:153–196, 2002.
3. E. Astesiano and G. Reggio. Formalism and method. In *Theoretical Computer Science*, pages 236(1–2), 2000.
4. b-novative GmbH. davinci presenter web site. http://www.b-novative.com/products/daVinci/.
5. K. Baclawski, M. K. Kokar, P. A. Kogut, L. Hart, J. Smith, W. S. Holmes III, J. Letkowski, and M. L. Aronson. Extending UML to support ontology engineering for the semantic Web. *Lecture Notes in Computer Science*, 2185:342–360, 2001.

6. H. Baumeister, M. Cerioli, A. Haxthausen, T. Mossakowski, P.D. Mosses, D. San-nella, and A. Tarlecki. CASL semantics. In P.D. Mosses, editor, CASL *Reference Manual.* [19], Part III.

7. J. Brunsmann, A. Homrighausen, H.-W. Six, and J. Voss. Assignments in a virtual university - the webassign-system. In *Proceedings of the 19th World Conference on Open Learning and Distance Education*, Vienna/Austria, June 1999.

8. E. M. Clarke and J. M. Wing. Formal methods: State of the art and future directions. In *ACM Computing Surveys 28*, pages 626–643, 1996.

9. M. Fröhlich. *Inkrementelles Graphlayout im Visualisierungssystem daVinci.* PhD thesis, Dissertation, Universität Bremen, 1998.

10. K. Guntermann and C. Spannagel. *PPower4 Manual.* TU Darmstadt, 2002.

11. Haskell web site. http://www.haskell.org/.

12. M. Kohlhase. OMDOC: Towards an internet standard for mathematical knowl-edge. In E. R. Lozano, editor, *Proceedings of Artificial Intelligence and Symbolic Computation, AISC'2000*, LNAI. Springer Verlag, 2001. See also http://www.mathweb.org/omdoc.

13. CoFILanguage Design Group, B. Krieg-Brückner and P.D. Mosses (eds.). CASL summary. In P.D. Mosses, editor, CASL *Reference Manual.* [19], Part I.

14. E. Melis, E. Andres, G. Goguadse, P. Libbrecht, M. Pollet, and C. Ullrich. Active-math: System description, 2001.

15. E. Melis and C. Ullrich G. Goguadse, P. Libbrecht. Wissensmodellierung und -nutzung in ACTIVEMATH. *KI*, to appear(1):12–18, 2003.

16. MMiSS web site. http://www.mmiss.de.

17. T. Mossakowski. CASL: From semantics to tools. In S. Graf and M. Schwartzbach, editors, *TACAS 2000*, volume 1785 of *Lecture Notes in Computer Science*, pages 93–108. Springer-Verlag, 2000.

18. P. D. Mosses and M. Bidoit. CASL — *the common algebraic specification language: User Manual.* Lecture Notes in Computer Science. Springer. To appear.

19. P. D. Mosses (ed.). CASL — *the common algebraic specification language: Reference Manual.* Lecture Notes in Computer Science. Springer. To appear.

20. OmDoc. http://www.openmath.org.

21. J. K. Ousterhout. *Tcl and the Tk Toolkit.* Addison Wesley, 1994.

22. A.W. Roscoe. *The theory and practice of concurrency.* Prentice Hall, 1998.

23. G. Russel and C. Lüth. Htk — graphical user interfaces for haskell programs. http://www.informatik.uni-bremen.de/htk/.

24. S.Autexier, D.Hutter, T.Mossakowski, and A.Schairer. The development graph manager MAYA (system description). In H. Kirchner and C. Reingeissen, edi-tors, *Algebraic Methodology and Software Technology, 2002*, volume 2422 of *Lecture Notes in Computer Science*, pages 495–502. Springer-Verlag, 2002.

25. Sleepycat Software. Berkeley DB. http://www.sleepycat.com/.

26. B. Steffen, T. Margaria, and V. Braun. The electronic tool integration platform: Concepts and design. In *International Journal on Software Tools for Technology Transfer (STTT) 1*, pages 9–30, 1997.

27. The Dublin Core Metadata Initiative. Dublin core metadata initiative - home page, 1998. http://purl.org/DC/.

28. M. Wallace and C. Runciman. Haskell and XML: Generic combinators or type-based translation? In *International Conference on Functional Programming ICFP'99*, pages 148– 159. ACM Press, 1999.

29. Xemacs web site. http://www.xemacs.org/.

Zero, Connected, Empty

Music by Ryoko Amadee Goguen
Lyrics by Joseph Amadee Goguen

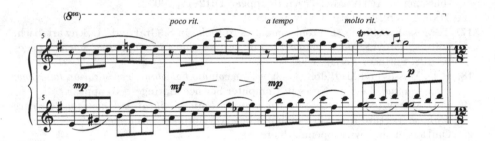

Copyright (C) JASRAC 2003

M. Wirsing, D. Pattinson, and R. Hennicker (Eds.): WADT 2002, LNCS 2755, pp. 118–126, 2003.
© Springer-Verlag Berlin Heidelberg 2003

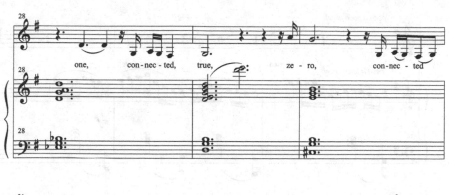

one,   con - nec - ted,   true,   ze - ro,   con - nec - ted

pow - er   e - lec - tric,   brain,   con - nec - ted

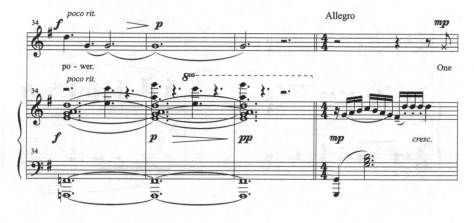

po - wer.   One

(keep pedal down till measure 72)

* played only on black keys

Zero, Connected, Empty:
An Essay after a Cantata

Joseph A. Goguen and Ryoko A. Goguen

Department of Computer Science & Engineering
University of California at San Diego, USA

Abstract. This brief essay attempts to explain some of the thinking behind the cantata *Zero, Connected, Empty*, composed by Ryoko Gougen,
with words by Joseph Gougen.

The cantata *Zer o, Connected, Empty* w as written for the banquet of the 16th
Workshop on Algebraic Development Techniques, a conference on theoretical
computer science, which was held in a monastery (more properly, a nunnery) on
the island of Frauenchiemsee, in a lake in the country about an hour by train
from Munich; it was performed there on 25 September 2002. It has 87 measures
and takes about 8 minutes to perform.

Artists, especially musicians, are often reluctant to "explain" their work, but
this is an exception, because the piece was written for a very specific situation,
and its creators are rather loquacious, though possibly obscure. The piece is a
brief history of Western culture from the Enlightenment to the present, told in
music and elliptical words, focusing on mathematics in a broad sense, or more
precisely , on the philosophy of mathematics, in a sense that includes computer
science. Its sections move from the eternal certainties and formal structures of the
classical period, through the emotional and cultural aggression of romanticism,
to contemporary confusions like postmodernism and computing. It is interesting
to recall that in the medieval universities, music was taught in the mathematics
facult y.

It is a cantata in the loose sense that modern composers use that term, rather
than in the traditional sense associated with Bach. It is written for voice and piano, preferably the same musician. Though originally performed with an electric
piano (which miraculously recov ered from a last minute surgery at the banquet),
it sounds much better on a concert grand, which allo ws the reverberations from
the cluster chords to circulate and decay properly.

The initial 8 measure Mozart-like introduction is the first theme from the first
mov ement of *Sonatina No. 1* for piano, by Ryok o. This leads into a 9 measure
recitativ e-like vocal section, expressing the philosophy of the classical period, exemplified by thinkers lik e Newton and Leibniz. This is follow ed b y an 8 measure
romantic treatment of the second theme from the same *Sonatina*, and then by another vocal section, with words reflecting the spirit of the romantic period (which
we take as lasting into the early 20th century), with its emphasis on conquering
and colonizing nature, and foreign territories in general. The w ord "power" is
meant in several senses, including those of physics, politics, electronics, and the

M. Wirsing, D. Pattinson, and R. Hennicker (Eds.): WADT 2002, LNCS 2755, pp. 127–128, 2003.

military. This is followed by a 9 measure counting section, featuring the natural numbers from 1 to 12, accompanied by a variation on the first *Sonatina* theme, using Beethovenesque chords. The music in measures 46 to 71 moves into the contemporary period, with more complex chords and rhythms, an accelerating tempo, and deconstructed quotations of familiar themes of Bach, Beethoven, and Mendelssohn, culminating in an explosive sequence of cluster chords.

The final vocal section in 12 measures mainly uses jazz modern hords. Its first phrase is a reference to the geography of Frauenchiemsee, but also, as the follo wing phrases suggest, to the buddhist philosophy of emptiness (*sunyata* in Sanskrit, *mu* in Japanese), which according for example to the 13th century Zen master Dogen Zenji [1], says that nothing has a "soul" or self-essence or ideal form, because everything arises together through mutual causation, i.e. through connectedness. The number zero gets recruited here as a symbol for the manifestation of emptiness as form. This section concludes with a brief recapitulation of the second *Sonatina* theme underneath its last 2 measures. The final 4 measures of the piece are an echo of emptiness, or emptiness ringing your door bell. (Some more general aspects of the relation betw een music and *sunyata* will probably be discussed in [3].) An element of non-linear coherence enters through the repeated reappearance of melodic and harmonic fragments in transformed guises.

A clever person in the audience pointed out that the repeated phrase "zero, connected," can be interpreted as a "generation constraint" (e.g, in the sense of [2], page 121) for the natural numbers; this shows how the audience can enrich an artist's comprehension of a piece.

In ternational copyrigh t is secured throughJASRAC, the Japanese equivalent of the American ASCAP, but there may not be a great demand from performers, because the piece requires an unusual vocal range, contemporary classical piano skills, and the ability to the interpret with shades of contemporary free jazz, and even a bit of pop.

We wish to thank Yumiko Morita, Koji Nakano, and Pei Xiang for their very generous help.

References

1. Dogen. *Shob ogenzo: Zen Essays by Dgen.* University of Haw aii, 1992.
2. Joseph Goguen and Rod Burstall. Institutions: Abstract model theory for specification and programming. *Journal of the Asso ciationfor Computing Machinery*, 39(1):95–146, January 1992.
3. Joseph Goguen and Erik Myin, editors. *A rt, Br ain and Consciousness.* Imprint Academic, to appear 2004.

Type Checking Parametrised Programs and Specifications in ASL+$_{\text{FPC}}$

David Aspinall

School of Informatics, University of Edinburgh, UK
David.Aspinall@ed.ac.uk
http://homepages.inf.ed.ac.uk/da

Abstract ASL+ [SST92] is a kernel specification language with higher-order parametrisation for programs and specifications, based on a dependently typed λ-calculus. ASL+ has an institution-independent semantics, which leaves the underlying programming language and specification logic unspecified. To complete the definition, and in particular, to study the type checking problem for ASL+, the language ASL+$_{\text{FPC}}$ was conceived. It is a modified version of ASL+ for $\mathcal{FPC}$, and institution based on the paradigmatic programming calculus FPC. The institution $\mathcal{FPC}$ is notable for including *sharing equations* inside signatures, reminiscent of so-called *manifest types* or *translucent sums* in type systems for programming language modules [Ler94,HL94]. This allows type equalities to be propagated when composing modules. This paper introduces $\mathcal{FPC}$ and ASL+$_{\text{FPC}}$ and their type checking systems.

1 Program Development with Institutions

A simple setup for program development with institutions [GB92] is to consider programs to be syntactic expressions denoting models from an institution $\mathcal{I}$, and specifications to be syntactic expressions denoting classes of models. More elaborate views are certainly possible (e.g., programming languages considered as institutions whose satisfaction relation is a function), but perhaps unnecessary.

One issue that must be resolved is the relationship between identifiers in the syntax, and their semantic equivalents. In particular, the possibility of *aliasing*, or as it is known in the context of modular programming, *sharing*, should be considered. While a real language may already include an understanding of sharing, the usual institutional semantics of a specification language such as ASL in equational logic $\mathcal{EQ}$ or first-order logic $\mathcal{FOL}$ does not, simply because there is no way to specify sharing in algebraic signatures. For example, given

$$\Sigma =_{\text{def}} \mathbf{sig}$$
$$\mathbf{sorts}\ s, t$$
$$\mathbf{opns}\ c : s,\ d : t$$
$$\mathbf{end}$$

the equation "c=d" is ill-typed because c and d have distinct sorts; so this equation is not in $\mathbf{Sen}(\Sigma)$. However, flexible ways of parameter passing can

M. Wirsing, D. Pattinson, and R. Hennicker (Eds.): WADT 2002, LNCS 2755, pp. 129–144, 2003.

mean that the same sort can be referred to via several different identifiers, so there are occasions when this equation *should* be considered well-typed. The classical example is the "diamond-import" situation [Mac86], illustrated by the Standard ML (SML) functor heading:

```
functor F (structure S1 : sig type intset ... end
           structure S2 : sig type intset ... end
           sharing type S1.intset = S2.intset) = ...
```

The parametrised program F has two parameter modules S1 and S2, but requires that any actual parameters have identical implementations of the `intset` type. This means that when type checking the body of the functor, the given type equation can be assumed. In the algebraic case, sometimes we may want to suppose that sorts s and t denote the same set, so $c = d$ *is* type-correct.

This issue may seem simple, but propagating type equalities properly lies at the heart of type-theoretic explanations of programming language module systems, an issue which researchers have worked on for well over a decade (contributions include [HMM90,HL94,Ler95,Ler96,Jon96,Rus99]). The design of a module type system is affected both by the type system of the underlying language, and by the flexibility of the module system: higher-order, first-class and recursive modules have all been considered. The work reported here is a first attempt to design a type system for a language which has higher-order parametrisation of both programs and specifications.

With an institution-based semantics, we have two ways to go:

Ignore sharing: e.g., by extending the satisfaction relation $\models$ to be three valued, so that $A \models \varphi \in \{\,\textbf{true}, \textbf{false}, \textbf{wrong}\,\}$. Then $\textbf{Sen}(\Sigma)$ is extended to contain all formulae which could possibly have a denotation. So now $c = d \in \textbf{Sen}(\Sigma)$, but if $A_c \neq A_d$, then $(A \models c = d) = \textbf{wrong}$. This is a bit like *dynamic type checking* in programming languages, and similarly unattractive: nonsensical sentences accidentally become meaningful.

Handle sharing: e.g., by adding information to signatures, to maintain the idea of *static type checking*. Then $\textbf{Sen}(\Sigma)$ consists only of formulae which have a denotation in the semantics, as usual. This approach seems desirable when we have languages that can be statically type-checked.

Following the second choice, there are two ways of handling sharing:

External sharing: resolve sharing outwith the institutional notion of signature. For example, we could maintain a map from "external identifiers" to "internal names," the latter being names in an algebraic signature. This is (a bit) like the 1990 SML semantics, and was suggested in the algebraic semantics sketched for Extended ML in [ST86].

Internal sharing: make sharing part of the notion of signatures in the institution somehow. For example, to handle type sharing, signatures could be equipped with an equivalence relation on sorts.

The advantage of external sharing is keeping our familiar institutions. The considerable disadvantage is that we break the institution-independent framework:

for example, specification building operators of ASL must be lifted to operate
on the "external" part of algebraic signatures, and general results must be re-
proved. (Nonetheless, this route has been followed for the semantics of CASL by
introducing *institutions with symbols* [Mos99].)

The internal sharing alternative means that we must modify the institution.
But after that, we can apply the general institution-independent framework. We
treat signatures as *static typing environments* which contain all that's needed to
type-check terms and formulae; **Sen**$(\varSigma)$ is exactly the set of well-typed formulae
in the abstract syntax over $\varSigma$. This is the route that we follow for ASL+$_{\text{FPC}}$.

ASL+$_{\text{FPC}}$ is an attempt to give a complete but small definition of a formal
development framework. We start from the fixed-point calculus FPC, which
is a prototypical expression language for higher-order functional programming.
Then we define syntax and semantics for a programming language, specification
language and logic, and fit these into a λ-calculus used for structuring, based on
ASL+ [SST92,Asp95b]. The syntax and semantics of each part are put together
in the same way:

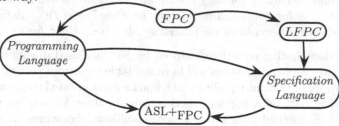

LFPC, the logic for FPC, is based on higher-order logic with an axiomatisation of the CPO order relation for the underlying fixed-point semantics of FPC.
The final result, ASL+$_{\text{FPC}}$, allows higher-order parameterisation of both programs and specifications, as well as the specification of parametrised programs,
as studied in the abstract setting of ASL+ [SST92,Asp95b,Asp97].

In Section 2, we give the definitions of the institution $\mathcal{FPC}$. Section 3 introduces syntax and semantics for $\mathcal{FPC}$ signatures and programs in context, and
Section 4 describes the full module language ASL+$_{\text{FPC}}$. Section 5 concludes.

2   An Institution for FPC

FPC [Plo85] is an extension of the simply-typed lambda calculus with products,
sums, and recursive types. The expressiveness of FPC is well-known: familiar
datatypes are built beginning from the empty type $\mu a.a$ and we can define a
fixed point operator for each function type $s -\rangle p$.

In practice, type expressions are too cumbersome to write out, so we need
type abbreviations. Similarly, real programming languages use definitions to
avoid repeating functions. So to the minimal FPC calculus we add type and term
constants (in algebraic terminology, these are the sort and operation names).

Let *TyVar* and *TyConst* be disjoint countable sets of *type variables* and *type
constants*. Types of FPC are given by the grammar:

$$t ::= c \mid a \mid t \rightarrow t \mid t \times t \mid t + t \mid \mu a.t$$

where $a \in$ *TyVar* and $c \in$ *TyConst*. Free and bound variables of a type are defined as usual, and α-convertible types are considered syntactically identical. Substitution of the type s for the type variable a in the type t is written $[s/a]t$. (Similar conventions and notation are used henceforth without note). Given a subset $Ty \subseteq$ *TyConst*, we write *ProgTypes*(Ty) for the set of closed types whose type constants are contained in the set Ty.

Terms of FPC are parametrised on a notion of signature, which is equipped with *sharing equations* for type constants. Let *TmConst* be a countable set of term constants.

Definition 1. *An FPC signature Σ is a triple $(Ty^\Sigma, Sh^\Sigma, Tm^\Sigma)$ where*

- *$Ty^\Sigma \subseteq$ TyConst*
- *$Sh^\Sigma : Ty^\Sigma \rightarrow Fin(ProgTypes(Ty^\Sigma))$*
- *$Tm^\Sigma : TmConst \rightharpoonup ProgTypes(Ty^\Sigma)$*

We let Σ stand variously for any component when no confusion would arise, and we write $=_\Sigma$ for the equality relation on ProgTypes(Ty^Σ) defined as the compatible closure of equalities introduced by Sh^Σ (i.e., $c = t$ for $t \in \Sigma(c)$).

The idea is that sharing equations induced by Sh^Σ are used during type checking. In practice, useful signatures will have unifiable equations (considering Ty^Σ as variables); we only want equalities which arise from abbreviations and aliased names. Having Sh^Σ as a partial function would suffice for our purposes, but defining $Sh^\Sigma(c)$ instead as a set of equations allows signatures to be put together easily. Checking that a finite signature is unifiable is a simple first-order unification problem (we do not unfold recursive types).

Terms in FPC are given by the grammar:

$$
\begin{aligned}
e ::= & \ v \mid x \mid \mathbf{fun}\,(x:t).\,e \mid e\,e \\
& \mid \langle e, e \rangle \mid \mathbf{fst}(e) \mid \mathbf{snd}(e) \\
& \mid \mathbf{inl}_{t+t}(e) \mid \mathbf{inr}_{t+t}(e) \mid \mathbf{case}\ e\ \mathbf{of}\ \mathbf{inl}(x) \Rightarrow e\ \mathbf{or}\ \mathbf{inr}(x) \Rightarrow e \\
& \mid \mathbf{intro}_{\mu a.t}(e) \mid \mathbf{elim}(e)
\end{aligned}
$$

where $v \in$ *TmConst* ranges over term constants and $x \in$ *TmVar* ranges over a set of term variables.

Terms are type-checked with the standard rules, together with a rule for typing term constants and a rule for using type equality:

$$\frac{\Sigma(v) = t}{G \triangleright^\Sigma v : t} \qquad \frac{G \triangleright^\Sigma e : s \qquad s =_\Sigma t}{G \triangleright^\Sigma e : t}$$

As usual, the type checking judgement $G \triangleright^\Sigma e : t$ uses a context G of type assignments $x : t$ giving types $t \in$ *ProgTypes*(Σ) to variables.

Definition 2. *A signature morphism $\sigma : \Sigma \rightarrow \Sigma'$ is a pair (Ty^σ, Tm^σ) where $Ty^\sigma : Ty^\Sigma \rightarrow Ty^{\Sigma'}$ and $Tm^\sigma : Dom(Tm^\Sigma) \rightarrow Dom(Tm^{\Sigma'})$ are functions such that for all $c \in \Sigma$, $t \in \Sigma(c) \Longrightarrow \sigma(c) =_{\Sigma'} \sigma(t)$ and for all $v \in \Sigma$, $\Sigma(v) = t \Longrightarrow Tm^{\Sigma'}(\sigma(v)) =_{\Sigma'} \sigma(t)$.*

A special case of signature morphism is the inclusion between a signature Σ and a richer one Σ' having more constants or equalities.

Definition 3 (FPC subsignatures and inclusions). *A signature Σ is a subsignature of Σ', written $\Sigma \subseteq \Sigma'$, if $Ty^\Sigma \subseteq Ty^{\Sigma'}$, $t \in \Sigma(c) \Longrightarrow c =_{\Sigma'} t$, and $\Sigma(v) = t \Longrightarrow Tm^{\Sigma'}(v) =_{\Sigma'} t$. If $\Sigma \subseteq \Sigma'$, then there is a canonical morphism $\iota_{\Sigma,\Sigma'} : \Sigma \hookrightarrow \Sigma'$, the inclusion of Σ in Σ', comprising the evident inclusions.*

Example 1. Define three signatures by:

$\Sigma_1 =_{\mathrm{def}}$	$\Sigma_2 =_{\mathrm{def}}$	$\Sigma_3 =_{\mathrm{def}}$
sig	**sig**	**sig**
type c	**type** c	**type** c
val $v : (c \times bool)$	**type** d	**type** d
$\times (c \times bool)$	**sharing** $d = c \times bool$	**val** $v : (c \times bool) \times d$
end	**val** $v : d \times d$	**end**
	end	

Then $\Sigma_1 \subseteq \Sigma_2$ and $\Sigma_3 \subseteq \Sigma_2$ but Σ_1 and Σ_3 are unrelated.

If $\Sigma \subseteq \Sigma'$ and $\Sigma' \subseteq \Sigma$ we consider Σ and Σ' semantically equivalent. Under this equivalence and a similar one on signature morphisms, we get the category **Sign**$^{\mathcal{FPC}}$. We usually work with particular concrete representatives.

2.1   FPC Algebras

FPC has a standard fixed-point semantics given using a universal domain (see e.g., [Gun92]). Using this we define interpretations for FPC types $\mathcal{M}[\![t]\!]_\iota$ in a type environment ι and well-typed FPC terms $\mathcal{M}[\![e : t]\!]_{\iota\rho}$ in a pair of environments ι, ρ. The details are routine, except to require that a Σ-type environment ι is defined on both type variables and constants, and respects the sharing relation in Σ (if $t \in \Sigma(c)$ then $\iota(c) = \mathcal{M}[\![t]\!]_\iota$). Similarly, we call ρ a (G, Σ, ι)-FPC environment if it maps term constants and variables to appropriate domains as required by G, Σ, and ι.

The interpretation of terms is preserved by signature change. This is the main part of the satisfaction condition.

Definition 4 (Environment reducts). *Let $\sigma : \Sigma \to \Sigma'$ be a signature morphism. Suppose ι is a Σ'-type environment and ρ is a $(\sigma(G), \Sigma', \iota)$-FPC environment. We define $\iota|_\sigma$ and $\rho|_\sigma$ by:*

$$\iota|_\sigma(c) = \begin{cases} \iota(\sigma(c)) & \text{for } c \in \Sigma, \\ \text{undefined} & \text{otherwise.} \end{cases} \qquad \rho|_\sigma(v) = \begin{cases} \rho(\sigma(v)) & \text{for } v \in \Sigma, \\ \text{undefined} & \text{otherwise.} \end{cases}$$
$$\iota|_\sigma(a) = \iota(a) \qquad\qquad\qquad\qquad \rho|_\sigma(x) = \rho(x)$$

for all c, a, v, x. It follows directly that $\iota|_\sigma$ is a Σ-type environment and $\rho|_\sigma$ is a $(G, \Sigma, \iota|_\sigma)$-FPC environment.

Proposition 1 (FPC meaning is preserved by signature change). *Let* $\sigma : \Sigma \to \Sigma'$ *be a signature morphism. Suppose ι is a Σ'-type environment and ρ is a $(\sigma(G), \Sigma', \iota)$-FPC environment. Then*

- $\mathcal{M}[\![t]\!]_{\iota|_\sigma} = \mathcal{M}[\![\sigma(t)]\!]_\iota$ *and*
- $\mathcal{M}[\![G \vartriangleright^\Sigma e : t]\!]_{\iota|_\sigma \rho|_\sigma} = \mathcal{M}[\![\sigma(G) \vartriangleright^{\Sigma'} \sigma(e) : \sigma(t)]\!]_{\iota\rho}.$

An *FPC Σ-algebra* A is now defined as a pair (ι_A, ρ_A) of a suitable type environment and term environment for an FPC signature Σ. We can define a model functor by setting $\mathbf{Mod}^{\mathcal{FPC}}(\Sigma)$ to be the discrete category of FPC Σ-algebras.

2.2   LFPC, a Logic for FPC

A suitable logic for FPC can be based on Gordon's HOL logic [GM93], which has equality, implication, and a choice operator as primitive; other connectives are definable. We add FPC types and the cpo order relation $\sqsubseteq$ inherited from the fixed-point semantics, to give the types and terms of LFPC:

$$\tau ::= t \mid \mathbf{prop} \mid \tau \to \tau$$
$$h ::= \Lambda z{:}\tau.\, h \mid h(h) \mid z \mid h = h \mid h \Longrightarrow h \mid \epsilon z{:}\tau.\, h \mid e \sqsubseteq e$$

where $t \in \mathit{ProgTypes}(\Sigma)$ and $z \in \mathit{LogVar} \supset \mathit{TmVar}$ ranges over a new countable set LogVar of logical variables. The typing judgement $G \vartriangleright^\Sigma e : \tau$ is defined for a fixed signature Σ and a context of bindings $z : \tau$.

Notice that the logical function space is distinct from the programming language one, and **prop** is distinct from any FPC type of booleans. No base type of individuals is necessary because FPC already includes types denoting countably infinite collections. Adding rules to axiomatise $\sqsubseteq$ gives us a higher-order logic of computable functions (similar to e.g., [MNOS99]). By the anti-symmetry of $\sqsubseteq$, FPC terms are automatically embedded in the logic, since we may express e as $\epsilon x{:}t.\, x \sqsubseteq e \bigwedge e \sqsubseteq x$. Because $\mathit{LogVar} \supset \mathit{TmVar}$, we can abstract over terms of the programming language inside the logic, but not vice-versa.

The semantics of LFPC is given using the standard set-theoretic construction for HOL. Each type denotes a non-empty set: **prop** is a two element set, $\tau \to \tau$ denotes a set of functions. The FPC type t is interpreted as the underlying set of the domain which interprets t. To define a sentence functor for $\mathcal{FPC}$, we set $\mathbf{Sen}^{\mathcal{FPC}}(\Sigma)$ to be the set of LFPC Σ-terms of type **prop**. The satisfaction condition is straightforward to verify, extending signature morphisms and Proposition 1 to terms of the logic.

Lemma 1 (Satisfaction condition for $\mathcal{FPC}$). *Let A be a Σ'-algebra and $\sigma : \Sigma \to \Sigma'$ a signature morphism. Then $A|_\sigma \models_\Sigma^{\mathcal{FPC}} \varphi$ iff $A \models_{\Sigma'}^{\mathcal{FPC}} \sigma(\varphi)$.*

3   Syntax for Signatures, Algebras, and Renamings

When writing parametrised programs and specifications, or using separate compilation of program parts, we have a *context* of declared programs and specifications; the context corresponds to the formal parameters or module interface. Working in a context, we use signature expressions which may not themselves be closed, but are closed when they are added to the context. Triples $(Ty^\Sigma, Sh^\Sigma, Tm^\Sigma)$ which have the same form as signatures but may not be closed are called *pre-signatures*. There is an inclusion between the signature of the context and the overall signature; if Σ_{ctx} is the former, we write $\Sigma_{ctx} \subseteq_\cup \Sigma$ to indicate that Σ is pre-signature such that $\Sigma_{ctx} \subseteq \Sigma_{ctx} \cup \Sigma$, where $\cup$ is "sequential" union of signatures. In this case, Σ is a signature-in-context. An algebra-in-context is then given by a function $f : \mathbf{Mod}(\Sigma_{ctx}) \to \mathbf{Mod}(\Sigma)$ which expands any Σ_{ctx}-algebra A to a Σ-algebra $f(A)$, so that $f(A)|_{\Sigma_{ctx}} = A$ (such an f is sometimes called a *persistent constructor* [ST88]).

Definition 5 (Signature morphism in context). *Given Σ_1, Σ_2 such that $\Sigma_{ctx} \subseteq_\cup \Sigma_1$ and $\Sigma_{ctx} \subseteq_\cup \Sigma_2$, a signature morphism in context Σ_{ctx} between them is defined to be an FPC signature morphism $\sigma : \Sigma_{ctx} \cup \Sigma_1 \to \Sigma_{ctx} \cup \Sigma_2$ such that*

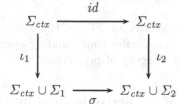

(i.e., the action of σ on Σ_{ctx} is the identity).

The grammar for syntactic signatures, signature morphisms, and algebras is:

$$
\begin{aligned}
S &::= \mathbf{sig}\ sdec^*\ \mathbf{end} \\
sdec &::= \mathbf{type}\ c \quad | \quad \mathbf{val}\ v : t \quad | \quad \mathbf{sharing}\ c = t \\
P &::= \mathbf{alg}\ pdec^*\ \mathbf{end} \\
pdec &::= \mathbf{type}\ c = t \quad | \quad \mathbf{val}\ v : t = e \\
s &::= [renam^*] \\
renam &::= c \mapsto c \quad | \quad v \mapsto v
\end{aligned}
$$

Each form has a type checking judgement:

$$
\begin{aligned}
\Sigma_{ctx} \triangleright S &\implies \Sigma & &\text{In } \Sigma_{ctx},\ S \text{ has pre-signature } \Sigma \\
\Sigma_{ctx} \triangleright P &\implies \Sigma & &\text{In } \Sigma_{ctx},\ P \text{ has pre-signature } \Sigma \\
\Sigma_{ctx} \triangleright s &\implies \Sigma \to \Sigma' & &\text{In } \Sigma_{ctx},\ s \text{ is a renaming from } \Sigma \text{ to } \Sigma'
\end{aligned}
$$

The first two judgements are inference judgements, since the pre-signature Σ is determined by the syntactic signature S or the program P. A renaming, on the other hand, does not determine its source or destination signature uniquely.

The typing rules are straightforward. They ensure that $\Sigma_{ctx} \cup \Sigma$ and $\Sigma_{ctx} \cup \Sigma'$ are proper signatures. The rule for adding a sharing equation is this:

$$
\frac{\Sigma_{ctx} \triangleright sdecs \implies \Sigma \qquad \begin{array}{c} t \in \mathit{ProgTypes}(\Sigma_{ctx} \cup \Sigma) \\ \mathit{Unifiable}(\Sigma_{ctx} \cup \Sigma \cup \{c = t\}) \end{array}}{\Sigma_{ctx} \triangleright sdecs\ \mathbf{sharing}\ c = t \implies \Sigma \cup \{c = t\}}
$$

The third premise ensures that the new equation is consistent with the equalities known so far. For typable phrases, it is easy to give a semantics.

Definition 6 (Interpretation of syntax in context).

- $[\![\Sigma_{ctx} \, \triangleright \, S \implies \Sigma]\!]$ *is the signature in context* $\Sigma_{ctx} \cup \Sigma$.
- $[\![\Sigma_{ctx} \, \triangleright \, s \implies \Sigma \to \Sigma']\!]$ *is the signature morphism in context* σ : $[\Sigma_{ctx}]\Sigma \to \Sigma'$ *determined by* s.
- $[\![\Sigma_{ctx} \, \triangleright \, P \implies \Sigma]\!]$ *is the functor* $f_P : \mathbf{Mod}(\Sigma_{ctx}) \to \mathbf{Mod}(\Sigma_{ctx} \cup \Sigma)$ *given by*

$$f_P(A) = \mathcal{P}[\![\Sigma_{ctx} \, \triangleright \, pdecs \implies \Sigma]\!]_{(\iota_A, \rho_A)}$$

where $P \equiv \mathbf{alg} \; pdecs \; \mathbf{end}$ *and* $\mathcal{P}[\![-]\!]_-$ *is defined by induction on the derivation of* $\Sigma_{ctx} \, \triangleright \, pdecs \implies \Sigma$, *extending* ι_A *and* ρ_A *in an obvious way.*

4   Modular Programs and Specifications in ASL+$_{\mathrm{FPC}}$

ASL+$_{\mathrm{FPC}}$ is based on the syntax for $\mathcal{FPC}$ of the previous sections. This syntax is combined using ASL-style specification building operators and a λ-calculus. There is a single syntactic category of pre-terms:

$$
\begin{aligned}
M ::= \; & X \quad | \quad P \quad | \quad S \\
& | \quad \mathbf{impose} \; \varphi \; \mathbf{on} \; M \quad | \quad \mathbf{derive \; from} \; M \; \mathbf{by} \; s : S \\
& | \quad \mathbf{translate} \; M \; \mathbf{by} \; s \quad | \quad \mathbf{enrich} \; M \; \mathbf{with} \; M \\
& | \quad \lambda X{:}M.\,M \quad | \quad M\,X \quad | \quad \varPi X{:}M.\,M \quad | \quad Spec\,(M) \\
& | \quad Let \; X = M \; in \; M : M
\end{aligned}
$$

Variables X range over a countable set $ModVar$. Meta-variables SP, A, M are all used to range over the set of pre-terms, with the hint that SP will denote a specification (collections of FPC algebras), and A some arbitrary collection.

Space precludes a complete motivation and explanation of the ASL+ calculus; we give only a brief overview. First, the ASL operators **impose**, etc, have their usual intentions in building specifications. The λ-calculus portion consists of λ-abstraction, application to variables, and $\varPi$-quantification for parametric (architectural) specifications. The $Spec\,(-)$ operator formalizes specification refinement: $SP' : Spec\,(SP)$ asserts that $SP \rightsquigarrow SP'$. This allows parametrised specifications and programs which accept any refinement of their formal parameter, written $\lambda X{:}\,Spec\,(SP).\,M$ (semantically, $Spec\,(-)$ is understood as a powerset operator). Finally, the let construct allows local definitions of modules, to relieve the restriction on function applications. It also imposes a signature or specification constraint: in $Let \; X = A \; in \; M : SP$, the constraining specification SP may be used to hide some details of the implementation M; in particular it *must* hide any mention of X from the result signature of M. (This prevents exporting a hidden symbol; module type systems solve this problem in varying ways).

Contexts for ASL+$_{\mathrm{FPC}}$ contain declarations and definitions for module variables. They may also directly include specifications, to allow "pervasive" datatypes of the language which are visible everywhere (BOOLEAN, INTEGER, etc).

$$\Gamma ::= \langle\rangle \quad | \quad \Gamma, X : A \quad | \quad \Gamma, X = M \quad | \quad \Gamma, SP$$

For type checking, we extract an FPC signature from a context Γ. This will include all of the pervasive elements, but also, any variables which stand for algebras will be included with their signatures, renamed using a "dot renaming" function to prefix identifiers with the module variable name. (We must assume the existence of suitable dot-renaming functions on $TyVar$, $TmVar$). Given a signature Σ, we write $X.\Sigma$ for the dot-renamed signature. Dot notation provides a way for programs to refer to components of modules. The notation can only be used on module variables because the syntax of FPC does not include ASL+$_{\text{FPC}}$ expressions; this restricts type propagation in the higher-order case.

4.1 Type Checking with Rough Types

Now we come to the main novelty in the development. ASL+ is equipped with two formal systems: one for proving satisfaction of a specification by a program, and the other for "rough" typing, which is designed to isolate the "static" type checking component of satisfaction. We follow the same plan in ASL+$_{\text{FPC}}$, except that rough types are improved to allow type equalities to be propagated from argument to result in parametrised programs. This generalisation is really the crux of the new system. Rough types have the syntax:

$$\kappa ::= \Sigma \quad | \quad \pi X : A.\kappa \quad | \quad P(\kappa)$$

where Σ ranges over pre-signatures. A program denoting a Σ-algebra will have rough type Σ; a Σ-specification expression will have type $P(\Sigma)$. The π-types classify functions. The main way that sharing information is propagated is through equations in pre-signatures that refer to the environment (e.g., $c = X.c$). The reason that the domain A of a type $\pi X : A.\kappa$ is a full ASL+ term is to account properly for sharing propagation between successive specification and program parameters; retaining a full term here allows rough types to be recalculated (see [Asp97] for further explanation).

There are three typing judgements:

$$\begin{array}{ll}
\Gamma \Longrightarrow_{\text{sig}} \Sigma_\Gamma & \Sigma_\Gamma \text{ is the underlying FPC signature of } \Gamma \\
\Gamma \rhd \kappa \leq \kappa' & \kappa \text{ is a subtype of } \kappa' \\
\Gamma \rhd M \Longrightarrow \kappa & M \text{ has rough type } \kappa
\end{array}$$

These judgements are defined in Figures 1–4, described in turn below.

Underlying signature (Figure 1). This judgement also serves to say that the context is well-formed. The underlying FPC signature is made by combining pre-signatures for the pervasive parts of the context, together with the dot-renamed components $X.\Sigma$ for variables X which range over Σ-algebras[1]. Module variables which have non-signature types (the rules assume κ is a non-signature) do not contribute to the FPC signature of the context; there is no way to use them directly in any FPC type or term.

---

[1] a sort of "flattening" operation, reminiscent of the way Java treats inner classes.

$$\langle\rangle \Longrightarrow_{\text{sig}} \emptyset$$

$$\frac{\Gamma \Longrightarrow_{\text{sig}} \Sigma_\Gamma \quad \Gamma \triangleright SP \Longrightarrow P(\kappa)}{\Gamma, X:SP \Longrightarrow_{\text{sig}} \Sigma_\Gamma}$$

$$\frac{\Gamma \Longrightarrow_{\text{sig}} \Sigma_\Gamma \quad \Gamma \triangleright SP \Longrightarrow P(\Sigma)}{\Gamma, SP \Longrightarrow_{\text{sig}} \Sigma_\Gamma \cup \Sigma}$$

$$\frac{\Gamma \Longrightarrow_{\text{sig}} \Sigma_\Gamma \quad \Gamma \triangleright M \Longrightarrow \Sigma}{\Gamma, X=M \Longrightarrow_{\text{sig}} \Sigma_\Gamma \cup X.\Sigma}$$

$$\frac{\Gamma \Longrightarrow_{\text{sig}} \Sigma_\Gamma \quad \Gamma \triangleright SP \Longrightarrow P(\kappa)}{\Gamma, SP \Longrightarrow_{\text{sig}} \Sigma_\Gamma}$$

$$\frac{\Gamma \Longrightarrow_{\text{sig}} \Sigma_\Gamma \quad \Gamma \triangleright M \Longrightarrow \kappa}{\Gamma, X=M \Longrightarrow_{\text{sig}} \Sigma_\Gamma}$$

$$\frac{\Gamma \Longrightarrow_{\text{sig}} \Sigma_\Gamma \quad \Gamma \triangleright SP \Longrightarrow P(\Sigma)}{\Gamma, X:SP \Longrightarrow_{\text{sig}} \Sigma_\Gamma \cup X.\Sigma}$$

Fig. 1. Underlying signature of a context

$$\frac{\Gamma \Longrightarrow_{\text{sig}} \Sigma_\Gamma \quad \Sigma_\Gamma \subseteq_\cup \Sigma \quad \Sigma_\Gamma \subseteq_\cup \Sigma' \quad (\Sigma_\Gamma \cup \Sigma') \subseteq_{\text{sh}} (\Sigma_\Gamma \cup \Sigma)}{\Gamma \triangleright \Sigma \leq \Sigma'}$$

$$\frac{\Gamma \triangleright A_1 \Longrightarrow \kappa \quad \Gamma \triangleright A_2 \Longrightarrow \kappa \quad \Gamma, X:A_2 \triangleright \kappa_1 \leq \kappa_2}{\Gamma \triangleright \pi X:A_1.\kappa_1 \leq \pi X:A_2.\kappa_2} \qquad \frac{\Gamma \triangleright \kappa \leq \kappa'}{\Gamma \triangleright P(\kappa) \leq P(\kappa')}$$

Fig. 2. Subtyping rules for rough types

Subtyping rules (Figure 2). We write $\Sigma_1 \subseteq_{\text{sh}} \Sigma_2$ if $\Sigma_1 \subseteq \Sigma_2$ but $Ty^{\Sigma_1} = Ty^{\Sigma_2}$ and $Tm^{\Sigma_1} = Tm^{\Sigma_2}$. In this case Σ_2 only differs from Σ_1 in having more sharing. The subtyping rules lift this relation to a relation on rough types. The rule for π-rough types appears as if it allows contravariancy in the domain; in fact, it does not because the rough types of A_1 and A_2 are required to be the same.

Programs and ASL terms (Figure 3). The rules for rough typing ASL terms, including FPC signatures and algebras, involve some signature calculation. The first two rules invoke the type checking system for the core-level from Section 3. The rule for **impose** checks that φ is a well-typed proposition.

The rules for **derive** and **translate** use renaming syntax, allowing some polymorphism. Arguments of **derive from** − **by** $s:S$ or of **translate** − **by** s can have any signature which fits suitably with s, according to the type checking rules for signature morphisms. The result signature of **derive** has to be given, but the result of **translate** is inferred, as the smallest image[2] of s. In fact, the rule for **translate** can be understood as constructing a pushout by propagating extra sharing; relying on the natural polymorphism of the syntax for renamings (as opposed to a semantic signature morphism in **Sign**$^{\mathcal{FPC}}$), this happens automatically. The rule for **derive**, by contrast, is provided with an explicit target signature, so any sharing in SP beyond that required by Σ' will be disregarded.

---

[2] This means that **translate** only uses surjective signature morphisms; but we can express translation along inclusions **translate** SP **by** $\iota : \Sigma \hookrightarrow \Sigma'$ using **enrich**.

$$\frac{\Gamma \implies_{sig} \Sigma_\Gamma \quad \Sigma_\Gamma \rhd S \implies \Sigma}{\Gamma \rhd S \implies P(\Sigma)} \qquad \frac{\Gamma \implies_{sig} \Sigma_\Gamma \quad \Sigma_\Gamma \rhd P \implies \Sigma}{\Gamma \rhd P \implies \Sigma}$$

$$\frac{\Gamma, SP \implies_{sig} \Sigma \quad \rhd^\Sigma \varphi : \textbf{prop}}{\Gamma \rhd \textbf{impose } \varphi \textbf{ on } SP \implies P(\Sigma)}$$

$$\frac{\Gamma \rhd SP \implies P(\Sigma') \quad \Gamma \implies_{sig} \Sigma_\Gamma \quad \begin{matrix} \Sigma_\Gamma \rhd S \implies \Sigma \\ \Sigma_\Gamma \rhd s \implies \Sigma \to \Sigma' \end{matrix}}{\Gamma \rhd \textbf{derive from } SP \textbf{ by } s : S \implies P(\Sigma)}$$

$$\frac{\Gamma \rhd SP \implies P(\Sigma) \quad \Gamma \implies_{sig} \Sigma_\Gamma \quad \Sigma_\Gamma \rhd s \implies \Sigma \to s(\Sigma)}{\Gamma \rhd \textbf{translate } SP \textbf{ by } s \implies P(s(\Sigma))}$$

$$\frac{\Gamma \rhd SP \implies P(\Sigma) \quad \Gamma, SP \rhd SP' \implies P(\Sigma')}{\Gamma \rhd \textbf{enrich } SP \textbf{ with } SP' \implies P(\Sigma \cup \Sigma')}$$

Fig. 3. Rough typing programs and ASL terms

The rule for **enrich** is similar to a rule for the dependent sum in type theory: just as x occurs bound in B in the term $\Sigma x{:}A.\,B$, so all the symbols of SP occur bound in SP' in the term **enrich** SP **with** SP'. This non-symmetry in **enrich** isn't revealed by its usual definition in terms of **translate** and **union**. The directly defined semantics of **enrich** SP **with** SP' also shows the dependency: models of the result are extensions of models of SP.

ASL+ terms (Figure 4). First, a variable which ranges over Σ-algebras is given a special *strengthened* type. The signature Σ/X is defined as Σ with the sharing equations augmented, so that $Sh^{\Sigma/X}(c) = Sh^\Sigma(c) \cup \{\, c = X.c \,\}$. This reflects the sharing of X with the context, since it denotes a projection on the $X.$-named part of the underlying environment. Strengthening was introduced by Leroy [Ler94] and a similar rule is present in most module type systems.

Rules for λ-abstractions and Π-abstractions are straightforward. Applications are restricted to variables; it may be necessary to rename the bound variable of the Π-type of the function to match the operand. The application rule is the crucial place where type identities are propagated. Subtyping here allows the actual parameter to have a richer type with more sharing equations than the type of the formal parameter A. Propagation of the type identities occurs because after application any mention of $X.c$ in the result type κ' will refer to a variable declared in the context, possibly having more sharing equations, rather than the bound variable of the Π-type.

The rule for a binding *Let* $X = M$ *in* $N : A$ allows N to be typed in the context extended by the typing of M and checks that the type of the constraint A is correct. The rough type of A is typed in Γ, so the dependency on X must be removed. The notation $\lceil \kappa \rceil$ in this rule embeds the rough type as a term of

$$\frac{\Gamma \implies_{\text{sig}} \Sigma_\Gamma \quad \Sigma_\Gamma \triangleright S \implies \Sigma}{\Gamma \triangleright S \implies P(\Sigma)} \qquad \frac{\Gamma \implies_{\text{sig}} \Sigma_\Gamma \quad \Sigma_\Gamma \triangleright P \implies \Sigma}{\Gamma \triangleright P \implies \Sigma}$$

$$\frac{\Gamma, SP \implies_{\text{sig}} \Sigma \quad \triangleright^\Sigma \varphi : \mathbf{prop}}{\Gamma \triangleright \mathbf{impose}\ \varphi\ \mathbf{on}\ SP \implies P(\Sigma)}$$

$$\frac{\Gamma \triangleright SP \implies P(\Sigma') \quad \Gamma \implies_{\text{sig}} \Sigma_\Gamma \quad \begin{array}{c} \Sigma_\Gamma \triangleright S \implies \Sigma \\ \Sigma_\Gamma \triangleright s \implies \Sigma \to \Sigma' \end{array}}{\Gamma \triangleright \mathbf{derive\ from}\ SP\ \mathbf{by}\ s : S \implies P(\Sigma)}$$

$$\frac{\Gamma \triangleright SP \implies P(\Sigma) \quad \Gamma \implies_{\text{sig}} \Sigma_\Gamma \quad \Sigma_\Gamma \triangleright s \implies \Sigma \to s(\Sigma)}{\Gamma \triangleright \mathbf{translate}\ SP\ \mathbf{by}\ s \implies P(s(\Sigma))}$$

$$\frac{\Gamma \triangleright SP \implies P(\Sigma) \quad \Gamma, SP \triangleright SP' \implies P(\Sigma')}{\Gamma \triangleright \mathbf{enrich}\ SP\ \mathbf{with}\ SP' \implies P(\Sigma \cup \Sigma')}$$

Fig. 4. Rough typing ASL+ terms

the calculus, defined by replacing Σ by its syntax, $\pi X : A.\kappa'$ by $\Pi X{:}A.\lceil \kappa' \rceil$ and $P(\kappa)'$ by $Spec(\kappa')$. This is simply a trick to avoid introducing a notion of rough-context (context with rough typing assumptions); when we project from the context, we get the rough type κ (or a strengthened version) again.

4.2 Brief Example

The following example shows how type equalities are propagated. We will build up a context of declarations step-by-step. First

$$\Gamma_1 =_{\text{def}} \mathbf{ELT} = \mathbf{sig}$$
$$\qquad\qquad \mathbf{type}\ \mathtt{elt}$$
$$\qquad \mathbf{end}$$

If the denotation of this expression is Σ_{ELT}, then we have the rough typing $\langle\rangle \triangleright \mathbf{ELT} \implies P(\Sigma_{ELT})$. Now we declare a parametrised program for building lists over some Σ_{ELT}-algebra:

$$\Gamma_2 =_{\text{def}} \Gamma_1, \mathtt{List} = \lambda\,\mathtt{Elt} : \mathbf{ELT}\,.\,\mathbf{alg}$$
$$\qquad\qquad\qquad\qquad \mathbf{type}\ \mathtt{elt} = \mathtt{Elt.elt}$$
$$\qquad\qquad\qquad\qquad \mathbf{type}\ \mathtt{list} = list_{\mathtt{elt}}$$
$$\qquad\qquad\qquad\qquad \mathbf{val}\ \mathtt{nil} : \mathtt{list} = \ldots$$
$$\qquad\qquad\qquad\qquad \mathbf{val}\ \mathtt{cons} : \mathtt{list} = \ldots$$
$$\qquad\qquad\qquad \mathbf{end}$$

($list_{\mathtt{elt}}$ is a type-expression in FPC which expresses the type of lists over the type $\mathtt{elt}$; the dots are filled with appropriate terms). This has the rough typing

$\Gamma \triangleright \texttt{List} \implies \Pi\texttt{Elt:ELT.}\, \Sigma_{LIST}[\texttt{Elt.elt}]$. The inferred signature of the algebra $\texttt{List}$ is $\Sigma_{LIST}[\texttt{Elt.elt}]$, where the square brackets are informal notation to indicate a dependency on $\texttt{Elt}$. To be more exact: this is the signature of lists extended with the equation $\texttt{elt} = \texttt{Elt.elt}$. Now we may apply the $\texttt{List}$ program to an algebra, for example:

$$\Gamma_3 =_{\text{def}} \Gamma_2, \texttt{Nat} =\textbf{alg}$$
$$\textbf{type } elt = nat$$
$$\textbf{end}$$

(where *nat* is an FPC type expression for natural numbers). Then $\Gamma_3 \triangleright \texttt{Nat} \implies \Sigma_{ELT}[nat]$ where $\Sigma_{ELT}[nat] = \Sigma_{ELT} \cup \{\, \texttt{elt} = nat \,\}$. Now we can derive $\Gamma_3 \triangleright \texttt{List Nat} \implies \Sigma_{LIST}[\texttt{Nat.elt}]$. Define $\Gamma_4 = \Gamma_3, \texttt{ListNat} = \texttt{List Nat}$. In the underlying FPC signature (such that $\Gamma_4 \implies_{\text{sig}} \Sigma_{\Gamma_4}$), we have the equation $\texttt{ListNat.elt} = nat$, which means that we can apply natural number functions to elements of $\texttt{ListNat}$ lists.

This very simple example demonstrates propagation of type equalities for the application of $\texttt{List}$. To prevent it, we could define an opaque version of list:

$$\texttt{OpaqueList} = \textit{Let } L = \texttt{List} \textit{ in } L : \Pi\texttt{Elt:ELT.}\, \Sigma_{LIST}$$

In a declaration $\texttt{OList} = \texttt{OpaqueList Nat}$ the type identity of the $\texttt{OList.elt}$ is unknown, so we could only pass elements of this list around.

4.3   Results and Further Developments

One important and non-trivial result is the decidability of rough type checking.

Theorem 1 (Decidability). *If all signatures are finite, each of the rough typing judgements is decidable.*

Proof. (Outline). First, observe that type checking in FPC and LFPC for finite signatures is decidable. For a slightly different formulation of the rough typing system viewed as an algorithm, we can give a measure on the inputs to each judgement which decreases from conclusions to premises of each rule. □

A set-theoretic semantics for ASL+$_{\text{FPC}}$ is given in [Asp97] together with a soundness proof. It interprets each of the typing judgements given above. However, the interpretation function is partial: rough type checking alone cannot guarantee that specifications are consistent, nor that actual arguments to parametrised programs or specifications meet the axiomatic requirements of their formal parameters.

To guarantee the well-definedness of an ASL+$_{\text{FPC}}$ term, we may need to do theorem proving. This is provided for with the satisfaction system, which incorporates ideas from other research into proof in structured specification.

5   Further Work, Related Work

The work here is mostly taken from Chapters 6 and 7 of my PhD thesis [Asp97], which contains additional results and full definitions. Theorem 1 is a new result.

The work began from the conception of adding type equations to algebraic signatures to explain sharing, an idea which occurred earlier to Tarlecki [Tar92]. The system here draws somewhat on later ideas of programming language researchers in investigating type systems for program modules, particularly those of Leroy [Ler95] and Harper and Lillibridge [HL94].

The most closely related and recent work in the algebraic specification community is on CASL's architectural specifications [SMT$^+$01]. This retains institution independence, but at the expense of complexity and for a language more restricted than ASL+.

Research is still highly active in the programming languages community in the quest to find more expressive type systems which are easier to understand and have good properties such as decidability (which failed for [HL94]). See e.g., [Sha98,Sha99,Jud97,Jon96,Rus99,DCH03]. Space precludes a detailed survey, but one aspect is worthy of note: several recent systems employ *singleton kinds* [SH00,DCH03] as an alternative to manifest types, as a way of succinctly internalising type equalities within the type system. The original version of ASL+ [SST92] in fact included a singleton construct (isolated in [Asp95a]) to allow a program to be turned into a trivial specification, and it was suggested how the dot notation could be expressed using this construct. (There are also connections here with the work of Cengarle [Cen94] who defined a syntax with an operator $Sig(-)$ for extracting the signature of an actual parameter; her work is an older relative of ASL+$_{\mathrm{FPC}}$).

In the end, it is a challenge to balance the various requirements and give a feasible system for type checking. The solution here is not ideal and has drawbacks outlined in [Asp97]. Typing modules for a specification language like ASL+ has different requirements to the programming case, and the system proposed here should be regarded only as a first attempt at a type-theoretic solution.

Towards Edinburgh CASL

One future venture we would like to undertake is the design and implementation of a CASL extension for a subset of Standard ML. While the specification constructs and CASL variations have received a great deal of attention, connection to specific programming languages remains relatively unexplored. A significant exception is the work at Bremen on HasCASL [SM02], which has parallels with what we want to do (and connections with work described above). We have early design ideas for a CASL extension called Edinburgh CASL, which is dedicated to specification for a subset of Standard ML, and constructed using a type-theoretic approach similar approach to ASL+$_{\mathrm{FPC}}$. We go beyond FPC in considering additional features of SML like polymorphism and pattern matching.

Since ASL+$_{\mathrm{FPC}}$ was invented, improvements to generic institutional technology were developed which may allow a more abstract approach (for example, using institutions with symbols and derived signature morphisms, instead of the concrete sharing relation in $\mathcal{FPC}$ signatures); however, these may not help with more advanced features such as first-class modules. And it remains important to verify that abstract constructions produce the desired result in different scenarios, by experimenting with ways of adding programming languages to CASL.

Acknowledgements

I'm grateful to Don Sannella and Andrzej Tarlecki for their guidance during development of this work, and the collaboration with Don since on ideas for Edinburgh CASL described above. Thanks are also due to the HasCASL team in Bremen (especially Till Mossakowski) for discussions.

References

[Asp95a]   David Aspinall. Subtyping with singleton types. In *Proc. Computer Science Logic, CSL'94, Kazimierz, Poland*, LNCS 933. Springer-Verlag, 1995.

[Asp95b]   David Aspinall. Types, subtypes, and ASL+. In *Recent Trends in Data Type Specification*, LNCS 906. Springer-Verlag, 1995.

[Asp97]   David Aspinall. *Type Systems for Modular Programs and Specification*. PhD thesis, Department of Computer Science, University of Edinburgh, 1997.

[Cen94]   María Victoria Cengarle. *Formal Specifications with Higher-Order Parameterisation*. PhD thesis, Institut für Informatik, Ludwig-Maximilians-Universität München, 1994.

[DCH03]   Derek Dreyer, Karl Crary, and Robert Harper. A type system for higher-order modules. In *Proceedings of POPL 2003, New Orleans*, 2003.

[GB92]   J. A. Goguen and R. M. Burstall. Institutions: abstract model theory for specification and programming. *Journal of the ACM*, 39:95–146, 1992.

[GM93]   M. J. C. Gordon and T. F. Melham. *Introduction to HOL*. Cambridge University Press, 1993.

[Gun92]   Carl A. Gunter. *Semantics of Programming Languages*. MIT Press, 1992.

[HL94]   Robert Harper and Mark Lillibridge. A type-theoretic approach to higher-order modules with sharing. In *Conference Record of the 21st ACM SIGPLAN-SIGACT Symposium on Principles of Programming Languages (POPL'94)*, pages 123–137, Portland, Oregon, January 17–21, 1994. ACM Press.

[HMM90]   Robert Harper, John C. Mitchell, and Eugenio Moggi. Higher-order modules and the phase distinction. In *Conference record of the Seventeenth Annual ACM Symposium on Principles of Programming Languages*, pages 341–354, San Francisco, CA, January 1990.

[Jon96]   Mark P. Jones. Using parameterized signatures to express modular structure. In *Conference Record of the 23rd ACM SIGPLAN-SIGACT Symposium on Principles of Programming Languages (POPL'96)*, St. Petersburg, Florida, 21–24, 1996. ACM Press.

[Jud97]   Judicaël Courant. An applicative module calculus. In *TAPSOFT*, Lecture Notes in Computer Science, pages 622–636, Lille, France, April 1997. Springer-Verlag.

[Ler94]   Xavier Leroy. Manifest types, modules, and separate compilation. In *Proc. 21st symp. Principles of Programming Languages*, pages 109–122. ACM press, 1994.

[Ler95]   Xavier Leroy. Applicative functors and fully transparent higher-order modules. In *Conference Record of the 22nd ACM SIGPLAN-SIGACT Symposium on Principles of Programming Languages (POPL'95)*, pages 142–153, San Francisco, California, January 22–25, 1995. ACM Press.

[Ler96]   Xavier Leroy. A syntactic theory of type generativity and sharing. *Journal of Functional Programming*, 6(5):667–698, 1996.

[Mac86]   David MacQueen. Using dependent types to express modular structure. In *Proceedings, Thirteenth Annual ACM Symposium on Principles of Programming Languages*, pages 277–286, St. Petersburg Beach, Florida, January 13–15, 1986. ACM SIGACT-SIGPLAN, ACM Press.

[MNOS99]   Olaf Müller, Tobias Nipkow, David von Oheimb, and Oskar Slotosch. HOLCF = HOL + LCF. *Journal of Functional Programming*, 9:191–223, 1999.

[Mos99]   Till Mossakowski. Specifications in an arbitrary institution with symbols. In *Proc. 14th WADT 1999*, volume LNCS 1827, pages 252–270, 1999.

[Plo85]   Gordon Plotkin. Denotational semantics with partial functions. Lecture at C.S.L.I. Summer School, 1985.

[Rus99]   Claudio V. Russo. Non-dependent types for standard ML modules. In *Principles and Practice of Declarative Programming*, pages 80–97, 1999.

[SH00]   Christopher A. Stone and Robert Harper. Deciding type equivalence in a language with singleton kinds. In *ACM Symposium on Principles of Programming Languages (POPL), Boston, Massachusetts*, pages 214–227, 19–21, 2000.

[Sha98]   Z. Shao. Parameterized signatures and higher-order modules, 1998. Technical Report YALEU/DCS/TR-1161, Dept. of Computer Science, Yale University, August 1998.

[Sha99]   Zhong Shao. Transparent modules with fully syntactic signatures. In *International Conference on Functional Programming*, pages 220–232, 1999.

[SM02]   Lutz Schröder and Till Mossakowski. HasCASL: Towards integrated specification and development of Haskell programs. In *Proceedings of AMAST 2002*, 2002.

[SMT⁺01]   Lutz Schröder, Till Mossakowski, Andrzej Tarlecki, Bartek Klin, and Piotr Hoffman. Semantics of Architectural Specifications in CASL. In *Proc. FASE 2001*, volume LNCS 2029, pages 253–268, 2001.

[SST92]   Donald Sannella, Stefan Sokołowski, and Andrzej Tarlecki. Toward formal development of programs from algebraic specifications: Parameterisation revisited. *Acta Informatica*, 29:689–736, 1992.

[ST86]   Donald Sannella and Andrzej Tarlecki. Extended ML: An institution-independent framework for formal program development. In David H. Pitts, editor, *Proc. Workshop on Category Theory and Computer Programming*, LNCS 240, pages 364–389. Springer-Verlag, 1986.

[ST88]   Donald Sannella and Andrzej Tarlecki. Toward formal development of programs from algebraic specifications: implementation revisited. *Acta Informatica*, 25:233–281, 1988.

[Tar92]   Andrzej Tarlecki. Modules for a model-oriented specification language: a proposal for MetaSoft. In *Proc. 4th European Symposium on Programming ESOP'92*, LNCS 582, pages 452–472. Springer-Verlag, 1992.

Pre-nets, Read Arcs and Unfolding:
A Functorial Presentation*

Paolo Baldan[1], Roberto Bruni[2], and Ugo Montanari[2]

[1] Dipartimento di Informatica, Università Ca' Foscari di Venezia, Italia
baldan@dsi.unive.it,
[2] Dipartimento di Informatica, Università di Pisa, Italia
{bruni,ugo}@di.unipi.it

Abstract. Pre-nets have been recently proposed as a means of providing a functorial algebraic semantics to Petri nets (possibly with read arcs), overcoming some previously unsolved subtleties of the classical model. Here we develop a functorial semantics for pre-nets following a sibling classical approach based on an unfolding construction. Any pre-net is mapped to an acyclic branching net, representing its behaviour, then to a prime event structure and finally to a finitary prime algebraic domain. Then the algebraic and unfolding view are reconciled: we exploit the algebraic semantics to define a functor from the category of pre-nets to the category of domains that is shown to be naturally isomorphic to the unfolding-based functor. All the results are extended to pre-nets with read arcs.

Introduction

P/T Petri nets [Rei85] are one of the most widely known models of concurrency. Since their introduction, almost fifty years ago [Pet62], the conceptual simplicity of the model and its intuitive graphical presentation have attracted the interest of both theoreticians and practitioners. Nevertheless, the concurrent semantics of Petri nets still presents several aspects that cannot be considered fully encompassed. The aim of this paper is to point out the missing fragments of the overall picture and to fill as many gaps as possible, providing neat mathematical constructions.

We concentrate on the semantic interpretation arising from the so-called *individual token philosophy* (ITPh) as opposed to the *collective token philosophy* (CTPh). The two terminologies have been introduced in [GP95] to distinguish the interpretation of tokens in the same place as anonymous and indistinguishable resources (CTPh), from the view of tokens as resources uniquely characterised by their histories and causal dependencies (ITPh). On the one hand, the CTPh, taking the paradigm of multiset rewriting to the extreme consequence is somehow simpler: repeated elements in a multiset are completely equivalent and cannot be distinguished one from the other. On the other hand, the CTPh is less amenable to the full variety of concurrent semantic frameworks that can be studied in the ITPh. Roughly these can be classified in process-oriented, unfolding, algebraic, and logical:

---

* Research supported by the FET-GC Project IST-2001-32747 AGILE and by the MIUR Project COFIN 2001013518 COMETA. The second author is also supported by an Italian CNR fellowship for research on Information Sciences and Technologies, and by the CS Department of the University of Illinois at Urbana-Champaign.

M. Wirsing, D. Pattinson, and R. Hennicker (Eds.): WADT 2002, LNCS 2755, pp. 145–164, 2003.

- The *process* approach focuses on non-sequential / concurrent models of computations and on their composition. Several notions of (deterministic) process have been proposed that rely on different abstractions in modelling resources, executed events and concurrent computations [BD87,GR83,DMM96,Sas98].
- The *unfolding approach* is built on top of nondeterministic processes to account for a broader view of computations, which includes concurrency, causality and conflict. Starting from the seminal work of Winskel [Win87], which focuses on the simpler class of *safe* nets, several authors have contributed to the generalisation of the approach to the full class of P/T Petri nets [MMS92,MMS97a,MMS96], showing that a chain of adjunctions (coreflections in the case of safe or semi-weighted nets) leads from **PTNets** to **PES**, for **PTNets** the category of P/T Petri nets and **PES** the category of prime event structures, which is equivalent to the category **Dom** of coherent finitary prime algebraic domains (for this reason, the unfolding approach is sometimes referred to as a *denotational semantics*).
- The *algebraic approach*, originally proposed in [MM90] for the CTPh under the statement "Petri nets are monoid", recasts the process approach in universal algebras: The idea is to characterise the concurrent model of computation as the initial model in a suitable algebra of decorated computations.
- The *logical view* tries to recast the algebraic approach into deduction theories, whose sentences denote concurrent execution strategies and whose theorems select admissible computations [BMMS01].

Category theory has been shown instrumental in all the above approaches: processes come naturally equipped with notion of a parallel and a sequential composition, which provides the structure of a *monoidal category*; adjunctions and coreflections are categorical notion used in the unfolding semantics to guarantee that all constructions are as good as possible; P/T Petri nets are essentially graphs with structured nodes, and, as such, can be naturally equipped with structure-preserving homomorphism, which can also be seen as simulation morphisms; initiality in the algebraic semantics is again a categorical notion for selecting the best candidate model; finally, the logical view exploits the fact that adjunctions between the categories of models of two theories, like the theory of Petri nets and the theory of concurrent models, can be more conveniently expressed as theory morphisms (whose existence is easier to prove).

When categories are involved, a central property of the semantic constructions, witnessing their appropriateness, is *functoriality*, i.e., the fact that simulation morphisms between nets are preserved at the level of computational, algebraic, logical and denotational models. A second crucial property is *universality*, in the sense of constructions expressed as adjunctions. In fact, we remind that when functors are left/right adjoints they preserve colimits/limits yielding good compositionality properties.

For the ITPh the unfolding approach is completely stable and satisfactory. Instead, the application of the algebraic approach to the ITPh presents several problems basically related to the fact that the monoidal operation on computations is commutative only up to a symmetry natural isomorphism. As a consequence, the construction proposed in [DMM96] fails to preserves some ordinary simulation morphisms between nets. The situation is improved in [MMS97b] up to a pseudo-functorial construction [Sas98]. Correspondingly, different notions of deterministic processes, which differ just in the decoration of minimal and maximal places have been proposed as "concrete" models. The lack of functoriality has also discouraged the formulation of a logical semantics.

The problem intuitively resides in the dichotomy between the multiset view of a state and the need of distinguishing uniquely its elements to track their causal history. A relevant advance of the theory has been the introduction of *pre-nets* [BMMS99] (see also [BMMS01] for an extensive discussion), a variant of ordinary nets where a total ordering is imposed on the places occurring in the pre- and post-set of transitions. Any pre-net can be seen as a concrete "implementation" of the Petri net obtained by forgetting about the ordering of places in pre- and post-sets. Using strings rather than multisets allows to uniquely characterise each element by its position. Thus pre-nets allow to obtain a satisfactory algebraic treatment, where the construction of the model of computation yields an adjunction between the category of pre-nets and the category of models (symmetric monoidal categories [Mac71]) and it can be expressed as a theory morphism, accounting for the algebraic and logical views. Notably, the construction of the model of computation for all pre-nets implementing the same Petri net yields the same result, hence we can define the semantics of a net as *the* algebraic semantics of any of its pre-net implementations. Still the picture is incomplete, since some classical approaches to the semantics of Petri nets have not been yet explored for pre-nets.

In this paper we complete the theory of pre-nets by showing that:

- Concrete notions of deterministic occurrence pre-nets and of pre-net processes can be defined in analogy with Petri nets. Finite processes form a symmetric monoidal category which turns out to be isomorphic via a symmetric monoidal functor to the algebraic model of computation, thus reconciling the process and algebraic view in a fully functorial construction (a result not possible for Petri nets). Moreover, a graphical presentation is introduced for pre-net processes.
- A domain semantics for pre-nets can be defined by generalising a construction proposed for ordinary nets in [MMS96]. Given a pre-net R, the comma category $\langle u \downarrow \mathcal{Z}(R) \rangle$, where u is the initial state of R and $\mathcal{Z}(R)$ its algebraic model, is a preorder whose ideal completion is a prime algebraic domain. Roughly this domain consists of the set of deterministic processes of the net, endowed with a kind of prefix ordering.
- An unfolding semantics can be defined which associates to any pre-net, first an acyclic pre-net representing all its possible computations in a single branching structure, then an event structure and finally a prime algebraic domain.
- Since the unfolding is essentially a nondeterministic process that completely describes the behaviour of a pre-net, a clear link between the unfolding and the algebraic approach is called for. The result showing that the domain originating from the algebraic model of computation and the one extracted from the unfolding are isomorphic, can now be stated in a satisfactory categorical framework: the two constructions can be expressed as *naturally isomorphic functors* (while the analogous result for ordinary Petri nets [MMS96] holds only at the level of objects).
- Finally, the pre-net and Petri net framework are reconciled by explaining how the domain semantics of a net and of its pre-net implementations are related.

We remark that, although in the case of pre-nets all the construction are functorial, one link is still missing, because the functor that abstracts the unfolding of a pre-net to a prime event structure is not characterised as a universal construction. Whether the mentioned construction can be defined as a right adjoint or not is a non-trivial question. We strongly conjecture that the answer is negative, but this is left as an open problem.

Along the years, Petri nets have been generalised in several ways to increase their expressivity. In the last part of the paper we focus on a mild but significant extension, i.e., the addition of *read arcs*, which allows to provide a faithful representation of read-only accesses to resources. Nets with read arcs, called *contextual nets* in [MR95], have been used to model a variety of applications and phenomena, such as transaction serializability in databases [DFMR94], concurrent constraint programming [MR94], asynchronous systems [Vog97], and analysis of cryptographic protocols [CW01].

Pre-nets have been already shown to be useful to define a neat algebraic semantics for contextual nets [BMMS02]. Here, relying on some previous work on the different semantic approaches for nets with read arcs, we discuss how the whole theory developed in this paper for ordinary pre-nets generalises in the presence of read arcs.

Synopsis. The rest of the paper is structured as follows. Section 1 reviews the basics of pre-nets and their algebraic semantics. Section 2 defines a process semantics for pre-nets and compares it to the algebraic semantics. Section 3 develops the unfolding semantics of pre-nets. Section 5 extends our results to nets with read arcs. Finally, Section 6 summarises the results in the paper and some open questions. We assume that the reader has some familiarity with P/T Petri net theory and category theory.

1   Pre-nets and Their Algebraic Semantics

In this section we recall the basics of pre-nets [BMMS99,BMMS01], discussing their algebraic semantics and the relation with ordinary P/T Petri nets.

Notation. Given a set X, we denote by $X^\otimes$ the free monoid over X (finite strings of elements of X) with the empty string ε as the unit, and by $X^\oplus$ the free commutative monoid over X (finite multisets over X) with unit the empty set $\emptyset$. We write $\mu : X^\otimes \to X^\oplus$ for the function mapping any string to the underlying multiset. Furthermore, given a function $f : X \to Y^\otimes$ we denote by $f^\otimes : X^\otimes \to Y^\otimes$ its obvious monoidal extension. Similarly, given $g : X \to Y^\oplus$ we denote by $g^\oplus : X^\oplus \to Y^\oplus$ its commutative monoidal extension. Given $u \in X^\otimes$ or $u \in X^\oplus$ we denote by $[u]$ the underlying subset of X defined in the obvious way. When set relations are used over string and multisets, we implicitly refer to the underlying set. E.g., for $u, v \in X^\otimes$ (or $X^\oplus$) by $x \in u$ we mean $x \in [u]$ and similarly $u \cap v$ means $[u] \cap [v]$.

Recall that a *P/T Petri net* is a tuple $N = (\partial_0, \partial_1, S, T)$, where S is a set of *places*, T is a set of *transitions*, and $\partial_0, \partial_1 : T \to S^\oplus$ are functions assigning multisets called source and target, respectively, to each transition. A *marked* net is a pair $\langle N, m \rangle$ where N is a P/T Petri net and $m \in S^\oplus$. A *Petri net morphism* $f = \langle f_s, f_t \rangle : N \to N'$ is a pair where $f_s : S^\oplus \to S'^\oplus$ is a monoid homomorphism, and $f_t : T \to T'$ is a function such that $\partial_i' \circ f_t = f_s \circ \partial_i$, for any $t \in T$ and $i \in \{0, 1\}$. The category of P/T Petri nets (as objects) and Petri net morphisms (as arrows) is denoted by **Petri**. A morphism of marked P/T nets $f : \langle N, m \rangle \to \langle N', m' \rangle$ is subject to the additional requirement of preservation of the

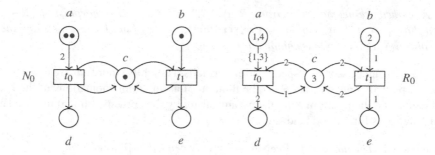

Fig. 1. The P/T Petri net N_0 and (one of) its pre-net implementation R_0.

initial marking, i.e., $f_s(m) = m'$. The category of marked P/T Petri nets (as objects) and marked Petri net morphisms (as arrows) is denoted by **Petri**$_*$.

A *pre-net* is roughly a Petri net where the resources (tokens in places) are linearly ordered. In other words, the state as well as the pre- and post-conditions of transitions are strings rather than multisets of places.

Definition 1 (pre-net). *A pre-net is a tuple* $R = (\zeta_0, \zeta_1, S, T)$, *where S is a set of* places, *T is a set of transitions, and* $\zeta_0, \zeta_1 : T \to S^\otimes$ *are functions assigning, respectively, source and target to transitions. A marked pre-net is a pair* $\langle R, u \rangle$ *with R a pre-net and* $u \in S^\otimes$.

The pictorial representation of Petri nets has certainly played an important role in their large diffusion as a specification framework. This graphical presentation (places represented as circles, transition as boxes, pre- and post-set multirelation as weighted arcs, tokens as black bullets) can be extended to pre-nets by adopting the following conventions: (1) weighted arcs are replaced by arcs labelled with the *ordered list of positions* in which the place appears in the pre- / post-set of the transition, with lists of length greater than one enclosed in curly brackets; (2) tokens are represented as numbers denoting their positions in the current state. An example of pre-net R_0 can be found in the right part of Fig. 1. It will be used throughout the paper to illustrate definitions and concepts. From the inscription $\{1, 3\}$ of the arc from a to t_0, we see that a firing of t_0 requires two tokens from a, to be taken as first and third consumed resources, while the second token to be consumed by t_0 must be taken from c, as imposed by the inscription 2 of the arc from c to t_0 (we remark that 2 denotes a position, not the number of tokens to be consumed). Moreover, from the inscriptions inside the circles for a, b and c, we note that the initial marking of R_0 is the string $u = abca$, i.e., that the a occurs in the first and fourth positions of u, b in the second, and c in the third.

As for P/T Petri nets, the notion of pre-net morphism naturally arises from an algebraic view, where places and transitions play the role of sorts and operators.

Definition 2 (pre-net morphism). *A pre-net morphism from* $R = (\zeta_0, \zeta_1, S, T)$ *to* $R' = (\zeta'_0, \zeta'_1, S', T')$ *is a pair* $f = \langle f_s, f_t \rangle$ *where* $f_s : S^\otimes \to S'^\otimes$ *is a monoid homomorphism, and* $f_t : T \to T'$ *is a function such that* $\zeta'_i \circ f_t = f_s \circ \zeta_i$, *for* $i \in \{0, 1\}$. *We denote by* **PreNet** *the category of pre-nets and their morphisms with the obvious composition.*

A marked pre-net morphism from $\langle R, u \rangle$ to $\langle R', u' \rangle$ is a pre-net morphism $f : R \to R'$ such that $f_s(u) = u'$. We denote by **PreNet**$_*$ the category of marked pre-nets and their morphisms with the obvious composition.

Pre-nets can be seen as a specification formalism (slightly) more concrete than Petri nets. In particular any pre-net R can be thought of as an "implementation" of the Petri net which is obtained from R replacing any string by the corresponding multiset. This construction is formalised below.

Definition 3. The functor $\mathcal{A} :$ **PreNet** $\to$ **Petri** is defined as follows:

- any pre-net $R = (\zeta_0, \zeta_1, S, T)$ is mapped to $\mathcal{A}(R) = (\partial_0, \partial_1, S, T)$, where $\partial_i(t) = \mu(\zeta_i(t))$ for each $t \in T$ and $i \in \{0, 1\}$;
- any pre-net morphism $f : R \to R'$ is mapped to $\mathcal{A}(f) = \langle g_s^{\oplus}, f_t \rangle$, where $g_s(s) = \mu(f_s(s))$ for each $s \in S$.

We denote by $\mathcal{A}_* :$ **PreNet**$_*$ $\to$ **Petri**$_*$ the obvious extension of $\mathcal{A}$ to marked nets.

For instance, referring to Fig. 1, the ordinary Petri net N_0 in the left part is implemented by R_0, i.e., we have $\mathcal{A}_*(R_0) = N_0$. The transition $t_0 : 2a \oplus c \to c \oplus d \in N_0$ is implemented as $t_0 : aca \to cd \in R_0$, and $t_1 : b \oplus c \to c \oplus e \in N_0$ as $t_1 : bc \to ec \in R_0$. Clearly alternative implementations would have been possible exploiting different linearizations.

Intuitively, a computation of a pre-net consists of "explicit" steps, namely firings of transitions which consume and produce resources, and of "implicit" steps which rearrange the order of the resources to allow the application of transitions. All the sequences of implicit steps that implement the same permutation of a given state are indistinguishable. Formally, the model of computation of a pre-net is the free symmetric strict monoidal category generated by the pre-net, the symmetries playing the role of the above mentioned implicit steps. Let **SSMC** be the category of symmetric strict monoidal categories (as objects) and symmetric monoidal functors (as arrows), and let **SSMC**$^{\otimes}$ denote the full subcategory containing only the categories whose monoid of objects is freely generated. Then the algebraic model of computation of a pre-net R is its image $\mathcal{Z}(R)$ through $\mathcal{Z} :$ **PreNet** $\to$ **SSMC**$^{\otimes}$, the left adjoint to the obvious forgetful functor from **SSMC**$^{\otimes}$ to **PreNet**. A more illustrative definition is given below.

Definition 4. Given a pre-net $R = (\zeta_0, \zeta_1, S, T)$, the model of computation $\mathcal{Z}(R)$ is a symmetric monoidal category whose objects are the elements of $S^{\otimes}$ and whose arrows are generated by the rules in Fig. 2, quotiented out by the axioms of monoidal categories and the coherence axioms making $\gamma_{-,-}$ the symmetry natural isomorphism (all axioms are collected in Fig. 3).

Recall that a *pointed* category is a pair $\langle \mathbf{C}, O_\mathbf{C} \rangle$, where $\mathbf{C}$ is a category and $O_\mathbf{C}$ is an object in $\mathbf{C}$. A *pointed functor* $F : \langle \mathbf{C}, O_\mathbf{C} \rangle \to \langle \mathbf{D}, O_\mathbf{D} \rangle$ is a functor $F : \mathbf{C} \to \mathbf{D}$ such that $F(O_\mathbf{C}) = O_\mathbf{D}$. The construction of the model of computation extends to marked pre-nets and to the category **SSMC**$^{\otimes}_*$ of pointed strictly symmetric monoidal categories.

Definition 5. Given a marked pre-net $\langle R, u \rangle$, the model of computation $\mathcal{Z}_*(\langle R, u \rangle)$ is the pointed category $\langle \mathcal{Z}(R), u \rangle$.

$$\frac{u \in S^{\otimes}}{id_u : u \to u \in \mathcal{Z}(R)} \qquad \frac{u, v \in S^{\otimes}}{\gamma_{u,v} : uv \to vu \in \mathcal{Z}(R)} \qquad \frac{t \in T \quad \zeta_0(t) = u \quad \zeta_1(t) = v}{t : u \to v \in \mathcal{Z}(R)}$$

$$\frac{\alpha : u \to v, \quad \alpha' : u' \to v' \in \mathcal{Z}(R)}{\alpha \otimes \alpha' : uu' \to vv' \in \mathcal{Z}(R)} \qquad\qquad \frac{\alpha : u \to v, \quad \beta : v \to w \in \mathcal{Z}(R)}{\alpha ; \beta : u \to w \in \mathcal{Z}(R)}$$

Fig. 2. Inference rules for $\mathcal{Z}(R)$.

For any $u, v, w \in S^{\otimes}$ and

for any $\alpha : u \to v, \beta : v \to w, \delta : w \to z, \alpha' : u' \to v', \beta' : v' \to w', \alpha'' : u'' \to v'' \in \mathcal{Z}(R)$:

UNIT:	$id_{\varepsilon} \otimes \alpha = \alpha = \alpha \otimes id_{\varepsilon},$	
ASSOCIATIVITY:	$(\alpha \otimes \alpha') \otimes \alpha'' = \alpha \otimes (\alpha' \otimes \alpha'')$	$(\alpha ; \beta) ; \delta = \alpha ; (\beta ; \delta)$
IDENTITIES:	$\alpha ; id_v = \alpha = id_u ; \alpha$	$id_u \otimes id_v = id_{uv}$
FUNCTORIALITY:	$(\alpha ; \beta) \otimes (\alpha' ; \beta') = (\alpha \otimes \alpha') ; (\beta \otimes \beta')$	
NATURALITY:	$(\alpha \otimes \alpha') ; \gamma_{v,v'} = \gamma_{u,u'} ; (\alpha' \otimes \alpha)$	
COHERENCE:	$\gamma_{u,vw} = (\gamma_{u,v} \otimes id_w) ; (id_v \otimes \gamma_{u,w})$	$\gamma_{u,v} ; \gamma_{v,u} = id_{uv}$

Fig. 3. Axioms for $\mathcal{Z}(R)$.

Notice that $\mathcal{Z}_*$ extends to a left adjoint functor from **PreNet**$_*$ to **SSMC**$_*^{\otimes}$.

Given a pre-net R and two states $u, v \in S^{\otimes}$ we say that v is reachable from u if there is an arrow $\alpha : u \to v$ in $\mathcal{Z}(R)$. If $\langle R, u \rangle$ is a marked pre-net we say that v is *reachable* if it is reachable from u. One can easily see that v is reachable in $\langle R, u \rangle$ if and only if $\mu(v)$ is reachable in $\mathcal{A}_*(\langle R, u \rangle)$. More generally, given any P/T net N, all its pre-net implementations have essentially the same behaviour, in the sense that they have isomorphic models of computation. Hence the semantics of N can be recovered by an arbitrarily chosen pre-net implementation.

Theorem 1. *For any pair of pre-nets R and R', if $\mathcal{A}(R) \simeq \mathcal{A}(R')$ then $\mathcal{Z}(R) \simeq \mathcal{Z}(R')$ via a symmetric monoidal functor.*

Moreover, the category $\mathcal{Z}(R)$ can be quotiented out by suitable axioms to recover all the algebraic computational models of $\mathcal{A}(R)$ in the literature (e.g. concatenable processes, commutative processes). Analogous results holds also in the marked case.

2   Concatenable Processes for Pre-nets

In this section we introduce a notion of (concatenable) process for pre-nets. A process is intended to provide a static representation of a concurrent computation, which makes explicit the events occurring in the computation and their causal dependencies. The appropriateness of our notion of pre-net process will be formalised by showing that for any pre-net the category of concatenable processes is isomorphic to its model of computation via a symmetric monoidal functor.

2.1 Safe and Occurrence Pre-nets

Let R be a pre-net. A state $u \in S^{\otimes}$ is called *safe* if any place occurs at most once in u, i.e., if $\mu(u)$ is a safe marking. A marked pre-net is called safe if the source and target of all transitions as well as all the reachable states are safe.

Definition 6 (causality, conflict, concurrency). *Let* $R = (\zeta_0, \zeta_1, S, T)$ *be a pre-net. The* causality relation *is the least transitive relation* $<_R \subseteq (S \cup T) \times (S \cup T)$ *such that*

$$(i) \text{ if } s \in \zeta_0(t) \text{ then } s <_R t; \quad (ii) \text{ if } s \in \zeta_1(t) \text{ then } t <_R s.$$

Given a place or transition $x \in S \cup T$, *we denote by* $\lfloor x \rfloor$ *the set of* causes *of* x *in* T, *defined as* $\lfloor x \rfloor = \{t \in T \mid t \leq_R x\} \subseteq T$, *where* $\leq_R$ *is the reflexive closure of* $<_R$.
    The conflict relation $\#_R \subseteq (S \cup T) \times (S \cup T)$ *is defined as the least relation such that*

$$(i) \text{ if } \zeta_0(t) \cap \zeta_0(t') \neq \emptyset \text{ then } t \#_R t'; \quad (ii) \text{ if } x \#_R x' \text{ and } x' \leq_R x'' \text{ then } x \#_R x''.$$

A set of places $X \subseteq S$ *is* concurrent, *written* $co(X)$ *if for any* $s, s' \in X$ *neither* $s < s'$ *nor* $s\#s'$, *and* $\bigcup_{x \in X} \lfloor x \rfloor$ *is finite.*

Definition 7 (occurrence pre-net). *An* occurrence pre-net *is a safe pre-net* R *such that (i) causality* $<_R$ *is a partial order and, for any transition* t, *the set of causes* $\lfloor t \rfloor$ *is finite; (ii) there are no backward conflicts, i.e., for any* $t \neq t'$, $\zeta_1(t) \cap \zeta_1(t') = \emptyset$; *(iii) conflict* $\#_R$ *is irreflexive. An occurrence pre-net is* deterministic *if it has no forward conflicts, i.e., for any* $t \neq t'$, $\zeta_0(t) \cap \zeta_0(t') = \emptyset$.
    We denote by **PreOcc**$_*$ *the full subcategory of* **PreNet**$_*$ *whose objects are occurrence pre-nets.*

It is immediate to verify that the relations of causality and conflict in a pre-net R are the same as in the implemented Petri net $\mathcal{A}(R)$. Hence R is a safe (occurrence) pre-net if and only if the corresponding Petri net $\mathcal{A}(R)$ is a safe (occurrence) net. This implies that $\mathcal{A}_*$ restricts to a well-defined functor from **PreOcc**$_*$ to **Occ**$_*$, the full subcategory of **Petri**$_*$ where objects are occurrence nets.

2.2 Processes of a Pre-net

An interesting feature of Petri nets is the fact that a net process can still be represented as a special Petri net (decorated with a morphism to the original net) [GR83]. This is true also for pre-nets.
    Let us call a pre-net morphism $f : R \to R'$ *elementary* if for any $s \in S$, $f_s(s) \in S'$ (places are sent to single places rather than to strings).

Definition 8 (process). *Let* $R = (\zeta_0, \zeta_1, S, T)$ *be a pre-net. A* process *of* R *is an elementary pre-net morphism* $\pi : O \to R$ *where* O *is an occurrence pre-net and for any* $t, t' \in T_O$, *if* $f_t(t) = f_t(t')$ *and* $\zeta_0(t) = \zeta_0(t')$ *then* $t = t'$ *(irredundancy).*
    The process π *is* finite / deterministic *if the underlying occurrence pre-net* O *is finite / deterministic. For a finite deterministic process* π *we denote by* $\min(\pi)$ *(resp.,* $\max(\pi)$*) the set of places of* O *which are minimal (resp., maximal) w.r.t.* $\leq_O$.

A concatenable process of a pre-net is a deterministic finite process of the net with explicit source and target states, i.e., with a total ordering in the minimal and maximal places of the underlying occurrence pre-net.

Definition 9 (concatenable process). *A concatenable process of a pre-net R is a triple* $\delta = \langle \sigma, \pi, \tau \rangle$, *where* π *is a deterministic finite process of R and* $\sigma, \tau \in S_O^{\otimes}$ *are string of places in* S_O *such that*

$$\mu(\sigma) = \min(\pi) \qquad and \qquad \mu(\tau) = \max(\pi).$$

We denote by $\zeta_0(\delta)$ *the string* $\pi_s^{\otimes}(\sigma)$ *and by* $\zeta_1(\delta)$ *the string* $\pi_s^{\otimes}(\tau)$.

An isomorphism of (concatenable) processes δ and δ' is an isomorphism of the underlying pre-nets consistent with the mapping to the original pre-net and with the linearizations of minimal and maximal places. The isomorphism class of a concatenable process δ is written $[\delta]$ and called an *abstract* concatenable process.

Concatenable processes $\delta = \langle \sigma, \pi, \tau \rangle$ of pre-nets can be graphically represented by slightly adjusting the visual modelling of ordinary Petri processes: (1) places (and transitions) are labelled by their images through π, (2) minimal and (resp. maximal) places carry also as superscript (resp., subscript) their position in σ (resp., τ); (3) arcs are labelled by the (unique) position in which the place appears in the pre- and post-set of the transition (again, we remark that arc labels stand for positions, not for weights).

In Fig. 4 some simple processes are illustrated (for our running example R_0) that correspond to single transitions, place identities and permutations.

Given two concatenable processes $\delta_1 = \langle \sigma_1, \pi_1, \tau_1 \rangle$ and $\delta_2 = \langle \sigma_2, \pi_2, \tau_2 \rangle$, such that $\zeta_1(\delta_1) = \zeta_0(\delta_2)$ their concatenation is defined as the process obtained by gluing the maximal places of π_1 and the minimal places of π_2 according to their orderings.

Definition 10 (sequential composition). *Let* $\delta_1 = \langle \sigma_1, \pi_1, \tau_1 \rangle$ *and* $\delta_2 = \langle \sigma_2, \pi_2, \tau_2 \rangle$ *be concatenable processes of a pre-net R such that* $\zeta_1(\delta_1) = \zeta_0(\delta_2)$. *Suppose* $T_1 \cap T_2 = \emptyset$ *and* $S_1 \cap S_2 = \max(\pi_1) = \min(\pi_2)$, *with* $\tau_1 = \sigma_2$. *In other words* δ_1 *and* δ_2 *overlap only on* $\max(\pi_1) = \min(\pi_2)$, *and such places carry the same ordering in the interfaces* τ_1 *and* σ_2. *Then their sequential composition* $\delta_1 ; \delta_2$ *is the concatenable process* $\delta = \langle \sigma_1, \pi, \tau_2 \rangle$, *where the process* π *is the (componentwise) union of* π_1 *and* π_2.

The above construction induces a well-defined operation of sequential composition between abstract concatenable processes. In particular, if $[\delta_1]$ and $[\delta_2]$ are abstract concatenable processes such that $\zeta_1(\delta_1) = \zeta_0(\delta_2)$ then we can always find $\delta_2' \in [\delta_2]$ such that $\delta_1 ; \delta_2'$ is defined. Moreover the result of the composition seen at abstract level, namely $[\delta_1 ; \delta_2']$, does not depend on the particular choice of the representatives.

Definition 11. *We denote by* $\mathcal{PP}(R)$ *the category having the elements of* $S^{\otimes}$ *as objects and abstract concatenable processes of R as arrows, with obvious composition as in Definition 10 and obvious identities.*

The category $\mathcal{PP}(R)$ is a symmetric strict monoidal category. In fact (1) parallel composition $\otimes$ is readily defined for processes $\delta_1 = \langle \sigma_1, \pi_1, \tau_1 \rangle$ and $\delta_2 = \langle \sigma_2, \pi_2, \tau_2 \rangle$ such that $T_1 \cap T_2 = S_1 \cap S_2 = \emptyset$, as $\delta_1 \otimes \delta_2 = \langle \sigma_1 \sigma_2, \pi, \tau_1 \tau_2 \rangle$, where π is the componentwise union of π_1 and π_2; (2) parallel composition induces a well-defined tensor

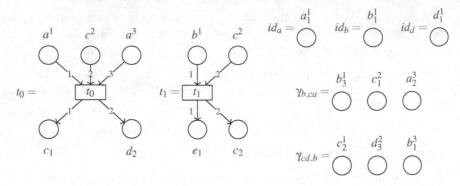

Fig. 4. Textual and graphical representation of simple pre-net processes.

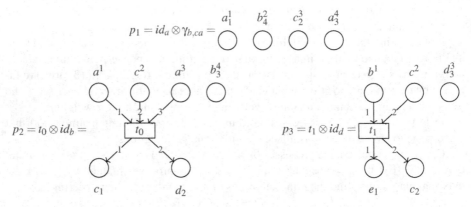

Fig. 5. Tensor product of simple processes.

product between abstract concatenable processes; (3) the tensor product is associative (but not commutative!) and it has the empty process $\langle \varepsilon, \pi_0, \varepsilon \rangle$ as unit; (4) the component $\gamma_{u,v}$ of the symmetry natural isomorphism is defined by the abstract class of processes $\langle \sigma_u \sigma_v, \pi, \sigma_v \sigma_u \rangle$ with no transitions and such that $\pi^\otimes(\sigma_u) = u$ and $\pi^\otimes(\sigma_v) = v$.

In Fig. 5 the processes of Fig. 4 are composed via tensor products in the larger processes $p_1 : abca \to acab$, $p_2 : acab \to cdb$ and $p_3 : bcd \to ecd$. Finally, in Fig. 6, the processes illustrated so far are composed sequentially in $p_4 : abca \to cdb$, $p_5 : cdb \to ecd$ and $p : abca \to ecd$.

The next theorem shows that pre-net processes provide an appropriate description of the concurrent computations of a pre-net R, in the sense that concatenable pre-net processes can be seen as concrete representatives of the arrows in $\mathcal{Z}(R)$.

Theorem 2. *The category $\mathcal{PP}(R)$ is isomorphic to the model of computation $\mathcal{Z}(R)$ via a symmetric monoidal functor.*

The theorem above is proved by observing that, being $\mathcal{PP}(R)$ a symmetric monoidal category, a functor from $F : \mathcal{Z}(R) \to \mathcal{PP}(R)$ can be easily defined by mapping generators to generators. A functor in the converse direction, is defined by identifying a normal

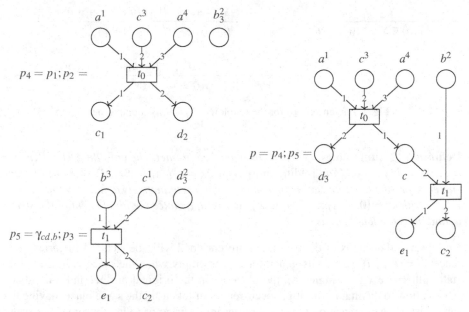

Fig. 6. Sequential composition of processes.

form for the processes in $\mathcal{PP}(R)$ which, roughly, corresponds to a maximally concurrent computation. As a technical remark, the proof is much simpler w.r.t. analogous ones for (concatenable, strongly concatenable) process categories associated to Petri nets, as we can (arbitrarily) fix the normal form expression in such a way that all isomorphic processes have exactly the same normal form (whereas in Petri nets the normal form can be fixed only up-to isomorphism).

3   Unfolding of Pre-nets

A deterministic process describes a single deterministic computation of the net. The unfolding approach, originally devised in [NPW81], associates to a system a single denotational structure representing, in an unambiguous way, all the events occurring in any possible computation and their dependencies. This structure expresses not only the causal ordering between the events, but also gives an explicit representation of the branching (choice) points of the computations.

In this section we develop a functorial unfolding semantics for pre-nets, discussing the difficulties which arise in trying to express this functor as a universal construction.

3.1   Unfolding Construction

Given a marked pre-net $\langle R, u \rangle$ the unfolding construction unwinds R into an occurrence pre-net, starting from the initial state u, firing transitions in all possible way and recording the corresponding occurrences.

$$\frac{1 \le i \le |u|}{u'_i = \langle \emptyset, u_i, i \rangle \in S' \quad \eta_s(u'_i) = u_i} \qquad \frac{v \in S'^\otimes \text{ safe} \quad co([v]) \quad t \in T \quad \eta_s^\otimes(v) = \zeta_0(t)}{t' = \langle v, t \rangle \in T' \quad \eta_t(t') = t \quad \zeta'_0(t') = v}$$

$$\frac{t' = \langle v, t \rangle \in T' \quad \zeta_1(t) = w_1 \ldots w_n}{w'_i = \langle \{t'\}, w_i, i \rangle \in S' \quad \eta_s(w'_i) = w_i \quad \zeta'_1(t') = w'_1 \ldots w'_n}$$

Fig. 7. Inference rules for the unfolding $\mathcal{U}_p(\langle R, u \rangle)$ of a pre-net R.

Definition 12 (unfolding). *Let $\langle R, u \rangle$ be a marked pre-net. The unfolding $\mathcal{U}_p(\langle R, u \rangle) = ((\zeta'_0, \zeta'_1, S', T'), u')$ and the folding morphism $\eta_R = \langle \eta_t, \eta_s \rangle : \mathcal{U}_p(R) \to R$ are the occurrence pre-net and (elementary) pre-net morphism inductively defined by the rules in Fig. 7, with $u' = \langle \emptyset, u_1, 1 \rangle \ldots \langle \emptyset, u_{|u|}, |u| \rangle$ (where u_i denotes the ith element of the string u, and $|u|$ is the length of u).*

Observe that items in the unfolding are enriched with their causal histories. Any place $s' = \langle x, w_i, i \rangle$ records its generator x (x is empty when the place is in the initial state, otherwise x is a singleton), the place w_i in the original pre-net and a number i which allow to distinguish multiple occurrences of tokens in the same place, having the same history. Any transition $t' = \langle v, t \rangle$ represents a firing of t that consumes the string of resources v.

The unfolding of our running example R_0, with initial state $abca$, is depicted in Fig. 8. The morphism $\eta_{R_0} : \mathcal{U}_p(\langle R_0, abca \rangle) \to R_0$ is implicitly represented by labelling each place and transition x with its image $\eta_{R_0}(x)$. For some items in the unfolding also the concrete identity is provided. For instance, $a_1 = \langle \emptyset, a, 4 \rangle$ represents the occurrence of a in the fourth position of the initial marking, $t'_0 = \langle a_4 c_3 a_1, t_0 \rangle$ represent an occurrence of t_0, which fires using the fourth, third and second resource in the initial state.

The unfolding construction can be characterised as a universal construction establishing a coreflection between the categories **PreOcc**$_*$ and **PreNet**$_*$.

Theorem 3. *The unfolding construction induces a functor $\mathcal{U}_p : \mathbf{PreNet}_* \to \mathbf{PreOcc}_*$, right adjoint to the inclusion $I_p : \mathbf{PreOcc}_* \to \mathbf{PreNet}_*$, with counit $\eta : I_p \circ \mathcal{U}_p \to 1$.*

3.2 Event Structure and Domain Semantics

The unfolding semantics for a pre-net can be naturally abstracted to a prime event structure semantics. *Prime event structures* (PES) are a simple event based model of (concurrent) computations in which events are considered as atomic and instantaneous steps, which can appear only once in a computation. An event can occur only after some other events (its *causes*) have taken place and the execution of an event can inhibit the execution of other events. This is formalised via two binary relations: *causality*, modelled by a partial order relation, and *conflict*, modelled by a symmetric and irreflexive relation, hereditary w.r.t. causality.

Definition 13 (prime event structures). *A prime event structure (PES) is a tuple $P = \langle E, \le, \# \rangle$, where E is a set of events and $\le$, $\#$ are binary relations on E called causality and conflict, respectively, such that:*

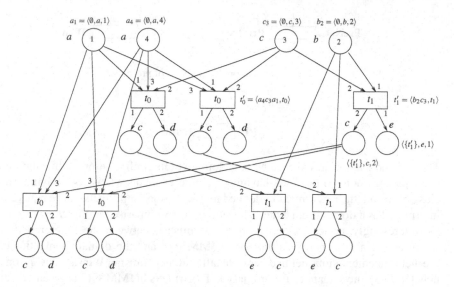

Fig. 8. The unfolding of $\langle R_0, abca \rangle$.

1. *the relation $\leq$ is a partial order and $\lfloor e \rfloor = \{e' \in E : e' \leq e\}$ is finite for all $e \in E$;*
2. *the relation $\#$ is irreflexive, symmetric and hereditary with respect to $\leq$, i.e., $e \# e'$ and $e' \leq e''$ implies $e \# e''$ for all $e, e', e'' \in E$;*

Let $P_0 = \langle E_0, \leq_0, \#_0 \rangle$ and $P_1 = \langle E_1, \leq_1, \#_1 \rangle$ be two PES's. A PES-morphism $f : P_0 \rightarrow P_1$ is a partial function $f : E_0 \rightarrow E_1$ such that for all $e_0, e'_0 \in E_0$, assuming that $f(e_0)$ and $f(e'_0)$ are defined:

1. $\lfloor f(e_0) \rfloor \subseteq f(\lfloor e_0 \rfloor)$;
2. (a) $f(e_0) = f(e'_0) \wedge e_0 \neq e'_0 \Rightarrow e_0 \#_0 e'_0$;   (b) $f(e_0) \#_1 f(e'_0) \Rightarrow e_0 \#_0 e'_0$;

The category of prime event structures and PES*-morphisms is denoted by* **PES**.

Given an occurrence pre-net the corresponding PES can be obtained by forgetting about the places, keeping the transitions and the dependency relations among them. The transformation is functorial since the transition component of a morphism between occurrence pre-nets satisfies the requirements to be a PES-morphism between the underlying PES's.

Definition 14 (from occurrence pre-nets to PES's). *Let $\mathcal{E}_p : \mathbf{PreOcc}_* \rightarrow \mathbf{PES}$ be the functor defined on objects by $\mathcal{E}_p(R) = \langle T, \leq_R, \#_R \rangle$ for any occurrence pre-net R and on arrows by $\mathcal{E}_p(f) = f_t$ for each occurrence pre-net morphism $f : R_0 \rightarrow R_1$.*

Winskel in his seminal work [Win87] shows that PES's are intimately connected with another classical semantical model, i.e., *prime algebraic, finitely coherent, finitary partial orders*, hereafter referred to simply as *domains* [Ber78]. Formally, an equivalence is established between the category **PES** of prime event structures and the category **Dom** of domains and additive, stable, immediate precedence-preserving functions:

$$\mathbf{PreNet}_* \xrightarrow[\mathcal{U}_p]{\overset{I_p}{\longleftarrow}} \mathbf{PreOcc}_* \xrightarrow{\mathcal{E}_p} \mathbf{PES} \xrightarrow[\mathcal{L}]{\overset{\mathcal{P}}{\longleftarrow}} \mathbf{Dom}$$

Fig. 9. Denotational semantics of pre-nets.

$$\mathbf{PES} \xrightarrow[\mathcal{L}]{\overset{\mathcal{P}}{\longleftarrow}} \mathbf{Dom}$$

The functor $\mathcal{L}$ associates to each PES the domain of its configurations, while the functor $\mathcal{P}$ maps each domain to a PES having its prime elements as events. Relying on this classical result, the PES semantics defined in this section for pre-nets can be equivalently interpreted as a domain semantics. The situation is summarised in Fig. 9.

Interestingly, a clear relation can be established between the functorial domain semantics of a Petri net N as defined in [MMS96] and the domain semantics of its pre-net implementations defined here. Recall that, generalising Winskel's work on safe nets [Win87], the semantics for ordinary P/T Petri nets in [MMS96] is given as a chain of adjunctions from the category of nets to the category of domains. The diagram below summarises these results.

$$\mathbf{PTNets}_* \xrightarrow[\mathcal{U}_d]{\overset{\longleftarrow}{\perp}} \mathbf{DecOcc}_*$$

$$\mathcal{F} \vdash \mathcal{D}$$

$$\mathbf{Safe}_* \xrightarrow[\mathcal{U}]{\overset{\longleftarrow}{\perp}} \mathbf{Occ}_* \xrightarrow[\mathcal{E}]{\overset{\mathcal{N}}{\longleftarrow}} \mathbf{PES} \xrightarrow[\mathcal{L}]{\overset{\mathcal{P}}{\longleftarrow}} \mathbf{Dom}$$

The domain associated to a Petri net by the above construction can be obtained from that of any of its pre-net implementations by equating all the events which correspond to occurrences of the same transition with different linearizations of the same resources (which may differ for the order of tokens in the same place). Formally this is expressed as a natural transformation between the two semantics:

Theorem 4. *There is a natural transformation* $\varsigma : \mathcal{L} \circ \mathcal{E}_p \circ \mathcal{U}_p \to \mathcal{L} \circ \mathcal{E} \circ \mathcal{F} \circ \mathcal{U}_d \circ \mathcal{A}$.

As a consequence (as it happens for the algebraic models of computation) the domains associated to the pre-net implementations of a given net are all isomorphic, i.e., for all R, R', if $\mathcal{A}(R) \simeq \mathcal{A}(R')$ then $\mathcal{L} \circ \mathcal{E}_p \circ \mathcal{U}_p(R) \simeq \mathcal{L} \circ \mathcal{E}_p \circ \mathcal{U}_p(R')$.

Unfortunately, in the case of pre-nets finding a left adjoint for the functor $\mathcal{E}_p$ appears to be quite problematic. Intuitively, the left adjoint should freely generate an occurrence pre-net from any PES in a way which guarantees the existence and uniqueness of a representation of PES-morphisms in **PreOcc**$_*$. Places could be freely generated as for ordinary Petri nets, but then it would be impossible to fix a linear order on the pre- and post-sets of transitions in a "universal" way. Our conjecture is that $\mathcal{E}_p$ is not a right adjoint functor.

An idea which seems promising in view of a universal characterisation of the mentioned construction is to abandon the purely algebraic view of pre-nets, considering an alternative notion of pre-net morphism, based on a weaker condition which requires the preservation of pre- and post-sets of transitions only up to a permutation. The permutation should be explicitly mentioned in the morphism itself, i.e., a morphism $f : R \to R'$ would be enriched with a family of permutations $\{\omega_t^0, \omega_t^1\}_{t \in T}$ such that $\omega_t : f_s^\otimes(\zeta_i(t)) \to \zeta_i(f_t(t))$ for any transition t in R.

4 Reconciling the Unfolding and Algebraic Semantics of Pre-nets

The unfolding of a marked pre-net can be seen as a maximal nondeterministic process, representing all its possible computations. Hence it is natural to expect that a tight relationship can be established between the unfolding and the algebraic / process approach. In this section we show that the domain produced through the unfolding construction can be obtained, equivalently, by means of a functorial construction based on the model of computation. The correspondence holds at categorical level, namely the functor $\mathcal{L} \circ \mathcal{E}_p \circ \mathcal{U}_p$ (see Fig. 9) and the new functor based on the algebraic semantics are naturally isomorphic. This improves the analogous result existing for Petri nets [MMS96], which only holds at the object level.

Let $\langle R, u \rangle$ be a marked pre-net and consider the comma category $\langle u \downarrow \mathcal{PP}(R) \rangle$ (which, by Theorem 2, is isomorphic to $\langle u \downarrow Z(R) \rangle$). Objects are concatenable processes of R with source in u, and an arrow exists from a process δ_1 to δ_2 if $\delta_2 = \delta_1; \delta$ for some process δ. It can be shown that $\langle u \downarrow \mathcal{PP}(R) \rangle$ is a preorder, i.e., in $\langle u \downarrow \mathcal{PP}(R) \rangle$ there is at most one arrow between any two objects. Let $\lesssim_R$ denote the corresponding preorder relation i.e., $\delta_1 \lesssim_R \delta_2$ if there exists δ such that $\delta_1; \delta = \delta_2$.

An alternative characterisation of $\lesssim_R$, enforces the intuitive idea that it is a generalisation of the prefix ordering over processes. First, we need to introduce the notion of left injection for concatenable processes.

Definition 15 (left injection). *Let* $\delta_i : u \to v_i$ *($i \in \{1, 2\}$) be two objects in* $\langle u \downarrow \mathcal{PP}(R) \rangle$, *with* $\delta_i = \langle \sigma_i, \pi_i, \tau_i \rangle$. *A left injection* $\iota : \delta_1 \to \delta_2$ *is a morphism of pre-nets* $\iota : R_{\pi_1} \to R_{\pi_2}$ *(where* R_{π_i} *is the pre-net underlying* π_i), *such that*

1. ι *preserves the ordering of minimal places, namely* $\sigma_2 = \iota_s^\otimes(\sigma_1)$;
2. ι *is rigid on transitions, namely for* t_2' *in* R_{π_2} *and* t_1 *in* R_{π_1}, *if* $t_2' \leq \iota(t_1)$ *then* $t_2' = \iota(t_1')$ *for some* t_1' *in* R_{π_1} *(the image of a lower set is a lower set).*

The name "injection" comes from the fact that any morphism ι between marked deterministic occurrence nets results to be injective on places and transitions. The word "left" is related to the fact that ι is required to preserve only the string of minimal places.

Lemma 1. *Let* $\delta_i : u \to v_i$ *($i \in \{1, 2\}$) be objects in* $\langle m \downarrow \mathcal{PP}(R) \rangle$, *with* $\delta_i = \langle \sigma_i, \pi_i, \tau_i \rangle$. *Then* $\delta_1 \lesssim_R \delta_2$ *iff there exists a left injection* $\iota : \delta_1 \to \delta_2$.

By exploiting the above characterisation and the fact that $\mathcal{U}_p$ is a right adjoint we can conclude that the ideal completion of the preorder $\langle u \downarrow \mathcal{PP}(R) \rangle$, denoted by $\mathrm{Idl}(\langle u \downarrow \mathcal{PP}(R) \rangle)$, is isomorphic to the domain $\mathcal{L}(\mathcal{E}_p(\mathcal{U}_p(R)))$ obtained from the unfolding of the pre-net R.

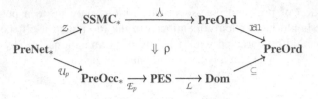

Fig. 10. Reconciling the algebraic and unfolding semantics of pre-nets.

To gain some intuition observe that the elements of the partial order induced by the preorder $\langle u \downarrow \mathcal{PP}(R) \rangle$ are classes of concatenable processes which are "left isomorphic", i.e., isomorphic via a left injection. Intuitively, the partial order consists of processes starting from a fixed initial state and ordered by prefix. Since processes are finite, taking the ideal completion of the partial order induced by the preorder $\langle u \downarrow \mathcal{PP}(R) \rangle$ (which produces the same result as taking directly the ideal completion of $\langle u \downarrow \mathcal{PP}(R) \rangle$) is necessary for moving from finite computations to arbitrary ones.

Theorem 5 (unfolding vs. concatenable processes). *Let $\langle R, u \rangle$ be a marked pre-net. Then* $\mathsf{Idl}(\langle u \downarrow \mathcal{PP}(R) \rangle)$ *is isomorphic to the domain* $\mathcal{L}(\mathcal{E}_p(\mathcal{U}_p(R)))$.

The above results admits a nice categorical formulation, since all the involved constructions can be seen as functors. Let **PreOrd** be the category of preorders and monotone functions, and let $\mathsf{Flat} : \mathbf{Cat} \to \mathbf{PreOrd}$ be the functor mapping any category to the underlying preorder (where $x \leq y$ if and only if there was an arrow $f : x \to y$ in the original category). Let $\wedge: \mathbf{SSMC_*} \to \mathbf{PreOrd}$ be the functor mapping any pointed symmetric strict monoidal category $\langle \mathbf{C}, O_C \rangle$ to $\mathsf{Flat}(\langle O_C \downarrow \mathbf{C} \rangle)$. Finally let $\mathbf{PreOrd} \to \mathbf{PreOrd}$ be the ideal completion functor, mapping any preorder to its ideal completion. Then the following result holds (see Fig. 10).

Theorem 6. *There is a natural isomorphism* $\rho : \mathsf{Idl} \circ \wedge \circ \mathcal{PP}_* \to \mathcal{L} \circ \mathcal{E}_p \circ \mathcal{U}_p$.

5   Adding Read Arcs

Several extensions of ordinary Petri nets have been proposed in the literature to enrich the expressiveness of the basic model. A mild generalisation which has been shown to be quite useful is the addition of the so-called *read arcs* which allow a transition to check for the presence of a token in a place without removing the token itself. Observe that a read arc cannot be safely replaced by a self-loop, since the former allows a greater amount of concurrency in the system: a resource can be read in parallel by several transitions at the same time, concurrently. For instance consider again the net N_0 in Fig. 1, and compare it to the net N_1 in Fig. 11, where place c is connected to transitions t_0 and t_1 by read arcs (denoted by undirected lines), meaning that c represent a resource accessed in a read-only manner. While in N_1 the transitions t_0 and t_1 can fire concurrently, in the net N_0 where read arcs are replaced by self-loops, the two transitions are serialised.

Formally a *contextual Petri net* is a tuple $N = \langle \partial_0, \partial_1, \partial_2, S, T \rangle$, where $\langle \partial_0, \partial_1, S, T \rangle$ is an ordinary Petri net and $\partial_2 : T \to S^\oplus$ associates to each transition its context. Notice

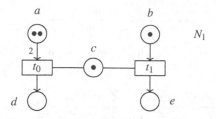

Fig. 11. Ordinary nets do not allow for concurrent read-only operations.

that, as a single token can be read concurrently by different transitions, it can be read also with multiplicity greater than 1 by the same transition. Hence a transition t can use, to fire, any marking ranging from $\partial_0(t) \oplus [\partial_2(t)]$ to $\partial_0(t) \oplus \partial_2(t)$, i.e., it is sufficient that any context place contains at least one token. The notion of *marked contextual net* and of (marked) contextual net morphism are defined as structure-preserving graph homomorphisms in the obvious way, yielding the categories **CPetri** and **CPetri**$_*$.

Correspondingly, we consider an extension of pre-nets with read arcs.

Definition 16 (contextual pre-net). *A contextual pre-net is a tuple* $R = \langle \zeta_0, \zeta_1, \zeta_2, S, T \rangle$ *such that* $\langle \zeta_0, \zeta_1, S, T \rangle$ *is a pre-net and* $\zeta_2 : T \to S^{\otimes}$ *is the* context *function.*

The notion of pre-net morphism can be extended to contextual pre-nets in the obvious way, requiring for any transition the preservation of the context, besides pre- and post-conditions. Also the extension to marked pre-nets is immediate. We denote by **CPreNet** and by **CPreNet**$_*$ the corresponding categories.

The algebraic semantics of contextual pre-nets has been developed in [BMMS02] taking as models the so-called *match-share categories*, a kind of symmetric monoidal category equipped with two additional (non-natural) transformations. Let $\mathcal{Z}_c(R)$ denote the model of computation for a contextual pre-net R, as defined in such paper (as a special case, in the absence of contexts, $\mathcal{Z}_c(R) = \mathcal{Z}(R)$). The construction can be expressed as an adjunction and it is defined in terms of theory morphisms between suitable equational theories. Due to space limitation, we cannot give full details here.

The results developed in this paper for Petri nets and pre-nets generalise to the contextual case. In the following we sketch the basic notions, constructions and the results involved in the extension.

First, as it happens for ordinary contextual nets [BCM01], the dependencies among events in a contextual pre-net computation cannot be captured completely by two binary relations representing causality and symmetric conflict. While *causality* can be defined essentially as in the ordinary case, due to the possibility of preserving part of the state in a step of computation, an *asymmetric* form of conflict arises between transitions. In fact let t, t' be transitions such that $\zeta_2(t) = s = \zeta_0(t')$. Then the firing of t' prevents t to be fired, since it consumes the shared resource in s. Instead the firing of t just reads a resource in s and thus t' can fire after t. This kind of dependency is represented by introducing an *asymmetric conflict* relation $\nearrow$ on transitions, which models the previous situation as $t \nearrow t'$. An ordinary symmetric conflict, arising when two transition t and t' have a common precondition, is represented as an asymmetric conflict in both

$$\frac{1 \leq i \leq |u|}{u'_i = \langle \emptyset, u_i, i \rangle \in S' \quad \eta_s(u'_i) = u_i}$$

$$\frac{v_p, v_c \in S'^{\otimes} \quad v_p \text{ safe} \quad v_p \cap v_c = \emptyset \quad co([v_p \cdot v_c]) \quad t \in T \quad \eta_s^{\otimes}(v_p) = \zeta_0(t) \quad \eta_s^{\otimes}(v_c) = \zeta_2(t)}{t' = \langle v_p, v_c, t \rangle \in T' \quad \eta_t(t') = t \quad \zeta_0'(t') = v_p \quad \zeta_2'(t') = v_c}$$

$$\frac{t' = \langle v, t \rangle \in T' \quad \zeta_1(t) = w_1 \dots w_n}{w'_i = \langle t', w_i, i \rangle \in S' \quad \eta_s(w'_i) = w_i \quad \zeta_1'(t') = w'_1 \dots w'_n}$$

Fig. 12. Inference rules for the unfolding $\mathcal{U}_c(\langle R, u \rangle)$ of a contextual pre-net R.

directions, i.e., $t \nearrow t' \nearrow t$. Finally, since $<$ represents a global order of execution, while $\nearrow$ determines an order of execution only locally to each computation, it is natural to impose $\nearrow$ to be an extension of $<$.

The notion of *concurrency* is updated to take into account the presence of asymmetric conflict: a set of places $X \subseteq S$ is *concurrent*, written $co(X)$, if for any $s, s' \in X$ it does not hold $s < s'$, $\lfloor X \rfloor$ is finite and $\nearrow$ is acyclic on $\lfloor X \rfloor$.

Then the contextual occurrence pre-nets can be naturally defined.

Definition 17 (occurrence contextual pre-net). *An* occurrence contextual pre-net *is a safe pre-net R such that (i) causality $<_R$ is a partial order; (ii) R has no backward conflicts; (iii) for any transition t, the set of causes $\lfloor t \rfloor$ is finite and asymmetric conflict $\nearrow_R$ is acyclic on $\lfloor t \rfloor$. An occurrence contextual pre-net is* deterministic *if it has no forward conflicts.*

We denote by **CPreOcc**$_*$ *full subcategory of* **CPreNet**$_*$ *having marked occurrence contextual pre-nets as objects.*

In the unfolding construction below just notice that the second rule takes a context v_c which is not required to be safe, consistently with the fact that single token can be read with multiplicity greater than 1.

Definition 18 (contextual unfolding). *Let $\langle R, u \rangle$ be a marked contextual pre-net. The unfolding $\mathcal{U}_c(\langle R, u \rangle) = ((\zeta_0', \zeta_1', \zeta_2', S', T'), u')$ and the folding morphism $\eta_R = \langle \eta_t, \eta_s \rangle :$ $\mathcal{U}_c(R) \to R$ are the occurrence pre-net and (elementary) contextual pre-net morphism inductively defined by the rules in Fig. 12, with $u' = \langle \emptyset, u_1, 1 \rangle \dots \langle \emptyset, u_{|u|}, |u| \rangle$.*

Also in this case the unfolding extends to a functor $\mathcal{U}_c :$ **CPreNet**$_* \to$ **CPreOcc**$_*$ which is right adjoint to the inclusion of **CPreOcc**$_*$ into **CPreNet**$_*$. The unfolding can be abstracted to an event based model, called *asymmetric event structure* (AES's), introduced in [BCM01] as a generalisation of Winskel PES's where conflict is allowed to be non-symmetric. As proved in the mentioned paper, the category of AES's coreflects into **Dom** allowing to recover a domain semantics. The situation is summarised in Fig. 13.

The algebraic and unfolding approach to the semantics of contextual pre-nets can be reconciled, along the same schema followed for pre-net, obtaining a commutative functorial diagram which generalises Fig. 10 in the presence of read arcs.

$$\mathbf{CPreNet}_* \xleftarrow[\;\;\mathcal{U}_c\;\;]{\begin{array}{c}\overbrace{\phantom{\mathcal{U}_c}}\\ \bot\end{array}} \mathbf{CPreOcc}_* \xrightarrow[\mathcal{E}_c]{} \mathbf{AES} \xleftarrow[\;\;\mathcal{L}\;\;]{\begin{array}{c}\mathcal{P}\\ \bot\end{array}} \mathbf{Dom}$$

Fig. 13. Denotational semantics of contextual pre-nets.

	Algebraic	Process	Logical	Unfolding	Reconciliation
P/T nets CTPh	**[MM90]**	**[BD87]**	**[BMMS01]**		
P/T nets ITPh	*[DMM96,Sas98]*	*[GR83,DMM96,Sas98]*		**[Win87,MMS97a]**	*[MMS96]*
pre-nets ITPh	**[BMMS01]**	**Section 2**	**[BMMS01]**	Section 3	Section 4

Fig. 14. Net semantics.

6 Conclusions

We have shown that a functorial unfolding semantics for pre-nets can be developed along the lines of the seminal work of Winskel. The semantics is expressed as a chain of functors leading from the category **PreNet**$_*$ to the category **Dom**, through **PreOcc**$_*$ and **PES**. A different construction of a domain for any pre-net can be defined by relying on the algebraic semantics of pre-nets, already defined in the literature. Differently from what happens for Petri nets, this latter construction can be expressed as a functor from **PreNet**$_*$ to **Dom**. The unfolding and algebraic constructions can be reconciled in a fully satisfactory categorical setting, by showing that the corresponding functors are naturally isomorphic. The proof relies on the introduction of a concrete notion of process for pre-nets, and on a characterisation of the algebraic semantics in terms such processes.

Figure 14 summarises our results, connecting them to the known (CTPh and ITPh) net semantics. Each column is devoted to a specific semantic flavour (see the classification in the Introduction). The last column refers to the possibility of relating the algebraic and unfolding views. The entries are either references to the literature where the corresponding construction has been presented, or pointers to the sections of our contribution. Empty cells stands for unfeasible constructions. Italic text refers to non-functorial constructions, i.e., constructions that are defined just at the object level, but cannot deal with simulation morphisms. Regular entries stands for functorial constructions, and bold entries for adjunctions. Note that, in the case of pre-nets, all constructions are feasible and functorial. Finally, we mention that all constructions and results for pre-nets are extended to work in the presence of read arcs.

References

[BCM01]   P. Baldan, A. Corradini, and U. Montanari. Contextual Petri nets, asymmetric event structures and processes. *Inform. and Comput.*, 1(171):1–49, 2001.

[BD87]   E. Best and R. Devillers. Sequential and concurrent behaviour in Petri net theory. *Theoret. Comput. Sci.*, 55:87–136, 1987.

[Ber78]   G. Berry. Stable models of typed lambda-calculi. In *Proceedings of ICALP'78*, vol. 62 of *Lect. Notes in Comput. Sci.*, pages 72–89. Springer-Verlag, 1978.

[BMMS99]   R. Bruni, J. Meseguer, U. Montanari, and V. Sassone. Functorial semantics for Petri nets under the individual token philosophy. In *Proceedings of CTCS'99*, vol. 29 of *Elect. Notes in Th. Comput. Sci.* Elsevier Science, 1999.

[BMMS01]   R. Bruni, J. Meseguer, U. Montanari, and V. Sassone. Functorial models for Petri nets. *Inform. and Comput.*, 170(2):207–236, 2001.

[BMMS02]   R. Bruni, J. Meseguer, U. Montanari, and V. Sassone. Functorial models for contextual pre-nets. Technical Report TR-02-09, University of Pisa, 2002.

[CW01]   F. Crazzolara and G. Winskel. Events in security protocols. In *Proceedings of CCS'01*, pages 96–105. ACM, 2001.

[DFMR94]   N. De Francesco, U. Montanari, and G. Ristori. Modeling concurrent accesses to shared data via Petri nets. In *Programming Concepts, Methods and Calculi*, vol. A-56 of *IFIP Transactions*, pages 403–422. North Holland, 1994.

[DMM96]   P. Degano, J. Meseguer, and U. Montanari. Axiomatizing the algebra of net computations and processes. *Acta Inform.*, 33(7):641–667, 1996.

[GP95]   R.J. van Glabbeek and G.D. Plotkin. Configuration structures. In *Proceedings of LICS'95*, pages 199–209. IEEE Computer Society Press, 1995.

[GR83]   U. Goltz and W. Reisig. The non-sequential behaviour of Petri nets. *Inform. and Comput.*, 57:125–147, 1983.

[Mac71]   S. MacLane. *Categories for the Working Mathematician*. Springer, 1971.

[MM90]   J. Meseguer and U. Montanari. Petri nets are monoids. *Inform. and Comput.*, 88:105–155, 1990.

[MMS92]   J. Meseguer, U. Montanari, and V. Sassone. On the semantics of Petri nets. In *Proceedings of CONCUR '92*, vol. 630 of *Lect. Notes in Comput. Sci.*, pages 286–301. Springer-Verlag, 1992.

[MMS96]   J. Meseguer, U. Montanari, and V. Sassone. Process versus unfolding semantics for Place/Transition Petri nets. *Theoret. Comput. Sci.*, 153(1-2):171–210, 1996.

[MMS97a]   J. Meseguer, U. Montanari, and V. Sassone. On the semantics of Place/Transition Petri nets. *Math. Struct. in Comput. Sci.*, 7:359–397, 1997.

[MMS97b]   J. Meseguer, U. Montanari, and V. Sassone. Representation theorems for Petri nets. In *Foundations of Computer Science: Potential - Theory - Cognition*, vol. 1337 of *Lect. Notes in Comput. Sci.*, pages 239–249. Springer, 1997.

[MR94]   U. Montanari and F. Rossi. Contextual occurrence nets and concurrent constraint programming. In *Graph Transformations in Computer Science*, vol. 776 of *LNCS*, pages 280–295. Springer, 1994.

[MR95]   U. Montanari and F. Rossi. Contextual nets. *Acta Inform.*, 32:545–596, 1995.

[NPW81]   M. Nielsen, G. Plotkin, and G. Winskel. Petri Nets, Event Structures and Domains, Part 1. *Theoret. Comput. Sci.*, 13:85–108, 1981.

[Pet62]   C.A. Petri. *Kommunikation mit Automaten*. PhD thesis, Schriften des Institutes für Instrumentelle Matematik, Bonn, 1962.

[Rei85]   W. Reisig. *Petri Nets: An Introduction*. EATCS Monographs on Theoretical Computer Science. Springer, 1985.

[Sas98]   V. Sassone. An axiomatization of the category of Petri net computations. *Math. Struct. in Comput. Sci.*, 8(2):117–151, 1998.

[Vog97]   W. Vogler. Efficiency of asynchronous systems and read arcs in Petri nets. In *Proceeding of ICALP'97*, vol. 1256 of *LNCS*, pages 538–548. Springer, 1997.

[Win87]   G. Winskel. Event Structures. In *Petri Nets: Applications and Relationships to Other Models of Concurrency*, vol. 255 of *LNCS*, pages 325–392. Springer, 1987.

Coreflective Concurrent Semantics
for Single-Pushout Graph Grammars*

Paolo Baldan[1], Andrea Corradini[2], Ugo Montanari[2], and Leila Ribeiro[3]

[1] Dipartimento di Informatica, Università Ca' Foscari di Venezia, Italia
`baldan@dsi.unive.it`
[2] Dipartimento di Informatica, Università di Pisa, Italia
`{andrea,ugo}@di.unipi.it`
[3] Instituto de Informática, Universidade Federal do Rio Grande do Sul, Brazil
`leila@inf.ufrgs.br`

Abstract. The problem of extending to graph grammars the unfolding semantics originally developed by Winskel for (safe) Petri nets has been faced several times along the years, both for the single-pushout and double-pushout approaches, but only partial results were obtained. In this paper we fully extend Winskel's approach to single-pushout grammars providing them with a categorical concurrent semantics expressed as a coreflection between the category of graph grammars and the category of prime algebraic domains.

Introduction

It belongs to the folklore that Graph Grammars [25] generalise Petri nets, in that they allow for a more structured representation of system states, modelled in terms of graphs rather than (multi)sets, and for a more general kind of state transformation, modelling also preservation of parts of the state, besides deletion and creation.

During the last years, a rich theory of concurrency for the *algebraic approaches* to graph transformation has been developed, including the generalisation of various classical Petri net concurrency models, like Goltz-Reisig process semantics [13] and Winskel's unfolding semantics [27].

Recall that, building on [22], the seminal work [27] gives the concurrent semantics of (safe) nets by means of a chain of coreflections leading from the category of safe Petri nets to the category of prime algebraic domains.

$$\text{Safe} \xleftarrow{\quad} \text{Occurrence} \xleftarrow{\;\mathcal{N}\;} \text{Prime Event} \xleftarrow{\;\mathcal{P}\;} \text{Domains}$$
$$\text{Nets} \xrightarrow[\;\mathcal{U}\;]{} \text{Nets} \xrightarrow[\;\mathcal{E}\;]{} \text{Structures} \xrightarrow[\;\mathcal{L}\;]{\sim}$$

The first step unfolds any (safe) net into an occurrence net, i.e., a branching acyclic net making explicit causality and conflict (nondeterministic choice point) between events in the net. The second step produces a *prime event structure*

---

* Research supported by the FET-GC Project IST-2001-32747 AGILE and by the MIUR Project COFIN 2001013518 COMETA.

M. Wirsing, D. Pattinson, and R. Hennicker (Eds.): WADT 2002, LNCS 2755, pp. 165–184, 2003.

(PES) abstracting away the state and recording only the events and the relationships between events. Finally, the last step maps any PES into the corresponding prime algebraic domain of configurations.

Some important steps have been taken in the direction of developing an analogous semantical framework for algebraic graph grammars, but a definitive answer has not been provided yet. More precisely, a number of constructions have been defined for algebraic, *double-pushout* (DPO) graph grammars [12, 9] by the first three authors (see, e.g., [1]), as summarised by the following diagram:

$$
\textbf{DPO Graph} \xleftarrow[\ \mathcal{U}_g\ ]{\perp} \textbf{Occurrence} \xrightarrow[\ \mathcal{E}_g\ ]{} \textbf{Inhibitor Event} \xleftarrow[\ \mathcal{L}_i\ ]{\overset{\mathcal{P}_i}{\perp}} \textbf{Domains}
$$
$$
\textbf{Grammars} \qquad\qquad \textbf{Grammars} \qquad\qquad \textbf{Structures}
$$

Even if at this level of abstraction it is not possible to see the relevant differences in the technical treatment of DPO grammars w.r.t. the much simpler case of Petri nets, still it is worth pointing at the evident differences between this chain of categories and the corresponding one for nets. Firstly, the category of PES's is replaced by that of *inhibitor event structures* (IES's), which, assuming conditional or-causality as a basic relation between events, are able to capture both the asymmetric conflicts between events arising from the capability of preserving part of the state and the inhibiting effects related to the presence of the application conditions for rules. The category of domains can be viewed as a coreflective subcategory of IES's (as shown by the last step of the chain) and thus one can also recover a semantics for DPO grammars in terms of domains and PES's. Secondly, the functor from the category of occurrence grammars to the category of IES's *does not admit* a left adjoint establishing a coreflection between IES's and occurrence grammars, and thus the whole semantic transformation is not expressed as a coreflection.

In this paper we concentrate on the *single-pushout* (SPO) approach [18, 11] to graph transformation. One of the main differences with respect to the DPO approach lies in the fact that there are no conditions on rule application, i.e., whenever a match is found the corresponding rule can always be applied. For SPO grammars an unfolding construction has been proposed in [24], corresponding to the first step in the above chains of coreflections.

Building on the results briefly summarised above, we provide a coreflective unfolding semantics for SPO graph grammars, defined through the following chain of coreflections:

$$
\textbf{SPO Graph} \xleftarrow[\ \mathcal{U}_s\ ]{\perp} \textbf{Occurrence} \xleftarrow[\ \mathcal{E}_s\ ]{\overset{N_s}{\perp}} \textbf{Asymmetric Event} \xleftarrow[\ \mathcal{L}_a\ ]{\overset{\mathcal{P}_a}{\perp}} \textbf{Domains}
$$
$$
\textbf{Grammars} \qquad\qquad \textbf{Grammars} \qquad\qquad \textbf{Structures}
$$

In particular, this construction differs from and improves that for DPO graph grammars, discussed above, because of the following facts:

- Due to the absence of application conditions for rules, a less powerful and more manageable kind of event structures called *asymmetric event structures* (introduced to deal with contextual nets in [4]), can be used to represent the dependency structure of SPO graph grammars.

– A novel construction, inspired by the work on contextual nets, allows to associate a canonical occurrence SPO graph grammar to any asymmetric event structure. This provides the lacking step, i.e., a left adjoint functor establishing a coreflection between the category of occurrence graph grammars and the category of asymmetric event structures.

An existing result [4] establishes a coreflection between asymmetric event structures and domains, so that we obtain a coreflective PES and domain semantics for SPO graph grammars.

These results do not extend immediately to the DPO approach because of the presence of application conditions for rules. However, as discussed in the conclusions, they can give some suggestions for improving the treatment of this more complex case.

The rest of the paper is structured as follows. In Section 1 we review the basics of single-pushout graph grammars and we define the notion of graph grammar morphism we shall work with. In Section 2 we discuss the kind of dependencies arising between events in SPO graph grammars and we introduce the notion of occurrence graph grammar. In Section 3 we briefly discuss the unfolding construction for SPO graph grammars and its characterisation as a universal construction. In Section 4 we complete the chain of coreflections from grammars to domains, showing how any occurrence grammar can be abstracted to an asymmetric event structure and, vice versa, how a canonical occurrence grammar can be associated to any asymmetric event structure. Finally Section 5 draws some conclusions.

1   Typed Graph Grammars and Their Morphisms

In this section we summarise the basics of graph grammars in the *single-pushout* (SPO) approach [18], an algebraic approach to graph rewriting alternative to the classical *double-pushout* (DPO) approach. The original SPO approach is adapted to deal with *typed graphs* [8, 19], which are, roughly, graphs labelled over a structure (the *graph of types*) that is itself a graph. Then some insights are provided on the relationship between typed graph grammars and Petri nets. Finally, the class of SPO typed graph grammars is turned into a category **GG** by defining a notion of grammar morphism, which recasts in this setting the morphisms for DPO grammars introduced in [3].

1.1   Typed Graph Grammars

Given a partial function $f : A \rightarrowtail B$ we will denote by $dom(f)$ its *domain*, i.e., the set $\{a \in A \mid f(a) \text{ is defined}\}$. Let $f, g : A \rightarrowtail B$ be two partial functions. We will write $f \leq g$ when $dom(f) \subseteq dom(g)$ and $f(x) = g(x)$ for all $x \in dom(f)$.

For a graph G we will denote by N_G and E_G the sets of *nodes* and *edges* of G, and by $s_G, t_G : E_G \rightarrow N_G$ its *source* and *target* functions.

Definition 1 (partial graph morphism). *A partial graph morphism* $f : G \rightarrowtail H$ *is a pair of partial functions* $f = \langle f_N : N_G \rightarrowtail N_H, f_E : E_G \rightarrowtail E_H \rangle$ *such that (see Fig. 1.(a)):*

$$s_H \circ f_E \leq f_N \circ s_G \quad and \quad t_H \circ f_E \leq f_N \circ t_G. \; (*)$$

We denote by **PGraph** *the category of (directed, unlabelled) graphs and partial graph morphisms. A morphism is called* total *if both components are total, and the corresponding full subcategory of* **PGraph** *is denoted by* **Graph**.

Notice that, according to condition (*), if f is defined over an edge then it must be defined both on its source and target nodes: this ensures that the domain of f is a well-formed graph. The inequalities in condition (*) ensure that *any* subgraph of a graph G can be the domain of a partial morphism $f : G \rightarrowtail H$. Instead, the stronger (apparently natural) conditions $s_H \circ f_E = f_N \circ s_G$ and $t_H \circ f_E = f_N \circ t_G$ would have imposed f to be defined over an edge whenever it is defined either on its source or on its target node.

Given a graph TG, a *typed graph* G over TG is a graph $|G|$, together with a total morphism $t_G : |G| \to TG$. A *partial morphism* between TG-typed graphs $f : G_1 \rightarrowtail G_2$ is a partial graph morphisms $f : |G_1| \rightarrowtail |G_2|$ consistent with the typing, i.e., such that $t_{G_1} \geq t_{G_2} \circ f$ (see Fig. 1.(b)). A typed graph G is called *injective* if the typing morphism t_G is injective. The category of TG-typed graphs and partial typed graph morphisms is denoted by TG-**PGraph**.

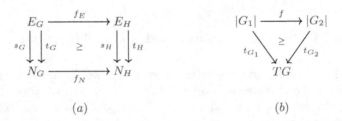

$$(a) \qquad\qquad\qquad (b)$$

Fig. 1. Diagrams for partial graph and typed graph morphisms.

Given a partial typed graph morphism $f : G_1 \rightarrowtail G_2$, we denote by $dom(f)$ the domain of f typed in the obvious way.

Definition 2 (graph production and direct derivation). *Fixed a graph TG of types, a (TG-typed graph) production q is an injective partial typed graph morphism $L_q \overset{r_q}{\rightarrowtail} R_q$. It is called* consuming *if the morphism is not total. The typed graphs L_q and R_q are called the* left-hand side *and the* right-hand side *of the production, respectively.*

Given a typed graph G and a match (i.e., a total morphism) $g : L_q \to G$, we say that there is a direct derivation *δ from G to H using q (based on g), written $\delta : G \Rightarrow_q H$, if the following is a pushout square in TG-**PGraph**.*

$$L_q \xrightarrow{r_q} R_q$$
$$g \downarrow \qquad \downarrow h$$
$$G \xrightarrow{d} H$$

Roughly speaking, the rewriting step removes from the graph G the image of the items of the left-hand side which are not in the domain of r_q, namely $g(L_q - dom(r_q))$, adding the items of the right-hand side which are not in the image of r_q, namely $R_q - dom(r_q)$. The items in the image of $dom(r_q)$ are "preserved" by the rewriting step (intuitively, they are accessed in a "read-only" manner).

A relevant difference with respect to the DPO approach is that here there is no *dangling condition* [9] preventing a rule to be applied whenever its application would leave dangling edges. In fact, as a consequence of the way pushouts are constructed in TG-**PGraph**, when a node is deleted by the application of a rule also all the edges having such node as source or target are deleted by the rewriting step, as a kind of *side-effect*. For instance, production q in the top row of Fig. 2, which consumes node B, can be applied to the graph G in the same figure. As a result both node B and the loop edge L are removed.

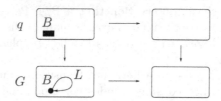

Fig. 2. Side-effects in SPO rewriting.

Even if the category **PGraph** has all pushouts, still we will consider a condition which corresponds to the *identification condition* of the DPO approach.

Definition 3 (valid match). *A match $g : L_q \to G$ is called* valid *when for any $x, y \in |L_q|$, if $g(x) = g(y)$ then $x, y \in dom(r_q)$.*

Conceptually, a match is not valid if it requires a single resource to be consumed twice, or to be consumed and preserved at the same time.

Definition 4 (typed graph grammar and derivation). *A (TG-typed) SPO graph grammar $\mathcal{G}$ is a tuple $\langle TG, G_s, P, \pi \rangle$, where G_s is the (typed) start graph, P is a set of production names, and π is a function which associates a production to each name in P. A graph grammar is* consuming *if all the productions in the range of π are consuming. A derivation in $\mathcal{G}$ is a sequence of direct derivations beginning from the start graph $\rho = \{G_{i-1} \Rightarrow_{q_{i-1}} G_i\}_{i \in \{1,\dots,n\}}$, with $G_0 = G_s$. A derivation is* valid *if so are all the matches in its direct derivations.*

In the paper we will consider only *consuming* graph grammars and *valid* derivations. The restriction to consuming grammars is essential to obtain a meaningful semantics combining concurrency and nondeterminism. In fact, the presence of non-consuming productions, which can be applied without deleting any item, would lead to an unbounded number of concurrent events with the same causal history. This would not fit with the approach to concurrency (see, e.g., [13,27]) where events in computations are identified with their causal history (formally, the unfolding construction would not work). On the other hand, considering valid derivations only, is needed to have a computational interpretation which is resource-conscious, i.e., where a resource can be consumed only once.

We denote by $Elem(\mathcal{G})$ the set $N_{TG} \cup E_{TG} \cup P$. We will assume that for each production name q the corresponding production $\pi(q)$ is $L_q \overset{r_q}{\rightarrowtail} R_q$. Without loss of generality, we will assume that the injective partial morphism r_q is a partial inclusion (i.e., that $r_q(x) = x$ whenever defined).

1.2   Relation with Petri Nets

The reader who is familiar with Petri net theory can gain a solid intuition about grammar morphisms and many other definitions and constructions in this paper, by referring to the relation between Petri nets and (SPO) graph grammars. The correspondence between these two formalisms (see, e.g., [6] and references therein) relies on the basic observation that a P/T Petri net is essentially a rewriting system on a restricted kind of graphs, namely discrete, labelled graphs (that can be identified with sets of tokens labelled by places), the productions being the net transitions.

For instance, Fig. 3 presents a Petri net transition t and the corresponding graph production r_t which consumes nodes corresponding to two tokens in s_0 and one token in s_1 and produces new nodes corresponding to one token in s_2 and one token in s_3. The domain of the rule morphism is empty, i.e., $r_t : L \rightarrowtail R$ is the empty function, since nothing is explicitly preserved by a net transition.

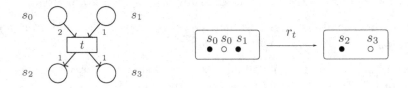

Fig. 3. A Petri net transition and a corresponding SPO production.

Note that, in this encoding of transitions into productions, the restriction to consuming graph grammars corresponds, in the theory of Petri nets, to the common requirement that transitions must have non-empty preconditions.

A tighter correspondence can be established with *contextual nets* [21], also called nets with test arcs in [5], activator arcs in [15] or read arcs in [26], an extension ordinary nets with the possibility of checking for the presence of tokens which are not consumed. Non-directed (usually horizontal) arcs are used to represent context conditions. For instance, transition t in the left part of Fig. 4 has place s as context, hence at least one token in s is needed for enabling t, and the firing of t does not affect such token.

As shown in Fig. 4, the context of a transition t in a contextual net corresponds to the graph $dom(r_t)$ of the corresponding SPO production $r_t : L \rightarrowtail R$. Thus, in general, a contextual net corresponds to an SPO graph grammar still acting on discrete graphs, but where a production may preserve some nodes, i.e., its domain might not be empty.

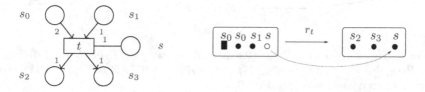

Fig. 4. A contextual Petri net transition and a corresponding SPO production.

1.3   Grammar Morphisms

The notion of SPO grammar morphism defined in this paper recasts in the setting of the SPO approach the notion introduced for DPO grammars in [7, 3], which in turn was a generalisation of Petri net morphisms. Recall that a Petri net morphism [27] consists of two components: a multirelation between the sets of places, and a partial function mapping transitions of the first net into transitions of the second one. Net morphisms are required to "preserve" the algebraic structure of a net in the sense that the pre- (post-)set of the image of a transition t must be the image of the pre- (post-)set of t.

Recall that, given two sets A and B, a *multirelation* $R : A \leftrightarrow B$ is a function $R : A \times B \to \mathbb{N}$. Intuitively, R relates elements $a \in A$ and $b \in B$ with multiplicity $R(a, b)$. As the items of the type graph of a graph grammar can be seen as generalisations of Petri net places and typed graphs as generalisations of multisets of places, the first component of a grammar morphism will be a span of total graph morphisms between the type graphs of the source and target grammars, arising as a categorical generalisation of the notion of multirelation. Here we give only some basic definitions. For an extensive discussion we refer the reader to [7, 1].

Definition 5 (spans). *Let* **C** *be a category. A (concrete) span in* **C** *is a pair of coinitial arrows* $f = \langle f^L, f^R \rangle$ *with* $f^L : x_f \to a$ *and* $f^R : x_f \to b$*. Objects* a *and* b *are called the source and the target of the span, written* $f : a \leftrightarrow b$*. The span* f *will be sometimes denoted as* $\langle f^L, x_f, f^R \rangle$*, explicitly giving the common source object* x_f*.*

Consider the equivalence $\sim$ *over the set of spans with the same source and target defined, for* $f, f' : a \leftrightarrow b$*, as* $f \sim f'$ *if there exists an isomorphism*

$k : x_f \rightarrow x_{f'}$ such that $f'^L \circ k = f^L$ and $f'^R \circ k = f^R$ (see Fig. 6.(a)). The isomorphism class of a span f is denoted by $[f]$ and called a semi-abstract span.

Fig. 5 gives two examples of multirelations in **Set**, with the corresponding span representations.

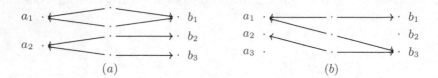

(a)   (b)

Fig. 5. The (semi-abstract) spans for the multirelations (a) $R_1(a_1, b_1) = 2$, $R_1(a_2, b_2) = 1$, $R_1(a_2, b_3) = 1$ and (b) $R_2(a_1, b_1) = 1$, $R_2(a_1, b_3) = 1$, $R_2(a_2, b_3) = 1$ (Pairs which are not mentioned are mapped to 0).

Definition 6 (category of spans). Let **C** be a category with pullbacks. Then the category **Span(C)** has the same objects of **C** and semi-abstract spans on **C** as arrows. More precisely, a semi-abstract span $[f]$ is an arrow from the source to the target of f. The composition of two semi-abstract spans $[f_1] : a \leftrightarrow b$ and $[f_2] : b \leftrightarrow c$ is the (equivalence class) of a span f constructed as in Fig. 6.(b) (i.e., $f^L = f_1^L \circ y$ and $f^R = f_2^R \circ z$), where the square is a pullback. The identity on an object a is the equivalence class of the span $\langle id_a, id_a \rangle$, where id_a is the identity of a in **C**.

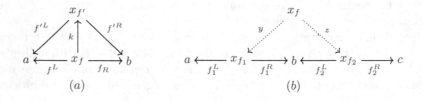

(a)   (b)

Fig. 6. Equivalence and composition of spans.

Relations can be identified with special multirelations $R : A \leftrightarrow B$ where multiplicities are bounded by one (namely $R(a, b) \leq 1$ for all $a \in A$ and $b \in B$). The corresponding condition on a span $f : A \leftrightarrow B$ is the existence of at most one path between any two elements $a \in A$ and $b \in B$. For instance, the span in Fig. 5.(a) is not relational, while that in Fig. 5.(b) is relational.

Definition 7 (relational span). Let **C** be a category. A span $f : a \leftrightarrow b$ in **C** is called relational if $\langle f^L, f^R \rangle : x_f \rightarrow a \times b$ is mono.

We can also find a categorical analogue of constructing the image of a multiset through a multirelation. The next definition is given for graphs, but it could be generalised to any category with pullbacks.

Definition 8 (pullback-retyping construction). *Let $[f_T] : TG_1 \leftrightarrow TG_2$ be a semi-abstract span in **Graph** and let G_1 be a TG_1-typed graph. Then G_1 can be "transformed" into a TG_2-typed graph as depicted in the diagram below, by first taking a pullback (in **Graph**) of the arrows $f_T^L : X_{f_T} \to TG_1$ and $t_{G_1} : |G_1| \to TG_1$, and then typing the pullback object over TG_2 by using the right part of the span $f_T^R : X_{f_T} \to TG_2$.*

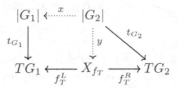

The TG_2-typed graph $G_2 = \langle |G_2|, f_T^R \circ y \rangle$ is determined only up to isomorphism. Sometimes we will write $f_T\{x, y\}(G_1, G_2)$ (or simply $f_T(G_1, G_2)$ if we are not interested in morphisms x and y) to express the fact that G_1 and G_2 are related in this way by the pullback-retyping construction induced by $[f_T]$.

We are now ready to define grammar morphisms. Besides the component specifying the relation between the type graphs, a morphism from $\mathcal{G}_1$ to $\mathcal{G}_2$ includes a (partial) mapping between production names. Furthermore a third component explicitly relates the (untyped) graphs underlying corresponding productions of the two grammars, as well as the graphs underlying the start graphs.

Definition 9 (grammar morphism). *Let $\mathcal{G}_i = \langle TG_i, G_{s_i}, P_i, \pi_i \rangle$ ($i \in \{1, 2\}$) be two graph grammars. A morphism $f : \mathcal{G}_1 \to \mathcal{G}_2$ is a triple $\langle [f_T], f_P, \iota_f \rangle$ where*

- $[f_T] : TG_1 \leftrightarrow TG_2$ *is a semi-abstract span in **Graph**, called the* type-span;
- $f_P : P_1 \to P_2 \cup \{\emptyset\}$ *is a total function, where $\emptyset$ is a new production name (not in P_2), with associated production $\emptyset \rightarrowtail \emptyset$ [1];*
- ι_f *is a family $\{\iota_f(q_1) \mid q_1 \in P_1\} \cup \{\iota_f^s\}$ of morphisms in **Graph** such that $\iota_f^s : |G_{s_2}| \to |G_{s_1}|$ and for each $q_1 \in P_1$, if $f_P(q_1) = q_2$, then $\iota_f(q_1)$ is pair*

$$\langle \iota_f^L(q_1) : |L_{q_2}| \to |L_{q_1}|, \iota_f^R(q_1) : |R_{q_2}| \to |R_{q_1}| \rangle.$$

such that the following conditions are satisfied:

1. Preservation of the start graph.
   There exists a morphism k such that $f_T\{\iota_f^s, k\}(G_{s_1}, G_{s_2})$, namely such that the diagram in Fig. 7.(a) commutes and the square is a pullback.

---

[1] Considering the empty production $\emptyset$ is technically preferable to the use of a partial mapping $f_P : P_1 \rightarrowtail P_2$.

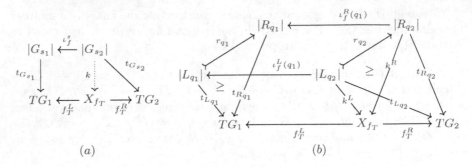

Fig. 7. Diagrams for SPO grammar morphisms.

2. Preservation of productions.

For each $q_1 \in P_1$, with $q_2 = f_P(q_1)$, there exist morphisms k^L and k^R such that the diagram in Fig. 7.(b) commutes, and $f_T\{\iota_f^Y(q_1), k^Y\}(Y_{q_1}, Y_{q_2})$ for $Y \in \{L, R\}$.

The morphism f is called relational *if the type component f_T is relational.*

As in [1, 7] one can show that grammar morphisms are "simulations", namely that every derivation ρ_1 in $\mathcal{G}_1$ can be transformed into a derivation ρ_2 in $\mathcal{G}_2$, related to ρ_1 by the pullback-retyping construction induced by the morphism.

2   Nondeterministic Occurrence Grammars

Analogously to what happens for Petri nets, occurrence grammars are "safe" grammars, where the dependency relations between productions satisfy suitable acyclicity and well-foundedness requirements. Nondeterministic occurrence grammars will be used to provide a static description of the computation of a given graph grammar, recording the events (production applications) which can appear in all possible derivations and the dependency relations among them.

While for nets it suffices to take into account only the causality and conflict relations, for grammars the fact that a production application not only consumes and produces, but also preserves a part of the state leads to a form of asymmetric conflict between productions. Quite interestingly, instead, as we shall discuss later there is no need of taking into account the dependencies between events related to the side-effects of rule applications (i.e., the deletion of an edge caused by the deletion of its source or target node).

The notion of safe graph grammar [8] generalises the one for P/T nets which requires that each place contains at most one token in any reachable marking.

Definition 10 ((strongly) safe grammar). *A grammar $\mathcal{G} = \langle TG, G_s, P, \pi \rangle$ is* (strongly) safe *if, for all H such that $G_s \Rightarrow^* H$, H is injective.*

In a safe grammar, each graph G reachable from the start graph is injectively typed, and thus we can identify it with the corresponding subgraph $t_G(|G|)$ of the type graph. With this identification, a production can only be applied to the subgraph of the type graph which is the image via the typing morphism of its left-hand side. Thus, according to its typing, we can safely think that a production *produces*, *preserves* or *consumes* items of the type graph. Using a net-like language, we speak of *pre-set* ${}^{\bullet}q$, *context* q and *post-set* $q^{\bullet}$ of a production q, defined in the obvious way. For instance, for grammar $\mathcal{G}$ in Fig. 8, ${}^{\bullet}q_1 = \{A\}$, $\underline{q_1} = \{B\}$ and $q_1^{\bullet} = \{L\}$, while ${}^{\bullet}B = \emptyset$, $\underline{B} = \{q_1, q_2\}$ and $B^{\bullet} = \{q_3\}$.

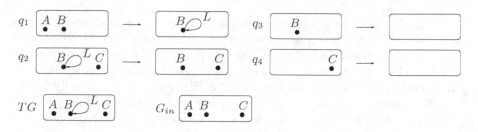

Fig. 8. A safe SPO graph grammar $\mathcal{G}$.

Although the notion of causal relation is meaningful only for safe grammars, it is technically convenient to define it for general grammars. The same holds for the asymmetric conflict relation introduced below.

Definition 11 (causal relation). *The* causal relation *of a grammar $\mathcal{G}$ is the binary relation $<$ over $Elem(\mathcal{G})$ defined as the least transitive relation satisfying: for any node or edge x in the type graph TG, and for productions $q, q' \in P$*

1. *if $x \in {}^{\bullet}q$ then $x < q$;*
2. *if $x \in q^{\bullet}$ then $q < x$;*
3. *if $q^{\bullet} \cap \underline{q'} \neq \emptyset$ then $q < q'$.*

As usual $\leq$ is the reflexive closure of $<$. Moreover, for $x \in Elem(\mathcal{G})$ we denote by $\lfloor x \rfloor$ the set of causes of x in P, namely $\{q \in P : q \leq x\}$.

Notice that the fact that an item is preserved by q and consumed by q', i.e., $\underline{q} \cap {}^{\bullet}q' \neq \emptyset$ (e.g., item $C \in \underline{q_2} \cap {}^{\bullet}q_4$ in grammar $\mathcal{G}$ of Fig. 8), does not imply $q < q'$. Actually, the dependency between the two productions is a kind of *asymmetric conflict* (see [2, 23, 17]). The application of q' prevents q from being applied, so that q can never follow q' in a derivation (or equivalently when both q and q' occur in a derivation then q must precede q'). But the converse is not true, since q *can* be applied before q'.

Definition 12 (asymmetric conflict). *The* asymmetric conflict relation *of a grammar $\mathcal{G}$ is the binary relation $\nearrow$ over the set of productions, defined by:*

1. *if $\underline{q} \cap {}^{\bullet}q' \neq \emptyset$ then $q \nearrow q'$;*
2. *if ${}^{\bullet}q \cap {}^{\bullet}q' \neq \emptyset$ and $q \neq q'$ then $q \nearrow q'$;*
3. *if $q < q'$ then $q \nearrow q'$.*

Condition 1 is justified by the discussion above. Condition 2 essentially expresses the fact that the ordinary symmetric conflict is encoded, in this setting, as an asymmetric conflict in both directions. Finally, since $<$ represents a global order of execution, while $\nearrow$ determines an order of execution only locally to each computation, it is natural to impose $\nearrow$ to be an extension of $<$ (Condition 3).

As already mentioned, the side-effects of production applications can be disregarded when analysing the dependency relations between events. In fact:

Causality. Assume that production q produces an edge e, and q' deletes e as sideeffect (because it deletes its source or target). At a first glance we could think that q' should causally depend on q. However, although q' consumes the resource e produced by q, the application of q is not necessary to make q' applicable, since q' does not explicitly require the presence of e. Hence q' does not causally depend on q. For instance, referring to grammar $\mathcal{G}$ in Fig. 8, the application of q_3 after q_1 deletes node B and edge L as side-effect. However q_3 does not depend on q_1 since it can be applied already to the start graph.

Asymmetric conflict. Also asymmetric conflict (called *weak conflict* in [24]) can be defined disregarding the mentioned side-effects. This is basically due to the fact that when a production uses (consumes or preserves) an edge, it must use necessarily the corresponding source and target nodes, and therefore dependencies related to side-effects can be detected by looking only at explicitly used items. E.g., consider again grammar $\mathcal{G}$ in Fig. 8. Observe that production q_3 prevents q_2 from being applied since it deletes, as side-effect, edge L which is consumed by q_2. However, to consume L, production q_2 must preserve or consume node B (actually, it consumes it) and thus the "ordinary" definition of asymmetric conflict already tells us that $q_2 \nearrow q_3$.

A *nondeterministic occurrence grammar* is an acyclic grammar which represents, in a branching structure, several possible computations beginning from its start graph and using each production at most once.

Definition 13 ((nondeterministic) occurrence grammar). *A* (nondeterministic) *occurrence grammar is a grammar* $\mathcal{O} = \langle TG, G_s, P, \pi \rangle$ *such that*

1. *its causal relation* $\leq$ *is a partial order, and, for any* $q \in P$, *the set* $\lfloor q \rfloor$ *is finite and the asymmetric conflict* $\nearrow$ *is acyclic on* $\lfloor q \rfloor$;
2. *the start graph* G_s *is the set* $Min(\mathcal{O})$ *of minimal elements of* $\langle Elem(\mathcal{O}), \leq \rangle$ *(with the graphical structure inherited from* TG *and typed by the inclusion);*
3. *any item* x *in* TG *is created by at most one production in* P, *namely* $|\, {}^\bullet x \,| \leq 1$;
4. *for each* $q \in P$, *the typing* t_{L_q} *is injective on the "consumed part"* $|L_q| - |dom(r_q)|$, *and* t_{R_q} *is injective on the "produced part"* $|R_q| - r_q(|dom(r_q)|)$.

We denote by **OGG** *the full subcategory of* **GG** *with occurrence grammars as objects.*

Since the start graph of an occurrence grammar $\mathcal{O}$ is determined by $Min(\mathcal{O})$, we often do not mention it explicitly. One can show that, by the defining conditions, each occurrence grammar is *safe*.

Intuitively, conditions (1)–(3) recast in the framework of graph grammars the analogous conditions of occurrence nets (actually of occurrence contextual nets [4]). In particular, in Condition (1), the acyclicity of asymmetric conflict on $\lfloor q \rfloor$ corresponds to the requirement of irreflexivity for the conflict relation in occurrence nets. Condition (4), instead, is closely related to safety and requires that each production consumes and produces items with multiplicity one. An example of occurrence grammar is $\mathcal{G}$ in Fig. 8.

As in the case of Petri nets, reachable states can be characterised in terms of a concurrency relation.

Definition 14 (concurrent graph). *Let* $\mathcal{O} = \langle TG, P, \pi \rangle$ *be an occurrence grammar. A subgraph G of TG is called* concurrent *if*

1. *$\nearrow_G$, the asymmetric conflict restricted to $\bigcup_{x \in G} \lfloor x \rfloor$, is acyclic and finitary;*
2. *$\neg(x < y)$ for all $x, y \in G$.*

It is possible to show that a subgraph G of TG is concurrent iff it is a subgraph of a graph reachable from the start graph by means of a derivation which applies all the productions in $\bigcup_{x \in G} \lfloor x \rfloor$ exactly once in any order compatible with $\nearrow$.

3   Unfolding of Graph Grammars

The unfolding construction, when applied to a consuming grammar $\mathcal{G}$, produces a nondeterministic occurrence grammar $\mathcal{U}_s(\mathcal{G})$ describing the behaviour of $\mathcal{G}$. The unfolding can be characterised as a universal construction for several interesting categories of algebraic graph grammars.

Intuitively, given a graph grammar $\mathcal{G}$, the construction consists of starting from the start graph of $\mathcal{G}$, then applying in all possible ways its productions to concurrent subgraphs, and recording in the unfolding each occurrence of production and each new graph item generated in the rewriting process, both enriched with the corresponding causal history. Due to space limitations we skip the details of the constructions, giving only a summary of the main results.

3.1   Unfolding of Semi-weighted Graph Grammars

As it has been done for ordinary (and other larger classes of) Petri nets [27, 20, 1], we first restrict to a full subcategory **SGG** of **GG** where objects satisfy conditions analogous to those defining semi-weighted P/T Petri nets. A graph grammar is semi-weighted if the start graph is injective and the right-hand side of each production is injective when restricted to produced items (namely, to items which are not in the codomain of the production morphism).

Theorem 1. *The unfolding construction can be expressed as a functor $\mathcal{U}_s : \mathbf{SGG} \to \mathbf{OGG}$, which is right adjoint to the inclusion $\mathcal{I}_s : \mathbf{OGG} \to \mathbf{SGG}$.*

3.2   Unfolding of General Grammars

The restriction to the semi-weighted case is essential for the universal characterisation of the unfolding construction when one uses general morphisms. However, suitably restricting graph grammar morphisms to still interesting subclasses (comprising, for instance, the morphisms of [24, 14]) it is possible to regain the categorical result for general, possibly non semi-weighted, grammars.

More specifically, the coreflection result can be obtained by limiting our attention to a (non full) subcategory $\widehat{\mathbf{GG}}$ of $\mathbf{GG}$, where objects are general graph grammars, but all morphisms have a relational span as type component. The naive solution of taking *all* relational morphisms as arrows of $\widehat{\mathbf{GG}}$ does not work because they are not closed under composition. A possible appropriate choice is instead given by the category $\mathbf{GG}^R$, where the arrows are grammar morphisms such that the *right* component of the type span is mono. It is easy to realize that these kinds of span corresponds to partial graph morphisms in the opposite direction. In fact, a partial graph morphism $g : TG_2 \rightarrowtail TG_1$ can be identified with the span

$$TG_1 \xleftarrow{\quad g \quad} dom(g) \xhookrightarrow{\quad\quad} TG_2$$

Theorem 2. *The unfolding construction can be turned into a functor* $\mathcal{U}_s^R : \mathbf{GG}^R \to \mathbf{OGG}^R$, *having the inclusion* $\mathcal{I}_s^R : \mathbf{OGG}^R \to \mathbf{GG}^R$ *as left adjoint, establishing a coreflection between the two categories.*

Alternatively, the result can be proved for the subcategory $\mathbf{GG}^L$ of $\mathbf{GG}$ where arrows are grammar morphisms having the *left* component of the type span which is mono (corresponding to partial graph morphisms with the same source and target of the span).

4   Event Structure Semantics for SPO Graph Grammars

In this section we show that asymmetric event structures, a generalisation of prime event structures introduced in [4], provide a suitable setting for defining an event structure semantics for SPO graph grammars. After reviewing the basics of asymmetric event structures, we show that any occurrence SPO grammar can be mapped to an asymmetric event structure via a functorial construction. Furthermore, a left adjoint functor, back from asymmetric event structures to occurrence grammars, can be defined, associating a canonical occurrence grammar to any asymmetric event structure.

4.1   Asymmetric Event Structures

Asymmetric event structures [4] are a generalisation of prime event structures where the conflict relation is allowed to be non-symmetric. As already mentioned, this is needed to give a faithful representation of dependencies between events in formalisms such as string, term, graph rewriting and contextual nets, where a rule may preserve a part of the state, in the sense that part of the state is necessary for applying the rule, but it is not affected by the application.

For technical reasons we first introduce pre-asymmetric event structures. Then asymmetric event structures will be defined as special pre-asymmetric event structures satisfying a suitable condition of "saturation".

Definition 15 (asymmetric event structure). *A pre-asymmetric event structure (pre-AES) is a tuple $\mathcal{A} = \langle E, \leq, \nearrow \rangle$, where E is a set of events and $\leq$, $\nearrow$ are binary relations on E called* causality *and* asymmetric conflict, *respectively, such that:*

1. *$\leq$ is a partial order and $\lfloor e \rfloor = \{e' \in E \mid e' \leq e\}$ is finite for all $e \in E$;*
2. *$\nearrow$ satisfies, for all $e, e' \in E$:*

   (a) $e < e' \;\;\Rightarrow\;\; e \nearrow e'$,   *(b) $\nearrow$ is acyclic in $\lfloor e \rfloor$,*

where, as usual, $e < e'$ means $e \leq e'$ and $e \neq e'$.

An asymmetric event structure *(AES) is a pre-AES which satisfies:*

3. *for any $e, e' \in E$, if $\nearrow$ is cyclic in $\lfloor e \rfloor \cup \lfloor e' \rfloor$ then $e \nearrow e'$.*

The asymmetric conflict relation $\nearrow$ determines an order of execution locally to each computation: if $e \nearrow e'$ and e, e' occur in the same computation then e must precede e'. Therefore a set of events $e_1 \nearrow e_2 \nearrow \ldots \nearrow e_n \nearrow e_1$ forming a cycle of asymmetric conflict can never occur in the same computation, a fact that can be naturally interpreted as a kind of conflict over sets of events. Condition (3) above ensures that, in an AES, this kind conflict is inherited through causality, a typical property also of PES's.

Any pre-AES can be "saturated" to produce an AES. More precisely, given a pre-AES $\mathcal{A} = \langle E, \leq, \nearrow \rangle$, its saturation, denoted by $\overline{\mathcal{A}}$, is the AES $\langle E, \leq, \nearrow' \rangle$, where $\nearrow'$ is defined as $e \nearrow' e'$ iff $(e \nearrow e')$ or $\nearrow$ is cyclic in $\lfloor e \rfloor \cup \lfloor e' \rfloor$.

Definition 16 (category of AES's). *Let $\mathcal{A}_0$ and $\mathcal{A}_1$ be two AES's. An AES-morphism $f : \mathcal{A}_0 \to \mathcal{A}_1$ is a partial function $f : E_0 \rightharpoonup E_1$ such that, for all $e_0, e_0' \in E_0$, assuming that $f(e_0)$ and $f(e_0')$ are defined,*

1. *$\lfloor f(e_0) \rfloor \subseteq f(\lfloor e_0 \rfloor)$;*
2. *(a) $f(e_0) \nearrow_1 f(e_0') \;\;\Rightarrow\;\; e_0 \nearrow_0 e_0'$;*
   (b) $(f(e_0) = f(e_0')) \wedge (e_0 \neq e_0') \;\;\Rightarrow\;\; e_0 \nearrow e_0'$.

We denote by **AES** *the category having asymmetric event structures as objects and AES-morphisms as arrows.*

The notion of configuration extends smoothly from PES's to AES's, the main difference being the fact that the computational order between configurations is not simply set-inclusion. In fact, a configuration C can be extended with an event e' only if for any event $e \in C$, it does not hold that $e' \nearrow e$ (since, in this case, e would disable e'). The set of configurations of an AES with such a computational order is a domain. The corresponding functor from **AES** to

Dom, the category of finitary prime algebraic domains, has a left adjoint which maps each domain to the corresponding prime event structure (each PES can be seen as a special AES where conflict is symmetric). Hence Winskel's equivalence between **PES,** the category of prime event structures, and **Dom** generalises to a coreflection between **AES** and **Dom.**

$$\mathbf{AES} \xleftarrow[\mathcal{L}_a]{\mathcal{P}_a} \mathbf{Dom}$$

4.2   From Occurrence Grammars to AES's

Given any occurrence grammar, the corresponding asymmetric event structure is readily obtained by taking the production names as events. Causality and asymmetric conflict are the relations defined in Definitions 11 and 12.

Definition 17 (AES for an occurrence grammar). *Let $\mathcal{O} = \langle TG, P, \pi \rangle$ be an occurrence grammar. The AES associated to $\mathcal{O}$, denoted $\mathcal{E}_s(\mathcal{O})$, is the saturation of the pre-AES $\langle P, \leq, \nearrow \rangle$, with $\leq$ and $\nearrow$ as in Definitions 11 and 12.*

The above construction naturally gives rise to a functor.

Proposition 1. *For any morphism $h : \mathcal{O}_0 \to \mathcal{O}_1$ between occurrence grammars, let $\mathcal{E}_s(h)(q) = h_P(q)$ if $h_P(q) \neq \emptyset$ and $\mathcal{E}_s(h)(q)$ undefined, otherwise. Then $\mathcal{E}_s : \mathbf{OGG} \to \mathbf{AES}$ is a well-defined functor.*

For instance, Fig. 9 shows the AES (and the prime algebraic domain of its configurations) associated to the occurrence grammar $\mathcal{G}$ in Fig. 8. In the AES straight and dotted arrows represent causality and asymmetric conflict, respectively. In any configuration the event corresponding to q_i is written as "i".

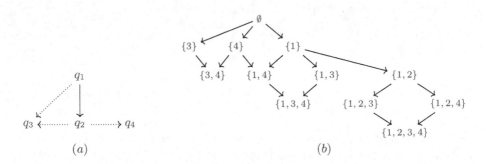

$$(a) \qquad\qquad\qquad (b)$$

Fig. 9. The (a) AES and (b) domain of configurations for $\mathcal{G}$ of Fig. 8.

4.3   From AES's to Occurrence Grammars

Any AES is identified with a canonical occurrence grammar, via a free construction that mimics Winskel's one. Given an asymmetric event structure $\mathcal{A}$, the corresponding grammar has the events of $\mathcal{A}$ as production names, while the graph items are freely generated in order to induce the right kind of dependencies between events. More specifically, first the graph nodes are freely generated according to the dependencies in $\mathcal{A}$. Then for any pair of nodes, edges connecting the two nodes are freely generated according to the dependencies in $\mathcal{A}$ and the specific restrictions of the SPO rewriting mechanism.

Definition 18. *Let $\mathcal{A} = \langle Ev, \leq, \nearrow \rangle$ be an AES. The corresponding SPO occurrence graph grammar, denoted by $\mathcal{N}_s(\mathcal{A}) = \langle TG, P, \pi \rangle$, is defined as follows:*

- *The* type graph *$TG = \langle N, E, s, t \rangle$ is defined as below, where $A, B, \ldots$ range over generic sets of events and x over sets of events of cardinality at most 1 (singletons or the empty set). Moreover by $x < e$, if $x = \{e'\}$ we mean that $e' < e$, while the relation trivially holds if $x = \emptyset$ (i.e. $\emptyset < e$, for any event e). Symmetrically, by $e < x$ with $x = \{e'\}$ we mean $e < e'$, while $e < \emptyset$ is intended to be always false.*

  - *Nodes:*
  $$N = \left\{ \langle x, A, B \rangle : \begin{array}{l} \forall e \in A \cup B.\ x < e, \\ \forall a \in A.\ \forall b \in B.\ a \nearrow b, \\ \forall b, b' \in B.\ b \neq b' \Rightarrow b \nearrow b' \end{array} \right\};$$

  - *Edges:*
  $$E = \left\{ \langle x, A, B, n_1, n_2 \rangle : \begin{array}{l} n_i = \langle x_i, A_i, B_i \rangle \in N, \\ \forall e \in A \cup B.\ x < e, \\ \forall a \in A.\ \forall b \in B.\ a \nearrow b, \\ \forall b, b' \in B.\ b \neq b' \Rightarrow b \nearrow b' \\[4pt] x_i \leq x \text{ for } i \in \{1, 2\} \\ A \subseteq A_1 \cap A_2 \\ B \subseteq (A_1 \cup B_1) \cap (A_2 \cup B_2) \\ \forall e_i \in B_i.\ \neg(e_i \leq x_j) \text{ for } i, j \in \{1, 2\},\ i \neq j \end{array} \right\};$$

  - *Source and target functions:*
  $$s(\langle x, A, B, n_1, n_2 \rangle) = n_1 \quad \text{and} \quad t(\langle x, A, B, n_1, n_2 \rangle) = n_2.$$

- *The set of productions $P = Ev$, and for any event $e \in Ev$ the corresponding production $\pi(e) = L_e \rightarrowtail R_e$ is defined as follows:*

  - $|L_e| = \{ n = \langle x, A, B \rangle,\ l = \langle x, A, B, n_1, n_2 \rangle \mid e \in A \cup B \}$
  - $|R_e| = \{ n = \langle x, A, B \rangle,\ l = \langle x, A, B, n_1, n_2 \rangle \mid e \in x \cup A \}$

  The typing and the (partial) inclusion of L_e in R_e are the obvious ones.

A node in the type graph TG is a triple $n = \langle x, A, B \rangle$. The set x might contain the event which generates the node n or might be empty if the node is in the start graph, A is the set of events which preserve the node n and B is the set of events which consume n. Clearly, the event in x, if any, must be a cause for every event in $A \cup B$, the events in A must be in asymmetric conflict with the events in B, and the events in B must be pairwise in conflict (represented as an asymmetric conflict in both directions, i.e., $b \nearrow b'$ and $b' \nearrow b$).

An edge in the type graph is a tuple $l = \langle x, A, B, n_1, n_2 \rangle$. The meaning of x, A, B is the same as for nodes. The components n_1 and n_2 are intended to represent the source and target nodes of edge l. They are subject to requirements which arise from the specific features of the SPO rewriting mechanism. First, $x_i \leq x$ since the event which produces an edge must produce or preserve the source/target nodes. Any event which preserves the edge must also preserve the source/target, hence $A \subseteq A_1 \cap A_2$. Any event which consumes the edge must preserve or consume the source/target nodes, hence $B \subseteq (A_1 \cup B_1) \cap (A_2 \cup B_2)$.

Finally the nodes n_1 and n_2 must be allowed to coexist: the requirement $x_i \leq x$ already ensures that n_1 and n_2 are not in conflict. Moreover each node is asked not to causally depend on the events which *consume* the other one.

We conclude with the main result, stating that the construction of the occurrence grammar associated to an AES is functorial and left adjoint to $\mathcal{E}_s$, establishing a coreflection between **OGG** and **AES**. For any AES $\mathcal{A}$, $\mathcal{E}_s(\mathcal{N}_s(\mathcal{A})) = \mathcal{A}$ and the component at $\mathcal{A}$ of the unit of the adjunction is the identity.

Theorem 3 (coreflection between OGG and AES). *The construction* $\mathcal{N}_s$ *extends to a functor that is left adjoint to* $\mathcal{E}_s$.

Roughly speaking, the proof shows that, given any AES $\mathcal{A}$ and occurrence graph grammar $\mathcal{O}$, all AES-morphisms $f : \mathcal{A} \to \mathcal{E}_s(\mathcal{O})$ uniquely extends to graph grammar morphisms $\hat{f} : \mathcal{N}_s(\mathcal{A}) \to \mathcal{O}$. The type span component of morphism $\hat{f}$ is $TG_{\mathcal{N}_s(\mathcal{A})} \overset{f_T}{\leftarrow} TG_{\mathcal{O}} \overset{id}{\to} TG_{\mathcal{O}}$, where f_T maps any item in $TG_{\mathcal{O}}$ to the only item in $TG_{\mathcal{N}_s(\mathcal{A})}$ which induces analogous dependencies among the events.

Summing up, Theorem 1 and Theorem 3 above give a chain of coreflections from the category **SGG** of semi-weighted SPO graph grammars to **AES** and **Dom**. The result can be extended to $\mathbf{GG}^R$, the category of general SPO grammars with restricted morphisms (having the *right* component in the type span which is mono), by exploiting Theorem 2 and observing that $\mathcal{N}_s$ restricts to a well-defined functor $\mathcal{N}_s^R : \mathbf{AES} \to \mathbf{OGG}^R$. The possibility of generalising the result to other categories of grammars with relational morphisms is still open.

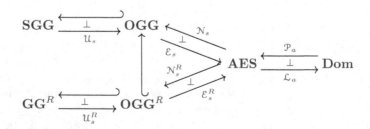

5 Conclusions

We have defined a functorial concurrent semantics for SPO graph grammars, expressed as a chain of coreflections leading from various categories of SPO grammars to the categories of AES's and domains. The approach originally proposed by Winskel in the setting of Petri nets has been fully extended to SPO graph grammars, improving the previous proposals where some steps of the construction were lacking, notably, in the case of the DPO approach, the functor from event structures to occurrence grammars.

A natural question regards the possibility of using these results for the DPO approach. We have already mentioned that for DPO graph grammars, due to the presence of application conditions for rules, a more complex kind of event structures, called *inhibitor event structures* [1] was introduced to obtain a functorial semantics. In this way a functor mapping any occurrence DPO grammar to an IES can be defined, which, however it does not admit a left adjoint. Still an idea could be to view asymmetric event structures as a coreflective subcategory of inhibitor event structures and then to devise a construction which associates a canonical DPO grammar to any asymmetric event structure.

The theory developed in this paper naturally suggests a notion of graph process for SPO grammars, which can be defined as a deterministic occurrence grammar with a morphism to the original grammar. We conjecture that these processes correspond exactly to the *concurrent derivations* of [16], which in turn were characterised as special classes of graph grammars in [24].

The analogies between the first steps of the constructions for the SPO and DPO approaches (the proper unfolding constructions) suggest the possibility of developing a general theory of unfolding in abstract categories (e.g., high level replacement systems [10]). Some parts of the construction are rather concrete and not easy to recast in an abstract categorical setting, but still this represents a challenging topic of further investigation.

References

1. P. Baldan. *Modelling concurrent computations: from contextual Petri nets to graph grammars*. PhD thesis, Department of Computer Science, University of Pisa, 2000. Available as technical report n. TD-1/00.
2. P. Baldan, A. Corradini, and U. Montanari. An event structure semantics for P/T contextual nets: Asymmetric event structures. *Proc. of FoSSaCS '98*, vol. 1378 of *LNCS*, pp. 63–80. Springer, 1998.
3. P. Baldan, A. Corradini, and U. Montanari. Unfolding of double-pushout graph grammars is a coreflection. *TAGT'98 Conference Proc.*, vol. 1764 of *LNCS*, pp. 145–163. Springer, 1999.
4. P. Baldan, A. Corradini, and U. Montanari. Contextual Petri nets, asymmetric event structures and processes. *Information and Computation*, 171(1):1–49, 2001.
5. S. Christensen and N. D. Hansen. Coloured Petri nets extended with place capacities, test arcs and inhibitor arcs. *Applications and Theory of Petri Nets*, vol. 691 of *LNCS*, pp. 186–205. Springer, 1993.

6. A. Corradini. Concurrent graph and term graph rewriting. *Proc. of CONCUR'96*, vol. 1119 of *LNCS*, pp. 438–464. Springer, 1996.

7. A. Corradini, H. Ehrig, M. Löwe, U. Montanari, and J. Padberg. The category of typed graph grammars and its adjunctions with categories of derivations. *Proc. of the 5th International Workshop on Graph Grammars and their Application to Computer Science*, vol. 1073 of *LNCS*. Springer, 1996.

8. A. Corradini, U. Montanari, and F. Rossi. Graph processes. *Fundamenta Informaticae*, 26:241–265, 1996.

9. A. Corradini, U. Montanari, F. Rossi, H. Ehrig, R. Heckel, and M. Löwe. Algebraic Approaches to Graph Transformation I: Basic Concepts and Double Pushout Approach. In Rozenberg [25], chapter 3.

10. H. Ehrig, A. Habel, H.-J. Kreowski, and F. Parisi-Presicce. Parallelism and concurrency in High-Level Replacement Systems. *Mathematical Structures in Computer Science*, 1:361–404, 1991.

11. H. Ehrig, R. Heckel, M. Korff, M. Löwe, L. Ribeiro, A. Wagner, and A. Corradini. Algebraic Approaches to Graph Transformation II: Single Pushout Approach and Comparison with Double Pushout Approach. In Rozenberg [25], chapter 4.

12. H. Ehrig, M. Pfender, and H.J. Schneider. Graph-grammars: an algebraic approach. In *Proc. of IEEE Conf. on Automata and Switching Theory*, pp. 167–180, 1973.

13. U. Golz and W. Reisig. The non-sequential behaviour of Petri nets. *Information and Control*, 57:125–147, 1983.

14. R. Heckel, A. Corradini, H. Ehrig, and M. Löwe. Horizontal and vertical structuring of graph transformation systems. *Mathematical Structures in Computer Science*, 6(6):613–648, 1996.

15. R. Janicki and M. Koutny. Semantics of inhibitor nets. *Information and Computation*, 123:1–16, 1995.

16. M. Korff. *Generalized graph structure grammars with applications to concurrent object-oriented systems*. PhD thesis, Technische Universität Berlin, 1996.

17. R. Langerak. *Transformation and Semantics for LOTOS*. PhD thesis, Department of Computer Science, University of Twente, 1992.

18. M. Löwe. Algebraic approach to single-pushout graph transformation. *Theoretical Computer Science*, 109:181–224, 1993.

19. M. Löwe, M. Korff, and A. Wagner. An Algebraic Framework for the Transformation of Attributed Graphs. *Term Graph Rewriting: Theory and Practice*, pp. 185–199. Wiley, London, 1993.

20. J. Meseguer, U. Montanari, and V. Sassone. On the semantics of Place/Transition Petri nets. *Mathematical Structures in Computer Science*, 7:359–397, 1997.

21. U. Montanari and F. Rossi. Contextual nets. *Acta Informatica*, 32(6), 1995.

22. M. Nielsen, G. Plotkin, and G. Winskel. Petri Nets, Event Structures and Domains, Part 1. *Theoretical Computer Science*, 13:85–108, 1981.

23. G. M. Pinna and A. Poigné. On the nature of events: another perspective in concurrency. *Theoretical Computer Science*, 138(2):425–454, 1995.

24. L. Ribeiro. *Parallel Composition and Unfolding Semantics of Graph Grammars*. PhD thesis, Technische Universität Berlin, 1996.

25. Grzegorz Rozenberg, editor. *Handbook of Graph Grammars and Computing by Graph Transformation. Vol. 1: Foundations*. World Scientific, 1997.

26. W. Vogler. Efficiency of asynchronous systems and read arcs in Petri nets. In *Proc. of ICALP'97*, vol. 1256 of *LNCS*, pp. 538–548. Springer, 1997.

27. G. Winskel. Event Structures. In *Petri Nets: Applications and Relationships to Other Models of Concurrency*, vol. 255 of *LNCS*, pp. 325–392. Springer, 1987.

Completeness Results for Fibred Parchments[*]
Beyond the Propositional Base

C. Caleiro, P. Gouveia, and J. Ramos

CLC – Departamento de Matemática, IST – UTL
Av. Rovisco Pais, 1049–001 Lisboa, Portugal

Abstract. In [6] it was shown that fibring could be used to combine institutions presented as c-parchments, and several completeness preservation results were established. However, their scope of applicability was limited to propositional-based logics. Herein, we extend these results to a broader class of logics, possibly including variables, terms and quantifiers. On the way, we need to consider an enriched notion of proof-calculus that deals explicitly with the substitution provisos that often appear in schematic inference rules. For illustration of the concepts, constructions and results, we shall adopt modal first-order logic as a working example.

1 Introduction

Working with several logics is the rule, in practice, to wit in knowledge representation and formal specification. Due to its intuitive simplicity and theoretical interest, the *fibring* mechanism for combining logics has deserved close attention [9, 3, 20, 25]. In [6], **c-***parchments* were proposed for bringing fibring to the realm of institutions [10, 15, 11, 24], as an alternative to other approaches for combining institutions [16–18]. A major strength of fibring is the possibility to establish general transfer results from the logics being combined to the resulting fibred logic. Soundness and completeness preservation for propositional-based logics was also obtained in [6]. Herein, we extend these results beyond the propositional base.

Recall that **c**-parchments, signature-indexed categories of **c**-rooms, are an evolution of [10, 17, 18] designed to promote a smooth characterization of fibring [20, 25, 5]. They differ from the *model-theoretic parchments* of [18] by endowing the algebras of truth-values with a Tarskian closure operation, rather than just a set of designated values. As shown in [6], fibred **c**-parchments appear as colimits in the corresponding category. The proof-theoretic counterpart of **c**-rooms in [6] was played by a notion of calculus with schematic inference rules, fit for representing the Hilbert-style axiomatizations of propositional-based logics. Since these logics are usually structural, every instantiation of a schematic inference rule was allowed. However, if we want to represent more complex logics, we need to gain control over these instantiations. A paradigmatic example are the provisos in some of the axioms of first-order logic, e.g., requiring that a variable is not

---

[*] This work was partially supported by FCT and FEDER, namely, via the Project FibLog POCTI/MAT/372 39/2001 of CLC.

M. Wirsing, D. Pattinson, and R. Hennicker (Eds.): WADT 2002, LNCS 2755, pp. 185–200, 2003.

free in a formula. The idea of making these side conditions explicit is not new [21], but the technique we shall use is improved along [7, 22]. This fine control of instantiations also has an impact on fibring, again characterizable by colimits. These aspects settled, we can study soundness and completeness transfer results in a broader context. As before [5], soundness preservation is immediate, by definition of fibring. For completeness, we capitalize on the notion of *fullness* [25] for guaranteeing that the logics at hand have a sufficient amount of models. Under reasonable assumptions on the logics being fibred, their syntactic constructors and the properties of their proof-calculi, we also generalize the completeness preservation results of [6]. We illustrate fibring by providing a detailed analysis of modal first-order logic as a fibring of propositional modal logic and first-order logic, considering various choices for its semantics, and clarifying the importance of provisos and the applicability of the completeness results.

In Section 2 we set up logic-parchments by recalling the details of **c**-parchments and introducing an improved version of proof-calculus. Section 3 is dedicated to fibring. After an overview of fibred semantics, we proceed to the categorial characterization of fibred logics, by understanding fibred deduction in the presence of provisos. A general soundness preservation result is also established, and the fundamental notion of fullness is introduced. In Section 4, we study completeness preservation under meaningful fullness requirements and reasonable assumptions on the syntactic constructors and the proof-calculi of the logics being fibred. We conclude by discussing the results obtained, their limitations and future work.

2   Rooms and Parchments

We consider, in turn, semantics, deduction, and finally logics.

2.1   Semantics

In the sequel, $\mathbf{AlgSig}_\phi$ is the category of many-sorted signatures $\Sigma = \langle S, O \rangle$, where S is a set (of *sorts*) and $O = \{O_u\}_{u \in S^+}$ is a family of sets (of *operators*) indexed by their type, with a distinguished sort $\phi \in S$ (for formulas) and morphisms preserving it. We denote by $\mathbf{Alg}(\Sigma)$ the category of Σ-algebras and homomorphisms, and by $\mathrm{cAlg}(\Sigma)$ the class of all *interpretation structures* $\langle \mathcal{A}, \mathbf{c} \rangle$ with $\mathcal{A}$ a Σ-algebra and $\mathbf{c}$ a closure operation on $|\mathcal{A}|_\phi$ (the carrier of sort ϕ, intuitively corresponding to the set of truth-values). Recall that a closure operation $\mathbf{c} : \wp(|\mathcal{A}|_\phi) \to \wp(|\mathcal{A}|_\phi)$ is extensive, monotonic and idempotent, i.e., $B \subseteq B^{\mathbf{c}}$, $B^{\mathbf{c}} \subseteq (B \cup B')^{\mathbf{c}}$ and $(B^{\mathbf{c}})^{\mathbf{c}} \subseteq B^{\mathbf{c}}$. We use $\mathcal{W}_\Sigma$ to denote the free Σ-algebra (the *word* algebra), Form_Σ to denote the set $|\mathcal{W}_\Sigma|_\phi$ of *formulas*, and $[\![\_]\!]_\mathcal{A}$ (for word *interpretation*) to denote the unique $\mathbf{Alg}(\Sigma)$-homomorphism from $\mathcal{W}_\Sigma$ to a given Σ-algebra $\mathcal{A}$. We use φ, ψ to denote formulas, and Φ, Ψ sets of formulas. Elements of $|\mathcal{W}_\Sigma|_s$ are referred to as *terms* and denoted by t. Every $\mathbf{AlgSig}_\phi$-morphism $h : \Sigma_1 \to \Sigma_2$ has an associated reduct functor $\_|_h : \mathbf{Alg}(\Sigma_2) \to \mathbf{Alg}(\Sigma_1)$. Note that $[\![t]\!]_{\mathcal{A}|_h} = [\![h(t)]\!]_\mathcal{A}$ for each $t \in |\mathcal{W}_{\Sigma_1}|_s$ and Σ_2-algebra $\mathcal{A}$. As usual, we overload the notation and write h for word *translation* instead of $[\![\_]\!]_{\mathcal{W}_{\Sigma_2}|_h}$ to denote the unique $\mathbf{Alg}(\Sigma_1)$-homomorphism from $\mathcal{W}_{\Sigma_1}$ to $\mathcal{W}_{\Sigma_2}|_h$.

Definition 1. A c-*room* is a pair $\mathcal{R} = \langle \Sigma, M \rangle$ with $\Sigma \in |\mathbf{AlgSig}_\phi|$ and $M \subseteq$ cAlg(Σ). A *morphism of* c-*rooms* from $\mathcal{R}_1 = \langle \Sigma_1, M_1 \rangle$ to $\mathcal{R}_2 = \langle \Sigma_2, M_2 \rangle$ is an $\mathbf{AlgSig}_\phi$-morphism $h : \Sigma_1 \to \Sigma_2$ such that $\langle \mathcal{A}|_h, \mathbf{c} \rangle \in M_1$ for every $\langle \mathcal{A}, \mathbf{c} \rangle \in M_2$.

Clearly, c-rooms and morphisms set up a category **CPRoom**, from which the category **CPar** of c-*parchments* is obtained by a simple Grothendieck construction (see [6]). Namely, a c-parchment is a functor from any given category **Sig** of abstract signatures to **CPRoom**. Building on the cocompleteness of **CPRoom**, **CPar** is cocomplete [6]. A c-room $\mathcal{R} = \langle \Sigma, M \rangle$ induces an *entailment*[1] relation defined by $\Phi \vDash_\mathcal{R} \psi$ if $[\![\psi]\!]_\mathcal{A} \in \{[\![\varphi]\!]_\mathcal{A} : \varphi \in \Phi\}^\mathbf{c}$ for every $\langle \mathcal{A}, \mathbf{c} \rangle \in M$, where $\Phi \cup \{\psi\} \subseteq \text{Form}_\Sigma$. The following property holds: if $h : \mathcal{R}_1 \to \mathcal{R}_2$ is a c-room morphism and $\Phi \vDash_{\mathcal{R}_1} \psi$ then $h(\Phi) \vDash_{\mathcal{R}_2} h(\psi)$. Given a c-parchment $R : \mathbf{Sig} \to \mathbf{CPRoom}$ and $\Omega \in |\mathbf{Sig}|$, we denote $\vDash_{R(\Omega)}$ simply by $\vDash_\Omega$.

Example 1. X is a fixed set of variables. The c-parchment of first-order logic with equality is the functor *FOLEq* defined from the category $\mathbf{Set}^{\mathbb{N}_0} \times \mathbf{Set}^{\mathbb{N}}$ of abstract signatures $\langle F, P \rangle$ of ranked function and predicate alphabets, by assigning to each $\langle F, P \rangle$ the c-room $\mathcal{R}_{FOLEq} = \langle \Sigma_{FOLEq}, M \rangle$ such that:

- $\Sigma_{FOLEq} = \langle \{\tau, \phi\}, O \rangle$ with $O_\tau = X \cup F_0$, $O_{\tau^n \tau} = F_n$ for $n > 0$, $O_{\tau^n \phi} = P_n$ for $n \neq 2$, $O_{\tau^2 \phi} = P_2 \cup \{\dot{=}\}$, $O_{\phi\phi} = \{\neg\} \cup \{\forall x : x \in X\}$, $O_{\phi^2 \phi} = \{\Rightarrow\}$;
- M contains all structures $\langle \mathcal{A}, \mathbf{c} \rangle$ obtained from $\langle F, P \rangle$-interpretations $\langle D, I \rangle$ with $D \neq \emptyset$, $f_I : D^n \to D$ for $f \in F_n$, $p_I \subseteq D^n$ for $p \in P_n$, by: $|\mathcal{A}|_\tau = D^{\text{Asg}(X, D)}$, $|\mathcal{A}|_\phi = \wp(\text{Asg}(X, D))$, where $\text{Asg}(X, D) = D^X$ is the set of assignments μ; $x_\mathcal{A}(\mu) = \mu(x)$ for $x \in X$, $f_\mathcal{A}(\langle e_i \rangle)(\mu) = f_I(\langle e_i(\mu) \rangle)$ for $f \in F$, $p_\mathcal{A}(\langle e_i \rangle) = \{\mu : \langle e_i(\mu) \rangle \in p_I\}$ for $p \in P$, $\dot{=}_\mathcal{A}(e_1, e_2) = \{\mu : e_1(\mu) = e_2(\mu)\}$, $\neg_\mathcal{A}(v) = \text{Asg}(X, D) \setminus v$, $\forall x_\mathcal{A}(v) = \{\mu : \mu[x/d] \in v$ for every $d \in D\}$, $\Rightarrow_\mathcal{A}(v_1, v_2) = (\text{Asg}(X, D) \setminus v_1) \cup v_2$; $\mathbf{c} : \wp(|\mathcal{A}|_\phi) \to \wp(|\mathcal{A}|_\phi)$ is the cut closure induced by set inclusion: given $V \in \wp(|\mathcal{A}|_\phi)$, $V^\mathbf{c} = \{v \in |\mathcal{A}|_\phi : (\bigcap V) \subseteq v\}$ is the principal ideal determined by $(\bigcap V)$ on the complete lattice $\langle \wp(|\mathcal{A}|_\phi), \supseteq \rangle$.

The denotation of a formula is the set of all assignments for which it holds. First-order logic without equality can be obtained by omitting $\dot{=}$.

Example 2. The c-parchment K of propositional modal logic is defined from the category **Set** of abstract signatures, by mapping each set PS of propositional symbols to the c-room $\mathcal{R}_K = \langle \Sigma_K, M \rangle$ such that:

- $\Sigma_K = \langle \{\phi\}, O \rangle$ with $O_\phi = \text{PS}$, $O_{\phi\phi} = \{\Box, \neg\}$, $O_{\phi^2 \phi} = \{\Rightarrow\}$;
- M contains every $\langle \mathcal{A}, \mathbf{c} \rangle$ obtained from Kripke models $\langle W, R, \vartheta \rangle$ with $W \neq \emptyset$, $R \subseteq W^2$, $\vartheta : \text{PS} \to \wp(W)$, by: $|\mathcal{A}|_\phi = \wp(W)$; $q_\mathcal{A} = \vartheta(q)$ for $q \in \text{PS}$, $\Box_\mathcal{A}(U) = \{w : \{w' : w R w'\} \subseteq U\}$, $\neg_\mathcal{A}(U) = W \setminus U$, $\Rightarrow_\mathcal{A}(U_1, U_2) = (W \setminus U_1) \cup U_2$; $\mathbf{c} : \wp(|\mathcal{A}|_\phi) \to \wp(|\mathcal{A}|_\phi)$ is the cut closure induced by $\supseteq$.

The denotation of a formula is the set of all worlds where it holds.

---

[1] This is a local entailment relation. A stronger, global, entailment can also be defined by letting $\Phi \vDash'_\mathcal{R} \psi$ if $[\![\psi]\!]_\mathcal{A} \in \emptyset^\mathbf{c}$ whenever $\{[\![\varphi]\!]_\mathcal{A} : \varphi \in \Phi\} \subseteq \emptyset^\mathbf{c}$ for every $\langle \mathcal{A}, \mathbf{c} \rangle \in M$. This terminology is borrowed from [5] and is reflected, below, by the separation of local and global rules in deduction-rooms.

2.2 Deduction

As noted in [21], to represent the Hilbert-calculi of non-propositional-based logics we need a notion of *schematic inference rule* with a *proviso* delimiting its possible instantiations. From now on ξ, with any decoration, stands for a *schema variable*. Given $\Sigma = \langle S, O \rangle$ and $s \in S$, we denote by Ξ_s the set $\{\xi_s^k : k \in \mathbb{N}_0\}$ and let $\Xi = \{\Xi_s\}_{s \in S}$. We define the set of *schema formulas* $\mathrm{Form}_\Sigma(\Xi)$ to be $|\mathcal{W}_\Sigma(\Xi)|_\phi$, the carrier of sort ϕ in the free algebra with generators Ξ. We use γ, δ to denote schema formulas, and Γ, Δ sets of schema formulas. Elements of $|\mathcal{W}_\Sigma(\Xi)|_s$ are *schema terms*, denoted by θ. A *schema substitution* is a family $\sigma = \{\sigma_s : \Xi_s \to |\mathcal{W}_\Sigma(\Xi)|_s\}_{s \in S}$, that extends freely to schema terms. We write $\theta\sigma$ for the corresponding instantiation. $\mathrm{SSub}(\Sigma)$ denotes the set of schema substitutions over Σ. If σ maps each schema variable to a term without schema variables, we call it a (ground) substitution and denote it by ρ. $\mathrm{Sub}(\Sigma)$ denotes the set of all substitutions. Given $h : \Sigma \to \Sigma'$, we use $h(\sigma)$ to denote $(h \circ \sigma) \in \mathrm{SSub}(\Sigma')$. In the sequel, $\mathbf{AlgSig}_\phi(\Sigma, \_)$ denotes the class of all morphisms with domain Σ.

Definition 2. A Σ-*proviso* is $\pi = \{\pi_h\}_{h \in \mathbf{AlgSig}_\phi(\Sigma, \_)}$, where $\pi_h \subseteq \mathrm{Sub}(\Sigma')$ for each $h : \Sigma \to \Sigma'$, such that $\rho \in \pi_h$ if and only if $h'(\rho) \in \pi_{h' \circ h}$.

Provisos make their behaviour explicit on signature changes, which is essential when inference rules are translated to a richer language [7]. We denote by *univ* the universal Σ-proviso, $univ_h = \mathrm{Sub}(\Sigma')$ for $h : \Sigma \to \Sigma'$. Given a Σ-proviso π we denote by π_Σ the component π_{id_Σ}. Given $h : \Sigma \to \Sigma'$, we denote by $h(\pi)$ the Σ'-proviso such that $h(\pi)_{h'} = \pi_{h' \circ h}$. Given $\sigma \in \mathrm{SSub}(\Sigma)$ we denote by $\pi\sigma$ the Σ-proviso defined by $(\pi\sigma)_h = \{\rho \in \mathrm{Sub}(\Sigma') : \rho \circ h(\sigma) \in \pi_h\}$. Note that for $\rho \in \mathrm{Sub}(\Sigma)$, $\pi\rho = univ$ if $\rho \in \pi_\Sigma$, and $\pi\rho = \emptyset$ if $\rho \notin \pi_\Sigma$. By analogy, we define $\sigma \in \pi_\Sigma$ if $\pi\sigma = univ$. Given Σ-provisos π_1 and π_2 we denote by $\pi_1 \cap \pi_2$ the Σ-proviso such that $(\pi_1 \cap \pi_2)_h = (\pi_1)_h \cap (\pi_2)_h$. We say that $\pi_1 \subseteq \pi_2$ iff $\pi_{1_h} \subseteq \pi_{2_h}$. A Σ-proviso π is said to be *insensitive* to ξ if for every $h : \Sigma \to \Sigma'$ and any $\rho, \rho' \in \mathrm{Sub}(\Sigma')$ that may only differ on ξ, $\rho \in \pi_h$ if and only if $\rho' \in \pi_h$.

Example 3. Recall Example 1 and let $x \in X$. We define the following provisos:

- $nfv(\xi_\phi^n, x)$: given h, $\rho \in nfv(\xi_\phi^n, x)_h$ if $h(x)$ does not occur free in $\rho(\xi_\phi^n)$[2];
- $fts(\xi_\phi^n, \xi_\phi^m, \xi_\tau^k, x)$: given h, $\rho \in fts(\xi_\phi^n, \xi_\phi^m, \xi_\tau^k, x)_h$ if $\rho(\xi_\tau^k)$ is free for $h(x)$ in $\rho(\xi_\phi^n)$ and $\rho(\xi_\phi^m)$ is $\rho(\xi_\phi^n)$ with all free occurrences of $h(x)$ replaced by $\rho(\xi_\tau^k)$;
- $eqrep(\xi_\phi^n, \xi_\phi^m, x, y)$: given h, $\rho \in eqrep(\xi_\phi^n, \xi_\phi^m, x, y)_h$ if $\rho(\xi_\phi^m)$ is $\rho(\xi_\phi^n)$ with some free occurrences of $h(x)$, out of the scope of $h(\forall y)$, replaced by $h(y)$.

Definition 3. A Σ-*rule* is a triple $\langle \Gamma, \delta, \pi \rangle$ with $\Gamma \cup \{\delta\} \subseteq \mathrm{Form}_\Sigma(\Xi)$ finite and π a Σ-proviso insensitive to all the schema variables not in $\Gamma \cup \{\delta\}$.

We represent $r = \langle \Gamma, \delta, \pi \rangle$ by $\frac{\Gamma}{\delta} : \pi$, or even $\frac{\gamma_1 \ldots \gamma_n}{\delta} : \pi$ if $\Gamma = \{\gamma_1, \ldots, \gamma_n\}$. If the set of premises Γ is empty the rule is identified with its conclusion δ and proviso π and called a *schema axiom*. The translation of r by $h : \Sigma \to \Sigma'$ is the Σ'-rule $h(r) = \langle h(\Gamma), h(\delta), h(\pi) \rangle$.

---

[2] We mean that $h(x)$ always occurs in the scope of $h(\forall x)$. If variables and quantifiers are maintained on translating via h, it means precisely that x occurs under $\forall x$.

As in [20, 25, 5, 6], we explicitly distinguish local from global inference rules (see Examples 4 and 5). The nature of this distiction shows up both on their diverse deductive roles and on their different soundness requirements, below.

Definition 4. A *deduction-room (d-room, for short)* is a triple $\mathcal{D} = \langle \Sigma, lR, gR \rangle$ where $lR \cup gR$ is a set of Σ-rules such that gR-rules have non-empty premises. A *morphism of d-rooms* from $\mathcal{D}_1 = \langle \Sigma_1, lR_1, gR_1 \rangle$ to $\mathcal{D}_2 = \langle \Sigma_2, lR_2, gR_2 \rangle$ is a morphism $h : \Sigma_1 \to \Sigma_2$ such that $h(lR_1) \subseteq lR_2$ and $h(gR_1) \subseteq gR_2$.

Deduction-rooms and morphisms set up a category **DRoom**, from which the category **DPar** of *deduction-parchments (d-parchments)* is obtained by a Grothendieck construction, its colimits built from colimits in **DRoom**, as in [6].

Each $\mathcal{D} = \langle \Sigma, lR, gR \rangle$ induces a *deducibility* relation, built on top of a notion of *theoremhood* (the global counterpart of (local) deducibility). Let $\Gamma \cup \{\delta\} \subseteq \mathrm{Form}_\Sigma(\Xi)$ and π be a Σ-proviso. We say that δ with proviso π is a *schema theorem* of $\mathcal{D}$ generated from Γ, $\Gamma \vdash_{\mathcal{D}}^{\mathrm{thm}} \delta : \pi$, if there exists a finite sequence $\langle \delta_1, \pi_1 \rangle, \ldots, \langle \delta_n, \pi_n \rangle$ with $\delta = \delta_n$ and $\pi \subseteq \pi_n$, such that for each i, either $\delta_i \in \Gamma$ and $\pi_i = univ$, or there exists a rule $\langle \Gamma_r, \delta_r, \pi_r \rangle \in lR \cup gR$ and $\sigma \in \mathrm{SSub}(\Sigma)$ with $\Gamma_r \sigma = \{\delta_{j_1}, \ldots, \delta_{j_k}\} \subseteq \{\delta_j : j < i\}$, $\delta_i = \delta_r \sigma$ and $\pi_i = (\pi_{j_1} \cap \cdots \cap \pi_{j_k}) \cap \pi_r \sigma$. To simplify, we write $\vdash_{\mathcal{D}}^{\mathrm{thm}} \delta : \pi$ if $\Gamma = \emptyset$, or $\Gamma \vdash_{\mathcal{D}}^{\mathrm{thm}} \delta$ if $\pi = univ$. Easily, for every $\sigma \in \mathrm{SSub}(\Sigma)$: if $\Gamma \vdash_{\mathcal{D}}^{\mathrm{thm}} \delta : \pi$ then $\Gamma\sigma \vdash_{\mathcal{D}}^{\mathrm{thm}} \delta\sigma : \pi\sigma$. In deductions, now, only lR-rules and schema theorems are allowed. We say that δ with proviso π is *deducible* in $\mathcal{D}$ from Γ, $\Gamma \vdash_{\mathcal{D}} \delta : \pi$, if there exist $\langle \delta_1, \pi_1 \rangle, \ldots, \langle \delta_n, \pi_n \rangle$ with $\delta = \delta_n$ and $\pi \subseteq \pi_n$, such that for each i, either $\delta_i \in \Gamma$ and $\pi_i = univ$, or $\vdash_{\mathcal{D}}^{\mathrm{thm}} \delta_i : \pi_i$, or there exists a rule $\langle \Gamma_r, \delta_r, \pi_r \rangle \in lR$ and $\sigma \in \mathrm{SSub}(\Sigma)$ such that $\Gamma_r \sigma = \{\delta_{j_1}, \ldots, \delta_{j_k}\} \subseteq \{\delta_j : j < i\}$, $\delta_i = \delta_r \sigma$ and $\pi_i = (\pi_{j_1} \cap \cdots \cap \pi_{j_k}) \cap \pi_r \sigma$. Simplified notation applies to $\vdash_{\mathcal{D}}$, and the following also holds: if $\Gamma \vdash_{\mathcal{D}} \delta : \pi$ then $\Gamma\sigma \vdash_{\mathcal{D}} \delta\sigma : \pi\sigma$. Given a d-parchment $D : \mathbf{Sig} \to \mathbf{DRoom}$, we denote each $\vdash_{D(\Omega)}$ by $\vdash_\Omega$. The next structurality result is straightforward.

Proposition 1. *Let $h : \mathcal{D} \to \mathcal{D}'$ be a d-room morphism. If $\Gamma \vdash_{\mathcal{D}}^{\mathrm{thm}} \delta : \pi$ then $h(\Gamma) \vdash_{\mathcal{D}'}^{\mathrm{thm}} h(\delta) : h(\pi)$, and if $\Gamma \vdash_{\mathcal{D}} \delta : \pi$ then $h(\Gamma) \vdash_{\mathcal{D}'} h(\delta) : h(\pi)$.*

Example 4. The d-parchment for first-order logic with equality is the functor $FOLEq : \mathbf{Set}^{\mathbb{N}_0} \times \mathbf{Set}^{\mathbb{N}} \to \mathbf{DRoom}$ that maps each $\langle F, P \rangle$ to the d-room $\mathcal{D}_{FOLEq} = \langle \Sigma_{FOLEq}, lR, gR \rangle$, with Σ_{FOLEq} defined as in Example 1 and:

lR: $\xi_\phi^1 \Rightarrow (\xi_\phi^2 \Rightarrow \xi_\phi^1) : univ$
$\quad (\xi_\phi^1 \Rightarrow (\xi_\phi^2 \Rightarrow \xi_\phi^3)) \Rightarrow ((\xi_\phi^1 \Rightarrow \xi_\phi^2) \Rightarrow (\xi_\phi^1 \Rightarrow \xi_\phi^3)) : univ$
$\quad (\neg\xi_\phi^2 \Rightarrow \neg\xi_\phi^1) \Rightarrow (\xi_\phi^1 \Rightarrow \xi_\phi^2) : univ$
$\quad (\forall x\, \xi_\phi^1) \Rightarrow \xi_\phi^2 : fts(\xi_\phi^1, \xi_\phi^2, \xi_\tau^1, x)$
$\quad (\forall x\, (\xi_\phi^1 \Rightarrow \xi_\phi^2)) \Rightarrow (\xi_\phi^1 \Rightarrow (\forall x\, \xi_\phi^2)) : nfv(\xi_\phi^1, x)$
$\quad \forall x\, (x \doteq x) : univ$
$\quad (x \doteq y) \Rightarrow (\xi_\phi^1 \Rightarrow \xi_\phi^2) : eqrep(\xi_\phi^1, \xi_\phi^2, x, y)$
$\quad \dfrac{\xi_\phi^1 \quad \xi_\phi^1 \Rightarrow \xi_\phi^2}{\xi_\phi^2} : univ;$

gR: $\dfrac{\xi_\phi^1}{\forall x\, \xi_\phi^1} : univ.$

The shape of the fourth axiom is unusual. The usual notation that replaces ξ_ϕ^2 by $\xi_\phi^1(x := t)$ and requires t to be free for x is fine, informally, but we make it precise with *fts*. Let us deduce $\forall x\, p(x) \vdash_{\mathcal{D}_{FOLEq}} p(x)$, with p a unary predicate.

1. $\langle \forall x\, p(x), univ \rangle$ Hypothesis
2. $\langle (\forall x\, p(x)) \Rightarrow p(x), univ \rangle$ Axiom 4
3. $\langle p(x), univ \rangle$ MP rule:1,2

In step 2, we used the fourth axiom with substitution $\rho_2(\xi_\phi^1) = \forall x\, p(x)$, $\rho_2(\xi_\phi^2) = p(x)$ and $\rho_2(\xi_\tau^1) = x$. Easily, $\rho_2 \in fts(\xi_\phi^1, \xi_\phi^2, \xi_\tau^1, x)_{\Sigma_{FOLEq}}$, and so $fts(\xi_\phi^1, \xi_\phi^2, \xi_\tau^1, x)\rho_2 = univ$. In step 3, we used MP with $\rho_3(\xi_\phi^1) = \forall x\, p(x)$ and $\rho_3(\xi_\phi^2) = (\forall x\, p(x)) \Rightarrow p(x)$.

Example 5. The d-parchment for modal logic is $K : \mathbf{Set} \to \mathbf{DRoom}$, mapping each PS to $\mathcal{D}_K = \langle \Sigma_K, lR, gR \rangle$, with Σ_K defined as in Example 2 and:

lR: $\xi_\phi^1 \Rightarrow (\xi_\phi^2 \Rightarrow \xi_\phi^1) : univ$

$\quad (\xi_\phi^1 \Rightarrow (\xi_\phi^2 \Rightarrow \xi_\phi^3)) \Rightarrow ((\xi_\phi^1 \Rightarrow \xi_\phi^2) \Rightarrow (\xi_\phi^1 \Rightarrow \xi_\phi^3)) : univ$

$\quad (\neg \xi_\phi^2 \Rightarrow \neg \xi_\phi^1) \Rightarrow (\xi_\phi^1 \Rightarrow \xi_\phi^2) : univ$

$\quad \Box(\xi_\phi^1 \Rightarrow \xi_\phi^2) \Rightarrow (\Box \xi_\phi^1 \Rightarrow \Box \xi_\phi^2) : univ$

$\quad \dfrac{\xi_\phi^1 \quad \xi_\phi^1 \Rightarrow \xi_\phi^2}{\xi_\phi^2} : univ$

gR: $\dfrac{\xi_\phi^1}{\Box \xi_\phi^1} : univ.$

2.3 Logics

Often, we shall consider a **c**-parchment $R : \mathbf{Sig} \to \mathbf{CPRoom}$ together with a d-parchment $D : \mathbf{Sig} \to \mathbf{DRoom}$ such that, for each $\Omega \in |\mathbf{Sig}|$, $R(\Omega)$ and $D(\Omega)$ share the same signature Σ_Ω. With $\Phi \cup \{\psi\}$ a set of formulas, we define:

- D is *sound* for R if $\Phi \vdash_\Omega \psi$ implies $\Phi \vDash_\Omega \psi$, for all Ω;
- D is *weakly complete* for R if $\vDash_\Omega \varphi$ implies $\vdash_\Omega \varphi$, for all Ω;
- D is *finitely complete* for R if Φ finite and $\Phi \vDash_\Omega \psi$ imply $\Phi \vdash_\Omega \psi$, for all Ω;
- D is *complete* for R if $\Phi \vDash_\Omega \psi$ implies $\Phi \vdash_\Omega \psi$, for all Ω.

A Σ-rule $r = \langle \Gamma, \delta, \pi \rangle$ is said to be *locally sound* for $\langle \mathcal{A}, \mathbf{c} \rangle \in cAlg(\Sigma)$ if $[\![\delta\rho]\!]_\mathcal{A} \in \{[\![\gamma\rho]\!]_\mathcal{A} : \gamma \in \Gamma\}^{\mathbf{c}}$, for every $\rho \in \pi_\Sigma$, and *globally sound* if $[\![\delta\rho]\!]_\mathcal{A} \in \emptyset^{\mathbf{c}}$ whenever $\{[\![\gamma\rho]\!]_\mathcal{A} : \gamma \in \Gamma\} \subseteq \emptyset^{\mathbf{c}}$, for every $\rho \in \pi_\Sigma$. If all the rules of a d-room $\mathcal{D}$ are sound for all the structures of a **c**-room $\mathcal{R}$, i.e., the rules in lR are locally sound and the rules in gR are globally sound, we say that the rules of $\mathcal{D}$ are sound for $\mathcal{R}$. Obviously, if the rules of $D(\Omega)$ are sound for $R(\Omega)$ for every Ω, then the d-parchment D is sound for the **c**-parchment R [20, 25, 5].

Definition 5. A *logic-room* (l-room, for short) is a tuple $\mathcal{L} = \langle \Sigma, M, lR, gR \rangle$ where $\mathcal{R}(\mathcal{L}) = \langle \Sigma, M \rangle$ is a **c**-room and $\mathcal{D}(\mathcal{L}) = \langle \Sigma, lR, gR \rangle$ is a d-room with rules sound for $\mathcal{R}(\mathcal{L})$. A *morphism of l-rooms* from $\mathcal{L}_1$ to $\mathcal{L}_2$ is an $\mathbf{AlgSig}_\phi$-morphism $h : \Sigma_1 \to \Sigma_2$ such that $h : \mathcal{R}(\mathcal{L}_1) \to \mathcal{R}(\mathcal{L}_2)$ is a morphism of **c**-rooms and $h : \mathcal{D}(\mathcal{L}_1) \to \mathcal{D}(\mathcal{L}_2)$ a morphism of d-rooms.

Logic-rooms and morphisms set up a cocomplete category **LRoom**. The category **LPar** of *logic-parchments (l-parchments)* is again obtained by a Grothendieck construction, and is also cocomplete (see [6]). We write $\vDash_{\mathcal{L}}$ and $\vdash_{\mathcal{L}}$ for $\vDash_{\mathcal{R}(\mathcal{L})}$ and $\vdash_{\mathcal{D}(\mathcal{L})}$, and say that $\mathcal{L}$ is (weakly/finitely) complete when $\mathcal{D}(\mathcal{L})$ is, for $\mathcal{R}(\mathcal{L})$. By definition, all l-rooms are sound. Given a l-parchment $L : \mathbf{Sig} \to \mathbf{LRoom}$ and $\Omega \in |\mathbf{Sig}|$, we denote $\vDash_{L(\Omega)}$ and $\vdash_{L(\Omega)}$ simply by $\vDash_{\Omega}$ and $\vdash_{\Omega}$.

In [20, 25] we have noted that fibring is very sensitive to the way logics are presented, leading sometimes to the so-called *collapsing problem* [23]. Deductively, this difficulty can be dealt with by an appropriate use of provisos. Semantically, a way to deal with the possible trivialization of fibred logics is to require certain *fullness* conditions, guaranteeing that the logics have "enough" models [25]. Given $\Sigma \in |\mathbf{AlgSig}_\phi|$, let $\mathcal{I} \subseteq \mathrm{cAlg}(\Sigma)$ be a class of *intended* structures.

Definition 6. A l-room $\mathcal{L} = \langle \Sigma, M, lR, gR \rangle$ is *full* for $\mathcal{I}$ if M contains every $\langle \mathcal{A}, \mathbf{c} \rangle \in \mathcal{I}$ for which the rules of $\mathcal{D}(\mathcal{L})$ are sound.

Although fullness seems to be a fairly strong requirement, making a l-room full for $\mathcal{I}$ is a well-behaved operation. Given $\mathcal{L} = \langle \Sigma, M, lR, gR \rangle$, its full version is $\overline{\mathcal{L}} = \langle \Sigma, \overline{M}, lR, gR \rangle$ with $\overline{M} = M \cup \{\langle \mathcal{A}, \mathbf{c} \rangle \in \mathcal{I} : \mathcal{D}(\mathcal{L}) \text{ is sound for } \langle \mathcal{A}, \mathbf{c} \rangle\}$. This definition easily extends to an endofunctor in **LPar**. More important is the fact that soundness and completeness carry over from $\mathcal{L}$ to $\overline{\mathcal{L}}$. In general, we have that $\vdash_{\mathcal{L}} = \vdash_{\overline{\mathcal{L}}} \subseteq \vDash_{\overline{\mathcal{L}}} \subseteq \vDash_{\mathcal{L}}$, meaning that $\vDash_{\overline{\mathcal{L}}}$ can be weaker than $\vDash_{\mathcal{L}}$ but if that happens $\overline{\mathcal{L}}$ is closer to being complete. Later, in the context of fibring, we shall consider several interesting choices of intended structures. For now, we just consider the unrestricted class of all interpretation structures.

Example 6. The l-parchment *FOLEq* of first-order logic with equality maps each $\langle F, P \rangle$ to the room $\mathcal{L}_{FOLEq} = \langle \Sigma_{FOLEq}, M, lR, gR \rangle$, with $\mathcal{R}_{FOLEq} = \langle \Sigma_{FOLEq}, M \rangle$ and $\mathcal{D}_{FOLEq} = \langle \Sigma_{FOLEq}, lR, gR \rangle$ as in Examples 1 and 4, well known to be sound and complete [14]. Its full unrestricted version $\overline{FOLEq}$ considers, instead, the room $\mathcal{L}_{\overline{FOLEq}} = \langle \Sigma_{FOLEq}, \overline{M}, lR, gR \rangle$ where $\overline{M}$ contains all structures making $\mathcal{D}_{FOLEq}$ sound, and not just the usual structures of first-order logic already in M. By construction, $\overline{FOLEq}$ is also sound and complete.

Example 7. The l-parchment K of modal logic assigns $\mathcal{L}_K = \langle \Sigma_K, M, lR, gR \rangle$, with $\mathcal{R}_K = \langle \Sigma_K, M \rangle$ and $\mathcal{D}_K = \langle \Sigma_K, lR, gR \rangle$ as in Examples 2 and 5, to each PS and is sound and complete [12]. The full unrestricted version $\overline{K}$ considers, instead, the room $\mathcal{L}_{\overline{K}} = \langle \Sigma_K, \overline{M}, lR, gR \rangle$ where $\overline{M}$ is the class of all structures for which $\mathcal{D}_K$ is sound. Besides the Kripke structures in M, $\overline{M}$ also contains, for instance, all modal algebras. $\overline{K}$ is also sound and complete.

3   Fibring

We already know that colimits of parchments are built from colimits of rooms. As in [6], we characterize fibring using colimits, and so we concentrate just on rooms. Obviously, all the characterizations to follow can be immediately lifted to parchments. When considering two rooms with signatures $\Sigma_1 = \langle S_1, O_1 \rangle$ and $\Sigma_2 = \langle S_2, O_2 \rangle$, we assume that their fibring is *constrained* by sharing the sorts and constructors in their largest common subsignature $\Sigma_0 = \langle S_0, O_0 \rangle$, with $S_0 = S_1 \cap S_2$ (it always includes ϕ) and $O_{0,u} = O_{1,u} \cap O_{2,u}$, for $u \in S_0^+$, according to the corresponding inclusions $h_1 : \Sigma_0 \to \Sigma_1$ and $h_2 : \Sigma_0 \to \Sigma_2$. Below, $\mathcal{R}_0 = \langle \Sigma_0, M_0 \rangle$ with $M_0 = \mathrm{cAlg}(\Sigma_0)$, $\mathcal{D}_0 = \langle \Sigma_0, \emptyset, \emptyset \rangle$ and $\mathcal{L}_0 = \langle \Sigma_0, M_0, \emptyset, \emptyset \rangle$.

The envisaged combined signature is $\Sigma_1 \circledast \Sigma_2 = \langle S, O \rangle$ such that $S = S_1 \cup S_2$, with inclusions $f_i : S_i \to S$, and $O_u = O_{1,u} \cup O_{2,u}$ if $u \in S_0^+$, $O_u = O_{i,u}$ if $u \in S_i^+ \setminus S_0^+$, with inclusions $g_i : O_i \to O$. Easily, $\Sigma_1 \circledast \Sigma_2$ is a pushout of $\{ h_i : \Sigma_0 \to \Sigma_i \}_{i \in \{1,2\}}$ in $\mathbf{AlgSig}_\phi$, with inclusions $\langle f_i, g_i \rangle : \Sigma_i \to \Sigma_1 \circledast \Sigma_2$. When $S_0 = \{\phi\}$ and $O_0 = \emptyset$ we say that the fibring is *unconstrained*, and the construction corresponds to a coproduct in $\mathbf{AlgSig}_\phi$. The fibring of two c-rooms $\mathcal{R}_1 = \langle \Sigma_1, M_1 \rangle$ and $\mathcal{R}_2 = \langle \Sigma_2, M_2 \rangle$ is $\mathcal{R}_1 \circledast \mathcal{R}_2 = \langle \Sigma_1 \circledast \Sigma_2, M_1 \circledast M_2 \rangle$, where $M_1 \circledast M_2$ is the class of all structures $\langle \mathcal{A}, \mathbf{c} \rangle \in \mathrm{cAlg}(\Sigma_1 \circledast \Sigma_2)$ such that both $\langle \mathcal{A}|_{\langle f_1, g_1 \rangle}, \mathbf{c} \rangle \in M_1$ and $\langle \mathcal{A}|_{\langle f_2, g_2 \rangle}, \mathbf{c} \rangle \in M_2$, i.e., $M_1 \circledast M_2$ is obtained by joining together $\langle \mathcal{A}_1, \mathbf{c}_1 \rangle \in M_1$ and $\langle \mathcal{A}_2, \mathbf{c}_2 \rangle \in M_2$ with $|\mathcal{A}_1|_s = |\mathcal{A}_2|_s = |\mathcal{A}|_s$ for $s \in S_0$, $o_{\mathcal{A}_1} = o_{\mathcal{A}_2} = o_\mathcal{A}$ for $o \in O_{0,u}$ with $u \in S_0^+$, and $\mathbf{c}_1 = \mathbf{c}_2 = \mathbf{c}$. The fibring $\mathcal{R}_1 \circledast \mathcal{R}_2$ is a pushout of $\{ h_i : \mathcal{R}_0 \to \mathcal{R}_i \}_{i \in \{1,2\}}$ in $\mathbf{CPRoom}$, as proved in [6], where a similar characterization for propositional-based proof-calculi, without provisos, was also given. To generalize the characterization, let $\mathcal{D}_1 = \langle \Sigma_1, lR_1, gR_1 \rangle$ and $\mathcal{D}_2 = \langle \Sigma_2, lR_2, gR_2 \rangle$ be d-rooms and $\mathcal{D}_1 \circledast \mathcal{D}_2$ their fibring.

Definition 7. $\mathcal{D}_1 \circledast \mathcal{D}_2 = \langle \Sigma_1 \circledast \Sigma_2, lR_1 \circledast lR_2, gR_1 \circledast gR_2 \rangle$ where $lR_1 \circledast lR_2 = \langle f_1, g_1 \rangle(lR_1) \cup \langle f_2, g_2 \rangle(lR_2)$ and $gR_1 \circledast gR_2 = \langle f_1, g_1 \rangle(gR_1) \cup \langle f_2, g_2 \rangle(gR_2)$.

Proposition 2. $\mathcal{D}_1 \circledast \mathcal{D}_2$ is a pushout of $\{ h_i : \mathcal{D}_0 \to \mathcal{D}_i \}_{i \in \{1,2\}}$ in $\mathbf{DRoom}$.

Fibred l-rooms capitalize on the characterizations above but, first, we need to note that soundness of rules is preserved. Let $\mathcal{L}_1 = \langle \Sigma_1, M_1, lR_1, gR_1 \rangle$ and $\mathcal{L}_2 = \langle \Sigma_2, M_2, lR_2, gR_2 \rangle$ be l-rooms and $\mathcal{L}_1 \circledast \mathcal{L}_2$ their fibring.

Theorem 1. The rules of $\mathcal{D}(\mathcal{L}_1) \circledast \mathcal{D}(\mathcal{L}_2)$ are sound for $\mathcal{R}(\mathcal{L}_1) \circledast \mathcal{R}(\mathcal{L}_2)$.

It is now safe to state the definition of fibred l-room.

Definition 8. $\mathcal{L}_1 \circledast \mathcal{L}_2 = \langle \Sigma_1 \circledast \Sigma_2, M_1 \circledast M_2, lR_1 \circledast lR_2, gR_1 \circledast gR_2 \rangle$.

Proposition 3. $\mathcal{L}_1 \circledast \mathcal{L}_2$ is a pushout of $\{ h_i : \mathcal{L}_0 \to \mathcal{L}_i \}_{i \in \{1,2\}}$ in $\mathbf{LRoom}$.

To see fibring interact with fullness, consider a system of intended structures $\mathcal{I}_1 \subseteq \mathrm{cAlg}(\Sigma_1)$, $\mathcal{I}_2 \subseteq \mathrm{cAlg}(\Sigma_2)$, $\mathcal{I} \subseteq \mathrm{cAlg}(\Sigma_1 \circledast \Sigma_2)$ satisfying a coherence requirement: if $\langle \mathcal{A}, \mathbf{c} \rangle \in \mathcal{I}$ then $\langle \mathcal{A}|_{\langle f_1, g_1 \rangle}, \mathbf{c} \rangle \in \mathcal{I}_1$ and $\langle \mathcal{A}|_{\langle f_2, g_2 \rangle}, \mathbf{c} \rangle \in \mathcal{I}_2$. The result below is an immediate consequence of the definition of fibring.

Proposition 4. If $\mathcal{L}_1$ and $\mathcal{L}_2$ are full for $\mathcal{I}_1$ and $\mathcal{I}_2$ then $\mathcal{L}_1 \circledast \mathcal{L}_2$ is full for $\mathcal{I}$.

Example 8. We obtain a l-room $\mathcal{L}_{\overline{FOLEq}} \circledast \mathcal{L}_{\overline{K}}$ for modal first-order logic by fibring the full versions of modal and first-order logic of Examples 7 and 6. The combined signature $\Sigma_{FOLEq} \circledast \Sigma_K = \langle S, O \rangle$ has:

$- S = \{\tau, \phi\}$, $O_\tau = X \cup F_0$, $O_{\tau^n \tau} = F_n$ for $n > 0$, $O_\phi = PS$, $O_{\tau^2 \phi} = P_2 \cup \{\doteq\}$,
$O_{\tau^n \phi} = P_n$ for $n > 0$ and $n \neq 2$, $O_{\phi\phi} = \{\neg, \Box\} \cup \{\forall x : x \in X\}$, $O_{\phi^2 \phi} = \{\Rightarrow\}$.

Consider the usual interpretation structures of modal first-order logic with constant domain and rigid interpretation of symbols, i.e., the first-order component does not change from one world to the other. It corresponds to considering both a *FOLEq* interpretation $\langle D, I \rangle$ and a Kripke model $\langle W, R, \vartheta \rangle$. It is easy to see that among the structures of $\mathcal{L}_{\overline{FOLEq}} \circledast \mathcal{L}_{\overline{K}}$ we can find the $\langle \mathcal{A}, c \rangle$ such that:

$- |\mathcal{A}|_\tau = D^{W \times \mathrm{Asg}(X,D)}$ and $|\mathcal{A}|_\phi = \wp(W \times \mathrm{Asg}(X, D))$; $x_\mathcal{A}(w, \mu) = \mu(x)$,
$f_\mathcal{A}(\langle e_i \rangle)(w, \mu) = f_I(\langle e_i(w, \mu) \rangle)$, $q_\mathcal{A} = \vartheta(q) \times \mathrm{Asg}(X, D)$, $p_\mathcal{A}(\langle e_i \rangle) = \{\langle w, \mu \rangle :$
$\langle e_i(w, \mu) \rangle \in p_I\}$, $\neg_\mathcal{A}(b) = (W \times \mathrm{Asg}(X, D)) \setminus b$, $\Rightarrow_\mathcal{A}(b_1, b_2) = ((W \times$
$\mathrm{Asg}(X, D)) \setminus b_1) \cup b_2$, $\forall x_\mathcal{A}(b) = \{\langle w, \mu \rangle : \langle w, \mu[x/d] \rangle \in b$ for every $d \in D\}$,
and $\Box_\mathcal{A}(b) = \{\langle w, \mu \rangle : \{\langle w', \mu \rangle : wRw'\} \subseteq b\}$; $c : \wp(|\mathcal{A}|_\phi) \to \wp(|\mathcal{A}|_\phi)$ is the cut closure operation induced by $\supseteq$.

This structure makes all the rules of $\mathcal{D}_{FOLEq} \circledast \mathcal{D}_K$ sound, but other usual modal first-order semantic structures could be considered (see Example 9). But now, the reason why we considered the full versions is obvious: the structure above is in $\mathcal{L}_{\overline{FOLEq}} \circledast \mathcal{L}_{\overline{K}}$ but certainly not in $\mathcal{L}_{FOLEq} \circledast \mathcal{L}_K$.

4   Completeness

Again we concentrate on l-rooms, since everything can be lifted to l-parchments. The results in this section generalize [6] and make thorough use of the notion of fullness. The first result applies to l-rooms full for the class of all structures.

Proposition 5. *If $\mathcal{L}$ is full for all structures then $\mathcal{L}$ is complete.*

Proof. The rules of $\mathcal{D}(\mathcal{L})$ are sound for $\langle \mathcal{W}_\Sigma, c \rangle$ with $c = \vdash_\mathcal{L}$. Thus, by fullness, the structure $\langle \mathcal{W}_\Sigma, c \rangle$ belongs to $\mathcal{L}$. Suppose now that $\Phi \not\vdash_\mathcal{L} \varphi$. To show that $\Phi \not\models_\mathcal{L} \varphi$ it is enough to note that $[\![\_]\!]_{\mathcal{W}_\Sigma}$ is the identity on formulas. $\square$

The system $\mathcal{I}_1 = \mathrm{cAlg}(\Sigma_1)$, $\mathcal{I}_2 = \mathrm{cAlg}(\Sigma_2)$ and $\mathcal{I} = \mathrm{cAlg}(\Sigma_1 \circledast \Sigma_2)$ of intended structures trivially satisfies the necessary coherence requirement.

Proposition 6. *Fullness for all structures is preserved by fibring.*

The first completeness transfer result follows from Propositions 5 and 6.

Theorem 2. *If $\mathcal{L}_1$ and $\mathcal{L}_2$ are full for all structures then $\mathcal{L}_1 \circledast \mathcal{L}_2$ is complete.*

Although this result is a bit too syntactic (see the structure in Proposition 5), its proof is enlightening. Reusing the technique of [25], we have shown that if a l-room has certain properties then it is complete (Proposition 5), and also that the relevant properties are preserved by fibring (Proposition 6). All the subsequent completeness preservation results follow the same pattern.

We start by considering a reasonable syntactic restriction. A signature $\Sigma = \langle S, O \rangle$ is said to be *plain* if for every $o \in O_{us}$ with $s \neq \phi$, $u \in (S \setminus \{\phi\})^*$. Plainhood of signatures prevents us from building terms using formulas.

Proposition 7. *If Σ_1 and Σ_2 are plain signatures then so is $\Sigma_1 \circledast \Sigma_2$.*

A closure operation $\langle A, \mathbf{c} \rangle$ is said to be *elementary* if for every $a_1, a_2 \in A$, $a_1 \in \{a_2\}^{\mathbf{c}}$ and $a_2 \in \{a_1\}^{\mathbf{c}}$ imply $a_1 = a_2$. Clearly, the structure in Proposition 5 is not elementary. Let us take as intended the class of all structures whose closure is elementary. The corresponding system of intended structures clearly fulfills the coherence requirement and fullness for this class is preserved by fibring.

Proposition 8. *Fullness for elementary structures is preserved by fibring.*

The following characterizations are a simple reformulation from [25, 6, 5]. A d-room $\mathcal{D}$ over $\Sigma = \langle S, O \rangle$ is said to have *implication* if there exists $\Rightarrow \in O_{\phi^2 \phi}$ satisfying: (i) $\vdash_{\mathcal{D}} \xi_\phi^1 \Rightarrow \xi_\phi^1$, (ii) $\xi_\phi^1, \xi_\phi^1 \Rightarrow \xi_\phi^2 \vdash_{\mathcal{D}} \xi_\phi^2$, (iii) $\xi_\phi^2 \vdash_{\mathcal{D}} \xi_\phi^1 \Rightarrow \xi_\phi^2$, (iv) for each $\langle \Gamma, \delta, \pi \rangle \in lR$ and ξ_ϕ^n not in $\Gamma \cup \{\delta\}$, $\{\xi_\phi^n \Rightarrow \gamma : \gamma \in \Gamma\} \vdash_{\mathcal{D}} \xi_\phi^n \Rightarrow \delta : \pi$.

In the sequel, $\mathcal{D}$ is said to be *formula-congruent* if it has implication and we have that, for every $o \in O_{s_1 \dots s_n \phi}$ and $s_i = \phi$, $\xi_\phi^i \Rightarrow \xi_\phi^0, \xi_\phi^0 \Rightarrow \xi_\phi^i \vdash_{\mathcal{D}}^{\text{thm}}$ $o(\xi_{s_1}^1, \dots, \xi_{s_{i-1}}^{i-1}, \xi_\phi^i, \xi_{s_{i+1}}^{i+1}, \dots, \xi_{s_n}^n) \Rightarrow o(\xi_{s_1}^1, \dots, \xi_{s_{i-1}}^{i-1}, \xi_\phi^0, \xi_{s_{i+1}}^{i+1}, \dots, \xi_{s_n}^n)$. The next result follows from Proposition 1.

Proposition 9. *If $\mathcal{D}_1$ and $\mathcal{D}_2$ are formula-congruent and share an implication then $\mathcal{D}_1 \circledast \mathcal{D}_2$ is formula-congruent.*

We are now able to present the following completeness result.

Proposition 10. *If $\mathcal{L}$ is full for elementary structures, has a plain signature, and $\mathcal{D}(\mathcal{L})$ is formula-congruent then $\mathcal{L}$ is complete.*
Proof. Let Σ be the signature of $\mathcal{L}$. Easily, $\equiv_\phi$ defined on $|\mathcal{W}_\Sigma|_\phi$ by $\varphi_1 \equiv_\phi \varphi_2$ if $\{\varphi_1\} \vdash_{\mathcal{L}} \varphi_2$ and $\{\varphi_2\} \vdash_{\mathcal{L}} \varphi_1$ is an equivalence. Since $\mathcal{D}(\mathcal{L})$ is formula-congruent and Σ is plain, $\equiv = \{\equiv_s\}_{s \in S}$ with $\equiv_s$ the identity if $s \neq \phi$ is a congruence on $\mathcal{W}_\Sigma$. Let us consider the Lindenbaum-Tarski structure $\langle \mathcal{W}_\Sigma / \equiv, \mathbf{c} \rangle$, with $\mathbf{c}$ defined by $\{[\psi] : \psi \in \Psi\}^{\mathbf{c}} = \{[\psi'] : \Psi \vdash_{\mathcal{L}} \psi'\}$. Clearly $\mathcal{D}(\mathcal{L})$ is sound for $\langle \mathcal{W}_\Sigma / \equiv, \mathbf{c} \rangle$, which is elementary and, by fullness, belongs to $\mathcal{L}$. Suppose that, $\Phi \nvdash_{\mathcal{L}} \varphi$. The structure just built shows that $\Phi \nvDash_{\mathcal{L}} \psi$. $\square$

Theorem 3. *If $\mathcal{L}_1$ and $\mathcal{L}_2$ are full for elementary structures, Σ_1 and Σ_2 are plain, and $\mathcal{D}(\mathcal{L}_1)$ and $\mathcal{D}(\mathcal{L}_2)$ are formula-congruent and share an implication then $\mathcal{L}_1 \circledast \mathcal{L}_2$ is complete.*

Let us try to improve this result. In logic, algebras of truth-values are often partially ordered. Every partial-order $\langle A, \leq \rangle$ induces two polarities $\mathrm{Upp}(B) = \{a \in A : b \leq a \text{ for every } b \in B\}$ and $\mathrm{Low}(B) = \{a \in A : a \leq b \text{ for every } b \in B\}$, and a cut closure operation on A defined by $B^{\mathbf{c}} = \mathrm{Upp}(\mathrm{Low}(B))$ [2]. *Partial-order* structures are precisely those whose closure operation fulfills this condition.

Proposition 11. *Fullness for partial-order structures is preserved by fibring.*

We can now present the following preservation results.

Proposition 12. *If $\mathcal{L}$ is full for partial-order structures, has a plain signature, and $\mathcal{D}(\mathcal{L})$ is formula-congruent then $\mathcal{L}$ is weakly complete.*

Proof. The proof is similar to Proposition 10, but with a different closure. Note that $[\varphi_1] \leq [\varphi_2]$ if $\vdash_{\mathcal{L}} \varphi_1 \Rightarrow \varphi_2$ defines a partial-order on $|\mathcal{W}_\Sigma/\equiv|_\phi$. Consider the structure $\langle \mathcal{W}_\Sigma/\equiv, \mathbf{c}\rangle$ where $\mathbf{c}$ is the cut closure induced by $\leq$. Let us check, in this less trivial case, that the structure makes the rules of $\mathcal{D}(\mathcal{L})$ sound. Consider a rule $r = \frac{\gamma_1 \ldots \gamma_n}{\delta} : \pi$ and fix a substitution $\rho \in \pi_\Sigma$. Assume that r is an l-rule. Since $[\![\sqcup]\!]_{\mathcal{W}_\Sigma/\equiv} = [\sqcup]$, we need to show that $[\delta\rho] \in \{[\gamma_1\rho], \ldots, [\gamma_n\rho]\}^{\mathbf{C}}$. Let φ be a formula such that $[\varphi] \leq [\gamma_i\rho]$ for $i = 1, \ldots, n$. This means that $\vdash_{\mathcal{L}} \varphi \Rightarrow \gamma_i\rho$ for each i. Using requirement (iv) of implication, $\{\xi_\phi^n \Rightarrow \gamma_i : i = 1, \ldots, n\} \vdash_{\mathcal{L}} \xi_\phi^n \Rightarrow \delta : \pi$ for ξ_ϕ^n not in r. Consider $\rho' \in \mathrm{Sub}(\Sigma)$ such that ρ' equals ρ, except that $\rho'(\xi_\phi^n) = \varphi$. Clearly, $\gamma_i\rho = \gamma_i\rho'$ and $\delta\rho = \delta\rho'$. Moreover, π is insensitive to ξ_ϕ^n and $\rho' \in \pi_\Sigma$. By the structurality of deducibility and the fact that $\pi\rho' = univ$, $\{\varphi \Rightarrow \gamma_i\rho : i = 1, \ldots, n\} \vdash_{\mathcal{L}} \varphi \Rightarrow \delta\rho$. So $\vdash_{\mathcal{L}} \varphi \Rightarrow \delta\rho$, or equivalently, $[\varphi] \leq [\delta\rho]$, and the l-rule is sound. Assume now that r is a g-rule. We need to show that $\{[\gamma_1\rho], \ldots, [\gamma_n\rho]\} \subseteq \emptyset^{\mathbf{C}}$ implies $[\delta\rho] \in \emptyset^{\mathbf{C}}$. Easily, $\emptyset^{\mathbf{C}}$ has precisely one element, the equivalence class of formulas ψ such that $\vdash_{\mathcal{L}} \psi$. If $\vdash_{\mathcal{L}} \gamma_i\rho$ for each i, then by using r, we conclude that $\vdash_{\mathcal{L}} \delta\rho$ and the g-rule is sound. By fullness, the structure belongs to $\mathcal{R}(\mathcal{L})$. So, if $\nvdash_{\mathcal{L}} \psi$ this structure shows that $\nvDash_{\mathcal{L}} \psi$.   □

Theorem 4. *If $\mathcal{L}_1$ and $\mathcal{L}_2$ are full for partial-order structures, Σ_1 and Σ_2 are plain, and $\mathcal{D}(\mathcal{L}_1)$ and $\mathcal{D}(\mathcal{L}_2)$ are formula-congruent and share an implication then $\mathcal{L}_1 \circledast \mathcal{L}_2$ is weakly complete.*

A little improvement is still possible. A d-room $\mathcal{D}$ over $\Sigma = \langle S, O\rangle$ is said to have *conjunction* if there exists $\wedge \in O_{\phi^2\phi}$ such that: (i) $\xi_\phi^1 \wedge \xi_\phi^2 \vdash_{\mathcal{D}} \xi_\phi^1$, (ii) $\xi_\phi^1 \wedge \xi_\phi^2 \vdash_{\mathcal{D}} \xi_\phi^2$, (iii) $\xi_\phi^1, \xi_\phi^2 \vdash_{\mathcal{D}} \xi_\phi^1 \wedge \xi_\phi^2$.

Proposition 13. *If $\mathcal{D}_1$ or $\mathcal{D}_2$ have conjunction then so has $\mathcal{D}_1 \circledast \mathcal{D}_2$.*

Proposition 14. *If $\mathcal{L}$ is full for partial-order structures, has a plain signature, and $\mathcal{D}(\mathcal{L})$ is formula-congruent and has conjunction then $\mathcal{L}$ is finitely complete.*

Proof. Consider the structure of Proposition 12, and suppose $\{\varphi_1, \ldots, \varphi_n\} \nvDash_{\mathcal{L}} \psi$. With $\varphi = \varphi_1 \wedge \ldots \wedge \varphi_n$, it is trivial that $\vdash_{\mathcal{L}} \varphi \Rightarrow \varphi_i$ for each i. Easily, it is also the case that $\nvdash_{\mathcal{L}} \varphi \Rightarrow \psi$ and the structure shows that $\{\varphi_1, \ldots, \varphi_n\} \nvDash_{\mathcal{L}} \psi$.   □

Theorem 5. *If $\mathcal{L}_1$ and $\mathcal{L}_2$ are full for partial-order structures, Σ_1 and Σ_2 are plain, $\mathcal{D}(\mathcal{L}_1)$ and $\mathcal{D}(\mathcal{L}_2)$ are formula-congruent, share an implication, and one of them has conjunction then $\mathcal{L}_1 \circledast \mathcal{L}_2$ is finitely complete.*

All the previous results are still valid if we concentrate only on structures providing a standard interpretation of equality, when it exists. A signature $\Sigma = \langle S, O\rangle \in |\mathbf{AlgSig}_\phi|$ is said to have a *system of equalities* if, for every $s \in S \setminus \{\phi\}$, there exists $\doteq \in O_{s^2\phi}$. The existence of equality symbols is preserved by fibring.

Proposition 15. *If Σ_1 and Σ_2 have systems of equalities then so has $\Sigma_1 \circledast \Sigma_2$.*

If Σ has a system of equalities, then $\langle \mathcal{A}, \mathbf{c}\rangle \in \mathrm{cAlg}(\Sigma)$ is said to be *standard* for equality if: (i) if $a_1 \doteq_\mathcal{A} a_2 \in \emptyset^{\mathbf{C}}$ then $a_1 = a_2$, (ii) for $T \subseteq |\mathcal{A}|_\phi$, the congruence $\equiv_T$ on $\mathcal{A}$ generated by $R_T = \{\langle a_1, a_2\rangle : a_1 \doteq_\mathcal{A} a_2 \in T^{\mathbf{C}}\}$ is such that $\equiv_{T,s} = R_T \cap (|\mathcal{A}|_s \times |\mathcal{A}|_s)$ for $s \neq \phi$. The conditions mean that $\mathbf{c}$ captures precisely the congruence imposed by the equalities.

Proposition 16. *Fullness for standard structures is preserved by fibring.*

Now, of course, we should require a similar standard treatment of equality at the deductive level. A d-room $\mathcal{D} = \langle \Sigma, lR, gR \rangle$ is said to have *equality* if Σ has a system of equality symbols and the following hold: (i) $\vdash_{\mathcal{D}} \xi_s^1 \doteq \xi_s^1$, (ii) $\xi_s^1 \doteq \xi_s^2 \vdash_{\mathcal{D}} \xi_s^2 \doteq \xi_s^1$, (iii) $\xi_s^1 \doteq \xi_s^2, \xi_s^2 \doteq \xi_s^3 \vdash_{\mathcal{D}} \xi_s^1 \doteq \xi_s^3$, (iv) for every $o \in O_{s_1 \dots s_n s}$ with $s \neq \phi$ and every $i \in \{1, \dots, n\}$ with $s_i \neq \phi$, $\xi_{s_i}^i \doteq \xi_{s_i}^0 \vdash_{\mathcal{D}}$ $o(\xi_{s_1}^1, \dots, \xi_{s_{i-1}}^{i-1}, \xi_{s_i}^i, \xi_{s_{i+1}}^{i+1}, \dots, \xi_{s_n}^n) \doteq o(\xi_{s_1}^1, \dots, \xi_{s_{i-1}}^{i-1}, \xi_{s_i}^0, \xi_{s_{i+1}}^{i+1}, \dots, \xi_{s_n}^n)$, and last but not least (v) for every $o \in O_{s_1 \dots s_n \phi}$ and every $i \in \{1, \dots, n\}$ with $s_i \neq \phi$, $\xi_{s_i}^i \doteq \xi_{s_i}^0, o(\xi_{s_1}^1, \dots, \xi_{s_{i-1}}^{i-1}, \xi_{s_i}^i, \xi_{s_{i+1}}^{i+1}, \dots, \xi_{s_n}^n) \vdash_{\mathcal{D}} o(\xi_{s_1}^1, \dots, \xi_{s_{i-1}}^{i-1}, \xi_{s_i}^0, \xi_{s_{i+1}}^{i+1}, \dots, \xi_{s_n}^n)$.

Proposition 17. *If $\mathcal{D}_1$ and $\mathcal{D}_2$ have equality then so has $\mathcal{D}_1 \circledast \mathcal{D}_2$.*

We can now state the following completeness result.

Proposition 18. *If $\mathcal{L}$ is full for standard elementary structures, has a plain signature, and $\mathcal{D}(\mathcal{L})$ is formula-congruent and has equality then $\mathcal{L}$ is complete.*

Proof. For each $s \in S \backslash \{\phi\}$, $\equiv_s$ such that $t_1 \equiv_s t_2$ if $\vdash_{\mathcal{L}} t_1 \doteq t_2$ is an equivalence on $|\mathcal{W}_\Sigma|_s$. Considering $\equiv = \{\equiv_s\}_{s \in S}$ with $\equiv_\phi$ as in Proposition 10, and noting that $\mathcal{D}(\mathcal{L})$ has equality, we conclude that $\equiv$ is a congruence on $\mathcal{W}_\Sigma$. Consider $\langle \mathcal{W}_\Sigma / \equiv, \mathbf{c} \rangle$, with $\mathbf{c}$ defined as in Proposition 10. The structure makes $\mathcal{D}(\mathcal{L})$ sound and is elementary. We now prove that it is standard for equality. Given $[t_1], [t_2] \in |\mathcal{W}_\Sigma / \equiv|_s$, if $[t_1] \doteq_{\mathcal{W}_\Sigma / \equiv} [t_2] \in \emptyset^{\mathbf{c}}$ then $\vdash_{\mathcal{L}} t_1 \doteq t_2$ and, by definition of $\equiv_s$, $t_1 \equiv_s t_2$ and $[t_1] = [t_2]$. Given $T = \{[\psi] : \psi \in \Psi\} \subseteq |\mathcal{W}_\Sigma / \equiv|_\phi$ and $R_T = \{\langle a_1, a_2 \rangle : a_1 \doteq_{\mathcal{W}_\Sigma / \equiv} a_2 \in T^{\mathbf{c}}\}$, let $R_{T,s} = R_T \cap (|\mathcal{A}|_s \times |\mathcal{A}|_s)$ for each $s \neq \phi$ and recall that $T^{\mathbf{c}} = \{[\psi'] : \Psi \vdash_{\mathcal{D}_1 \circledast \mathcal{D}_2} \psi'\}$. Consider the congruence $\equiv_T$ generated by R_T. By definition, $R_{T,s} \subseteq \equiv_{T,s}$. Since $\mathcal{D}(\mathcal{L})$ has equality, it is easy to see that $R_{T,s}$ is an equivalence and $\langle [o(t_1, \dots, t_i, \dots, t_n)], [o(t_1, \dots, t_i', \dots, t_n)] \rangle \in R_{T,s}$ for each $o \in O_{s_1 \dots s_n s}$ with $s \neq \phi$ (and, since Σ is plain, each $s_i \neq \phi$) whenever $\langle [t_i], [t_i'] \rangle \in R_{T,s_i}$. So, $\equiv_{T,s} \subseteq R_{T,s}$, $\langle \mathcal{W}_\Sigma / \equiv, \mathbf{c} \rangle$ is standard and belongs to $\mathcal{L}$, by fullness. As before, if $\Phi \nvdash_{\mathcal{L}} \varphi$, $\langle \mathcal{W}_\Sigma / \equiv, \mathbf{c} \rangle$ clearly shows that $\Phi \nvdash_{\mathcal{L}} \psi$.   □

Completeness preservation results for elementary, or partial-order, structures, assuming herein that they are also standard for equality, easily follow.

Theorem 6. *If $\mathcal{L}_1$ and $\mathcal{L}_2$ are full for standard elementary structures, Σ_1 and Σ_2 are plain, and $\mathcal{D}(\mathcal{L}_1)$ and $\mathcal{D}(\mathcal{L}_2)$ are formula-congruent, with equality and a shared implication then $\mathcal{L}_1 \circledast \mathcal{L}_2$ is complete.*

Theorem 7. *If $\mathcal{L}_1$ and $\mathcal{L}_2$ are full for standard partial-order structures, Σ_1 and Σ_2 are plain, and $\mathcal{D}(\mathcal{L}_1)$ and $\mathcal{D}(\mathcal{L}_2)$ are formula-congruent, with equality and a shared implication then $\mathcal{L}_1 \circledast \mathcal{L}_2$ is weakly complete.*

Theorem 8. *If $\mathcal{L}_1$ and $\mathcal{L}_2$ are full for standard partial-order structures, Σ_1 and Σ_2 are plain, $\mathcal{D}(\mathcal{L}_1)$ and $\mathcal{D}(\mathcal{L}_2)$ are formula-congruent, with equality, a shared implication, and one of them has conjunction then $\mathcal{L}_1 \circledast \mathcal{L}_2$ is finitely complete.*

Example 9. In Example 8 we obtained a system of modal first-order logic by fibring full versions of propositional modal logic and first-order logic. It is well known (e.g., [12]) that the structures considered therein make both the Barcan formula $(\forall x(\Box\xi_\phi^n)) \Rightarrow (\Box(\forall x\xi_\phi^n))$ and its converse $(\Box(\forall x\xi_\phi^n)) \Rightarrow (\forall x(\Box\xi_\phi^n))$ sound. However, although the latter is deducible from $\mathcal{D}_{FOLEq} \circledast \mathcal{D}_K$, the former is not. Since our completeness preservation results apply, $\mathcal{L}_{\overline{FOLEq}} \circledast \mathcal{L}_{\overline{K}}$ is complete and must contain structures where the Barcan formula fails. This is the case for the expanding domains interpretations of [12], with an extra component $Q : W \to \wp(D)$ that assigns to each world a domain of interpretation such that if wRw' then $Q(w) \subseteq Q(w')$. We denote by D_w the set $Q(w)$, by $\text{Asg}(X, D)_w$ the set D_w^X of assignments in D_w, and by U the set $\{\langle w, \mu\rangle : w \in W \text{ and } \mu \in \text{Asg}(X, D)_w\}$. Then, $\langle \mathcal{A}, \mathbf{c}\rangle$ is defined just as in Example 8, considering $\mathcal{A}_\tau = D^U$, $\mathcal{A}_\phi = \wp(U)$ and $\forall x_\mathcal{A}(b) = \{\langle w, \mu\rangle : \{\langle w, \mu[x/d]\rangle : d \in D_w\} \subseteq b\}$.

One important aspect of the structures considered so far is that the interpretation of symbols is rigid. However, if we consider flexible symbols, we must proceed with caution. As noted for instance in [22], some axioms of *FOLEq* do not behave well in the presence of flexible symbols. Consider the following instance of the fourth *FOLEq* axiom, $(\forall x ((c \doteq x) \Rightarrow \Diamond(c > x))) \Rightarrow ((c \doteq c) \Rightarrow \Diamond(c > c))$, where c is a flexible symbol and $>$ is an irreflexive ordering. It is easy to find a structure that falsifies the formula. The problem arises when we try to replace a variable by a flexible term (in this case c) in the scope of a modality. One way to avoid this problem is to strengthen the proviso as follows:

- $fts(\xi_\phi^n, \xi_\phi^m, \xi_\tau^k, x)$ is such that, given $h : \Sigma_{FOLEq} \to \Sigma'$, $\rho \in fts(\xi_\phi^n, \xi_\phi^m, \xi_\tau^k, x)_h$ iff $\rho(\xi_\tau^k)$ free for $h(x)$ in $\rho(\xi_\phi^n)$ and $\rho(\xi_\phi^m)$ results from $\rho(\xi_\phi^n)$ by replacing by $\rho(\xi_\tau^k)$ the free occurrences of $h(x)$. Plus, if $\rho(\xi_\tau^k)$ is a Σ_{FOLEq}-term, then no free occurrence of $h(x)$ can appear in the scope of a $\Sigma' \setminus h(\Sigma_{FOLEq})$-symbol.

In our example, this means that no term of Σ_{FOLEq} may be replaced in the scope of a modality. This change has no impact whatsoever on first-order logic *per se*, but makes a huge difference when we combine it with modal logic. Likewise, the fifth axiom in $\mathcal{D}_{FOLEq}$ must also be changed to prevent $\rho(\xi_\phi^n)$ to contain modalities. If we consider equality, *eqrep* must also be changed so that [13]: "if x occurs free in the scope of a modal operator, then either all or no occurrence of x may be replaced by y". With these changes, the corresponding fibred system includes structures $\langle \mathcal{A}, \mathbf{c}\rangle$ defined from $\langle W, R, \vartheta, D, Q, I\rangle$ with $f_I = \{f_{I,w} : D_w^n \to D_w\}_{w \in W}$ for $f \in F_n$, and $p_I = \{p_{I,w}\}_{w \in W}$ with $p_{I,w} \subseteq D_w^n$ for $p \in P_n$, by letting:

- $f_\mathcal{A}(\langle e_i\rangle)(w, \mu) = f_{I,w}(\langle e_i(w, \mu)\rangle)$, $p_\mathcal{A}(\langle e_i\rangle) = \{\langle w, \mu\rangle : \langle e_i(w, \mu)\rangle \in p_{I,w}\}$.

However, after changing the provisos, the converse Barcan formula is no longer deducible (see [13]). According to the completeness results, the class of fibred models must now contain structures falsifying it. The structures of [22] are general enough to provide such counterexamples. Consider $\langle W, R, \vartheta, D, I\rangle$ where $R = \{R_\mu\}_{\mu \in D^X}$ with each $R_\mu \subseteq W^2$, $f_I = \{f_{I,w} : D^n \to D\}_{w \in W}$ for $f \in F_n$, and $p_I = \{p_{I,w}\}_{w \in W}$ with $p_{I,w} \subseteq D^n$ for $p \in P_n$. Letting $U = W \times \text{Asg}(X, D)$, and for each $b \subseteq U$, $b_w = \{w : \langle w, \mu \rangle \in b \text{ for some } \mu\}$ and $b_\mu = \{\mu : \langle w, \mu \rangle \in b \text{ for some } w\}$ we define $\langle \mathcal{A}, \mathbf{c} \rangle$ by:

- $|\mathcal{A}|_\tau = D^U$, $|\mathcal{A}|_\phi = \wp(U)$; $x_\mathcal{A}(w, \mu) = \mu(x)$, $f_\mathcal{A}(\langle e_i \rangle)(w, \mu) = f_{I,w}(\langle e_i(w, \mu) \rangle)$, $q_\mathcal{A} = \vartheta(q) \times \mathrm{Asg}(X, D)$, $p_\mathcal{A}(\langle e_i \rangle) = \{\langle w, \mu \rangle : \langle e_i(w, \mu) \rangle \in p_{I,w}\}$, $\forall x_\mathcal{A}(b) = \bigcup_{w \in b_w} \{\langle w, \mu \rangle : \mu[x/d] \in b_\mu \text{ for every } d \in D\}$, $\square_\mathcal{A}(b) = \bigcup_{\mu \in b_\mu} \{\langle w, \mu \rangle : \{w' : wR_\mu w'\} \subseteq b_w\}$.

Of course, these are just examples of structures obtained in the fibred system. Due to fullness, it should contain many others.

5   Conclusion

We have extended the restricted propositional-based setting of [6] to the fibring of logics also admitting variables, terms and quantifiers. Along with the semantic dimension provided by c-parchments, we have adopted an improved notion of Hilbert-style calculus with explicit control of schema rule instantiations, following [7, 22]. Besides a detailed account of modal first-order logic as a fibring, we have also reused the technique of fullness from [25] to provide a smooth generalization of the completeness preservation results of [6] to this more general context. The techniques used include congruences and Lindenbaum-Tarski algebras, together with assumptions on the existence of suitable logic constructors.

A word is due on the relationship between our work and Pawlowski's [19]. Indeed, the context information provided by his inference systems presentations can be seen as an alternative way to achieve the same kind of control over schema rule instantiations, leading to a setting where schema variable substitutions are restricted accordingly. Pawlowski's approach is certainly more abstract and systematic, namely in the sense that it tends to treat logical variables and schema variables in a uniform way. However, his claim that, thanks to context information, he can express and manipulate inference rules "without referring to binding operators or requirements" is a little misleading. Context information is certainly present in his framework from the very beginning, but his inference rules are still decorated with additional relevant context information regarding schema variables. This information is strongly related to our provisos. Of course, the provisos seem to be more complex since they have to carry over on signature changes. However, this is due to the fact that all the context information is placed exactly where we really need it: the inference rules. Moreover, note that our provisos are sufficiently general to take into account the scope of modal operators (c.f., Example 9). Binding operators like modalities, very different in nature from quantifiers (namely on the absence of any explicit reference to logical variables), seem to be a good challenge to Pawlowski's notion of context, which is directly built around variables. Last but not least, the incomplete deductive system for first-order logic with equality that he obtains by combining first-order logic and equational logic is certainly to be expected and does not contradict the completeness preservation results presented herein. Note that, on the one hand, equational logic does not come with an implication connective (thus barring the application of Theorems 3 to 8), and on the other hand, fullness would require considering also semantic structures for equational logic whose interpretation of equality would not be the standard (also ruling out Theorem 2). In fact, we could as well have mentioned this example to motivate the difficulties involved in preserving completeness, and to stress the importance of obtaining non-trivial sufficient conditions for completeness preservation as the ones we have presented.

Despite all the results obtained so far, the challenge of combining logics is still far from over. Regarding fibring, specifically, one interesting line of research to pursue is a comprehensive comparison to Diaconescu's Grothendieck institutions [8]. Another important subject that needs further investigation is the *collapsing problem*. In this paper we have avoided the problem by making a careful use of fullness. However, in general, fibring logics of very distinct nature can give rise to trivialities. In [23] *modulated fibring* was presented as a first solution to this problem, using adjunctions between orders on truth-values. Work already in progress aims at solving the same problem using simpler machinery, via the novel notion of *cryptofibring*. Other interesting lines of research require a deep understanding of the process of algebraization of logics, putting in context the notion of fullness and the role that it plays in the completeness results, bringing us closer to the rich field of algebraic logic [4, 1]. We are also interested in studying the representation of fibring in logical frameworks, by capitalizing on the theory of general logics [15]. Finally, future work must also cover transfer results for other relevant properties, like decidability, complexity or interpolation.

References

1. H. Andréka, I. Németi, and I. Sain. Algebraic logic. In *Handbook of Philosophical Logic, 2nd Edition*, volume 2. Kluwer Academic Publishers, 2001.
2. G. Birkhoff. *Lattice Theory*. AMS Colloquium Publications, 1967.
3. P. Blackburn and M. Rijke. Why combine logics? *Studia Logica*, 59(1):5–27, 1997.
4. W. Blok and D. Pigozzi. *Algebraizable Logics*, volume 77. Memoires AMS, 1989.
5. C. Caleiro. *Combining Logics*. PhD thesis, Universidade Técnica de Lisboa, 2000.
6. C. Caleiro, P. Mateus, J. Ramos, and A. Sernadas. Combining logics: Parchments revisited. In *Recent Trends in Algebraic Development Techniques*, volume 2267 of *LNCS*, pages 48–70. Springer, 2001.
7. M.E. Coniglio, A. Sernadas, and C. Sernadas. Fibring logics with topos semantics. *Journal of Logic and Computation*, in print.
8. R. Diaconescu. Grothendieck institutions. *App. Cat. Str.*, 10(4):383–402, 2002.
9. D. Gabbay. Fibred semantics and the weaving of logics: part 1. *Journal of Symbolic Logic*, 61(4):1057–1120, 1996.
10. J. Goguen and R. Burstall. A study in the foundations of programming methodology: specifications, institutions, charters and parchments. In *Category Theory and Computer Programming*, volume 240 of *LNCS*, pages 313–333. Springer, 1986.
11. J. Goguen and R. Burstall. Institutions: abstract model theory for specification and programming. *Journal of the ACM*, 39(1):95–146, 1992.
12. G. Hughes and M. Cresswell. *A New Introduction to Modal Logic*. Routledge, 1996.
13. M. Kracht and O. Kutz. The semantics of modal predicate logic I: Counter-part frames. In *Advances in Modal Logic, Volume 3*. CSLI, 2002.

14. E. Mendelson. *Introduction to Mathematical Logic.* D. van Nostrand, 1979.
15. J. Meseguer. General logics. In *Proceedings of the Logic Colloquium'87*, pages 275–329. North-Holland, 1989.
16. T. Mossakowski. Using limits of parchments to systematically construct institutions of partial algebras. In *Recent Trends in Data Type Specification*, volume 1130 of *LNCS*, pages 379–393. Springer, 1996.
17. T. Mossakowski, A. Tarlecki, and W. Pawłowski. Combining and representing logical systems. In *Category Theory and Computer Science 97*, volume 1290 of *LNCS*, pages 177–196. Springer, 1997.
18. T. Mossakowski, A. Tarlecki, and W. Pawłowski. Combining and representing logical systems using model-theoretic parchments. In *Recent Trends in Algebraic Development Techniques*, volume 1376 of *LNCS*, pages 349–364. Springer, 1998.
19. W. Pawłowski. Presenting and combining inference systems. In *Recent Trends in Algebraic Development Techniques*, LNCS. Springer, this volume.
20. A. Sernadas, C. Sernadas, and C. Caleiro. Fibring of logics as a categorial construction. *Journal of Logic and Computation*, 9(2):149–179, 1999.
21. A. Sernadas, C. Sernadas, C. Caleiro, and T. Mossakowski. Categorial fibring of logics with terms and binding operators. In *Frontiers of Combining Systems 2*, pages 295–316. Research Studies Press, 2000.
22. A. Sernadas, C. Sernadas, and A. Zanardo. Fibring modal first-order logics: Completeness preservation. *Logic Journal of the IGPL*, 10(4):413–451, 2002.
23. C. Sernadas, J. Rasga, and W.A. Carnielli. Modulated fibring and the collapsing problem. *Journal of Symbolic Logic*, 67(4):1541–1569, 2002.
24. A. Tarlecki. Moving between logical systems. In *Recent Trends in Data Type Specification*, volume 1130 of *LNCS*, pages 478–502. Springer, 1996.
25. A. Zanardo, A. Sernadas, and C. Sernadas. Fibring: Completeness preservation. *Journal of Symbolic Logic*, 66(1):414–439, 2001.

Use of Patterns in Formal Development: Systematic Transition from Problems to Architectural Designs

Christine Choppy[1] and Maritta Heisel[2]

[1] LIPN, Institut Galilée – Université Paris XIII, France
Christine.Choppy@lipn.univ-paris13.fr
[2] Otto-von-Guericke-Universität Magdeburg, Fakultät für Informatik,
Institut für Verteilte Systeme, D-39016 Magdeburg, Germany
heisel@cs.uni-magdeburg.de

Abstract. We present a pattern-based software lifecycle and a method that supports the systematic execution of that lifecycle. First, *problem frames* are used to develop a formal specification of the problem to be solved. In a second phase, *architectural styles* are used to construct an architectural specification of the software system to be developed. That specification forms the basis for fine-grained design and implementation.

1 Elaborating the Software Development Process

Experience has shown that problems and bugs in software systems take their source mainly in the early phases of the software development process[1]. Hence, a software development lifecycle that derives the design of the software directly from the requirements and then passes on to the implementation cannot be regarded as satisfactory. The step between requirements and design is too large.

An additional phase should be introduced between the requirements and the design. One idea that has been accepted for some time now is that some kind of *specification* should be set up on the basis of the requirements, so that the requirements are transformed into documents useful for developers. Specifications lead to a deeper understanding of the problems to be solved, and they can be used to support other development activities (e.g. coding, testing, maintenance). However, producing appropriate specifications often turns out to be difficult for practitioners. For instance, finding an appropriate starting point for the formal specification process is a very common problem.

M. Jackson [Jac95,Jac01] proposes the use of *problem frames* for presenting and understanding software development problems. A problem frame is a characterization of a class of problems in terms of their main components and the connections between these components. A set of typical solution methods is associated to each problem frame. The basic idea is that once an appropriate problem frame for a given problem is found, we also have good proposals for constructing a solution to that problem. We think this idea

---

[1] See for example http://www.standishgroup.com/sample_research

M. Wirsing, D. Pattinson, and R. Hennicker (Eds.): WADT 2002, LNCS 2755, pp. 201–215, 2003.
© Springer-Verlag Berlin Heidelberg 2003

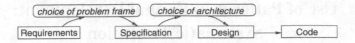

Fig. 1. Lifecycle using problem frames and architectures

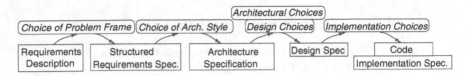

Fig. 2. Complete lifecycle using problem frames and architectural styles

is useful, but it provides only a coarse structure of the problem. Hence, problem frames should be supplemented by means that allow for a finer structuring.

Architectural styles [SG96,BCK98] are a means to structure a software system, i.e. to choose its architecture. Since architectural styles are used to construct designs, they should not be used right at the beginning of the development process, but only after the problem has been fully understood and specified. Figure 1 shows how to bridge the gap between the requirements and the design of a software system. It is possible to elaborate the software development lifecyle further, as suggested in Figure 2. Here, several phases are introduced between the requirements and the design of a software system.

Problem frames and architectural styles are both forms of *patterns*. While problem frames are concerned with *problems*, architectural styles are concerned with *solutions*. Hence, with Figures 1 and 2, we propose a pattern-based software lifecycle. Patterns should be used systematically and on different levels of abstraction.

In the following, we show how the steps from an informal requirements description to an architectural specification shown in Figure 2 can be carried out in a systematic way. This work further elaborates the approach by Choppy and Reggio [CR00], where problem frames are used to structure formal specifications.

We first discuss how patterns can be used on different abstraction levels and in different phases of the software development process in Section 2. Section 3 presents a method to carry out pattern based formal development in a systematic way. The application of that method is illustrated by the case study of a robot simulation in Section 4. In Section 5, we summarize our work and also discuss related work that aims at methodological support for developing formal specifications.

2 Patterns for Different Software Development Activities

Patterns are a means to reuse software development knowledge on different levels of abstraction. Patterns classify sets of software development problems or solutions that share the same structure.

Patterns have been introduced on the level of detailed object oriented design [GHJV95]. Today, patterns are defined for different activities. *Problem Frames* [Jac01] are patterns that classify software development *problems*. *Architectural styles* are pat-

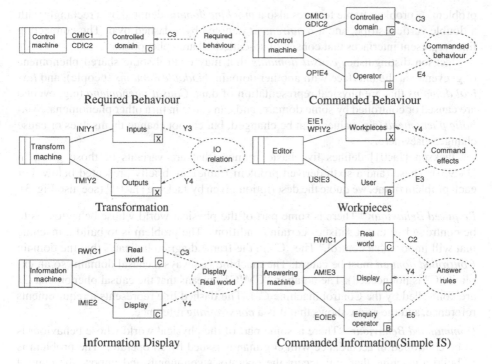

Fig. 3. Frame diagrams

terns that characterize software architectures [SG96,BCK98]. They are sometimes called "architectural patterns". *Design Patterns* are referred to as "micro-architectures", while *frameworks* are considered as less abstract, more specialized. Finally, *idioms* are low-level patterns related to specific programming languages [BMR+96], and are sometimes called "code patterns".

Using patterns, we can hope to construct software in a systematic way, making use of a body of accumulated knowledge, instead of starting from scratch each time. In the following, we briefly introduce problem frames and architectural styles, which will be used in our method.

2.1 Problem Frames

Jackson [Jac01] describes problem frames as follows:

> A problem frame is a kind of pattern. It defines an intuitively identifiable problem class in terms of its context and the characteristics of its domains, interfaces and requirement.

For each problem frame, a *frame diagram* is set up (cf. Figure 3), which contains the different parts involved. Plain rectangles denote *application domains*. The characteristics of these domains play an important role in the application of a problem frame to a problem. A problem frame features also a *machine domain* denoted by a rectangle with a double vertical stripe, and a *requirement* denoted by a dashed oval. The connecting lines represent interfaces that consist of so-called "shared phenomena".

Jackson distinguishes *causal domains* that may control some shared phenomena (e.g. events) at the interface with another domain, *biddable domains* (people), and *lexical domains* that are physical representation of data. *Causal phenomena* (e.g. events) are caused or controlled by some domain, and can cause in turn other phenomena. *Symbolic phenomena* (e.g. values) can be changed, but cannot change themselves or cause changes elsewhere.

Jackson [Jac01] defines five basic frames (that are variants of those given in [Jac95]). These (and a sixth derived problem frame) are briefly presented below. For each problem frame, we quote the description given by Jackson [Jac01] (see also Fig. 3).

Required Behaviour "There is some part of the physical world whose behaviour is to be controlled so that it satisfies certain conditions. The problem is to build a machine that will impose that control." The "C" in the frame diagram indicates that the domain *Controlled domain* must be causal. The machine is always a causal domain (so an explicit "C" is not needed). The notation "CM!C1" means that the causal phenomena *C1* are controlled by the Control machine *CM*. The dashed line represents a requirements reference, and the arrow shows that it is a *constraining* reference.

Commanded Behaviour "There is some part of the physical world whose behaviour is to be controlled in accordance with commands issued by an operator. The problem is to build a machine that will accept the operator's commands and impose the control accordingly." The "B" indicates that the domain *Operator* is a biddable domain, and the phenomena *E4* are the operator commands.

Transformation "There are some computer-readable input files whose data must be transformed to give certain required output files. The output data must be in a particular format, and it must be derived from the input data according to certain rules. The problem is to build a machine that will produce the required outputs from the inputs." The "X" indicates that *Inputs* and *Outputs* are lexical (inert) domains.

Workpieces "A tool is needed to allow a user to create and edit a certain class of computer processable text or graphic objects, or similar structures, so that they can be subsequently copied, printed, analysed or used in other ways. The problem is to build a machine that can act as this tool."

Information Display "There is some part of the physical world whose states and behaviour is continually needed. The problem is to build a machine that will obtain this information from the world and present it at the required place in the required form." Here, the purpose of the machine is to display things that happen in the real world. Both domains are causal. *Y4* are symbolic requirement phenomena.

Commanded Information is derived from the Simple IS frame [Jac95]. There is some part of the physical world whose states and behavior are needed upon requests from an operator. The problem is to build a machine that will obtain this information from the world and present it at the required place in the required form.

Let us note that these problem frames do not cover every conceivable problem class. Some more problem frames have been identified by Souquières and Heisel [SH00].

2.2   Architectural Styles

According to Bass, Clements, and Kazman [BCK98],

the software architecture of a program or computing system is the structure or structures of the system, which comprise software components, the externally visible properties of those components, and the relationships among them.

Architectural styles are patterns for software architectures. A style is characterized by [BCK98]:

- a set of component types (e.g., data repository, process, procedure) that perform some function at runtime,
- a topological layout of these components indicating their runtime interrelationships,
- a set of semantic constraints (for example, a data repository is not allowed to change the values stored in it),
- a set of connectors (e.g., subroutine call, remote procedure call, data streams, sockets) that mediate communication, coordination, or cooperation among components.

Important architectural styles are the following:

- **Data-Centered** with substyles *Repository* and *Blackbord*
- **Data Flow** with substyles *Batch Sequential* and *Pipe-and-Filter*
- **Virtual Machine** with substyles *Interpreter* and *Rule-Based Systems*
- **Call-and-Return** with substyles *Main Program and Subroutine*, *Layered*, *Object-Oriented* or *Abstract Data Types*
- **Independent Components** with substyles *Communicating Processes* and *Event Systems* (implicit/explicit invocation)

When choosing an architecture for a system, usually several architectural styles are possible, which means that all of them could be used to implement the functional requirements. Which architectural style is the most appropriate must then be decided using *non-functional* criteria such as efficiency, scalability, or modifiability. How such a choice is made is illustrated in Section 4.

2.3   Design Patterns

Design patterns [GHJV95] are used on a lower level of abstraction than problem frames or architectural styles. They provide concrete means to combine objects, or classes, respectively. In our overall software lifecyle, they would be used after an architectural style has been chosen. This step is beyond the scope of this paper.

3   An Agenda for Pattern-Based Specification and Design

We now present our method for carrying out a pattern-based software lifecycle as shown in Figures 1 and 2. As a means of presentation, we use the *agenda* concept [Hei98]. An agenda is a list of steps or phases to be performed when carrying out some task in the

Table 1. Agenda for pattern-based specification

No.	Description	Result	Validation
1.	Fit the problem into an appropriate problem frame.	Instantiated frame diagram	All important issues of the problem must be treated adequately, see also [Jac01].
2.	Set up a formal specification for each domain of the instantiated frame diagram (including the machine domain) and the requirements.	Set of formal specifications	– The specification must be coherent with the instantiated problem frame diagram. – The shared phenomena must belong to the interfaces of all domains where they are visible. – Control of phenomena must be taken into account. – The specification S of the machine domain (in combination with the domain knowledge D) must suffice to satisfy the requirements R, i.e., $S \wedge D \rightarrow R$ must hold.
3.	Choose an appropriate architectural style for structuring the machine domain and instantiate it.	Architectural diagram and informal text	The chosen architecture must be able to satisfy the machine specification.
4.	Set up a formal specification of all components obtained in Step 3 and of the overall system (i.e., specify how the components cooperate).	Set of formal specifications	– The formal specification must correspond to the architectural diagram. – The overall specification must be a *refinement* of the machine specification developed in Step 2. – The constraints imposed by the chosen architectural style must be satisfied.

context of software engineering. The result of the task will be a document expressed in some language. Agendas contain informal descriptions of the steps, which may depend on each other. Agendas are not only a means to guide software development activities. They also support quality assurance, because the steps may have validation conditions associated with them. These validation conditions state necessary semantic conditions that the developed artifact must fulfill in order to serve its purpose properly.

Table 1 shows an agenda that precisely describes how to carry out and validate the first steps of the lifecycle proposed in Figure 2. A precondition for the applicability of the agenda is that the problem is sufficiently small that it may be fitted into one problem frame. Complex problems have to be decomposed first, for example by projection, as described by [Jac01].

Step 1 of the agenda is performed in principle as described by Jackson [Jac01]. To find the right problem frame, the structure of the frame diagram and the domain characteristics as described in Section 2.1 must be taken into account. However, this is not as straightforward as it might seem, because we first need to choose between possibly different viewpoints on the problem. For instance, the choice of taking into account a user/operator influences the choice of problem frame, and it also changes the

Table 2. Problem frames and related architectural styles

Problem Frame	Architectural Style
Required Behaviour Commanded Behaviour	Communicating Processes, Event/Action, Process Control
Transformation Workpieces	Repository, Batch Sequential, Pipe and Filter, Virtual Machine, Layered, ADT/OO, Event Systems
Information Display Commanded Information	Repository, Blackboard

characteristics of the domains and phenomena. We think it is worthwhile to examine for each problem frame whether we find a meaningful instantiation of it or else a clear reason why not.

Once the choice of a problem frame is made, we rely on the structure provided by the problem frame to proceed and establish a corresponding formal specification [CR00].

Step 2 uses the instantiated frame diagram from Step 1 that determines the structure of the formal specification to be set up. For each box in the instantiated frame diagram, a specification must be given. The validation condition "coherence of the instantiated frame diagram and specification" means that the phenomena at the interfaces of the requirements box must be used in expressing the requirements. Moreover, the shared phenomena that are given in the instantiated problem frame must belong to the interfaces of the respective domain specifications. A domain which is in control of a shared phenomenon must be able to produce that phenomenon as an output, and a domain which is able to observe a phenomenon of which it is not in control must be able to take the phenomenon as an input. The domain knowledge D mentioned in the last validation condition of Step 2 refers to the specification of the application domains, i.e. the domains of the instantiated problem frame other than the machine domain.

Step 3 uses the specification of the machine domain developed in Step 2. This specification describes the machine to be developed, whose structure will be determined by the architectural style. Several possible architectural styles should be explored and assessed according to those non-functional criteria that are regarded to be important for the given problem.

Table 2 gives heuristics for performing Step 3. It has been developed from the general characteristics of the involved problem frames and architectural styles as well as by conducting several case studies. It shows rules of thumb giving hints which architectural styles to consider first.

As can be seen, there are several architectural styles associated to each problem frame. Which one is finally chosen depends on *non-functional* requirements. It remains to make these explicit in order to really guide the transition from a problem frame to an architectural style.

For the problem frames Transformation and Workpieces, we have quite a number of architectural styles to consider. This is due to the fact that these problem frames cover most of the "classical" software development problems and that they are less constraining than the other frames. For Required Behaviour and Commanded Behaviour, we

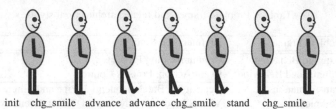

init   chg_smile   advance   advance   chg_smile   stand   chg_smile

Fig. 4. The movements of the robot

should consider architectural styles that are well suited for reactive systems, and for Information Display and Commanded Information, it seems natural to choose data-centered architectures.

Step 4 uses the architectural style instantiated in Step 3 to develop a specification that formally describes the chosen architecture. That instantiation determines a set of components that structure the system to be developed. It also shows the cooperations between these components. For each component a formal specification must be given. Furthermore, it must be specified how the components cooperate. Such an architectural specification is the basis for detailed design and implementation. The most important validation condition associated with Step 4 of the agenda is to show that the chosen architecture indeed correctly implements the machine specified in Step 2, i.e., that the architectural specification refines the machine specification. Because we use formal specifications, this validation condition can be demonstrated in a rigorous or formal way.

In the following section, we demonstrate the application of the agenda by means of a concrete example.

4   Case Study: Robot Simulation

This case study is taken from [HL97], where it was used to illustrate different architectural styles. Here, we demonstrate how the most suitable architectural style can be found in a systematic manner, performing the steps of the agenda presented in Table 1.

The task is to build a system simulating a simple robot. This robot can make the movements shown in Figure 4: it can advance by moving its right or its left leg; it can stand still; and it can smile or not. The robot can be modeled as an automaton with three states: standing, left_up and right_up as shown in Figure 5. To each state a boolean value is associated indicating whether the robot is smiling or not. The initial state is standing and smiling.

The robot is defined by the abstract data type ROBOT where the states are defined as constants and the movements as transitions from one state to another, except for smiling, which is defined by a boolean value: true for smiling. For each state a predicate is defined deciding if the robot is in this state.

The input for the system to be built is a list of commands to be executed by the robot, i.e., a list consisting of the elements stand, advance and chg_smile. The output is a list of pairs, where the first component of each pair is the current state of the

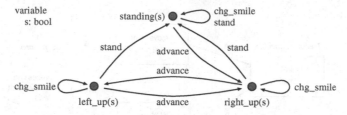

Fig. 5. The robot automaton

robot, and the second component of each pair is the list of commands not yet executed. Each command must be executed, and the intermediate states entered during execution of the command list must be given as an output.

Step 1: Choice and Instantiation of a Problem Frame. We consider the problem frames (cf. Figure 3) one by one and give reasons for each problem frame why it is rejected or accepted.

Required Behaviour The "C" says that the *Controlled Domain* must be causal. Since the robot is defined by an abstract data type, the domain corresponding to the robot is not causal but lexical. Moreover, the problem frame Required Behaviour does not let us distinguish between the input domain (being a list of commands) and the output domain (being a list of pairs). Hence, we reject this problem frame.

Commanded Behaviour This problem frame must be rejected for the same reasons as before. Moreover, we cannot find a domain corresponding to the *Operator* domain.

Information Display Here, the purpose of the machine is to display things that happen in the real world. Both domains are causal, which does not fit well with the robot problem.

Commanded Information This frame must be rejected, because we cannot find an *Enquiry operator* and because the domains involved in the robot problem are not causal.

Workpieces This problem frame is more promising than the ones considered before, because we have a lexical domain here. The workpieces are the robot's state, together with the current command list. However, we cannot find an instantiation for the *User* domain, because command lists are not biddable. Hence, we finally reject the Workpieces frame.

Transformation It is this frame that we finally choose for our problem. A lexical input list is transformed into a lexical output list. The relation between the two lists is given by the robot automaton. Figure 6 shows the instantiated frame diagram.

Step 2: Structured Requirements Specification. Having chosen a problem frame for the robot problem, we must give a specification of all the domains involved and of the requirements. As a specification language, we use LOTOS [BB87], because LOTOS is one of the specification languages allowing us to define software architectures, and especially the interaction of different components, in a suitable way.

Specification of the *Inputs* **Domain.** As shown in Figure 6, the input domain is a list of movement commands.

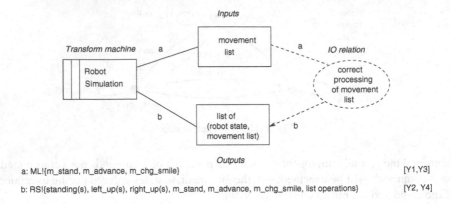

Fig. 6. The robot problem fitted into the Transformation problem frame

The movements are defined by the type `MVT` with three constants `m_stand`, `m_advance` and `m_chg_smile`.

The robot will be asked to execute several movements collected in a list. This list is defined by an abstract data type `M_LIST` whose definition is straightforward.

Specification of the *Outputs* **Domain.** The output consists of a list of pairs, whose first element is the current state of the robot and whose second element is the list of movements yet to be performed.

The definition of the abstract data type `ROBOT` reflects exactly the automaton given in Figure 5.

To define the *Outputs* domain `O_LIST`, a data type `VALUE` must be defined as the Cartesian product (with constructor `make`) of the two types `ROBOT` and `M_LIST`. The type `O_LIST` of lists of elements of type `VALUE` is then defined in much the same way as the type `M_LIST`.

Specification of the *IO relation.* The IO relation says that, given a list of commands, the robot simulation must execute that list of commands one by one and output the current state of the robot after execution of each command, together with the commands yet to be executed.

For example, if the input command list has the form $(m_1, m_2, m_3, \ldots m_k)$ then the output list has the form

$$((m_1(init\ of\ robot), (m_2, m_3, \ldots m_k)), (m_2(m_1(init\ of\ robot)), (m_3, \ldots m_k)),$$
$$\ldots, (m_k(\ldots (m_3(m_2(m_1(init\ of\ robot)))) \ldots), empty))$$

where $m(r)$ denotes the robot state that is reached from state r by executing movement m. This requirement is defined by a predicate `is_correct` which takes a movement list and and output list as its arguments. This predicate is defined in a type `IO_REL`.

Specification of the Machine Domain *Robot Simulation.* For each input list, the robot simulation must produce an output list in such a way that the two lists are in the relation `is_correct`.

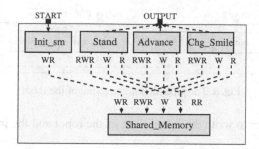

Fig. 7. The repository architecture for the robot

```
type ROBOT_SIMULATION
is   IO_REL
     opns  robsim : m_list  -> o_list
     eqns
           forall ml : m_list
           ofsort bool
             is_correct(ml, robsim(ml)) = true
endtype
```

Steps 3 and 4: Architectural Design of the Robot. We will explore several possibilities to structure the machine domain specified in Step 2. The non-functional criteria for assessing the different architectures will be efficiency and simplicity. Moreover, we give a specification of the top-level behavior for each considered architecture. For reasons of space, we cannot give the specifications of the different components.

All architectures we will consider in the following have the same interface. This interface consists of an input channel START and an output channel OUTPUT, where START corresponds to interface *a* and OUTPUT corresponds to interface *b* of Figure 6.

The list of movements to be processed is given in one step. The simulation must show the intermediate states of the robot when processing the input list. Hence, instead of producing the output list at once, the machine will produce the elements of the output list one by one. Then, the correctness condition required to be proven in Step 4 of the agenda is that the sequence of events occurring on gate OUTPUT is an output list that is in relation IO_rel with the input list.

The gate START is used to start the simulation, yielding in the following top-level behavior:

```
         START  !make(init of robot,input_list); exit
| [START] | (behav_expr)
```

The different architectures will result in different definitions of *behav_expr*.

The Repository Architecture. The basic idea is to use a repository that contains the current state of the robot and the list of commands still to be executed. There are three components, one for each command. These components change the state according to the automaton and discard the first element of the command list.

Figure 7 illustrates the repository architecture, where channel names R, W and RW denote the read, write and read/write access to the repository, respectively. The compo-

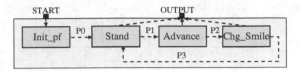

Fig. 8. The pipe/filter architecture of the robot

nent *Init_sm* serves to write the initial state of the robot and the initial command list
into the repository.

The components try to access the shared memory in parallel in order to execute the
movement they are responsible for. Each of them first reads the list of movements. If
the first movement is the one it is responsible for, the movement is executed, the robot
state changed, and the new state and the rest of the movement list is written back in the
shared memory. If the movement cannot be executed by the component that has been
granted access, it writes back the unchanged state in order to unlock the shared memory.
The top-level behavior of this architecture is as follows:

```
START  !make(init of robot,init_list); exit
|[START]|
(
hide   RR, R, WR, W, RWR in
    SM [RR, R, WR, W, RWR]
            (init of shared_memory,false,for_nobody)
    |[ RR, R, WR, W, RWR ]|
    (     Init_sm [START, W, WR]
       ||| Stand [OUTPUT, R, W, RWR]
       ||| Chg_Smile [OUTPUT, R, W, RWR]
       ||| Advance [OUTPUT, R, W, RWR] ))
```

This architecture has the disadvantage that the system implementation must guarantee
fairness, i.e. each component must be given the chance to access the shared memory.
Otherwise, an infinite number of unsuccessful accesses is possible, and the system does
not terminate (*live-lock*).

The Pipe-and-Filter Architecture. In the pipe/filter modeling, we can make sure that
each component is given the possibility to execute its movement if required. The idea
is to have a line of filters. Each filter inspects the movement list. If it can execute the
movement, it does so and hands the new robot state and the new movement list to the
next filter. Otherwise, it passes on the unchanged data. Again, we need an initializing
component, called here Init_pf. The architecture is shown in Figure 8. The top-level
behavior of this architecture is as follows:

```
hide P0, P1, P2, P3 in
    ( Init_pf [START, P0]
          |[ P0 ]|
      Stand [P0, P1, P3, OUTPUT]
          |[ P1, P3 ]|
      Advance [P1, P2, OUTPUT]
          |[ P2 ]|
      Chg_Smile [P2, P3, OUTPUT]    )
```

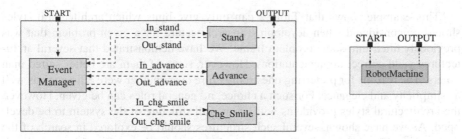

Fig. 9. The event system and the virtual machine architectures for the robot

This solution is better than the repository architecture because it always terminates. It is not ideal, however, because each component must inspect the data, even if it cannot process them.

The Event System Architecture. The event system style can be used to overcome the disadvantages of the previous two architectures. An event manager inspects the movement list and passes the data only to the component that can process them. The initial state of the robot and the movement list are given to the event manager. No initialization component is required. This architecture is shown on the left-hand side of Figure 9. We have the following overall behavior:

```
hide In_stand, Out_stand, In_chg_smile, Out_chg_smile,
    In_advance, Out_advance in
    Event_Manager [START, In_stand, Out_stand, In_chg_smile,
                   Out_chg_smile, In_advance, Out_advance]
        |[In_stand, Out_stand, In_chg_smile, Out_chg_smile,
          In_advance, Out_advance]|
    ( Stand [OUTPUT, In_stand, Out_stand]
        |||
      Advance [OUTPUT, In_advance, Out_advance]
        |||
      Chg_Smile [OUTPUT, In_chg_smile, Out_chg_smile] )
```

The components executing the movements are much simpler now than in the other architectures.

The Virtual Machine Architecture. The architecture can be improved once more. We should not have three components that can only execute a single command, but a virtual machine that can execute all three commands. This architectural style seems to be the most natural one, because virtual machines are well suited for simulation tasks. This architecture is shown on the right-hand side of Figure 9. It is quite simple:

```
process Robot [START, OUTPUT] : exit :=
      START ? v: value; RobotMachine[OUTPUT](v)
endproc
```

where the process RobotMachine just recursively processes the given movement list contained in v.

This example shows that Table 2 can only give hints which architectural styles should be considered when developing an architecture for a given problem that was previously fitted into some problem frame. We have demonstrated that several architectures yield correct implementations. However, some of them are better suited than others. The reasons for preferring one architecture over another were efficiency as well as simplicity and elegance. For such a choice, no general rules can be given. However, the architectural styles provide us with an overall structure of the system to be developed. As we have shown, several such structures should be explored in search of the optimal one. The structure finally chosen is the starting point of the subsequent development steps.

For further validation of our approach, we have also carried out other case studies using CASL [CH03].

5   Conclusions

Methodological issues in writing specifications are many, and we would like to point to related work that addresses issues complementary to ours. Roggenbach and Mossakowski [RM02] address the writing of readable specifications in CASL, avoiding semantic pitfalls (these concerns are also addressed in the CASL reference manual [BM02]). Bidoit, Hennicker and Kurz [BHK02] explore the use of observability concepts which are found to be useful and relevant for writing specifications. Blanc [Bla02] proposes guidelines for the iterative and incremental development of specifications.

In this paper, we have introduced a methodology for formal specification that is systematic and that stresses reuse of previously acquired knowledge. Both patterns and agendas are a means to represent knowledge. Patterns are abstractions of the *products* developed during the software lifecycle, and reuse is achieved by instantiating a pattern. Agendas, on the other hand, are explicit representations of process knowledge. Both concepts are orthogonal, and in order to base the software development process as much as possible on previously acquired knowledge, the two concepts should be used in combination. In particular, the contributions of this paper are:

- We have elaborated a software lifecycle where patterns play an important and well-defined role.
- We have developed an agenda that gives guidance how to perform this pattern-based software lifecycle in a systematic way.
- We have shown how to combine problem frames, architectural styles and formal specifications. So far, these three were considered in isolation; no explicit connection between them has previously been established.

In the future we will provide methodological support also for the subsequent development steps of the software lifecycle proposed in Figure 2. In particular, this will involve the application of design patterns. Furthermore, we will investigate problem decomposition and multiframe problems in more detail.

Acknowledgments

We thank Thomas Santen, Carsten von Schwichow and an anonymous referee for their helpful comments on this paper.

References

[BB87]   T. Bolognesi and E. Brinksma. Introduction to the ISO specification language LO-TOS. *Computer Networks and ISDN Systems, North-Holland*, 14:25–59, 1987.

[BCK98]   L. Bass, P. Clements, and R. Kazman. *Software Architecture in Practice*. Addison-Wesley, 1998.

[BHK02]   M. Bidoit, R. Hennicker, and A. Kurz. On the Integration of Observability and Reachability Concepts. In *Proc. 5th Int. Conf. Foundations of Software Science and Computation Structures (FOSSACS'2002)*, LNCS 2303, pages 21–36. Springer Verlag, 2002.

[Bla02]   B. Blanc. *Prise en compte de principes architecturaux lors de la formalisation des besoins - Proposition d'une extension en CASL et d'un guide méthodologique associé*. Thèse de Doctorat, ENS Cachan, 2002.

[BM02]   M. Bidoit and P. Mosses. *CASL User Manual*, 2002. http://www.brics.dk/Projects/CoFI/.

[BMR+96]   F. Buschmann, R. Meunier, H. Rohnert, P. Sommerlad, and M. Stal. *Pattern-Oriented Software Architecture: A System of Patterns*. John Wiley & Sons, 1996.

[CH03]   C. Choppy and M. Heisel. Systematic transition from problems to architectural designs. Technical report, Université Paris Nord, 2003. To appear.

[CR00]   C. Choppy and G. Reggio. Using CASL to Specify the Requirements and the Design: A Problem Specific Approach. In *Recent Trends in Algebraic Development Techniques*, LNCS 1827, pages 104–123. Springer Verlag, 2000.

[GHJV95]   E. Gamma, R. Helm, R. Johnson, and J. Vlissides. *Design Patterns – Elements of Reusable Object-Oriented Software*. Addison Wesley, Reading, 1995.

[Hei98]   M. Heisel. Agendas – a concept to guide software development activites. In R. N. Horspool, editor, *Proc. Systems Implementation 2000*, pages 19–32. Chapman & Hall London, 1998.

[HL97]   M. Heisel and N. Lévy. Using LOTOS patterns to characterize architectural styles. In M. Bidoit and M. Dauchet, editors, *Proceedings TAPSOFT'97*, LNCS 1214, pages 818–832. Springer-Verlag, 1997.

[Jac95]   M. Jackson. *Software Requirements & Specifications: a Lexicon of Practice, Principles and Prejudices*. Addison-Wesley, 1995.

[Jac01]   M. Jackson. *Problem Frames. Analyzing and structuring software development problems*. Addison-Wesley, 2001.

[RM02]   M. Roggenbach and T. Mossakowski. What is a good CASL specification, 2002. WADT.

[SG96]   M. Shaw and D. Garlan. *Software Architecture. Perspectives on an Emerging Discipline*. Prentice-Hall, 1996.

[SH00]   J. Souquières and M. Heisel. Structuring the first steps of requirements elicitation. Technical Report A00-R-123, LORIA, Nancy, France, 2000.

Conditional Circular Coinductive Rewriting with Case Analysis

Joseph A. Goguen[1], Kai Lin[1], and Grigore Roşu[2]

[1] Department of Computer Science & Engineering
University of California at San Diego, USA
[2] Department of Computer Science
University of Illinois at Urbana-Champaign, USA

Abstract. We argue for an algorithmic approach to behavioral proofs, review the hidden algebra approach, develop circular coinductive rewriting for conditional goals, extend it with case analysis, and give some examples.

1 Introduction

A natural extension of algebraic specification distinguishes *visible* sorts for data from *hidden* sorts for states, with states *behaviorally equivalent* iff they are indistinguishable under a given set of experiments; we have formalized this as *hidden algebra*, originating in [9] and further developed in [11, 20, 18, 19] and other papers. While standard equational proof techniques like induction are suitable for data, *coinduction* or *context induction* is generally needed for non-trivial behavioral properties, typically requiring extensive human intervention. This is not surprising, since behavioral satisfaction is Π_2^0-hard [5], so that no algorithm can prove [or disprove] all behaviorally true [or false] statements. However, successful technology transfer requires placing less demand on users, as illustrated by the success of model checking. Hence our recent research concerns coinduction algorithms that require no human intervention.

The languages we know that support automated behavioral reasoning are Spike [2], CafeOBJ [6], and BOBJ [10], the first based on context induction, and the other two on forms of coinduction. The powerful coinduction algorithm now in BOBJ has developed through several stages. The first restricts ordinary term rewriting to behavioral rewriting [18], by allowing rules to apply only in certain contexts (since standard equational reasoning may be unsound for behavioral satisfaction); this can be used like ordinary rewriting for ordinary equational reasoning, to check behavioral equivalence by computing and comparing behavioral normal forms. Circular coinductive rewriting (CCRW) [10] attempts to prove that a hidden equation holds (in the sense that the two terms cannot be distinguished by any context) by applying behavioral rewriting to both sides, allowing also application of the goal in even more restricted contexts than for rewriting, and generating new goals when rewriting fails to show equivalence; forms of this algorithm have been in BOBJ for more than three years. Conditional circular coinductive rewriting (CCCRW) generalizes CCRW to proving (sets

M. Wirsing, D. Pattinson, and R. Hennicker (Eds.): WADT 2002, LNCS 2755, pp. 216–232, 2003.
© Springer-Verlag Berlin Heidelberg 2003

of) conditional equations (although conditional axioms were already allowed by CCRW, conditional equations could only be proved by implication elimination, which we show is quite limited). Finally, conditional circular coinductive rewriting with case analysis (CCCCRW, or C4RW) adds case analysis, and seems to be the most powerful automated proof technique now available for behavioral equivalence.

This paper only discusses details for a simplified version of the C4RW algorithm, showing its correctness by relating its steps to sound inference rules given in Section 3. Some more sophisticated extensions have been implemented in BOBJ, and are briefly sketched in Section 5, but details are left for future papers. These extensions make the algorithm much more powerful in practice, and are needed, for example, in our recent proofs for the alternating bit protocol and the Petersen mutual exclusion algorithm.

BOBJ's C4RW algorithm takes as input a behavioral specification and a set of hidden sorted conditional equations, and it returns `true`, or `failed` (which may mean that algorithm could not prove the goal, or that the goal is false, depending on the specification), or else goes into an infinite loop. Here is a simple example, illustrating case analysis in a coinductive proof of a conditional equation:

Example 1. <u>Sets with insertion</u> The behavioral theory SET has one hidden sort, Set, one hidden constant for the empty set, and operations for element membership and insertion. The case definition separates the situation where X equals Y from that where it does not; this split is applied only when a subterm of the term being reduced matches the pattern, eq(X,Y). BOBJ allows case definitions to be named, reused, and combined with other such definitions.

```
bth SET is sort Set .
  pr NAT .
  op empty : -> Set .
  op _in_ : Nat Set -> Bool  .
  op insert : Nat Set -> Set .
  vars M N : Nat .  var S : Set .
  eq N in empty = false .
  eq N in insert(M, S) = eq(N,M) or N in S .
end
cases CASES for SET is
  vars X Y : Nat .
  context eq(X,Y) .
  case eq X = Y .
  case eq eq(X,Y) = false .
end
cred with CASES  insert(N, S) == S if N in S .
```

BOBJ's C4RW algorithm is called by the `cred` command; notice that the goal here is a conditional equation. An algorithm of [21] first determines that {in} is a *cobasis* for set-sorted terms, i.e., that two terms are behaviorally equivalent iff they are indistinguishable by experiments with in. Next, the condition of the goal is added to the specification as a new equation, with its variables replaced by new constants (see the Condition Elimination rule in Section 3). Then BOBJ

attempts to prove the goal by reducing each side to its behavioral normal form
and checking for equality; since this fails, the goal is added to the specification
as a *circularity*, that can only be applied in a restricted way; then the cobasis is
applied to each side, and behavioral rewriting produces

    eq(M,n) or M in s = M in s

where n and s are the new constants for N and S, respectively. After that, case
analysis is applied, and since both cases reduce to true, the result is proved.
All this takes just a few milliseconds, before BOBJ returns true. Note that the
circularity is not actually used in this proof (but we soon give an example that
does use a circularity). □

We have found case analysis essential for larger applications, such as our recent
proofs of the alternating bit protocol and the Petersen mutual exclusion algorithm; in addition, we reduced the proof score for liveness of a real-time asynchronous data transmission protocol done in CafeOBJ by Futatsugi and Ogata
[8], by a factor of about ten. Circularities have also been essential for many
non-trivial examples, but here is a simple example, proving an identity that is
familiar in functional programming, and also illustrating BOBJ's parameterized
module capability:

Example 2. iter and map Here DATA defines the interface to STREAM, that is,
we consider streams of elements from an arbitrary data structure having some
monadic operation f defined on its elements. These streams have the usual head
and tail operations, plus _&_, which appends an element to the head of a stream.
Its most interesting operations are map and iter, which respectively apply f to
all the elements of stream, and create a steam of iterations of f applied to its
argument.

```
th DATA is sort Elt .
  op f_ : Elt -> Elt .
end
bth STREAM[X :: DATA] is sort Stream .
  op head_ : Stream -> Elt .
  op tail_ : Stream -> Stream .
  op _&_ : Elt Stream -> Stream .
  op map_ : Stream -> Stream .
  op iter_ : Elt -> Stream .
  var E : Elt . var S : Stream .
  eq head(E & S) = E.
  eq tail(E & S) = S .
  eq head map S = f head S .
  eq tail map S = map tail S .
  eq head iter E = E .
  eq tail iter E = iter f E .
end
cred map iter E == iter f E .
```

The equation to be proved, map iter E = iter f E, often appears in proofs
about streams in functional programs. Pure behavioral rewriting fails to prove
the goal, so circular coinduction is invoked, with the goal added to the specification in a form that limits its application. The cobasis is determined to consist

of head and tail, and so new goals are produced by applying these operations to the original goal. The goal generated by head is directly proved by rewriting, but the goal generated by tail is reduced by behavioral rewriting to

    map iter f E = iter f f E

By applying the circularity at the top level, reduces the left side to iter f f E, which is the same as the behavioral normal form of the right side of the goal. □

2   Specification and Satisfaction

We first briefly review the basics of algebraic specification, and then the version of hidden algebra implemented in BOBJ [10], which drops several common but unnecessary restrictions, including that operations have just one hidden argument, that equations have just one hidden variable, and that all operations preserve behavioral equivalence.

2.1   Preliminaries

The reader is assumed familiar with basic equational logic and algebra. Given an S-sorted signature Σ and an S-indexed set of variables Z, let $T_\Sigma(Z)$ denote the Σ-term algebra over variables in Z. If $V \subseteq S$ then $\Sigma{\upharpoonright}_V$ is a V-sorted signature consisting of all those operations in Σ with sorts entirely in V. We may let $\sigma(X)$ denote the term $\sigma(x_1, ..., x_n)$ when the number of arguments of σ and their order and sorts are not important. If only one argument is important, then to simplify writing we place it at the beginning; for example, $\sigma(t, X)$ is a term having σ as root with only variables as arguments except one, and we do not care which one, which is t. $Der(\Sigma)$ is the derived signature of Σ, which contains all Σ-terms, viewed as operations. If t is a Σ-term and A is a Σ-algebra, then $A_t \colon A^{var(t)} \to A$ is the interpretation of t in A, defined as follows: given $\theta \colon var(t) \to A$, let $A_t(\theta)$ be $\theta(t)$, the evaluation of t in A with variables replaced by the values given by θ. If one variable of t, say $\star$, is of special importance, then we may view the evaluation of t in two steps, as $A_t \colon A \to (A^{(var(t)-\{\star\})} \to A)$ with the obvious meaning.

2.2   Behavioral Specification and Satisfaction

We generalize the hidden algebra of [9, 11] to include variants such as observational logic [1, 3, 14] and coherent hidden algebra [6, 7]. See [19] for a detailed presentation of variants, history, many other concepts, and proofs for some results mentioned here. Two important variants of behavioral logic are the *fixed data* and the *loose data*, depending on whether or not the data universe is fixed (i.e, "built-in"). Due to space limitations, our exposition focuses on the loose data version, but all results also hold for the fixed data version. (However, validity of case analysis often depends on having a suitable fixed data algebra; for example, the above proof for SET requires that sort Bool have the usual two truth values.)

Definition 1. *Given disjoint sets V, H called* **visible** *and* **hidden sorts**, *a* **hidden (V, H)-signature** *is a many sorted $(V \cup H)$-signature. A* **hidden subsignature of Σ** *is a hidden (V, H)-signature Γ with $\Gamma \subseteq \Sigma$ and $\Gamma|_V = \Sigma|_V$. The* **data signature** *is $\Sigma|_V$, which may be denoted Ω. A visible sorted operation not in Ω is called an* **attribute**, *and a hidden sorted operation is called a* **method**.

Unless otherwise stated, the rest of this paper assumes fixed a hidden signature Σ with a fixed subsignature Γ. Informally, Σ-algebras are universes of possible states of a system, i.e., "black boxes," where one is only concerned with behavior under experiments with operations in Γ, where an experiment is an observation of a system attribute after perturbation; this is formalized below, where the symbol $\star$ is a placeholder (i.e., a variable) for the state being experimented upon.

Definition 2. *A* **Γ-context for sort** $h \in H$ *is a term in $T_\Gamma(\{\star : h\} \cup Z)$ with one occurrence of $\star$, where Z is an infinite set of special variables. A* **data context for sort** $v \in V$ *is a term in $T_\Omega(\{\star : v\} \cup Z')$ with one occurrence of $\star$, where Z' is an infinite set of special visible variables. A Γ-context with visible result sort is called a* **Γ-experiment**. *If c is a context for sort h and $t \in T_{\Sigma,h}$ then $c[t]$ denotes the term obtained from c by substituting t for $\star$; we may also write $c[\star]$ for the context itself.*

Definition 3. *Given a hidden Σ-algebra A with a hidden subsignature Γ, for sorts $s \in (V \cup H)$, we define* **Γ-behavioral equivalence** *of $a, a' \in A_s$ by $a \equiv_\Sigma^\Gamma a'$ iff $A_c(a)(\theta) = A_c(a')(\theta)$ for all Γ-experiments c and all $(V \cup H)$-sorted maps $\theta : var(c) \to A$; we may write $\equiv$ instead of $\equiv_\Sigma^\Gamma$ when Σ and Γ can be inferred from context, and we write $\equiv_\Sigma$ when $\Sigma = \Gamma$. Given an $(V \cup H)$-equivalence $\sim$ on A, an operation σ in $\Sigma_{s_1...s_n,s}$ is* **congruent**[1] *for $\sim$ iff $A_\sigma(a_1, ..., a_n) \sim A_\sigma(a'_1, ..., a'_n)$ whenever $a_i \sim a'_i$ for $i = 1...n$. An operation σ is* **Γ-behaviorally congruent for A** *iff it is congruent for $\equiv_\Sigma^\Gamma$. We often write just* **congruent** *for behaviorally congruent (a similar notion is given by Padawitz [17]). A* **hidden Γ-congruence on A** *is a $(V \cup H)$-equivalence on A which is the identity on visible sorts and for which each operation in Γ is congruent.*

Behavioral equivalence is the identity on visible sorts, since the trivial contexts $\star : v$ are experiments for all $v \in V$. The following foundation for coinduction and other important results generalizes [11] to operations not behavioral and/or with more than one hidden argument; [20, 19] give proofs. Since final algebras need not exist in this setting, Γ-behavioral equivalence cannot use them as coalgebra does [22, 16, 15].

Theorem 1. *Given a hidden subsignature Γ of Σ and a hidden Σ-algebra A, then Γ-behavioral equivalence is the largest hidden Γ-congruence on A.*

Behavioral satisfaction of conditional equations can now be naturally defined in terms of behavioral equivalence:

---

[1] This is called "coherent" in [7], where the concept originated.

Definition 4. *A hidden Σ-algebra A Γ-**behaviorally satisfies** a Σ-equation $(\forall X)\, t = t'$ if $t_1 = t'_1, ..., t_n = t'_n$, say e, iff for each $\theta: X \to A$, if $\theta(t_i) \equiv^{\Gamma}_{\Sigma} \theta(t'_i)$ for all $1 \le i \le n$, then $\theta(t) \equiv^{\Gamma}_{\Sigma} \theta(t')$; in this case we write $A \models^{\Gamma}_{\Sigma} e$. If E is a set of Σ-equations we then write $A \models^{\Gamma}_{\Sigma} E$ when A Γ-behaviorally satisfies each Σ-equation in E. We may omit Σ and/or Γ from $\models^{\Gamma}_{\Sigma}$ when they are clear.*

An elegant formulation of case analysis adds a new kind of sentence, which can also be used in specifications:

Definition 5. *Given a hidden signature Σ, a Σ-**case sentence over variables** X is a nonempty set $\{C_1, C_2, ..., C_n\}$, written $(\forall X) \bigvee^{n}_{i=1} C_i$, where each C_i for $1 \le i \le n$ is a set of pairs of Σ-terms over variables in X. For A a hidden Σ-algebra, $A \models (\forall X) \bigvee^{n}_{i=1} C_i$ iff for any $\theta: X \to A$ there is some $1 \le i \le n$ such that $\theta(t) \equiv^{\Gamma}_{\Sigma} \theta(t')$ for each $t = t'$ in C_i.*

Definition 6. *A behavioral (or **hidden**) Σ-**specification** (or -**theory**) is a triple (Σ, Γ, E) where Σ is a hidden signature, Γ is a hidden subsignature of Σ, and E is a set of Σ-sentences (equations or cases). Non-data Γ-operations (i.e., in $\Gamma - \Sigma\lceil_V$) are called **behavioral**. A Σ-algebra A **behaviorally satisfies** (or **is a model of**) a behavioral specification $\mathcal{B} = (\Sigma, \Gamma, E)$ iff $A \models^{\Gamma}_{\Sigma} E$, in which case we write $A \models \mathcal{B}$; also $\mathcal{B} \models e$ iff $A \models \mathcal{B}$ implies $A \models^{\Gamma}_{\Sigma} e$. An operation $\sigma \in \Sigma$ is **behaviorally congruent for** $\mathcal{B}$ iff σ is Γ-behaviorally congruent for each A such that $A \models \mathcal{B}$.*

Interesting examples of non-Γ-behaviorally congruent operations arise in programming language semantics. For example, considering two programs in a language equivalent iff both terminate with the same output, then a Γ can be defined to enforce this behavioral equivalence relation, but an adequate behavioral specification requires operations that do not preserve this behavioral equivalence, such as observing the execution environment (e.g., two programs may declare a variable x, one instantiate it to 0 and the other to 1, and then never use that variable).

Proposition 1. *For any behavioral specification $\mathcal{B} = (\Sigma, \Gamma, E)$, all operations in Γ and all hidden constants are behaviorally congruent for $\mathcal{B}$.*

Of course, E may be such that other operations are also congruent. An easy criterion for congruence is given in [20] and is generalized in [4, 21]; [20] also shows that congruent operations can be added to or removed from Γ as desired when no equation in E has hidden sorted equalities in its condition (which is a common situation).

3   Behavioral Inference

This section introduces five sound rules for behavioral equivalence, beyond the usual reflexivity, symmetry, transitivity and substitution; they all work on conditional equations. We let $\Vdash$ denote the relation being defined, for deduction from a specification to an equation. Also, if $\mathcal{B}$ is a behavioral specification and Y a set of variables, let $\mathcal{B}(Y)$ denote $\mathcal{B}$ with Y adjoined to the signature of $\mathcal{B}$; similarly, if E is a set of equations, let $\mathcal{B}\{E\}$ denote $\mathcal{B}$ with E adjoined to the equations of $\mathcal{B}$. This notation allows writing things like $\mathcal{B}(Y)\{E\}$ and $\mathcal{B}(Y)\{E, E'\}$.

As discussed in the paragraph before Proposition 1, operations are not always congruent. This implies that the congruence rule of equational deduction is not always sound for behavioral reasoning. However, there are important situations where it is sound: a) when applied at the top of a visible term; and b) when applied to behaviorally congruent operations. The reason for the first is that behavioral equivalence is the identity on visible sorts, and for the second is that behaviorally congruent operations preserve the behavioral equivalence relation. Thus we have:

$$
\text{Congruence}:
\begin{cases}
a) & \dfrac{\mathcal{B} \ \Vdash\ (\forall X)\ t = t'\ \texttt{if}\ c,\ sort(t, t') \in V}{\mathcal{B}\ \Vdash\ (\forall X, W)\ \sigma(W, t) = \sigma(W, t')\ \texttt{if}\ c,\ \text{for all } \sigma \in Der(\Sigma)} \\[2em]
b) & \dfrac{\mathcal{B}\ \Vdash\ (\forall X)\ t = t'\ \texttt{if}\ c,\ sort(t, t') \in H}{\mathcal{B}\ \Vdash\ (\forall X, W)\ \delta(\overline{s}, t) = \delta(\overline{s}, t')\ \texttt{if}\ c,\ \text{for all } \delta \in \Gamma \text{ and } \overline{s} \in T_\Sigma(W)}
\end{cases}
$$

where $\overline{s}$ is an appropriate string of Σ-terms over variables in W. Deduction with this plus reflexivity, symmetry and transitivity satisfies an important property given in Proposition 2 below, using the following:

Definition 7. *A Σ-context γ is **behavioral** iff all operations on the path to $\star$ in γ are behavioral, and is **safe** iff either it is behavioral or there is some behavioral experiment γ' (of visible result) such that $\gamma = \gamma''[\gamma']$ for some appropriate γ''.*

Proposition 2. *If $\mathcal{B}\ \Vdash\ (\forall X)\ t = t'\ \texttt{if}\ c$ then $\mathcal{B}\ \Vdash\ (\forall X, W)\ \gamma[t] = \gamma[t']\ \texttt{if}\ c$ for any appropriate safe Σ-context γ, where W gives the variables of γ.*

The *deduction theorem* says that to prove an implication $p \to q$ one can add p to the set of axioms and then prove q. In equational logics, since universal quantifiers bind both the condition and the conclusion of a conditional equation, to make the deduction theorem work one must first unbind the variables in the condition. This is typically done via the "theorem of constants," which adds a new constant to the signature for each variable of interest. Here is a behavioral rule combining these two:

$$
\text{Condition Elimination}: \frac{\mathcal{B}(Y)\{E(c)\}\ \Vdash\ (\forall X - Y)\ t = t'}{\mathcal{B}\ \Vdash\ (\forall X)\ t = t'\ \texttt{if}\ c}
$$

where Y is the set of variables occurring in c, and $E(c)$ is the set of ground unconditional equations contained in c (arising since c is a conjunction of equalities). In the lower part of this rule, t, t', c are all defined over the signature of $\mathcal{B}$ and use variables in X, while in the upper part, t and t' still use variables in X but all their variables in Y are replaced by new constants, thus giving a new behavioral specification $\mathcal{B}(Y)$, where each variable in Y is regarded as a new constant.

A case sentence can be used to derive new equations by providing a substitution from the case statement's variables into terms over the equation's variables. Formally, let $\mathcal{B}$ be a behavioral specification containing the case statement $(\forall Y) \bigvee_{i=1}^{n} C_i$, let φ be a map $Y \to T_\Sigma(X)$, and let $V_i = var(\varphi(C_i))$. Then

$$
\text{Case Analysis}: \frac{\mathcal{B}(V_i)\{(\forall \emptyset)\varphi(C_i)\}\ \Vdash\ (\forall X - V_i)\ t = t'\ \texttt{if}\ c,\ \text{for } 1 \le i \le n}{\mathcal{B}\ \Vdash\ (\forall X)\ t = t'\ \texttt{if}\ c}
$$

This says that to prove $(\forall X)\ t = t'$ if c by case analysis using $(\forall Y)\ \bigvee_{i=1}^{n} C_i$, one must provide a substitution instantiating the case statement to the context of the proof task and then, for each case C_i, prove the equational sentence using that case as an axiom, after the relationship between the case's variables and the equation's variables is made explicit by replacing them by constants.

Unrestricted use of the Case Analysis rule can be very expensive, even non-terminating, and it can also be hard to find appropriate substitutions φ. Since our main goal is an automatic and relatively efficient procedure for proving behavioral equivalence, we have developed a simple mechanism to tell BOBJ both when to perform a case analysis and with what substitution. In BOBJ, each case statement comes with a *pattern*, usually denoted by p, which is just a Σ-term with variables. The Case Analysis rule is enabled only if the pattern p matches a subterm of t or t', and then the substitution also comes for free.

Our most powerful rule is Circular Coinduction, but first we recall the important notion of cobasis, originating in [20] and later simplified in [18, 12, 21]. For this paper, a **cobasis** Δ is a subset of Γ that generates enough experiments, in the sense that no Γ-experiment can distinguish two states that cannot be distinguished by these experiments. Finding a minimal cobasis seems undecidable, but there are *cobasis criteria* that work well in practice [20, 4, 21], and are implemented in BOBJ. Users can also declare their own cobases in BOBJ.

Intuition for circular coinduction can be enhanced by considering its duality with structural induction. Inductive proofs show equality of terms $t(x), t'(x)$ over a given variable x (seen as a constant) by showing $t(\sigma(x))$ equals $t'(\sigma(x))$ for all σ in a basis, while circular coinduction shows terms t, t' behaviorally equivalent by showing equivalence of $\sigma(t)$ and $\sigma(t')$ for all σ in a cobasis. Moreover, coinduction applies cobasis operations at the top, while structural induction applies basis operations at the bottom. Both induction and circular coinduction assume some "frozen" instances of t, t' equal when checking the inductive/coinductive step: for induction, the terms are frozen at the bottom by replacing the induction variable by a constant, so that no other terms can be placed beneath the induction variable, while for coinduction, the terms are frozen at the top, so that they cannot be used as subterms of other terms (with some important but subtle exceptions, treated by the Special Context inference rule below).

Freezing terms at the top is elegantly handled by a simple trick. Suppose every specification has a special visible sort b, and for each (hidden or visible) sort s in the specification, a special congruent operation $[\_] : s \to b$. No equations are assumed for these operations and no user defined sentence can refer to them; they are there for technical reasons. Thus, with the inference rules introduced so far, for any behavioral specification $\mathcal{B}$ and any conditional equation $(\forall X)\ t = t'$ if c, it is necessarily the case that $\mathcal{B} \Vdash (\forall X)\ t = t'$ if c iff $\mathcal{B} \Vdash (\forall X)\ [t] = [t']$ if c. The rule below preserves this property. Let Δ be a cobasis for $\mathcal{B}$, and let the sort of t, t' be hidden. Then

Circular Coinduction :
$$\frac{\mathcal{B}\{(\forall X)\ [t] = [t']\ \text{if}\ c\} \Vdash (\forall X, W)\ [\delta(t, W)] = [\delta(t', W)]\ \text{if}\ c,\ \text{for all appropriate}\ \delta \in \Delta}{\mathcal{B} \Vdash (\forall X)\ t = t'\ \text{if}\ c}$$

We call the equation $(\forall X)\ [t] = [t']$ if c added to $\mathcal{B}$ a **circularity**; it could just as well have been called a coinduction hypothesis or a co-hypothesis, but we find the first name more intuitive because from a coalgebraic point of view, coinduction is all about finding circularities. Another way to look at circular coinduction is through the lense of context induction [13]. To clarify this discussion, we replace the operations $[\star]$ by $C[\star]$. Then our rule says that to show $(\forall X)\ t = t'$ if c, one can assume $(\forall X)\ C[t] = C[t']$ if c and then show $(\forall X, W)\ C[\delta(t, W)] = C[\delta(t'), W]$ if c for each $\delta \in \Delta$, which, if one thinks of $C[\delta(\star, W)]$ as $(C; \delta)(\star, W)$, is just an induction scheme on contexts. In fact, this is how we prove soundness of this rule (in Theorem 2 below).

Unfortunately, restricting application of circularities to the top of proof goals using the operations $[\star]$ excludes many important situations (e.g., see Example 5). We begin considering when this restriction can be lifted with the following:

Definition 8. *A $\Gamma - \Delta$ context $\gamma[\star]$ is called **special** iff for any Δ-experiment $C[\star]$, there exists a Δ-experiment $D[\star]$ such that $C[\gamma[\star]] = D[\star]$ and the size of $D[\star]$ is not bigger than the size of $C[\star]$.*

The next rule allows using circularities inside special contexts:

$$\text{Special} \quad \dfrac{\mathcal{B} \Vdash (\forall X)\ [t] = [t']\ \text{if}\ c}{\mathcal{B} \Vdash (\forall X, W)\ [\gamma(t, W)] = [\gamma(t', W)]\ \text{if}\ c, \text{ when } \gamma \text{ is a special context}}$$
$$\text{Context}$$

We now describe two kinds of special contexts, but will consider others in the future.

Definition 9. *An operation f in $\Gamma - \Delta$ is **context collapsed** iff for any Δ-experiment $C[\star]$, one of the following two conditions holds:*

1. *there exists an attribute g in Δ and a data context $D[\star]$ such that $C[f(W)] = D[g(W)]$.*
2. *$C[f(W)] = t$ where t is a data (i.e., Ω) term.*

It is not hard to see that f is context collapsed if both of the following are satisfied:

1. For any attribute g in Δ and any variable x of hidden sort, $g(f(x, V), W) = t$, where t is a data term, or $t = D[g'(x, W)]$ where g' is another attribute in Δ and $D[\star]$ is a data context.
2. For any non-attribute operation g in Δ and any variable x on a hidden sort, $g(f(x, V), W) = x$, or else $g(f(x, V), W) = C[x]$ where C is a context made from context collapsed operations.

Definition 10. *An operation f in $\Gamma - \Delta$ is **context preserved** iff both of the following are satisfied:*

1. *For any attribute g in Δ and any variable x of hidden sort, $g(f(x, V), W) = t$ where t is a data term, or $t = D[g'(x, W)]$ where g' is an attribute in Δ and $D[\star]$ is a data context.*

2. *For any non-attribute g in Δ and variable x on a hidden sort,*

$$g(f(x, V), W) = C[\, f(g'(x, V, W), V, W)]\ or$$
$$g(f(x, V), W) = C[f(x, V, W)] \qquad\qquad or$$
$$g(f(x, V), W) = C[g'(x, V, W)] \qquad\quad or$$
$$g(f(x, V), W) = x$$

where $C[\star]$ is a context of context collapsed operations and g' is in Δ.

Circularities can be applied under contexts having just context collapsed operations and context preserved operations, since they satisfy the Special Context rule.

One should be extremely careful in checking that contexts are special. For example, the cobasis operations in Δ cannot be special, because otherwise the Circular Coinduction rule would prove everything. As shown by Example 3, even behavioral operations in $\Gamma - \Delta$ can fail to be special. The examples and theory in this paper show that the issues involved are more subtle than we realized in [18], which described an overly optimistic algorithm. The following is the main result of this paper; its soundness assertions will be used to justify the C4RW algorithm.

Theorem 2. *The usual equational inference rules,* Reflexivity, Symmetry, Transitivity, *and* Substitution, *as well as the new rules above,* Congruence, Condition Elimination, Case Analysis, Circular Coinduction *and* Special Context, *are all sound. By soundness here we mean that if $\mathcal{B} \ \Vdash\ (\forall X)\ t = t'$ if c and $sort(t, t') \neq b$, or if $\mathcal{B} \ \Vdash\ (\forall X)\ [t] = [t']$ if c, then $\mathcal{B} \models (\forall X)\ t = t'$ if c.*

4 Behavioral Rewriting

Behavioral rewriting is to the first five rules of Theorem 2 as ordinary term rewriting is to equational logic: it provides a simple, efficient and automatic procedure for checking equalities. To simplify the exposition, we treat only unconditional rewriting, but the generalization to conditional rewriting is similar to that for ordinary term rewriting. Behavioral rewriting must differ from ordinary term rewriting because operations are not necessarily behaviorally congruent.

Definition 11. *A Σ-rewrite rule is a triple $(\forall Y)\ l \to r$ where $l, r \in T_\Sigma(Y)$. A behavioral Σ-rewriting system is a triple (Σ, Γ, R) where Σ is a hidden signature, Γ is a hidden subsignature of Σ, and R is a set of Σ-rewrite rules.*

Definition 12. *The behavioral (term) rewriting relation associated to a behavioral rewriting system $\mathcal{R}$ is the smallest relation $\Rightarrow$ such that:*

1. *$\theta(l) \Rightarrow \theta(r)$ for each $(\forall Y)\ l \to r$ in $\mathcal{R}$ and $\theta: Y \to T_\Sigma(X)$;*
2. *if $t \Rightarrow t'$ and $sort(t, t') \in V$ then $\sigma(W, t) \Rightarrow \sigma(W, t')$ for all $\sigma \in Der(\Sigma)$ and all appropriate variables W; and*
3. *if $t \Rightarrow t'$ and $sort(t, t') \in H$ then $\delta(W, t) \Rightarrow \delta(W, t')$ for all $\delta \in \Gamma$ and all appropriate variables W.*

When $\mathcal{R}$ is important, we write $\Rightarrow_\mathcal{R}$ instead of $\Rightarrow$.

Behavioral rewriting applies a rule to a hidden redex if only behavioral[2] operations occur on the path from that redex towards the root until a visible sort is found; if no visible operation is found, rewriting is still allowed if all operations on the path from the redex to the root are behavioral. We can formulate an equivalent definition of behavioral rewriting with the following:

Proposition 3. $t \Rightarrow t'$ *iff there is a rewrite rule* $(\forall Y) \; l \to r$ *in* R, *a safe* Σ-*context* γ, *and a substitution* θ *such that* $t = \gamma[\theta(l)]$ *and* $t' = \gamma[\theta(r)]$.

Behavioral rewriting is implemented in CafeOBJ [6] and BOBJ [10]. Confluence and termination of behavioral rewriting are interesting subjects for research, but we do not focus on them here, except to notice that termination of ordinary rewriting produces termination of behavioral rewriting, because $\Rightarrow$ is a subrelation of the usual term rewriting relation; therefore any termination criterion for ordinary rewriting applies to behavioral rewriting. Many classical results also generalize:

Proposition 4. *If* $\mathcal{R} = (\Sigma, \Gamma, R)$ *is a behavioral* Σ-*rewriting system and* $\mathcal{B} = (\Sigma, \Gamma, E)$ *is its associated behavioral specification, i.e., if* $E = \{(\forall Y) \; l = r \mid (\forall Y) \; l \to r \in R\}$, *and if* $\Rightarrow$ *and* $\equiv_{Eq}$ *are the behavioral rewriting and equational derivability (using the first five rules in Theorem 2) relations on* $\mathcal{R}$ *and* $\mathcal{B}$, *respectively, then*

1. $\Rightarrow \; \subseteq \; \equiv_{Eq}$,
2. *If* $\Rightarrow$ *is confluent then* $\equiv_{Eq} \; = \overset{*}{\Rightarrow}; \overset{*}{\Leftarrow}$, *and*
3. *If* $\Rightarrow$ *is canonical then* $t \equiv_{Eq} t'$ *iff* $bnf_{\mathcal{B}}(t) = bnf_{\mathcal{B}}(t')$, *where* $bnf_{\mathcal{B}}(u)$ *is the behavioral normal form of a* Σ-*term* u.

We now extend behavioral rewriting to take account of special contexts, for use in the algorithm of the next section:

Definition 13. $\Rightarrow^{\sharp}$ *is defined for behavioral rewriting systems extended with the special sorts* b *and operations* $[\_]$, *by extending* $\Rightarrow$ *minimally such that if* $[t] \Rightarrow^{\sharp} [t']$ *then* $[\gamma(t, W)] \Rightarrow^{\sharp} [\gamma(t', W)]$ *for each special context (see Definition 8). Given a behavioral rewriting system* $\mathcal{R}$ *with its associated behavioral specification* $\mathcal{B}$, *let* $bnf_{\mathcal{R}}^{\sharp}(t)$ *denote the normal form of a term* t *under the rewriting relation* $\Rightarrow^{\sharp}$.

Soundness of $\Rightarrow^{\sharp}$ follows from Proposition 4 and soundness of the Special Context rule (Theorem 2, which also says what we mean by soundness in the context of the special sort b and operations $[\_]$).

5   The C4RW Algorithm

A simple way to automate behavioral reasoning is just to behaviorally rewrite the two terms to normal forms, and then compare them, as suggested by Proposition 4. Although this is too weak to prove most interesting properties, the C4RW

---

[2] We recommend declaring as many operations as possible behavioral, and in particular, all congruent operations [20]; those who don't like this may substitute "behavioral or congruent" for "behavioral" through the rest of this paper.

algorithm combines it with circular coinduction and case analysis in a harmonious and automatic way, for proving hidden properties, which are usually the most interesting and difficult. Intuitively, to prove a hidden conditional equation $(\forall X)\ t = t'$ if c, one applies the circular coinduction rule and tries to prove the resulting apparently more complex properties. The algorithm maintains a list of goals, which is reduced when a goal is proved, and is increased when new goals are generated by the coinduction or case analysis rules. The algorithm terminates when either a visible proof task cannot be proved, in which case failed is returned, or when the set of goals becomes empty, in which case true is returned. The proof goals are stored in bracketed form to control their application.

We first describe the main procedure, C4RW, which has a set, $\mathcal{G}$, of hidden equations with visible conditions as initial goals. The loop at step 1 processes each goal in $\mathcal{G}$, and terminates when $\mathcal{G}$ becomes empty or when a failed is returned at step 9. Step 2 removes goals from $\mathcal{G}$, and step 3 puts them (in frozen or bracketed form) into the specification. These frozen versions of goals can then be used in subsequent proofs, much as induction hypotheses are used in inductive proofs by induction, but "at the top" instead of "at the bottom" (see the discussion in Section 3).

---

procedure C4RW$(\mathcal{B}, \Delta, \mathcal{G})$ (can modify its $\mathcal{B}$ and $\mathcal{G}$ arguments)
INPUT: - a behavioral theory $\mathcal{B} = (\Sigma, \Gamma, E)$
           - a cobasis Δ of $\mathcal{B}$
           - a set $\mathcal{G}$ of hidden Σ-equations of visible conditions (in bracket form)
OUTPUT: true if a proof of $\mathcal{B} \models \mathcal{G}$ is found; otherwise failed or non-terminating
  1. **while** there is some $e := (\forall X)\ [t] = [t']$ if c in $\mathcal{G}$ **do**
  2.   **let** $\mathcal{G}$ **be** $\mathcal{G} - \{e\}$
  3.   **let** $\mathcal{B}$ **be** $\mathcal{B}\{e\}$
  4.   **let** θ **be** a substitution on X assigning new constants to the variables in c;
         add the new constants to $\mathcal{B}$
  5.   **let** $E_{\theta(c)}$ **be** the set of visible ground equations in $\theta(c)$
  6.   **for** each $\delta \in \Delta$ appropriate for e **do**
  7.   **if** PROVEEQ$(\mathcal{B}, bnf^{\sharp}_{\mathcal{B}\{E_{\theta(c)}\}}([\delta(t, W)]), bnf^{\sharp}_{\mathcal{B}\{E_{\theta(c)}\}}([\delta(t', W)]), \theta(c), E_{\theta(c)}) \neq$ true
  8.   **then if** δ is an attribute **then return** failed
  9.   **else let** $\mathcal{G}$ **be** $\mathcal{G} \cup \{(\forall X, W)\ bnf^{\sharp}_{\mathcal{B}}([\delta(t, W)]) = bnf^{\sharp}_{\mathcal{B}}([\delta(t', W)])$ if $c\}$
10.   **endfor   endwhile**
11. **return** true

Steps 4 and 5 prepare for applying the Circular Coinduction rule. Since it generates new conditional proof obligations, each with the same condition, and since all these will later be subject to Condition Elimination, for efficiency step 4 first generates new constants for the variables in the condition, and then step 5 calculates the set of ground unconditional equations that will later be added to the specification by Condition Elimination. Steps 6–11 apply the Circular Coinduction rule. For each appropriate operation δ in the cobasis Δ, step 7 tries to prove that $[\delta(t, W)]$ equals $[\delta(t', W)]$, by first applying the Condition Elimination rule $(\mathcal{B}\{E_{\theta(c)}\})$, then using behavioral rewriting on both terms, and finally checking equality with the procedure PROVEEQ, which is explained below. Notice that behavioral rewriting can use the frozen equation; more precisely, the frozen equation is applied (as a rewrite rule) if the term $[\delta(t, W)]$ reduces to an instance (via a substitution) of $[t]$ (of course, if the condition holds). This is equivalent to saying that the equation $(\forall X)\ t = t'$ **if** c can only be applied on the top when reducing the terms $\delta(t, W)$ generated by circular coinduction. If the procedure PROVEEQ does not return **true**, meaning that it was not able to prove the two b-sorted terms equal, then the algorithm returns **failed** if the cobasis operation was visible (step 8), or else it adds a new (hidden) goal to $\mathcal{G}$, as required by the Circular Coinduction rule.

We next discuss the procedure PROVEEQ, which takes as arguments a behavioral specification, b-sorted terms u, u', and the ground version of the equation's condition (its variables replaced by new constants) together with the set of ground equations it generates; it returns **true**, or **failed**, or loops forever. Step 1 returns **true** if the two terms are equal, and steps 2–6 check whether any case statement in $\mathcal{B}$ can be applied. Remember that the Case Analysis rule requires a substitution of the variables in the case statement into terms over the variables of the equation to be proved, and that we use a pattern in BOBJ which automatically selects a substitution τ (step 3). Step 4 checks whether the case analysis L can show the two terms equivalent, using the procedure CASEANALYSIS described below. If no case sentence can show the terms u, u' equivalent, then step 7 returns **failed**.

---

procedure PROVEEQ$(\mathcal{B}, u, u', \theta(c), E_{\theta(c)})$
INPUT: - a behavioral theory $\mathcal{B}$
          - two terms u and u' of visible sort b
          - a ground visible condition $\theta(c)$ and its ground equations $E_{\theta(c)}$
OUTPUT: **true** if a proof of $\mathcal{B} \models (\forall var(u, u'))\ u = u'$ **if** $\theta(c)$ is found;
          otherwise **failed** or non-terminating
1. **if** $u = u'$ **then return true**
2. **for** each case sentence (p, L) in $\mathcal{B}$ **do**
3.   **if** p matches a subterm of u or u' with substitution τ
4.   **then if** CASEANALYSIS$(L, \tau, \mathcal{B}, u, u', \theta(c), E_{\theta(c)})$
5.   **then return true   endfor**
6. **return failed**

The CASEANALYSIS procedure just applies the Case Analysis rule. For each case C, it first adds a new constant for each variable in C (step 2) and generates the ground equations of C (step 3). Steps 4–5 check the top derivation in the CASEANALYSIS rule: step 4 checks whether the condition of the equational sentence became false (to keep the presentation short we have not introduced an inference rule for false conditions), and if this is not the case, then step 5 recursively checks whether u and u' became equal under the new assumptions; since this recursion may not terminate, some care may be required when defining case statements.

---

procedure CASEANALYSIS($L, \tau, \mathcal{B}, u, u', \theta(c), E_{\theta(c)}$)

1. **for** each case C in L **do**
2.   **let** η **be** τ with a new constant substituted for each variable in C
3.   **let** $E_{\eta(C)}$ **be** the set of visible ground equations in $\eta(C)$
4.   **if** $bnf_{\mathcal{B}\{E_{\eta(C)}\}}(\theta(c)) \neq$ false **and**
5.   PROVEEQ($\mathcal{B}\{E_{\eta(C)}\}, bnf^{\sharp}_{\mathcal{B}\{E_{\eta(C)}, E_{\theta(c)}\}}(u), bnf^{\sharp}_{\mathcal{B}\{E_{\eta(C)}, E_{\theta(c)}\}}(u'), \theta(c), E_{\theta(c)}) \neq$ true
6.   **then return** failed   **endfor**
7. **return** true

To take full advantage of behavioral rewriting, one must carefully orient the new equations added at step 10 as rewrite rules; the success of C4RW depends on how well this is done. The following orientation procedure has worked very well in practice: If both directions are valid as rewrite rules (i.e., both sides have the same variables), then orient so that the left side has less symbols than the right side; if the terms have the same number of symbols, then orient with right side smaller than the left side under the lexicographic path ordering induced by the order in which operations are declared. The following is a more precise description:

1. if only one direction is valid, then use it;
2. if both directions are valid, but one of $\delta(t, W)$ and $\delta(t', W)$, say t_1 has more symbols than the other, say t_2, then add the rule $[t_1] \to [t_2]$ if c to $\mathcal{G}$;
3. if both directions are valid and have the same number of symbols, but $t_1 \gg t_2$, then add $[t_1] \to [t_2]$ if c to $\mathcal{G}$, where $\gg$ is the lexicographic path ordering induced by the operation declaration ordering, defined by
   $f(t_1, ..., t_n) \gg t_i$ for all $1 \leq i \leq n$,
   $f(t_1, ..., t_n) \gg f(s_1, ..., s_n)$ if $\langle t_1, ..., t_n \rangle \gg \langle s_1, ..., s_n \rangle$ in lexicographic order,
   $f(t_1, ..., t_n) \gg g(s_1, ..., s_m)$ if $f > g$ and $f(t_1, ..., t_n) \gg s_i$ for all $1 \leq i \leq m$.

Theorem 3. *The procedure C4RW described above is correct. More precisely, for any behavioral theory $\mathcal{B}$ and any correct cobasis Δ for it, C4RW($\mathcal{B}, \Delta, \mathcal{G}$) returns true if and only if $\mathcal{B} \models \mathcal{G}$.*

Proof. Since C4RW is just a discipline for applying the behavioral inference rules, its correctness follows from Theorem 2.

Example 3. <u>An invalid coinduction</u> This shows how the unrestricted use of circularities can give rise to incorrect results. Notice that odd is a congruent operation not in the cobasis, which for streams, consists of just head and tail.

```
bth FOO is pr STREAM[NAT] .
  op odd_ : Stream -> Stream .
  var S : Stream .
  eq head odd S = head S .
  eq tail odd S = odd tail tail S .
  ops a b : -> Stream .
  eq head a = head b .
  eq tail a = odd a .
  eq tail tail b = odd b .
end
cred odd b == a .
```

The "proof" goes as follows, using the cobasis {head, tail} as usual:

    head odd b == head a

follows by behavioral reduction, and

    tail odd b == tail a

reduces to

    odd odd b == odd a

which follows (illegitimately) by applying the circularity inside the context odd. To show that the result really is false, one may take a to be the stream which begins with 001 and then has all 2s, and b to be the sequence which also begins with 001, and then continues with all 0s. (Of course, BOBJ's C4RW algorithm also fails to prove it; in fact, it goes into an infinite loop during behavioral rewriting when given this input.) □

Example 4. Two definitions of Fibonacci, plus evenness Here is a not so trivial example. The goal of the first invocation of C4RW via cred is to show equality of two different definitions of the stream of all Fibonacci numbers; for this, the algorithm generates an unending stream of new circularities, thus illustrating how C4RW itself can fail to terminate. The zip function interleaves two streams. The second goal involves both a conditional goal and a circularity, and it succeeds.

```
bth 2FIBO is pr STREAM[NAT] .
  ops fib fib' : Nat Nat -> Stream .
  vars N N' : Nat . vars S S' : Stream .
  eq head fib(N, N') = N .
  eq tail fib(N, N') = fib(N', N + N') .
  eq head fib'(N, N') = N .
  eq head tail fib'(N, N') = N' .
  op zip : Stream Stream -> Stream .
  eq head zip(S, S') = head S .
  eq tail zip(S, S') = zip(S', tail S) .
  op add_ : Stream -> Stream .
  eq tail tail fib'(N, N') = add zip(fib'(N, N'), tail fib'(N, N')).
  eq head add S = head S + head tail S .
  eq tail add S = add tail tail S .
end
cred fib(N, N') == fib'(N, N') .
bth EVENNESS is pr 2FIBO + STREAM[BOOL] * (sort Stream to BStream) .
  op all-true : -> BStream .
  eq head all-true = true .
  eq tail all-true = all-true .
  op even?_ : Nat -> Bool .
  op even?_ : Stream -> BStream .
  vars M N : Nat . var S : Stream .
  eq even? 0 = true .
  eq even? s 0 = false .
  eq even? s s N = even? N .
  eq head even? S = even? head S .
  eq tail even? S = even? tail S .
  eq even?(M + N) = true if even?(M) and even?(N) .
end
cred even? fib(M, N) == all-true if even?(M) and even?(N) .
```

The last equation is really a lemma that should be proved by induction; it is needed in the proof that if `fib` is given two even arguments, then all its values are even. □

Example 5. Two definitions for iteration The goal is proving equivalence of two ways to produce a stream of increasingly many copies of a function applied to an element.

```
bth MAP-ITER [X :: DATA] is pr STREAM[X] .
  ops (iter1_) (iter2_) : Elt -> Stream .
  var E : Elt . var S : Stream .
  eq head iter1 E = E .
  eq tail iter1 E = iter1 f E .
  eq head iter2 E = E .
  eq tail iter2 E = map iter2(E) .
end
cred iter1 E == iter2 E .
```

The C4RW algorithm generates two circularities, one of which is applied in a special context, i.e., not at the top, in fact, under `map`. Hence this example actually requires special contexts. □

6   Conclusions and Future Research

We believe that the C4RW algorithm, especially with its use of special contexts, is the most powerful algorithm now available for proving behavioral properties of complex systems. However, much can be done to improve it. First, the conditions for contexts to be special in Section 3 are only the beginning of what could be a long journey, parallel to the one followed in research on automatic induction algorithms. In fact, it would probably be useful to combine the C4RW algorithm with some automatic induction methods. In any case, we will consider more powerful conditions for special contexts in future publications.

Another topic that seems worth exploring is adding conditions to case statements; the idea is that after the pattern is matched, the case split would only be applied if the condition is satisfied. This could make the application of case splits more precise, as well as reduce the computation needed for some large examples.

Finally, more should be done on the duality between induction and circular coinduction. In particular, since we are talking about sophisticated algorithms that generate new hypotheses, not just about basic forms of induction and coinduction, the very notion of duality may need some careful explication.

References

1. G. Bernot, M. Bidoit, and T. Knapik. Observational specifications and the indistinguishability assumption. *Theoretical Computer Science*, 139:275–314, 1995.
2. N. Berregeb, A. Bouhoula, and M. Rusinowitch. Observational proofs with critical contexts. In *Fundamental Approaches to Software Engineering*, volume 1382 of *LNCS*, pages 38–53. Springer, 1998.

3. M. Bidoit and R. Hennicker. Behavioral theories and the proof of behavioral properties. *Theoretical Computer Science*, 165(1):3–55, 1996.
4. M. Bidoit and R. Hennicker. Observer complete definitions are behaviourally coherent. In K. Futatsugi, J. Goguen, and J. Meseguer, editors, *OBJ/CafeOBJ/Maude at Formal Methods '99*, pages 83–94. Theta, 1999.
5. S. Buss and G. Roşu. Incompleteness of behavioral logics. In H. Reichel, editor, *Proceedings of Coalgebraic Methods in Computer Science*, volume 33 of *Electronic Notes in Theoretical Computer Science*, pages 61–79. Elsevier Science, 2000.
6. R. Diaconescu and K. Futatsugi. *CafeOBJ Report: The Language, Proof Techniques, and Methodologies for Object-Oriented Algebraic Specification*. World Scientific, 1998. AMAST Series in Computing, volume 6.
7. R. Diaconescu and K. Futatsugi. Behavioral coherence in object-oriented algebraic specification. *Journal of Universal Computer Science*, 6(1):74–96, 2000.
8. K. Futatsugi and K. Ogata. Rewriting can verify distributed real-time systems – how to verify in CafeOBJ. In Y. Toyama, editor, *Proc. Int. Workshop on Rewriting in Proof and Computation*, pages 60–79. Tohoku University, 2001.
9. J. Goguen. Types as theories. In G.M. Reed, A.W. Roscoe, and R.F. Wachter, editors, *Topology and Category Theory in Computer Science*, pages 357–390. Oxford, 1991.
10. J. Goguen, K. Lin, and G. Roşu. Circular coinductive rewriting. In *Proceedings, Automated Software Engineering*, pages 123–131. IEEE, 2000.
11. J. Goguen and G. Malcolm. A hidden agenda. *Theoretical Computer Science*, 245(1):55–101, 2000.
12. J. Goguen and G. Roşu. Hiding more of hidden algebra. In *Proceeding, FM'99*, volume 1709 of *LNCS*, pages 1704–1719. Springer, 1999.
13. R. Hennicker. Context induction: a proof principle for behavioral abstractions. *Formal Aspects of Computing*, 3(4):326–345, 1991.
14. R. Hennicker and M. Bidoit. Observational logic. In *Proceedings, AMAST'98*, volume 1548 of *LNCS*, pages 263–277. Springer, 1999.
15. B. Jacobs. Mongruences and cofree coalgebras. In M. Nivat, editor, *Algebraic Methodology and Software Technology (AMAST95)*, pages 245–260. Springer, 1995. LNCS 936.
16. B. Jacobs and J. Rutten. A tutorial on (co)algebras and (co)induction. *Bulletin of European Association for Theoretical Computer Science*, 62:222–259, 1997.
17. P. Padawitz. Towards the one-tiered design of data types and transition systems. In *Proceedings, WADT'97*, volume 1376 of *LNCS*, pages 365–380. Springer, 1998.
18. G. Roşu. Behavioral coinductive rewriting. In K. Futatsugi, J. Goguen, and J. Meseguer, editors, *OBJ/CafeOBJ/Maude at Formal Methods '99*, pages 179–196. Theta, 1999.
19. G. Roşu. *Hidden Logic*. PhD thesis, University of California, San Diego, 2000.
20. G. Roşu and J. Goguen. Hidden congruent deduction. In R. Caferra and G. Salzer, editors, *Automated Deduction in Classical and Non-Classical Logics*, volume 1761 of *LNAI*, pages 252–267. Springer, 2000.
21. G. Roşu and J. Goguen. Circular coinduction. In *International Joint Conference on Automated Reasoning (IJCAR'01)*. 2001.
22. J. Rutten. Universal coalgebra: a theory of systems. *Theoretical Computer Science*, 249:3–80, 2000.

Verifying Generative CASL Architectural Specifications

Piotr Hoffman

Warsaw University, Institute of Informatics,
Banacha 2, 02-097 Warszawa, Poland
piotrek@mimuw.edu.pl

Abstract. We present a proof-calculus for architectural specifications, complete w.r.t. their generative semantics. Architectural specifications, introduced in the CASL specification language, are a formal mechanism for expressing implementation steps in program development. They state that to implement a needed unit, one may implement some other units and then assemble them in the prescribed manner; thus they capture modular design steps in the development process. We focus on developing verification techniques applicable to full CASL architectural specifications, which involves, inter alia, getting around the lack of amalgamation in the CASL institution.

Introduction

The formal development of a program consists of three phases: writing a requirements specification, implementing it, and proving the correctness of the implementation with respect to the specification (see [ST97]). Architectural specifications [BST99] are meant to be a tool aiding the developer in breaking down the implementation task into independent subtasks. These subtasks can be either further broken down, or directly coded in a chosen programming language. The benefits of dividing a programming task into independent subtasks are obvious: work on parts of the project can proceed in parallel, the project as a whole is easier to maintain and comprehend. Also, the correctness proof can now be constructed from the correctness proofs for the individual subtasks and a correctness proof of the subdivision (i.e., architectural specification) itself.

Our goal is to provide one crucial ingredient necessary for architectural specifications to be used in practice in the formal development of software, namely verification of their correctness. What we do is reduce the correctness problem to the problem of proving semantic consequence in an institution (usually the underlying institution of the architectural specification at issue).

In Sect. 1 we define the syntax and formal semantics of architectural specifications in an institution-independent fashion. In Sect. 2 we then devise a calculus for proving properties of architectural specifications and state its correctness and completeness w.r.t. a generative semantics. This calculus in many respects builds on what has been presented in [Hof01].

M. Wirsing, D. Pattinson, and R. Hennicker (Eds.): WADT 2002, LNCS 2755, pp. 233–252, 2003.
© Springer-Verlag Berlin Heidelberg 2003

Sect. 3 and 4 (together with Sect. 6) form the novel part of the paper. Here, we develop techniques for discharging proof-obligations which arise when using the proof-calculus of Sect. 2. These techniques are institution-independent, i.e., they work in any institution having certain, abstractly formulated, expressibility properties. To extend the applicability of these techniques to institutions without amalgamation, a representation of the institution of interest in some "better" institution is introduced.

In Sect. 5 we define the full CASL logic [CASL00,CASL], including subsorting and structured specifications, and construct a representation of this logic in a many-sorted logic. Finally, in Sect. 6 we prove that this many-sorted logic and the used representation of CASL have all the properties required in order to apply the techniques presented in Sect. 3 and 4.

1   Preliminaries

Some notation: by $f\,[a/x]$ we denote the function which extends f with the mapping $x \mapsto a$; at the same time, if X is a syntactic object, then $X[B/A]$ substitutes all As in X by Bs; for any notion $\mathcal{N}$, we write $\mathcal{N}^*$ for the same notion applied to tuples.

The definition of architectural specifications will be parametrized by an underlying logical system, formalized as an *institution* (see [GB92]). An institution **I** is a quadruple $(Sig, \mathbf{Mod}, Sen, \models)$, where:

- Sig is a category of *signatures*;
- $\mathbf{Mod} : Sig^{op} \to \mathcal{CAT}$ is a functor into the quasi-category of all categories;
- $Sen : Sig \to \mathbf{Set}$ is a functor into the category of all small sets (we sometimes drop the functoriality requirement);
- $\models$ is a family $\{\models_\Sigma\}_{\Sigma \in \mathbf{Ob}(Sig)}$, where $\models_\Sigma$ is a relation on $\mathbf{Mod}(\Sigma) \times Sen(\Sigma)$.

These data are subject to the so-called *satisfaction condition*, i.e., for any $\sigma : \Sigma \to \Delta$ in Sig, $\phi \in Sen(\Sigma)$ and $M \in \mathbf{Mod}(\Delta)$ the following equivalence holds:

$$M \models_\Delta Sen(\sigma)(\phi) \iff \mathbf{Mod}(\sigma)(M) \models_\Sigma \phi$$

Objects of $\mathbf{Mod}(\Sigma)$ are called Σ-*models* and elements of $Sen(\Sigma)$ are called Σ-*sentences*. We usually denote the functor $\mathbf{Mod}(\sigma)$ by $\cdot|_\sigma$ and call it the $(\sigma\text{-})$ *reduct*. If $N|_\sigma = M$, then N is a σ-*extension* of M (and M is σ-*extendible*). For any $\sigma : \Sigma \to \Delta$ in Sig and $\phi \in Sen(\Sigma)$, by $\sigma(\phi)$ we denote $Sen(\sigma)(\phi)$ and we call it the *translation* of ϕ along σ. We will overload the symbols $\mathbf{Mod}$ and $\models$ as follows. For any set Φ of Σ-sentences, $\mathbf{Mod}_\Sigma(\Phi)$ denotes the class of all Σ-models satisfying all sentences from Φ. If Ψ is a set of Σ-sentences, then we write $\Phi \models_\Sigma \Psi$ if $\mathbf{Mod}_\Sigma(\Phi) \subseteq \mathbf{Mod}_\Sigma(\Psi)$.

By Pres(**I**) we denote the category of *presentations*, i.e., pairs (Σ, Φ), where Φ is a finite set of Σ-sentences, a morphism $\sigma : (\Sigma, \Phi) \to (\Delta, \Psi)$ being a signature morphism $\sigma : \Sigma \to \Delta$ such that $\sigma(\Phi) \models_\Delta \Psi$. By $Sig[\_]$ and $Ax[\_]$ we denote the respective components of a presentation.

$ASP ::= \textbf{units } UD_1 \ldots UD_n \textbf{ result } UE$
$UD ::= A : USP$
$UE ::= \bullet T \mid \lambda A : SP \bullet T$
$T ::= A \mid A(T) \mid T \textbf{ and } T \mid \textbf{reduce } T \textbf{ by } \sigma \mid \textbf{let } A = T \textbf{ in } T$

Fig. 1. Language of architectural specifications.

We parametrize the definition of architectural specifications by an underlying institution $\mathbf{I} = (Sig, \mathbf{Mod}, Sen, \models)$ and a subcategory of Sig called the *inclusion subcategory*, which we require to be a partial order on the objects of Sig. An inclusion of Σ into Δ is denoted $\iota_{\Sigma,\Delta}$. If this inclusion exists, we say that Σ is a subsignature of Δ. The language of architectural specifications is described by the grammar in Figure 1 (here, the As come from an infinite set of identifiers, the σs are signature morphisms in $\mathbf{I}$, and the SPs are objects in $\mathrm{Pres}(\mathbf{I})$; USPs are introduced below).

Syntactic differences aside, this language of architectural specifications differs somewhat from what is used in CASL. On the one hand, the reduct is more restrictive in CASL. On the other hand, we have disallowed a number of constructs available in CASL: multi-parameter units, unit definitions, fitting morphisms in applications, declarations with imports, and translation. Except for the imports, this does not constitute a big difference, it just makes the analysis and presentation more clear. Also, we have chosen our specifications to be presentations. As we will see in Sect. 5, this is not a real restriction either, since structured specifications fit well into such a framework.

In order to introduce a formal semantics of architectural specifications, we need a few definitions. A *parametric signature* is a pair $P\Sigma = (\Sigma, \Delta)$ such that Σ is a subsignature of Δ. A *unit signature* $U\Sigma$ is either a (regular) signature or a parametric signature. A *parametric unit* over a parametric signature (Σ, Δ) is a partial function U from $\mathbf{Mod}(\Sigma)$ to $\mathbf{Mod}(\Delta)$ which is *persistent*, i.e., for any $M \in \mathrm{dom}(U)$ we have $U(M)|_{\iota_{\Sigma,\Delta}} = M$. A *unit* over a unit signature is a model over that signature, if it is a (regular) signature, or a parametric unit over that signature, if it is a parametric signature. A *unit specification* USP is either a presentation (Σ, Φ), with $[\![USP]\!] = \mathbf{Mod}_\Sigma(\Phi)$, or a *parametric unit specification* $(\Sigma, \Phi) \to (\Delta, \Psi)$, where Σ is a subsignature of Δ, with $[\![USP]\!]$ being the class of all parametric units U over (Σ, Δ) such that $\mathrm{dom}(U) = \mathbf{Mod}_\Sigma(\Phi)$ and for all $M \in \mathrm{dom}(U)$, $U(M) \in \mathbf{Mod}_\Delta(\Psi)$. A *static environment* δ is a pair (B, P), where B maps identifiers to signatures and P maps identifiers to parametric signatures. The domains of B and P are required to be finite and disjoint. A *unit environment* fitting a static environment (B, P) is a map e, sending any identifier $X \in \mathrm{dom}(B)$ to a model over $B(X)$ and any identifier $X \in \mathrm{dom}(P)$ to a parametric unit over $P(X)$. A *unit context* fitting a static environment δ is any set E of environments fitting δ. A *model function of type* (E, Σ) is a total function F from the context E into $\mathbf{Mod}(\Sigma)$. *Unit functions of type* $(E, U\Sigma)$, denoted UF, are defined analogically.

To deal with the **and**-construct, we need two notions. First, we define the *sum* of signatures Σ, Δ to be their least upper bound $\Gamma = \Sigma \cup \Delta$ with respect to the inclusion order, if it exists. Then for any models M over Σ and N over Δ we define their *amalgamation*, denoted $M \oplus N$, to be a Γ-model P satisfying $P|_{\iota_{\Sigma,\Gamma}} = M$ and $P|_{\iota_{\Delta,\Gamma}} = N$, if it exists and is unique. Models are called *amalgamable* if their amalgamation exists.

We use the natural semantics style, with judgments written $\_ \vdash \_ \rhd \_$. We always implicitly add premises stating that objects denoted by δ are static environments, in any pair δ, E, E fits δ, etc.

The semantics in Fig. 2and 3 assigns:

- to a unit declaration, a static environment δ and a unit context E fitting δ;
- to a term, given a static environment δ and a unit context E fitting δ, a signature Σ and a model function F of type (E, Σ);
- to a unit expression, given a static environment δ and a unit context E fitting δ, a unit signature $U\Sigma$ and a unit function UF of type $(E, U\Sigma)$;
- to an architectural specification, a static environment δ, a unit signature $U\Sigma$ and a unit function UF of type $(\text{dom}(UF), U\Sigma)$, where $\text{dom}(UF)$ fits δ.

This semantics, though considerably simplified, is consistent with the formal semantics of CASL, found in [CASL00].

2   Proving Properties of Architectural Specifications

An architectural specification describes the decomposition of a programming task. This may be expressed by a statement of the form $\vdash ASP :: USP$, which says "*ASP* describes a correct procedure of building a unit $U \in [\![USP]\!]$". This means that, first of all, the procedure itself is correct, i.e., *ASP* has a denotation $\delta, U\Sigma, UF$; and, second, the procedure gives the required result, i.e., any unit in the image of UF is in $[\![USP]\!]$.

A *context* consists of a finite number of declarations of two types:

- $A :_\Sigma \Phi$, where (Σ, Φ) is a presentation; we then say that $A \in \text{dom}(\Gamma)$ and $\Gamma(A) = \Sigma$,
- $\sigma : A \to B$, where $\Gamma(A) = \text{dom}(\sigma)$ and $\Gamma(B) = \text{cod}(\sigma)$.

A *parametric context* Γ_{par} consists of a finite number of declarations of the form $A :_{\Sigma \to \Delta} \Phi \to \Psi$, where $(\Sigma, \Phi) \to (\Delta, \Psi)$ is a unit specification. For both kinds of contexts, if Γ_1 and Γ_2 coincide on the intersection of their domains, then $\Gamma_1 \cup \Gamma_2$ is defined naturally.

We say that a *model family* $\{M_X\}_{X \in \text{dom}(\Gamma)}$ is *consistent* with a context Γ, written $\{M_X\}_{X \in \text{dom}(\Gamma)} \models \Gamma$, if:

- $X :_\Sigma \Phi$ in Γ implies $M_X \in \mathbf{Mod}_\Sigma(\Phi)$,
- $\sigma : X \to Y$ in Γ implies $M_X = M_Y|_\sigma$.

A calculus for deriving statements $\vdash ASP :: USP$ is given in Fig. 4 and 5. This calculus is correct, but its completeness depends on additional assumptions. A proof by induction over the structure of (the **result**-term of) the architectural specification (see [Hof01] for a few further hints) is omitted.

$$\boxed{\vdash ASP \rhd \delta, U\Sigma, UF}$$

$$\vdash UD_1 \rhd (B_1, P_1), E_1 \ \ldots \ \vdash UD_n \rhd (B_n, P_n), E_n$$
$$\delta = (B_1 \cup \cdots \cup B_n, P_1 \cup \cdots \cup P_n) \text{ is an } n\text{-element static environment}$$
$$E = \{\ e_1 \cup \cdots \cup e_n \mid e_1 \in E_1, \ldots, e_n \in E_n\ \}$$
$$\delta, E \vdash UE \rhd U\Sigma, UF$$

$$\overline{\vdash \textbf{units } UD_1 \ldots UD_n \textbf{ result } UE \rhd \delta, U\Sigma, UF}$$

$$\boxed{\vdash UD \rhd \delta, E}$$

$$\overline{\vdash A : (\Sigma, \Phi) \rhd (A \mapsto \Sigma, \emptyset), \{\ A \mapsto M \mid M \in \textbf{Mod}_\Sigma(\Phi)\ \}}$$

$$\overline{\vdash A : SP \to SP' \rhd (\emptyset, A \mapsto Sig[SP] \to Sig[SP']), \{A \mapsto U \mid U \in [\![SP \to SP']\!]\ \}}$$

$$\boxed{\delta, E \vdash UE \rhd U\Sigma, UF}$$

$$\frac{(B, P), E \vdash T \rhd \Sigma, F}{(B, P), E \vdash \bullet T \rhd \Sigma, F}$$

$$(B\,[\Sigma/A]\,, P), \{\,e\,[M/A] \mid e \in E, \ M \in \textbf{Mod}_\Sigma(\Phi)\,\} \vdash T \rhd \Delta, F$$
$$\Sigma \text{ is a subsignature of } \Delta$$
$$\text{for all } e \in E \text{ and } M \in \textbf{Mod}_\Sigma(\Phi), \ F(e\,[M/A])|_{\iota_{\Sigma,\Delta}} = M$$

$$\overline{\begin{array}{c}(B, P), E \vdash \lambda A : (\Sigma, \Phi) \bullet T \rhd \\ (\Sigma, \Delta), \lambda e \in E \cdot \lambda M \in \textbf{Mod}_\Sigma(\Phi) \cdot F(e\,[M/A])\end{array}}$$

$$\boxed{\delta, E \vdash T \rhd \Sigma, F}$$

$$\frac{A \in \text{dom}(B)}{(B, P), E \vdash A \rhd B(A), \lambda e \in E \cdot e(A)}$$

$$A \in \text{dom}(P) \text{ and } P(A) = (\Sigma, \Delta)$$
$$(B, P), E \vdash T \rhd \Sigma, F$$
$$\text{for all } e \in E, \ F(e) \in \text{dom}(e(A))$$

$$\overline{(B, P), E \vdash A(T) \rhd \Delta, \lambda e \in E \cdot e(A)(F(e))}$$

$$\delta, E \vdash T_1 \rhd \Sigma_1, F_1 \text{ and } \delta, E \vdash T_2 \rhd \Sigma_2, F_2$$
$$\Delta = \Sigma_1 \cup \Sigma_2$$
$$\text{for all } e \in E, \ F_1(e) \oplus F_2(e) \text{ exists}$$

$$\overline{\delta, E \vdash T_1 \textbf{ and } T_2 \rhd \Delta, \lambda e \in E \cdot F_1(e) \oplus F_2(e)}$$

Fig. 2. The semantics of architectural specifications.

$$\frac{\delta, E \vdash T \vartriangleright \varSigma', F}{\delta, E \vdash \mathbf{reduce}\ T\ \mathbf{by}\ \sigma : \varSigma \to \varSigma' \vartriangleright \varSigma, \lambda e \in E \cdot F(e)|_\sigma}$$

$$\frac{\begin{array}{c}(B, P), E \vdash T \vartriangleright \varSigma, F \\ A \notin \mathrm{dom}(B) \cup \mathrm{dom}(P) \\ (B\,[\varSigma/A]\,, P), \{\, e\,[F(e)/A] \mid e \in E\,\} \vdash T' \vartriangleright \varSigma', F'\end{array}}{(B, P), E \vdash \mathbf{let}\ A = T\ \mathbf{in}\ T' \vartriangleright \varSigma', \lambda e \in E \cdot F'(e\,[F(e)/A])}$$

Fig. 3. The semantics of architectural specifications, continued.

Theorem 1. *For any architectural specification ASP and unit specification USP, if* $\vdash ASP :: USP$, *then* $\vdash ASP \vartriangleright \delta, U\varSigma, UF$ *for some* $\delta, U\varSigma, UF$ *such that* $UF(e) \in [\![USP]\!]$ *for all* $e \in \mathrm{dom}(UF)$. $\qquad\square$

Theorem 2. *For any architectural specification ASP and unit specification USP, if* $\vdash ASP \vartriangleright \delta, U\varSigma, UF$ *for some* $\delta, U\varSigma, UF$ *such that* $UF(e) \in [\![USP]\!]$ *for all* $e \in \mathrm{dom}(UF)$ *and:*

- *no parametric unit in ASP is applied more than once, and*
- *no parametric unit declaration in ASP is inconsistent,*

then $\vdash ASP :: USP$. $\qquad\square$

The first of the requirements of Th. 2 effectively means that our calculus is complete w.r.t. a *generative* semantics, i.e., one in which two applications of the same parametric unit to the same argument may yield different results (the semantics of Fig. 2 and 3 is non-generative, but both coincide if no unit is applied more than once). There are also methodological reasons for choosing a generative semantics (e.g., SML is generative [Pau96]; for a discussion of generativity see [Hof01], Sect. 4 *in fine*). The second of the above requirements is purely technical and of little importance, since an inconsistent declaration makes an architectural specification useless in any case.

Of course, the above theorems do not solve our problem fully, since we still need some method for discharging the following types of premises:

type I "for any model family $\{M_X\}_{X \in \mathrm{dom}(\varGamma)} \models \varGamma$, we have $M_A \models_{\varGamma(A)} \phi$", and
type II "for any model family $\{M_X\}_{X \in \mathrm{dom}(\varGamma)} \models \varGamma$, $M_A \oplus M_B$ exists".

(the premise $\phi \models_\varSigma \psi$ is merely a special case of a type I premise).

If the signature category has coproducts, then a type II premise may be transformed into a type IIa premise: "for any model family $\{M_X\}_{X \in \mathrm{dom}(\varGamma)} \models \varGamma$, M_C has a unique η-extension", where η is the universal morphism from a coproduct $\varGamma(A) \sqcup \varGamma(B)$ to the sum $\varGamma(A) \cup \varGamma(B)$.

3   Type I Premises

In this section we develop a technique for checking premises of the form "for any model family $\{M_X\}_{X \in \mathrm{dom}(\varGamma)} \models \varGamma$, we have $M_A \models_{\varGamma(A)} \phi$". It should be

$$\boxed{\vdash ASP :: USP}$$

$$\vdash UD_1 :: \Gamma_{par}^1, \Gamma^1 \ \dots \ \vdash UD_n :: \Gamma_{par}^n, \Gamma^n$$

$$\mathrm{dom}(\Gamma_{par}^i) \cap \mathrm{dom}(\Gamma_{par}^j) = \emptyset \text{ and } \mathrm{dom}(\Gamma^i) \cap \mathrm{dom}(\Gamma^j) = \emptyset \text{ for all } 1 \le i < j \le n$$

$$\Gamma_{par} = \Gamma_{par}^1 \cup \cdots \cup \Gamma_{par}^n, \ \Gamma = \Gamma^1 \cup \cdots \cup \Gamma^n \text{ and } \mathrm{dom}(\Gamma_{par}) \cap \mathrm{dom}(\Gamma) = \emptyset$$

$$\Gamma_{par}, \Gamma \vdash UE :: USP$$

$$\overline{\vdash \textbf{units } UD_1 \dots UD_n \textbf{ result } UE :: USP}$$

$$\boxed{\vdash UD :: \Gamma_{par}, \Gamma}$$

$$\overline{\vdash A : (\Sigma, \Phi) :: \emptyset, \{A :_{\Sigma} \Phi\}} \qquad \overline{\vdash A : (\Sigma, \Phi) \to (\Delta, \Psi) :: \{A :_{\Sigma \to \Delta} \Phi \to \Psi\}, \emptyset}$$

$$\boxed{\Gamma_{par}, \Gamma \vdash UE :: USP}$$

$$\Gamma_{par}, \Gamma \vdash T :: \Gamma', A$$

$$\frac{\text{for any family } \{M_X\}_{X \in \mathrm{dom}(\Gamma')} \models \Gamma', \text{ we have } M_A \models_{\Gamma'(A)} \Phi}{\Gamma_{par}, \Gamma \vdash T :: (\Sigma, \Phi)}$$

$$\Phi_0 \models_{\Sigma} \Phi \text{ and } \Phi \models_{\Sigma} \Phi_0$$

$$\Gamma_{par}, \Gamma \cup \{A :_{\Sigma} \Phi\} \vdash T :: \Gamma', B$$

$$\Sigma \text{ is a subsignature of } \Gamma'(B)$$

$$\text{for any model family } \{M_X\}_{X \in \mathrm{dom}(\Gamma')}, \ M_A \oplus M_B \text{ exists}$$

$$\frac{\text{for any model family } \{M_X\}_{X \in \mathrm{dom}(\Gamma')} \models \Gamma', \text{ we have } M_B \models_{\Gamma'(B)} \Psi}{\Gamma_{par}, \Gamma \vdash \lambda A : (\Sigma, \Phi_0) \bullet T :: (\Sigma, \Phi) \to (\Gamma'(B), \Psi)}$$

$$\boxed{\Gamma_{par}, \Gamma \vdash T :: \Gamma', A}$$

$$A \in \mathrm{dom}(\Gamma)$$

$$\overline{\Gamma_{par}, \Gamma \vdash A :: \Gamma, A}$$

$$\Gamma_{par}, \Gamma \vdash T :: \Gamma', A'$$

$$A :_{\Gamma'(A') \to \Delta} \Phi \to \Psi \text{ in } \Gamma_{par} \text{ and } B \notin \mathrm{dom}(\Gamma')$$

$$\frac{\text{for any model family } \{M_X\}_{X \in \mathrm{dom}(\Gamma')} \models \Gamma', \text{ we have } M_{A'} \models_{\Gamma'(A')} \Phi}{\Gamma_{par}, \Gamma \vdash A \,(\, T \,) :: \Gamma' \cup \{B :_{\Delta} \Psi, \ \iota_{\Gamma'(A') \subseteq \Delta} : A' \to B\}, B}$$

$$\Gamma_{par}, \Gamma \vdash T_1 :: \Gamma_1, A_1 \text{ and } \Gamma_{par}, \Gamma \vdash T_2 :: \Gamma_2, A_2$$

$$\Delta = \Gamma_1(A_1) \cup \Gamma_2(A_2)$$

$$\mathrm{dom}(\Gamma_1) \cap \mathrm{dom}(\Gamma_2) = \mathrm{dom}(\Gamma) \text{ and } B \notin \mathrm{dom}(\Gamma_1) \cup \mathrm{dom}(\Gamma_2)$$

$$\frac{\text{for any model family } \{M_X\}_{X \in \mathrm{dom}(\Gamma_1 \cup \Gamma_2)} \models \Gamma_1 \cup \Gamma_2, \ M_{A_1} \oplus M_{A_2} \text{ exists}}{\Gamma_{par}, \Gamma \vdash T_1 \textbf{ and } T_2 :: \Gamma_1 \cup \Gamma_2 \cup \{B :_{\Delta} \emptyset, \ \iota_{\Gamma_1(A_1) \subseteq \Delta} : A_1 \to B, \ \iota_{\Gamma_2(A_2) \subseteq \Delta} : A_2 \to B\}, B}$$

Fig. 4. A proof-calculus for architectural specifications.

$$\frac{\begin{array}{c}\Gamma_{par}, \Gamma \vdash T :: \Gamma', A \\ B \notin \mathrm{dom}(\Gamma')\end{array}}{\Gamma_{par}, \Gamma \vdash \mathbf{reduce}\ T\ \mathbf{by}\ \sigma : \Sigma \to \Gamma'(A) :: \Gamma' \cup \{B :_\Sigma \emptyset, \sigma : B \to A\}, B}$$

$$\frac{\begin{array}{c}\Gamma_{par}, \Gamma \vdash T :: \Gamma', B \\ \Gamma_{par}, \Gamma' \cup \{A :_{\Gamma'(A)} \emptyset, \mathrm{id}_{\Gamma'(A)} : A \to B\} \vdash T' :: \Gamma'', E \\ A \notin \mathrm{dom}(\Gamma')\ \text{and}\ D \notin \mathrm{dom}(\Gamma'')\end{array}}{\Gamma_{par}, \Gamma \vdash \mathbf{local}\ A = T\ \mathbf{within}\ T' :: \Gamma''[D/A], E[D/A]}$$

Fig. 5. A proof-calculus for architectural specifications, continued.

noted that this problem (and this applies to type IIa premises just as well) is so complex since we are not allowed to assume that the underlying institution has logical amalgamation (see below). The reason is the CASL institution lacking logical amalgamation. As suggested in [SMTKH01], to circumvent this problem, we represent the underlying institution in a second institution which enjoys the logical amalgamation property. In the sequel, we will work with two institutions, $\mathbf{I} = (Sig, \mathbf{Mod}, Sen, \models)$ and $\mathbf{I}' = (Sig', \mathbf{Mod}', Sen', \models')$.

An institution $\mathbf{I}$ has *logical amalgamation* if for any context Γ one may compute a sink $\{\alpha_X : \Gamma(X) \to \Sigma\}_{X \in \mathrm{dom}(\Gamma)}$ and a finite set of Σ-sentences Φ such that the set $\{\ \{M|_{\alpha_X}\}_{X \in \mathrm{dom}(\Gamma)} \mid M \in \mathbf{Mod}_\Sigma(\Phi)\ \}$ is equal to the set of all model families consistent with Γ. The pair $(\{\alpha_X\}_{X \in \mathrm{dom}(\Gamma)}, \Phi)$ is called an *amalgamating sink*. The idea here is that with logical amalgamation we are able to replace quantification of the form "for any model family $\{M_X\}_{X \in \mathrm{dom}(\Gamma)} \models \Gamma, \dots$" by "for any model $M \models_\Sigma \Phi, \dots$", with references to M_A in the former replaced by $M|_{\alpha_A}$ in the latter.

A *representation of* $\mathbf{I}$ *in* $\mathbf{I}'$ is a triple $R = (p, s, m)$, where:

- $p : Sig \to \mathrm{Pres}(\mathbf{I}')$ is functor,
- $s : Sen \to p; Sig[\_]; Sen'$ is a natural transformation,
- $m : p^{op}; \mathbf{Mod}' \to \mathbf{Mod}$ is a natural isomorphism.

These data must satisfy the following *representation condition* for all $\Sigma \in \mathbf{Ob}(Sig)$, $M' \in \mathbf{Mod}'(p(\Sigma))$, $\phi \in Sen(\Sigma)$:

$$M' \models'_{Sig[p(\Sigma)]} s_\Sigma(\phi) \iff m_\Sigma(M') \models_\Sigma \phi$$

The above-defined form of representation is in fact very strong, effectively implying that $\mathbf{I}$ is a "subinstitution" of $\mathbf{I}'$ definable in terms of $\mathbf{I}'$-sentences.

For any set Φ of Σ-sentences over $\mathbf{I}$, by $\overline{s_\Sigma}(\Phi)$ we denote the set $Ax[p(\Sigma)] \cup s_\Sigma(\Phi)$ of $Sig[p(\Sigma)]$-sentences over $\mathbf{I}'$. For any $\mathbf{I}$-context Γ, by $R(\Gamma)$ we denote the context obtained by mapping any declaration $A :_\Sigma \Phi$ in Γ to $A :_{Sig[p(\Sigma)]} \overline{s_\Sigma}(\Phi)$, and any declaration $\sigma : A \to B$ in Γ to $p(\sigma) : A \to B$.

Lemma 1. *Let Γ be an $\mathbf{I}$-context and $R = (p, s, m) : \mathbf{I} \to \mathbf{I}'$ a representation. Then the map taking any model family $\{M_X\}_{X \in \mathrm{dom}(\Gamma)}$ consistent with $R(\Gamma)$ to* *the model family* $\{m_{\Gamma(X)}(M_X)\}_{X \in \text{dom}(\Gamma)}$ *is a well defined bijection onto the set of all model families consistent with* Γ. $\qquad\qquad\square$

Theorem 3. *Assume that* $R : \mathbf{I} \to \mathbf{I}'$ *is a representation and that* $\mathbf{I}'$ *has logical amalgamation. Then, for any* $\mathbf{I}$-*context* Γ *with* $\Gamma(X) = \Sigma_X$ *for all* $X \in \text{dom}(\Gamma)$, *an amalgamating sink for* $R(\Gamma)$ *may be computed. Moreover, for any such sink* $(\{\tau_X\}_{X \in \text{dom}(\Gamma)}, \Phi)$, *for any* $A \in \text{dom}(\Gamma)$ *and any* Σ_A-*sentence* ϕ, *the following conditions are equivalent:*

1. *for any model family* $\{M_X\}_{X \in \text{dom}(\Gamma)} \models \Gamma$, *we have* $M_A \models_{\Sigma_A} \phi$,
2. $\Phi \models'_\Sigma \tau_A(s_{\Sigma_A}(\phi))$.

Proof. $((1) \Rightarrow (2))$ Take $Q \models'_\Sigma \Phi$. By definition of amalgamating sink, we have $\{ Q|_{\tau_X} \mid X \in \text{dom}(\Gamma) \} \models' R(\Gamma)$. Hence, $\{ m_{\Sigma_X}(Q|_{\tau_X}) \mid X \in \text{dom}(\Gamma) \} \models \Gamma$. Thus, by (1), $m_{\Sigma_A}(Q|_{\tau_A}) \models_{\Sigma_A} \phi$. This implies that $Q|_{\tau_A} \models'_{Sig[p(\Sigma_A)]} s_{\Sigma_A}(\phi)$, and so $Q \models'_{Sig[p(\Sigma_A)]} \tau_A(s_{\Sigma_A}(\phi))$ as required.

$((2) \Rightarrow (1))$ Take a model family $\{M_X\}_{X \in \text{dom}(\Gamma)} \models \Gamma$. Then by the lemma $\{m_{\Sigma_X}^{-1}(M_X)\}_{X \in \text{dom}(\Gamma)} \models' R(\Gamma)$. By definition of amalgamating sink, there exists a model $Q \models'_\Sigma \Phi$ such that $m_{\Sigma_X}^{-1}(M_X) = Q|_{\tau_X}$ for all $X \in \text{dom}(\Gamma)$. By (2), we have $Q \models'_\Sigma \tau_A(s_{\Sigma_A}(\phi))$. We may infer that $m_{\Sigma_A}^{-1}(M_A) = Q|_{\tau_A} \models'_{Sig[p(\Sigma_A)]} s_{\Sigma_A}(\phi)$, and so $M_A \models_{\Sigma_A} \phi$. $\qquad\qquad\square$

4   Type IIa Premises

We now develop techniques for checking the second type of premise, i.e., "for any model family $\{M_X\}_{X \in \text{dom}(\Gamma)} \models \Gamma$, M_A is uniquely η-extendible", where η comes from a certain class $\mathcal{E}$ of signature morphisms.

We say that in $\mathbf{I}$ *morphism equalities are expressible* if for any $\alpha, \beta : \Sigma \to \Delta$ one may compute a finite set Φ of Δ-sentences such that for any finite set of Δ-sentences Φ_0 we have $\Phi_0 \models_\Delta \Phi$ iff for any $M \models_\Delta \Phi_0$ the equality $M|_\alpha = M|_\beta$ holds. We then say that Φ *expresses* $\alpha = \beta$. Observe that we require much less than $M \models_\Delta \Phi$ iff $M|_\alpha = M|_\beta$ (such a Φ would usually not exist).

Let $\mathcal{E}$ be a class of signature morphisms in $\mathbf{I}$ and $R = (p, s, m) : \mathbf{I} \to \mathbf{I}'$ a representation. Then $\mathcal{E}$-*extendability is expressible in* R if for any $\eta : \Sigma \to \Sigma'$ from $\mathcal{E}$, setting $p(\Sigma) = (\Delta, \Phi)$ and $p(\Sigma') = (\Delta', \Phi')$, one may compute:

1. pairs $(\alpha_1, \beta_1), \ldots, (\alpha_n, \beta_n)$, where $\alpha_i, \beta_i : \Delta_i \to \Delta$ are signature morphisms in $\mathbf{I}'$, such that for any Δ-model M, M is $p(\eta)$-extendible iff $M|'_{\alpha_1} = M|'_{\beta_1}$, $\ldots$, $M|'_{\alpha_n} = M|'_{\beta_n}$, and
2. a finite set Ψ' of Δ'-sentences such that for any Δ-model N:
   (a) if N has a unique $p(\eta)$-extension satisfying Φ', then any of its $p(\eta)$-extensions satisfies Ψ', and
   (b) if N has a $p(\eta)$-extension satisfying Ψ', then it has a unique $p(\eta)$-extension satisfying Φ'.

We then say that $(\alpha_1, \beta_1), \ldots, (\alpha_n, \beta_n)$ and Ψ' express η-extendability. The idea behind this definition is as follows. Condition (1) captures $p(\eta)$-extendability. Condition (2) is designed to transform this into unique η-extendability: point (b) ensures unique η-extendability, while point (a) checks that Ψ' is not overly restrictive.

Theorem 4. *Let $R : \mathbf{I} \to \mathbf{I}'$ be a representation and assume that $\mathbf{I}'$ has logical amalgamation, $\mathcal{E}$-extendability is expressible in R and morphism equalities are expressible in $\mathbf{I}'$. For any $\mathbf{I}$-context Γ with $\Gamma(X) = \Sigma_X$ for all $X \in \mathrm{dom}(\Gamma)$, for any $A \in \mathrm{dom}(\Gamma)$ and any $\eta \in \mathcal{E}$ with $\mathrm{dom}(\eta) = \Sigma_A$ and $\mathrm{cod}(\eta) = \Delta$ the following objects may be computed:*

(A) an amalgamating sink $(\{\tau_X\}_{X \in \mathrm{dom}(\Gamma)}, \Phi)$ for $R(\Gamma)$,
(B) pairs $(\alpha_1, \beta_1), \ldots, (\alpha_n, \beta_n)$ and a set Ψ expressing η-extendability,
(C) sets $\Phi_1, \ldots, \Phi_n$ expressing the morphism equalities $(\alpha_1; \tau_A = \beta_1; \tau_A), \ldots, (\alpha_n; \tau_A = \beta_n; \tau_A)$, respectively,
(D) an amalgamating sink $(\{\sigma_X\}_{X \in \mathrm{dom}(R\Gamma')}, \Phi')$ for $R\Gamma' = R(\Gamma) \cup \{B :_{\mathrm{Sig}[p(\Delta)]} \emptyset, \ p(\eta) : A \to B\}$, where $B \notin \mathrm{dom}(\Gamma)$.

Moreover, for any such objects, the following conditions are equivalent:

1. for any model family $\{M_X\}_{X \in \mathrm{dom}(\Gamma)} \models \Gamma$, M_A has a unique η-extension,
2. (a) $\Phi \models'_\Sigma \Phi_1, \ldots, \Phi \models'_\Sigma \Phi_n$, and
*   (b) $\Phi' \models' \sigma_B(\Psi)$.*

Proof. $((1) \Rightarrow (2a))$ Take $1 \le i \le n$ and arbitrary $M \models'_\Sigma \Phi$. By (A), we have $\{ M|'_{\tau_X} \mid X \in \mathrm{dom}(\Gamma) \} \models' R(\Gamma)$. Thus $\{ m_{\Sigma_X}(M|'_{\tau_X}) \mid X \in \mathrm{dom}(\Gamma) \} \models \Gamma$, and, by condition (1), there exists a (unique) model N such that $N|_\eta = m_{\Sigma_A}(M|'_{\tau_A})$, which implies that $m_{\Sigma_A}^{-1}(N)|'_{p(\eta)} = M|'_{\tau_A}$. This proves that $M|'_{\tau_A}$ is $p(\eta)$-extendible, and so by (B) we have $M|'_{\alpha_i; \tau_A} = M|'_{\beta_i; \tau_A}$.

By (C), this proves that $\Phi \models'_\Sigma \Phi_i$.

$((1) \Rightarrow (2b))$ Take any model $M \models' \Phi'$. By (D), we have $\{ M|'_{\sigma_X} \mid X \in \mathrm{dom}(\Gamma') \} \models' R\Gamma'$, and so $\{ M|'_{\sigma_X} \mid X \in \mathrm{dom}(\Gamma) \} \models' R(\Gamma)$. Hence, we have $\{ m_{\Sigma_X}(M|'_{\sigma_X}) \mid X \in \mathrm{dom}(\Gamma) \} \models \Gamma$, and, by condition (1), $m_{\Sigma_A}(M|'_{\sigma_A})$ has a unique η-extension P.

This implies that $M|'_{\sigma_A}$ has a unique $p(\eta)$-extension satisfying Φ'. First, $m_{\Sigma_B}^{-1}(P)$ is such an extension, since $m_{\Sigma_B}^{-1}(P)|'_{p(\eta)} = m_{\Sigma_A}^{-1}(P|_\eta) = m_{\Sigma_A}^{-1}(m_{\Sigma_A}(M|'_{\sigma_A})) = M|'_{\sigma_A}$ and by the definition of m we have $m_{\Sigma_B}^{-1}(P) \in \mathbf{Mod}'(p(\Delta))$. Second, if $Q|'_{p(\eta)} = M|'_{\sigma_A}$ and $Q \in \mathbf{Mod}'(p(\Delta))$, then $m_{\Sigma_B}(Q)$ is well-defined, and $m_{\Sigma_B}(Q)|_\eta = m_{\Sigma_A}(Q|'_{p(\eta)}) = m_{\Sigma_A}(M|'_{\sigma_A})$, so by uniqueness of P we get $P = m_{\Sigma_B}(Q)$, i.e., $Q = m_{\Sigma_B}^{-1}(P)$.

Thus, by (B), we may conclude that any $p(\eta)$-extension of $M|'_{\sigma_A}$ satisfies Ψ', in particular $M|'_{\sigma_B} \models'_\Delta \Psi'$, i.e., $M \models' \sigma_B(\Psi')$.

$((2) \Rightarrow (1))$ Let $\{M_X\}_{X \in \mathrm{dom}(\Gamma)}$ be a model family consistent with Γ.

We will first prove that for any $1 \le i \le n$ we have $m_{\Sigma_A}^{-1}(M_A)|'_{\alpha_i} = m_{\Sigma_A}^{-1}(M_A)|'_{\beta_i}$. Consider the model family $\{m_{\Sigma_X}^{-1}(M_X)\}_{X \in \mathrm{dom}(\Gamma)}$. It is consistent with

$R(\Gamma)$, so, by (A), there exists a model $Q \models'_\Sigma \Phi$ such that for all $X \in \mathrm{dom}(\Gamma)$ we have $Q|'_{\tau_X} = m_{\Sigma_X}^{-1}(M_X)$. From condition (2a) we infer that $Q \models'_\Sigma \Phi_i$ for $1 \le i \le n$. Now (C) gives us $Q|'_{\alpha_i;\tau_A} = Q|'_{\beta_i;\tau_A}$ and, hence, $m_{\Sigma_A}^{-1}(M_A)|'_{\alpha_i} = m_{\Sigma_A}^{-1}(M_A)|'_{\beta_i}$ for $1 \le i \le n$.

From (B) we know that there exists a model N with $N|'_{p(\eta)} = m_{\Sigma_A}^{-1}(M_A)$.

We now prove that $N \models'_{Sig[p(\Delta)]} \Psi$. Define $N_X = m_{\Sigma_X}^{-1}(M_X)$ for $X \in \mathrm{dom}(\Gamma)$ and $N_B = N$. Then $\{N_X\}_{X \in \mathrm{dom}(R\Gamma')} \models' R\Gamma'$. Thus, by (D), there exists a model $Q \models' \Phi'$ such that $Q|'_{\sigma_X} = N_X$ for all $X \in \mathrm{dom}(R\Gamma')$. By condition (2b) we then have $Q \models' \sigma_B(\Psi)$, and so $N = N_B = Q|'_{\sigma_B} \models'_{Sig[p(\Delta)]} \Psi$.

By (B) we may infer that there exists a unique model $M' \in \mathbf{Mod}'(p(\Delta))$ such that $M'|'_{p(\eta)} = N|'_{p(\eta)}$.

We claim that $P = m_\Delta(M')$ is the unique model satisfying $P|_\eta = M_A$. First, it is well defined, since $M' \in \mathbf{Mod}'(p(\Delta))$. Second, $P|_\eta = m_\Delta(M')|_\eta = m_{\Sigma_A}(M'|'_{p(\eta)}) = m_{\Sigma_A}(N|'_{p(\eta)}) = M_A$. Third, for any P with $P|_\eta = M_A$ we have $m_\Delta^{-1}(P)|'_{p(\eta)} = m_{\Sigma_A}^{-1}(P|_\eta) = m_{\Sigma_A}^{-1}(M_A) = N|'_{p(\eta)}$, and by the uniqueness of M' we have $M' = m_\Delta^{-1}(P)$, which means $P = m_\Delta(M')$. $\square$

5   The Logic of CASL

In this section we define the institution of CASL logic and prove some basic facts about it (this definition is consistent with what may be found in the CASL semantics [CASL00]). We do this in three steps.

In step one, we construct the institution $\mathrm{CASL}_0 = (Sig_0, \mathbf{Mod}_0, Sen_0, \models^0)$. Its features include multi-sorted first-order logic, both total and partial function symbols, predicates, and sort-generation constraints. It is defined as follows:

- a signature Σ in Sig_0 is a tuple (S, TF, PF, P), where:
  - S is a finite set of *sorts*,
  - $TF = \{TF_{w,s}\}_{w \in S^*, s \in S}$ is a family of finite sets of *names of total function symbols*,
  - $PF = \{PF_{w,s}\}_{w \in S^*, s \in S}$ is a family of finite sets of *names of partial function symbols*,
  - $P = \{P_w\}_{w \in S^*}$ is a family of finite sets of *names of predicates* .

  We additionally require that always $TF_{w,s} \cap PF_{w,s} = \emptyset$. A function symbol $f \in TF_{w,s} \cup PF_{w,s}$ is denoted $f_{w,s}$. A predicate $b \in P_w$ is denoted b_w.
- a signature morphism $\sigma : (S, TF, PF, P) \to (S', TF', PF', P')$ in Sig_0 is a triple $(\sigma^S, \sigma^F, \sigma^P)$, where:
  - $\sigma^S : S \to S'$ is a function,
  - $\sigma^F = \{\sigma^F_{w,s}\}_{w \in S^*, s \in S}$ is a family of functions, $\sigma^F_{w,s} : TF_{w,s} \cup PF_{w,s} \to TF'_{(\sigma^S)^*(w),\sigma^S(s)} \cup PF'_{(\sigma^S)^*(w),\sigma^S(s)}$ preserving the totality of symbols,
  - $\sigma^P = \{\sigma^P_w\}_{w \in S^*}$ is a family of functions, with $\sigma^P_w : P_w \to P'_{(\sigma^S)^*(w)}$.

– for any signature Σ, $\mathbf{Mod_0}(\Sigma)$ is the category of models having:
  - as models M, functions taking:
    * any sort s in Σ to a non-empty carrier set $M[cr, s]$,
    * any function symbol $f_{w,s}$ in Σ, where $w = s_1 \ldots s_n$, to a partial function $M[fn, f_{w,s}] : M[cr, s_1] \times \cdots \times M[cr, s_n] \to M[cr, s]$, the function being total if $f_{w,s}$ is,
    * any predicate b_w in Σ, where $w = s_1 \ldots s_n$, to a set $M[pr, b_w] \subseteq M[cr, s_1] \times \cdots \times M[cr, s_n]$.
  - as (homo)morphisms $h : M \to M'$, families $\{h_s\}_{s \in S^*}$ of functions $h_s : M[cr, s] \to M'[cr, s]$ such that for any tuple of sorts $w = s_1 \ldots s_n$ in Σ and any tuple $(x_1, \ldots, x_n) \in M[cr, s_1] \times \cdots \times M[cr, s_n]$:
    * for any sort s and function symbol $f_{w,s}$ in Σ, if $h_s(M[fn, f_{w,s}](x_1, \ldots, x_n))$ is defined, then so is $M'[fn, f_{w,s}](h_{s_1}(x_1), \ldots, h_{s_n}(x_n))$ and they are both equal,
    * for any predicate b_w in Σ, if $(x_1, \ldots, x_n) \in M[pr, b_w]$, then $(h_{s_1}(x_1), \ldots, h_{s_n}(x_n)) \in M'[pr, b_w]$.
– for any signature morphism $\sigma = (\sigma^S, \sigma^F, \sigma^P) : \Sigma = (S, TF, PF, P) \to \Sigma' = (S', TF', PF', P')$, the reduct functor $\cdot|_\sigma : \mathbf{Mod_0}(\Sigma') \to \mathbf{Mod_0}(\Sigma)$ takes:
  - any model $M' \in \mathbf{Mod_0}(\Sigma')$ to the model $M \in \mathbf{Mod_0}(\Sigma)$ defined by:
    * for all sorts s in Σ, $M[cr, s] = M'[cr, \sigma^S(s)]$,
    * for all function symbols $f_{w,s}$ in Σ, $M[fn, f_{w,s}] = M'[fn, \sigma^F_{w,s}(f)_{(\sigma^S)^*(w), \sigma^S(s)}]$,
    * for all predicates b_w in Σ, $M[pr, b_w] = M'[pr, \sigma^P_w(b)_{(\sigma^S)^*(w)}]$.
  - any Σ'-homomorphism $h' = \{h'_{s'}\}_{s' \in S'} : M' \to N'$ to the Σ-homomorphism $h = h'|_\sigma : M'|_\sigma \to N'|_\sigma$ defined by $h = \{h'_{\sigma^S(s)}\}_{s \in S}$.
– for any signature Σ, $Sen_0(\Sigma)$ is the set of all closed first-order formulas, with atomic formulas built using variables of sorts in Σ, function symbols and predicates in Σ, and equality. Additionally, triples (S_0, F_0, θ), called *sort-generation constraints*, are also in this set, provided that:
  - $\theta : \Sigma_0 \to \Sigma$ is a signature morphism,
  - S_0 is the set of *generated sorts*, i.e., a set of sorts from Σ_0,
  - F_0 is the set of *constructors*, i.e., a set of function symbols from Σ_0.
– for any signature morphism $\sigma : \Sigma \to \Sigma'$, the function $Sen_0(\sigma) : Sen_0(\Sigma) \to Sen_0(\Sigma')$ is defined naturally on first-order sentences and by the rule that $\sigma((S_0, F_0, \theta)) = Sen_0(\sigma)((S_0, F_0, \theta)) = (S_0, F_0, \theta; \sigma)$ on sort-generation constraints.
– the satisfaction relation $M \models^0_\Sigma \phi$ for first-order ϕ is defined as normal first-order satisfaction, with $=$ interpreted strongly and predicates on non-defined values being false. For sort-generation constraints we have $M \models^0_\Sigma (S_0, F_0, \theta)$ if for any $s_0 \in S_0$ and any $x_0 \in (M|_\theta)[cr, s_0]$ there exists a term t_0 over $dom(\theta)$ of sort s_0 built using function symbols from F_0 and with no variables of sorts from S_0, and a valuation v_0 into $M|_\theta$ such that t_0 under valuation v_0 has value x_0.

Proposition 1. CASL$_0$ *is an institution.*   □

For any morphism $\sigma : \Sigma \to \Delta$, the "image" of Σ is denoted $\sigma(\Sigma)$, and its complement $\Delta \setminus \sigma(\Sigma)$. We say that σ is an *inclusion* if $\sigma : \Sigma \to \sigma(\Sigma)$ is the identity.

In the second step we construct the institution $\text{CASL}_{\leq} = (Sig_{\leq}, \mathbf{Mod}_{\leq}, Sen_{\leq}, \models^{\leq})$. Here, signatures are additionally equipped with a subsort preorder which, on the model level, is interpreted by injective mappings embedding a subsort, say *Zero*, in a supersort, say *NonNeg* or *NonPos*. By means of an overloading relation it is then ensured that if two function symbols use the same name, say $f : NonNeg \to Elem$ and $f : NonPos \to Elem$, then both are interpreted so that they "coincide" on the carrier of *Zero* (the common subsort). The formal definition follows:

- a signature Σ in $Sig_{\leq}$ is a tuple $(S, TF, PF, P, \leq)$ such that (S, TF, PF, P) is a signature in Sig_0 and $\leq$ is a reflexive-transitive *subsort preorder* on $S \times S$; for such a signature the *overloading relations* $\sim_F$ and $\sim_P$ are defined by:
  - $f_{w_1, s_1} \sim_F f_{w_2, s_2}$ if there exist w and s with $w \leq^* w_1, w_2$ and $s_1, s_2 \leq s$,
  - $b_{w_1} \sim_P b_{w_2}$ if there exists w with $w \leq^* w_1, w_2$.
- a signature morphism $\sigma : \Sigma = (S, TF, PF, P, \leq) \to \Sigma' = (S', TF', PF', P', \leq')$ in $Sig_{\leq}$ is a signature morphism $\sigma = (\sigma^S, \sigma^F, \sigma^P) : (S, TF, PF, P) \to (S', TF', PF', P')$ in Sig_0 such that:
  - $s \leq s'$ implies $\sigma^S(s) \leq' \sigma^S(s')$ for all sorts s, s' in Σ,
  - $f_{w_1, s_1} \sim_F f_{w_2, s_2}$ implies $\sigma^F_{w_1, s_1}(f)_{(\sigma^S)^*(w_1), \sigma^S(s_1)} \sim'_F \sigma^F_{w_2, s_2}(f)_{(\sigma^S)^*(w_2), \sigma^S(s_2)}$ for all function symbols $f_{w_1, s_1}, f_{w_2, s_2}$ in Σ,
  - $b_{w_1} \sim_P b_{w_2}$ implies $\sigma^P_{w_1}(b)_{(\sigma^S)^*(w_1)} \sim'_P \sigma^P_{w_2}(b)_{(\sigma^S)^*(w_2)}$ for all predicates b_{w_1}, b_{w_2} in Σ.
- for any signature $\Sigma = (S, TF, PF, P, \leq)$ in $Sig_{\leq}$, by $\Sigma^{\#}$ we denote the Sig_0 signature (S, TF', PF', P'), where:
  - $TF'_{w,s} = TF_{w,s} \uplus \{em_{s' \leq s} | w = s' \leq s\}$ (the *embeddings*),
  - $PF'_{w,s} = PF_{w,s} \uplus \{prj_{s' \geq s} | s \leq s' = w\}$ (the *projections*),
  - $P'_w = P_w \uplus \{in_{s,s'} | s' \leq s = w\}$ (the *membership predicates*).
This naturally extends to a functor $(\_)^{\#} : Sig_{\leq} \to Sig_0$.

By $\Sigma^{\#\#}$ we denote the presentation $(\Sigma^{\#}, \Phi)$, where Φ consists of the *defining sentences*, which state that:
1. an embedding of s into s is the identity,
2. an embedding of s into s' is injective,
3. the composition of embeddings of s into s' and s' into s'' is equal to the embedding of s into s',
4. a projection of s' onto s is the least right-sided inverse of the embedding of s into s',
5. a membership predicate $in_{s,s'}$ holds on an element x of sort s iff the projection of s onto s' is defined on x,
6. if $f_{w_1, s_1} \sim_F f_{w_2, s_2}$, $w \leq^* w_1, w_2$ and $s_1, s_2 \leq s$, then for any tuple x^* of elements of sorts w, embedding x^* in w_1, applying f_{w_1, s_1} and embedding that in s gives the same result, as embedding x^* in w_2, applying f_{w_2, s_2} and embedding that in s,
7. if $b_{w_1} \sim_P b_{w_2}$ and $w \leq^* w_1, w_2$, then for any tuple x^* of elements of sorts w, embedding x^* in w_1 and checking, whether it is in b_{w_1}, gives the same result as embedding x^* in w_2 and checking, whether it is in b_{w_2}.

Again, this naturally extends to a functor $(\_)^{\#\#} : Sig_\le \to \mathrm{Pres}(\mathrm{CASL}_0)$.

- the model functor $\mathbf{Mod}_\le$ is defined as the composition $((\_)^{\#\#})^{op}; \mathbf{Mod}_0$,
- the sentence functor $Sen_\le$ is defined as the composition $(\_)^{\#}; Sen_0$,
- the satisfaction relation is defined by $M \models^\le_\Sigma \phi$ if $M \models^0_{\Sigma\#} \phi$, for $M \in \mathbf{Mod}_\le(\Sigma)$ and $\phi \in Sen_\le(\Sigma)$.

Proposition 2. *The triple R_0 consisting of:*

- $(\_)^{\#\#} : Sig_\le \to \mathrm{Pres}(\mathrm{CASL}_0)$,
- $\{\mathrm{id}_{Sen_\le(\Sigma)}\}_{\Sigma \in \mathbf{Ob}(Sig_\le)} : Sen_\le \to (\_)^{\#}; Sen_0$,
- $\{\mathrm{id}_{\mathbf{Mod}_\le(\Sigma)}\}_{\Sigma \in \mathbf{Ob}(Sig_\le)} : ((\_)^{\#\#})^{op}; \mathbf{Mod}_0 \to \mathbf{Mod}_\le$,

is a representation of $\mathrm{CASL}_\le$ in CASL_0. $\qquad\qquad\square$

We now present a construction developed in [ST88]. Given an institution $\mathbf{I} = (Sig, \mathbf{Mod}, Sen, \models)$ and a class $\mathcal{D}$ of signature morphisms, define the institution of *structured specifications over* $\mathbf{I}$, denoted $Str_\mathcal{D}(\mathbf{I}) = (Sig, \mathbf{Mod}, Sen', \models')$, by structural induction. For any signature Σ, a sentence in $Sen'(\Sigma)$ is:

- a presentation (Σ, Φ) over $\mathbf{I}$; we define $M \models'_\Sigma (\Sigma, \Phi)$ if $M \models_\Sigma \Phi$,
- **translate** ϕ **by** σ, with $\sigma : \Delta \to \Sigma$ and ϕ a Δ-sentence over $Str_\mathcal{D}(\mathbf{I})$; we define $M \models'_\Delta$ **translate** ϕ **by** σ if $M|_\sigma \models'_\Sigma \phi$,
- $\phi_1 \cup \phi_2$, where ϕ_1 and ϕ_2 are Σ-sentences over $Str_\mathcal{D}(\mathbf{I})$; we define $M \models'_\Sigma \phi_1 \cup \phi_2$ if $M \models'_\Sigma \phi_1$ and $M \models'_\Sigma \phi_2$,
- **derive** σ **from** ϕ, where $\sigma : \Sigma \to \Delta$ comes from $\mathcal{D}$ and ϕ is a Δ-sentence over $Str_\mathcal{D}(\mathbf{I})$; we define $M \models'_\Delta$ **derive** σ **from** ϕ if M has a σ-extension $N \models'_\Sigma \phi$,
- **free** ϕ **along** σ, with $\sigma : \Delta \to \Sigma$ and ϕ a Σ-sentence over $Str_\mathcal{D}(\mathbf{I})$; we define $M \models'_\Delta$ **free** ϕ **along** σ if $M \models'_\Delta \phi$ and for any model $N \models'_\Delta \phi$ and any homomorphism $h : M|_\sigma \to N|_\sigma$ there exists a unique homomorphism $g : M \to N$ such that $g|_\sigma = h$.

The translation of a Σ-sentence ϕ over $Str_\mathcal{D}(\mathbf{I})$ along a signature morphism $\sigma : \Sigma \to \Delta$ is defined to be **translate** ϕ **by** σ.

Now, the *full* CASL *institution*, denoted CASL_1, is simply $Str_\mathcal{I}(\mathrm{CASL}_\le)$, where $\mathcal{I}$ is the class of all signature morphisms σ, which are inclusions.

The following proposition allows us to lift representations of institutions to representations of structured specifications over them:

Proposition 3. *For any class $\mathcal{D}$ of signature morphisms in $\mathbf{I}$ and representation $R = (p, s, m) : \mathbf{I} \to \mathbf{I'}$ the following is a representation $R' = (p', s', m') : Str_\mathcal{D}(\mathbf{I}) \to Str_{\mathcal{D'}}(\mathbf{I'})$, where $\mathcal{D'} = \{p(\sigma) \mid \sigma \in \mathcal{D}\}$:*

- $p'(\Sigma) = (Sig[p(\Sigma)], \{p(\Sigma)\})$,
- $m' = m$,

– s' *is defined by structural induction:*
- $s'_\Sigma((\Sigma, \Phi)) = (Sig[p(\Sigma)], \overline{s_\Sigma}(\Phi))$,
- $s'_\Sigma(\text{translate } \phi \text{ by } \sigma) = \left(\text{translate } s'_{\text{dom}(\sigma)}(\phi) \text{ by } p(\sigma)\right) \cup p(\text{cod}(\Sigma))$,
- $s'_\Sigma(\phi_1 \cup \phi_2) = s'_\Sigma(\phi_1) \cup s'_\Sigma(\phi_2)$;
- $s'_\Sigma(\text{derive } \sigma \text{ from } \phi) = \text{derive } p(\sigma) \text{ from } s'_{\text{cod}(\sigma)}(\phi)$,
- $s'_\Sigma(\text{free } \phi \text{ along } \sigma) = \text{free } s'_{\text{cod}(\sigma)}(\phi) \text{ along } p(\sigma)$. $\square$

Let $\text{STRCASL}_0 = Str_\mathcal{I}(\text{CASL}_0)$. The *standard representation* $R = (p, m, s)$: $\text{CASL}_1 \to \text{STRCASL}_0$ is the representation obtained via the above proposition from the representation $R_0 : \text{CASL}_\le \to \text{CASL}_0$. Note that any CASL structured specification, as defined in the official semantics, may be easily represented as a sentence over CASL_1; the institution STRCASL_0, on the other hand, is equivalent to CASL structured specifications in which no subsorting is ever used.

6   Applying the Theorems to CASL

In this section we apply Th. 3 and 4 to CASL and the standard representation.

From [Hof01] (Prop. 1 and Cor. 1) we know that if the signature category is finitely cocomplete, the colimits are computable, and the model functor preserves finite limits, then the institution has logical amalgamation. The category Sig_0 is indeed finitely cocomplete [Mos98], the colimits being computable. The functor $\mathbf{Mod}_0$ preserves finite limits, too (cf. [GB84] for the first-order case). Thus:

Proposition 4. *The institution* STRCASL_0 *has logical amalgamation.* $\square$

The same is of course true for CASL_0. However, neither $\text{CASL}_\le$, nor CASL_1 have logical amalgamation (see [SMTKH01]), and this is the reason for introducing the representation R.

We now take a look at the assumptions of Th. 4. We will need the following auxiliary fact:

Proposition 5. *In the institution* STRCASL_0 *isomorphic models are elementarily equivalent, i.e., they satisfy precisely the same sentences.* $\square$

Proposition 6. *In* STRCASL_0 *morphism equalities are expressible.*

Proof. Let $\alpha, \beta : \Sigma \to \Delta$ be arbitrary signature morphisms. If there exists a sort s in Σ with $\alpha^S(s) \ne \beta^S(s)$, then let ϕ be the sentence *"false"*. Otherwise let ϕ be the conjunction of:

– $"\forall x^* : (\alpha^S)^*(w) \cdot \alpha^S_{w,s}(f)_{(\alpha^S)^*(w), \alpha^S(s)}(x^*) = \beta^S_{w,s}(f)_{(\alpha^S)^*(w), \alpha^S(s)}(x^*)"$, for all $f_{w,s}$ in Σ,
– $"\forall x^* : (\alpha^S)^*(w) \cdot \alpha^P_w(b)_{(\alpha^S)^*(w)}(x^*) \iff \beta^P_w(b)_{(\alpha^S)^*(w)}(x^*)"$, for b_w in Σ.

We claim that for any finite set Φ_0 of Δ-sentences, we have $\Phi_0 \models_\Delta \phi$ iff for any $M \models_\Delta \Phi_0$ we have $M|_\alpha = M|_\beta$.

The proof "$\Rightarrow$" is trivial. For "$\Leftarrow$", take Φ_0 satisfying the premise and let $M \models_\Delta \Phi_0$.

There exists a model N with all carriers distinct (here we use the non-emptyness of carriers), isomorphic to M. By Prop. 5, M and N satisfy precisely the same Δ-sentences. Thus $N \models_\Delta \Phi_0$ and therefore $N|_\alpha = N|_\beta$. This implies that for any sort s in Σ $N[cr, \alpha^S(s)] = N[cr, \beta^S(s)]$, and so $\alpha^S(s) = \beta^S(s)$.

Thus, ϕ is a conjunction of sentences, as defined above. We only need to prove that this conjunction holds in M. But this is obvious, since $M|_\alpha = M|_\beta$, so for any $f_{w,s}$ in Σ we must have

$$M[fn, \alpha^F_{w,s}(f)_{(\alpha^S)^*(w), \alpha^S(s)}] = M[fn, \beta^F_{w,s}(f)_{(\beta^S)^*(w), \beta^S(s)}],$$

which implies that M must satisfy the appropriate element of the conjunction; a similar argument applies to predicates.   □

Let $\mathcal{E}$ be the class of morphisms $\eta : \Sigma \to \Sigma'$ in CASL_1 such that η^S, η^F and η^P are surjective functions and the subsort preorder in Σ' is the least preorder generated by the η^S-image of the subsort preorder of Σ. Note that $\eta \in \mathcal{E}$ does not imply that $p(\eta)$ is an epimorphism. We do have:

Proposition 7. *Any universal morphism* $\sigma : \Sigma_1 \sqcup \Sigma_2 \to \Sigma_1 \cup \Sigma_2$ *in* CASL_1 *is in* $\mathcal{E}$.   □

Proposition 8. $\mathcal{E}$-*extendability is expressible in* R.

Proof. Let $\eta : \Sigma \to \Sigma'$ be in $\mathcal{E}$; set $p(\Sigma) = (\Delta, \Phi)$ and $p(\Sigma') = (\Delta', \Phi')$. Let $\theta = p(\eta) : \Delta = (S, TF, PF, P) \to \Delta' = (S', TF', PF', P')$. By $C_\emptyset$ we denote a function constantly equal $\emptyset$ and with an appropriate domain.

To prove condition (1) of the definition we now define three families of pairs of morphisms:

- for any sorts s, s' in Δ such that $\theta^S(s) = \theta^S(s')$, let $\alpha, \beta : (\{s\}, C_\emptyset, C_\emptyset, C_\emptyset) \to \Delta$ be such that α takes s to s and β takes s to s',
- for any function symbols $f_{w,s}$ and $f'_{w',s'}$ in Δ, where $w = s_1 \ldots s_n$ and $w' = s'_1 \ldots s'_n$, such that $\theta^F_{w,s_0}(f) = \theta^F_{w',s'_0}(f')$, let $\alpha, \beta : (\{t_1, \ldots, t_n, t_0\}, C_\emptyset, C_\emptyset[\{f\}/(t_1 \ldots t_n, t_0)], C_\emptyset) \to \Delta$ be such that α takes f to f and t_i to s_i, and β takes f to f' and t_i to s'_i (for $0 \le i \le n$; the t_i are chosen distinct),
- for any predicates b_w and $b'_{w'}$ in Δ, where $w = s_1 \ldots s_n$ and $w' = s'_1 \ldots s'_n$, such that $\theta^P_w(b) = \theta^P_{w'}(b')$, let $\alpha, \beta : (\{t_1, \ldots, t_n\}, C_\emptyset, C_\emptyset, C_\emptyset[\{b\}/t_1 \ldots t_n]) \to \Delta$ be such that α takes b to b and t_i to s_i, and β takes b to b' and t_i to s'_i (for $1 \le i \le n$; the t_i again chosen distinct).

We claim that a Δ-model M is θ-extendible iff for all of the above-defined pairs (α, β) we have $M|'_\alpha = M|'_\beta$.

The "only if" part is fairly obvious: if we have a Δ'-model N with $N|_\theta = M$ and (α, β) is one of the above pairs, then, since $\alpha; \theta = \beta; \theta$, we have $M|'_\alpha = (N|'_\theta)|_\alpha = N|'_{\alpha;\theta} = N|'_{\beta;\theta} = (N|'_\theta)|_\beta = M|'_\beta$.

For the "if" part, suppose that a Δ-model M indeed satisfies all of the required equations. We define a Δ'-model N:

- for any sort s' in Δ', $N[cr, s'] = M[cr, s]$, for any sort s in Δ with $s' = \theta^S(s)$ (since $\eta \in \mathcal{E}$, such an s must exist),
- for any function symbol $f'_{w',s'}$ in Δ', $N[fn, f'_{w',s'}]$ is:
  - $M[fn, f_{w,s}]$ if $w' = (\theta^S)^*(w)$, $s' = \theta^S(s)$ and $f' = \theta^F_{w,s}(f)$ for some function symbol $f_{w,s}$ in Δ,
  - any function of the appropriate type, otherwise,
- for any predicate $b'_{w'}$ in Δ', $N[pr, b'_{w'}]$ is:
  - $M[pr, b_w]$ if $w' = (\theta^S)^*(w)$ and $b' = \theta^P_w(b)$ for some predicate b_w in Δ,
  - the empty relation, otherwise.

This definition is correct precisely because for any of the above pairs of morphisms (α, β) we have $M|'_\alpha = M|'_\beta$. Directly from the definition it follows that $M = N|'_\theta$. This proves condition (1) of the definition. Thus, we move to condition (2).

For any sorts s, t in Δ', an s, t-path q is a sequence $s_1 \leq \cdots \leq s_n$ ($n > 1$) such that $s = \theta^S(s_1)$, $t = \theta^S(s_n)$ and that $s_i = s_j$, $i < j$ imply $i = 1$, $j = n$. For any Δ'-term T of sort s, by $q(T)$ we denote the "composition-of-embeddings" term $em_{\theta^S(s_{n-1}) \leq \theta^S(s_n)}(\ldots em_{\theta^S(s_1) \leq \theta^S(s_2)}(T)\ldots)$. These notions are further extended to tuples w of sorts and q^* of sequences (the sequence at each coordinate may be of different length). Observe that since $\eta \in \mathcal{E}$, for any sorts $s \leq' t$ in Δ' there exists an s, t-path. Also, the set of all paths is finite.

As the set Ψ' from condition (2) we take $\theta(\Phi)$ plus:

(A) for any s, t-paths q_1 and q_2, the sentence "$\forall x : s \cdot q_1(x) = q_2(x)$",
(B) for any $f_{w_1,s_1} \sim'_F f_{w_2,s_2}$ in Σ', any w, w_1-paths q_1^*, w, w_2-paths q_2^*, s_1, s-path q_1 and s_2, s-path q_2, the sentence "$\forall x^* : w \cdot q_1(f_{w_1,s_1}(q_1^*(x^*))) = q_2(f_{w_2,s_2}(q_2^*(x^*)))$",
(C) for any $b_{w_1} \sim'_P b_{w_2}$ in Σ', any w, w_1-paths q_1^* and w, w_2-paths q_2^*, the sentence "$\forall x^* : w \cdot b_{w_1}(q_1^*(x^*)) \iff b_{w_2}(q_2^*(x^*))$".

Let N be a Δ-model. We will prove that:

1. if N has a θ-extension satisfying Φ', then any of its θ-extensions satisfies Ψ',
2. if N has a θ-extension satisfying Ψ', then it has a unique θ-extension satisfying Φ'.

Observe that by definition of Ψ', for Δ'-models P, Q, if $P|'_\theta = Q|'_\theta$, then $P \models'_{\Delta'} \Psi'$ iff $Q \models'_{\Delta'} \Psi'$. Also, $\Phi' \models'_{\Delta'} \Psi'$. Now, if $N' \models'_{\Delta'} \Phi'$ and $N = N'|'_\theta = N''|'_\theta$, then $N' \models'_{\Delta'} \Psi'$ and $N'|'_\theta = N''|'_\theta$, hence $N'' \models'_{\Delta'} \Psi'$. This proves point (1).

As for point (2), take any model N and its θ-extension N' satisfying Ψ'. Define a Δ'-model M' as follows:

- carriers and function symbols (embeddings, etc., inclusive) and predicates in $\theta(\Delta)$ are interpreted in M' as in N',
- for any embedding $em_{s \leq t}$ in $\Delta' \setminus \theta(\Delta)$ we set $M'[fn, em_{s \leq t}] = N'[fn, em_{\theta^S(s_1) \leq \theta^S(s_2)}]; \ldots; N'[fn, em_{\theta^S(s_{n-1}) \leq \theta^S(s_n)}]$, where $q = s_1 \leq \cdots \leq s_n$ is an s, t-path,
- for any projection $prj_{t \geq s}$ in $\Delta' \setminus \theta(\Delta)$ we set $M'[fn, prj_{t \geq s}]$ to be the least right-sided inverse of $M'[fn, em_{s \leq t}]$,

- for any predicate $in_{s,t}$ in $\Delta' \setminus \theta(\Delta)$ we set $M'[pr, in_{s,t}] = \{x \in M'[cr, s] \mid M'[fn, prj_{s \geq t}](x)$ is defined $\}$.

The fact that N' satisfies the sentences from point (A) ensures that this definition is correct. It is also clear that M' is a θ-extension of N. We now check that $M' \models'_{\Delta'} \Phi'$.

Defining sentence no. 1 (defining sentences have been introduced in the previous section) holds, since all self-embeddings are in $\theta(\Delta)$. Defining sentences 3, 4 and 5 hold by definition of M'. Sentence 2 is a consequence of 3 and of the original embeddings being injective in N. Sentences 6 and 7 are consequences of the sentences from points (B) and (C).

Since it is obvious that any θ-extension of N satisfying defining sentences 1-7 is forced to be defined as above, this completes the proof.   □

Props. 4, 6, 7 and 8 allow us to state the following:

Corollary 1. *Theorems 3 and 4 apply to* CASL_1 *and the standard representation* $R : \text{CASL}_1 \to \text{STRCASL}_0$.   □

Actually, even more may be said. The cited propositions together with Theorems 3 and 4 define define an algorithm, which reduces the problem of checking whether a CASL_1 architectural specification has a denotation (i.e., is correct), to the problem of proving semantic consequence in the STRCASL_0 institution, that is, in the institution of structured specifications over multi-sorted CASL. This statement is somewhat weakened by the fact that, for this reduction to work, the architectural specification at issue must contain no inconsistent parametric unit declarations and no parametric unit in it may be applied more than once.

This algorithm is of particular value, because proving semantic consequence between structured specifications has gained a lot of attention; recent papers to be mentioned include [BCH99,Borz98,MAH01].

7   Conclusion

In this paper a system – correct and complete w.r.t. a generative semantics – for proving properties of architectural specifications has been presented. Of course this completeness is relative to the calculus used for proving semantic consequence. This system is general enough to include the full CASL logic, even though it does not enjoy the amalgamation property.

This work builds on the work of [Hof01]. There, techniques for the relatively simple case of first-order logic were introduced. Here, we provide abstract, institution-independent formulations of needed properties and of theorems. Moreover, it is proven that the full CASL logic satisfies those properties, thus giving us a verification algorithm for CASL architectural specifications, and at the same time showing to what a wide scope of logics our techniques apply.

It should, however, be borne in mind, that the algorithm of [Hof01] was complete w.r.t. a non-generative semantics, while the algorithm presented here is complete w.r.t. a generative semantics (a generative semantics easily reduces to a non-generative one).

A clear aim is to get all the advantages of the current approach while retaining completeness w.r.t. a non-generative semantics, and thus fully solve the verification problem for CASL architectural specifications. This would also require taking a closer look at the remaining CASL architectural constructs (imports may cause additional trouble). What might also be interesting is using a representation into a different institution, namely that of *enriched* CASL [SMTKH01,SMT01,SMTKH]; this way, one would not remove any information from specifications, as is the case with the standard representation (in enriched CASL a subsort category replaces the subsort preorder).

Acknowlegdements

This research has been partially supported by KBN grant no. 7 T11C 002 21 and by EU AGILE project IST-2001-32747.

References

[BCH99] M. Bidoit, M. V. Cengarle, R. Hennicker: Proof Systems for Structured Specifications and Their Refinements. In E. Astesiano, H.-J. Kreowski, B. Krieg-Brückner (eds.): Algebraic Foundations of Systems Specification; pp. 385–434. Springer, 1999.

[Borz98] T. Borzyszkowski: Completeness of a logical system for structured specifications. In F. Parisi Presicce (ed.): Proc. WADT'97, LNCS 1376, pp. 107–121 (1998).

[BST99] M. Bidoit, D. Sannella, A. Tarlecki. Architectural Specifications in CASL. In: Proc. AMAST'98, LNCS 1548, pp. 341–357 (1999); final ver. to appear in Formal Aspects of Computing.

[CASL] E. Astesiano, M. Bidoit, H. Kirchner, B. Krieg-Brückner, P. Mosses, D. Sannella, A. Tarlecki. CASL: The Common Algebraic Specification Language. TCS (to appear).

[CASL00] CASL – Summary and Semantics (note S-9), ver. 1.0. In [CoFI] (2000).

[CoFI] CoFI, The Common Framework Initiative for Algebraic Specification and Development. Documents accessible at http://www.brics.dk/Projects/CoFI.

[GB84] J. Goguen, R. Burstall: Some fundamental algebraic tools for the semantics of computation. Part 1: Comma categories, colimits, signatures, and theories. TCS 31(2), pp. 175–209 (1984)

[GB92] J. Goguen, R. Burstall: Institutions: abstract model theory for specification and programming. Journal of the ACM 39, pp. 95–146 (1992).

[Hof01] P. Hoffman: Verifying Architectural Specifications. In M. Cerioli, G. Reggio (eds.): Proc. WADT'01, LNCS 2267, pp. 152–175 (2001).

[MAH01] T. Mossakowski, S. Autexier, D. Hutter: Extending Development Graphs With Hiding. In H. Hussmann (ed.): Proc. FASE'01, LNCS 2029, pp. 269–283 (2001).

[Mos98] T. Mossakowski: Cocompleteness of the CASL signature category. In [CoFI], note S-7 (1998).

[Pau96] L. Paulson: ML for the Working Programmer. Cambridge Univ. Press, 1991.

[SMT01] L. Schröder, T. Mossakowski, A. Tarlecki: Amalgamation in CASL via Enriched Signatures. In F. Orejas, P. Spirakis, J. van Leeuwen (eds.): Proc. ICALP'01, LNCS 2076, pp. 993–1004 (2001).

[SMTKH01] L. Schröder, T. Mossakowski, A. Tarlecki, B. Klin, P. Hoffman: Semantics of Architectural Specifications in CASL. In H. Hussmann (ed.): Proc. FASE'01, LNCS 2029, pp. 253–268 (2001).

[SMTKH] L. Schröder, T. Mossakowski, A. Tarlecki, B. Klin, P. Hoffman: Amalgamation in the Semantics of CASL. TCS (to appear).

[ST88] D. Sannella, A. Tarlecki: Specifications in an Arbitrary Institution. Information and Computation vol. 76, 2/3, pp. 165–210 (1988).

[ST97] D. Sannella, A. Tarlecki: Essential Concepts of Algebraic Specification and Program Development. Formal Aspects of Computing 9, pp. 229–269 (1997).

Algebraic Higher-Order Nets: Graphs and Petri Nets as Tokens

Kathrin Hoffmann[1] and Till Mossakowski[2]

[1] Institute for Software Engeneering and Theroretical Computer Science
Technical University Berlin
[2] BISS, Department of Computer Science
University of Bremen

Abstract. Petri nets and Algebraic High-Level Nets are well-known to model parallel and concurrent systems. In this paper, we introduce the concept of Algebraic Higher-Order Nets, which allow to have dynamical tokens like graphs or (ordinary low-level) Petri nets. For this purpose, we specify graphs and Petri nets in the higher-order algebraic specification language HASCASL such that graphs and Petri nets become first-class citizens, i.e. members of algebras (rather than algebras themselves). As an example, we model hospital therapeutic processes by a single higher-order net. Individual care plans for each patient are tokens modeled by low-level nets.

1 Introduction

Petri nets are well established to support the modeling and simulation of parallel and concurrent systems and represent a well-known and widely used formalism. The combination of algebraic specification and Petri nets, called Algebraic High-Level Nets [EPR94,PER95] give rise to a formal and well defined description of the dynamic behaviour of concurrent and distributed systems. This formalism is adequate in application domains where tokens are simple data elements, that is, objects are modeled by basic sorts. In other application domains it is also desirable to use higher-order objects as tokens like graphs, which hardly can be realized by basic sorts. Furthermore, in the context of Petri nets there are interesting applications to consider dynamical tokens, like Petri nets themselves [Val98,Val00]. We propose the concept of Algebraic Higher-Order Nets [Hof00] to obtain higher-order objects as tokens. Roughly spoken Algebraic Higher-Order Nets are Algebraic High-Level Nets, where the algebraic specification is an higher-order specification.

The focus of the paper is on the high modeling capability given by Algebraic Higher-Order Nets (AHO-nets). We present an extension of Petri nets including the concept of the higher-order algebraic specification language HASCASL [SM02]. We sketch the semantical model of higher-order processes [EHP+02], while detailed definitions and resulting theorems are out of scope of the paper. Furthermore, we introduce a specification of graphs in HASCASL. But contrary to the first-order algebraic approach [EHKP91], where a graph is considered as

M. Wirsing, D. Pattinson, and R. Hennicker (Eds.): WADT 2002, LNCS 2755, pp. 253–267, 2003.
© Springer-Verlag Berlin Heidelberg 2003

a two sorted algebra with two unary operations, a graph is related to a higher-order type. Petri nets are bipartite graphs, therefore we transfer the specification of graphs to Petri nets. The main result of this paper is the specification of Petri nets in HASCASL leading to AHO-nets with Petri nets as tokens. On the one hand, these higher-order objects can be transported through the system. But on the other hand, also the structure of objects can be changed during transition firing. This is a new approach in the area of Petri nets.

The paper is organized in the following way: After a general introduction and motivation we sketch the specification of graphs. We give the specification of Petri nets and illustrate the main idea of AHO-nets with Petri nets as tokens. We apply AHO-nets with Petri nets as tokens to hospital therapeutic processes, which is a restricted version of an informal case study as proposed in [Han97]. We demonstrate that our approach is adequate in the application domain of business processes using higher-order processes as a semantical model. Further aspects concerning the data type part like the flexible modification of objects are discussed. Finally, we point out the relevance of our construction in other application domains like agent modeling.

2   Motivation and Related Work

AHO-nets are associated with the higher-order algebraic specification language HASCASL [SM02]. HASCASL-specifications are appropriate since they combine the simplicity of algebraic specification with higher-order features; the latter being needed for graphs and Petri nets as first-class citizens. The formalism of AHO-nets provides a two level modeling technique: Objects are considered as higher-order tokens, while the system reflects the organizational structure and describes how objects are processed.

In the area of Petri nets, the modeling of higher-order tokens is a hot topic of research. In [Val98,Val00] objects and systems are defined as condition/event systems. So, a simple notion of Petri nets as tokens is achieved, such that most principles of elementary net theory are respected and extended (e.g. processes as formal semantics). However, this approach enforces the use of the same formalism for both, the Petri net tokens and the overall net. By contrast, we distinguish between the object level and the system level by using different formal frameworks. Petri net tokens are encoded in an appropriately defined HASCASL-specification, which is used for the data type part of AHO-Nets. We specify operations for changing the structure of Petri net tokens, while the system structure is left unchanged. The advantage is a more flexible modeling technique.

There are approaches that combine object oriented modeling and Petri nets (see [ADCR01] for an overview). Their definition of Petri nets comes with a definition of object classes using an object oriented language, and tokens are instances of these classes. By contrast, we do not incorporate all the features of object oriented modeling like inheritance and encapsulation as we concentrate on such properties that can be expressed on the level of algebraic specification languages.

In Higher-Order Nets [Han97] interface places are used to integrate Petri net tokens as resources into the model. This approach is not formal but more application oriented and leads to an adequate informal modeling technique of business processes. While the basic idea is somehow similar to the presented approach, our work is more general in the sense that Petri net tokens are not restricted to special places.

Essentially, the idea of AHO-nets is to handle higher-order features in the data type part, e.g. higher-order types of graph and Petri net tokens consisting of one set and two functions. The main benefit of our approach are specific operations to change the structure of graph and Petri net tokens and to simulate the local behaviour of Petri net tokens.

We motivate our approach by some semi-formal examples. In Fig. 1, we consider a specification of graphs denoted by the type inscription *Graph* (see Section 3). We use zoom lines to illustrate the general distinction between the system level and the object level. The zoom enlarges the higher-order object *graph* that is a token in the respective system model. The box in the lower part of Fig. 1 is a closer view to a graph, an element of the graph carrier set in the corresponding HASCASL-model. Obviously, there may be different objects for the same system.

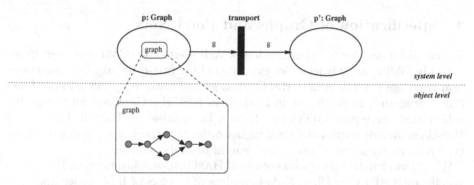

Fig. 1. Graphs as tokens

To demonstrate the semantics of the AHO-net in Fig. 1 we first assign the token *graph* to the variable *g* denoted in the arc inscription of the transition *transport*. Then, the follower marking is computed as follows: The token *graph* is consumed from the place *p*. Subsequently, the token *graph* is added to the place *p'*.

In general, the data type part of AHO-nets is not restricted to graphs. According to other application domains, objects can be specified in quite different ways. In Fig. 2, we use the specification of Petri nets (see Section 3) instead that of graphs leading to the notion of AHO-nets with Petri nets as tokens.

Petri nets have their own firing behaviour. Therefore, we can think about the autonomous activity of objects, i.e. the follower marking is computed on both levels, the system level and the object level, respectively. We assume, that the initial marking of the AHO-net in Fig. 2 consists of the Petri net *net1*. In *net1* the transition $t1$ is enabled since the current marking of $net1$ contains the token *start*. By the firing of the transition *follower marking*, on the one hand, the Petri net $net1$ is consumed from place $p1$. On the other hand, the marking of the Petri net $net1$ is changed in the sense that the follower marking is computed and the resulting Petri net $net1'$ is added to the place $p2$.

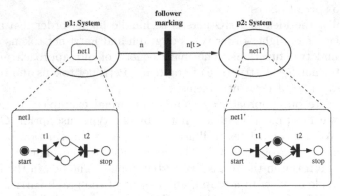

Fig. 2. Petri nets as tokens

3   Specification of Graphs and Petri Nets

We use HASCASL for the specification of AHO-Nets. HASCASL has been introduced in [SM02] as a higher-order extension of the first-order algebraic specification language CASL [CAS99]. HASCASL is geared towards specification of functional programs, in particular in Haskell; in fact, HASCASL has an executable subset that corresponds quite closely to a large subset of Haskell. Features of HASCASL include partial and total higher-order types, subtypes, polymorphism, type constructors, type classes, and general recursive functions.

We will work with a simplified version of HASCASL-specifications. A HASCASL-specification SPEC $= (S, \leq, F, Ax)$ consists of a set S of basic types and type constructors, a subsort relation $\leq$ on these, a set F of constants (being typed in higher-order types over S, including predicate types), and a set Ax of higher-order formulas.

Specifications of sets, partial maps and multisets can be found in [BH]. We here just mention some important operations and predicates. For example, the ternary predicate $f :: s \longrightarrow t$ indicates, for $f : S->?T$, $s : Set\ S$ and $t : Set\ T$ that f, when restricted to s, actually yields results in t. Here, the types S and T serve as "universes" in which sets s and t live (as subsets). This will be important when considering graphs and Petri nets as first-class citizens, i.e. as elements of some algebra carrier set (rather than as algebras, as it is often done).

Based on the specification of sets and partial maps, Fig. 3 specifies directed graphs, using a set of nodes and a set of arcs, the latter being implicitly given by the domain of the source and target functions. We use the ternary predicate $f :: s \longrightarrow t$ to ensure that the codomain actually is contained in the set of nodes. This yields a subtype of all tuples $(n, source, target)$.

---

spec GRAPH = SET **then**
    sort N, E
    type $Graph = \{(n, source, target) : Set\ N \times (E \rightarrow?\ N) \times (E \rightarrow?\ N)$ •
                    $dom\ source = dom\ target$
                    $\wedge\ source :: dom\ source \longrightarrow n$
                    $\wedge\ target :: dom\ target \longrightarrow n\}$

Fig. 3. Specification of directed graphs in HASCASL

---

spec PETRINET = SET **and** MAP **then**
    sorts P, T
    type $Net = \{(p, pre, post) : Set\ P \times (T \rightarrow?\ Set\ P) \times (T \rightarrow?\ Set\ P)$
              • $dom\ pre = dom\ post$
              $\wedge\ pre :: dom\ pre \rightarrow p$
              $\wedge\ post :: dom\ post \rightarrow p\}$

spec PETRISYSTEM = MULTISET **and** PETRINET**then**
    type $Marking = MultiSet\ P$
    type $System = \{(n, m) : Net \times Marking$
                    • $let\ (p, pre, post) = n$
                    $in\ \forall x : P$ • $x\ isIn\ m \Rightarrow x\ isIn\ p\}$
    ops $marking : System \rightarrow Marking;$
         $net : System \rightarrow Net;$
         $\_[\_> : System \times T \rightarrow?\ System;$

    forall $sys, sys1, sys2 : System;\ n : Net;\ x;\ P;\ m : Marking;\ t : T$
        • $net\ sys = let\ (n, m) = sys\ in\ n$
        • $marking\ sys = let\ (n, m) = sys\ in\ m$
        • $def\ sys[t> \Leftrightarrow$
            $(t\ isIn\ dom\ pre \wedge\ \forall x : P$ • $x\ isIn\ pre(t) \Rightarrow x\ isIn\ marking(sys))$
        • $def\ sys[t> \Rightarrow sys[t> = (net(sys),$
            $marking(sys) - setToMultiSet(pre(t)) + setToMultiSet(post(t)))$
            $as\ System$

Fig. 4. Specification of petri nets and systems in HASCASL

Fig. 4 introduces a specification of Petri nets. For simplicity, we assume that there are no multi-arcs between transitions and places. Hence, it suffices to use two functions *pre* and *post* that map each transition into its pre- and post-domain of places. Markings are just multisets of places, and a system is a net with a marking (such that the marking actually is only using the places of the net). The firing operation $\_[\_>$ is defined only if the pre domain of the transition to be fired is contained in the current marking, and in this case, it just subtracts the pre domain and adds the post domain to the current marking.

Fig. 5 now specifies workflow systems. These are systems equipped with an "start" and an "stop" place. The latter easily allow to paste together two workflow systems by uniting them disjointly while identifying the stop place of the

first system with the start place of the second one. We have to be a bit careful when performing constructions such as disjoint union and identification (i.e.
quotiening). When disjointly uniting two sets s_1 and s_2 both living in type S,
with the usual construction, we would end up with a disjoint union living in type
$S \times \{0, 1\}$. However, we want to end up in S again. Therefore, we assume that
type S comes with enough infrastructure to internally represent constructions
like products, disjoint union and quotiening (e.g. by embedding $S \times \{0, 1\}$ into
s somehow). This infrastructure is required by the specification SETCONSTRUC-
TIONS from the appendix. There we also show that it is actually possible to
provide such infrastructure for the natural numbers — and this should be easily
done for other domains as well.

We also provide operations *initNet* (yielding a net with two places and one
transition), *loop* (yielding a net with a single place and a "loop" transition) and
par. *par* performs a "parallel composition" of nets, identifying both start and
stop places. (We omit the easy specifications of *initNet* and *loop*.) Indeed, with
these operations, all finite workflow nets can be generated from scratch.

4   Petri Nets as Tokens

In this section we give the formal basis of AHO-nets as introduced in Section 2.
First, we review the basic concepts of AHO-nets as given in [Hof00] in order to
define the semantical model. In general, an AHO-net consists of a Petri net with
inscriptions of an algebraic higher-order specification.

Definition 1. *An* AHO-net,

$$N = (\text{SPEC}, A, P, T, pre, post, cond, type)$$

consists of

- *the* HASCASL-*specification* $\text{SPEC} = (S, \leq, F, Ax)$,
- *a* SPEC-*algebra* A,
- *sets* P *and* T *of places and transitions*,
- *pre- and post-domain functions*

$$pre, post : T \rightarrow (T_{\text{SPEC}}(X) \otimes P)^{\oplus}$$

assigning to each transition $t \in T$ *the pre- and post-domains* $pre(t)$ *and*
post(t), respectively. By $T_{\text{SPEC}}(X)$ *we denote the set of terms with variables*
X *over the specification* SPEC. *The set of all type-consistent arc inscriptions*
$T_{\text{SPEC}}(X) \otimes P$ *is defined by*

$$T_{\text{SPEC}}(X) \otimes P = \{(term, p) | term \in T_{\text{SPEC}}(X)_{type(p)}, p \in P\}$$

and $(T_{\text{SPEC}}(X) \otimes P)^{\oplus}$ *is the free commutative monoid over this set.*

spec WORKFLOWNET[SETCONSTRUCTIONS **with** $S \mapsto P$]
  [SETCONSTRUCTIONS **with** $S \mapsto T$] =
  PETRISYSTEM
and ABSTRACTSETCONSTRUCTIONS[SETCONSTRUCTIONS **with** $S \mapsto P$ **fit** $S \mapsto P$]
and ABSTRACTSETCONSTRUCTIONS[SETCONSTRUCTIONS **with** $T \mapsto P$ **fit** $T \mapsto P$]
then

> **type** $WFNet = \{(sys, start, stop) : System \times P \times P \bullet$
> $let\ ((p, pre, post), m) = sys\ in$
> $start\ isIn\ p$
> $\wedge\ stop\ isIn\ p$
> $\wedge\ \neg(start\ isIn\ range(post))$
> $\wedge\ \neg(stop\ isIn\ range(pre))\}$
> **ops** $initNet, loop : WFNet;$
> $\_paste\_, \_par\_ : WFNet \times WFNet \to WFNet$
> **forall** $w, w1, w2 : WFNet;\ sys : System$
>   • $w1\ paste\ w2 =$
> $let\ (((p1, pre1, post1), m1), start1, stop1) = w1$
> $(((p2, pre2, post2), m2), start2, stop2) = w2$
> $r = \lambda x, y \bullet x = y \vee (x = inl\ stop1 \wedge y = inr\ start2)$
> $\vee(y = inl\ stop1 \wedge x = inr\ start2)$
> $p = (p1\ coproduct\ p2)\ factor\ r$
> $q = coeq\ r$
> $t = (dom\ pre1)\ coproduct\ (dom\ pre2)$
> $pre\ t = if\ def\ left\ t\ \textbf{then}\ image(q\ o\ inl)(pre1(left\ t))$
> $else\ if\ def\ right\ t\ \textbf{then}\ image(q\ o\ inr)(pre2(right\ t))$
> $else\ undefined$
> $post\ t = if\ def\ left\ t\ \textbf{then}\ image(q\ o\ inl)(post1(left\ t))$
> $else\ if\ def\ right\ t\ \textbf{then}\ image(q\ o\ inr)(post2(right\ t))$
> $else\ undefined$
> $m = map(q\ o\ inl)m1 + map(q\ o\ inr)m2$
> $in\ (((p, pre, post), m), start1, stop2)\ as\ WFNet$
>   • $w1\ par\ w2 =$
> $let\ (((p1, pre1, post1), m1), start1, stop1) = w1$
> $(((p2, pre2, post2), m2), start2, stop2) = w2$
> $r = \lambda x, y \bullet x = y \vee (x = inl\ start1 \wedge y = inr\ start2)$
> $\vee(y = inl\ start1 \wedge x = inr\ start2)$
> $\vee(x = inl\ stop1 \wedge y = inr\ stop2)$
> $\vee(y = inl\ stop1 \wedge x = inr\ stop2)$
> $p = (p1\ coproduct\ p2)\ factor\ r$
> $q = coeq\ r$
> $t = (dom\ pre1)\ coproduct\ (dom\ pre2)$
> $pre\ t = if\ def\ left\ t\ \textbf{then}\ image(q\ o\ inl)(pre1(left\ t))$
> $else\ if\ def\ right\ t\ \textbf{then}\ image(q\ o\ inr)(pre2(right\ t))$
> $else\ undefined$
> $post\ t = if\ def\ left\ t\ \textbf{then}\ image(q\ o\ inl)(post1(left\ t))$
> $else\ if\ def\ right\ t\ \textbf{then}\ image(q\ o\ inr)(post2(right\ t))$
> $else\ undefined$
> $m = map(q\ o\ inl)m1 + map(q\ o\ inr)m2$
> $in\ (((p, pre, post), m), start1, stop1)\ as\ WFNet$

Fig. 5. Workflow systems and how to paste them together

– *the firing condition function*

$$cond : T \to \mathcal{P}_{fin}(EQNS(\text{S,F,X}))$$

assigning to each transition $t \in T$ a finite set $cond(t)$ of equations over the specification SPEC,
– *the type function*

$$type : P \to S$$

assigning to each place $p \in P$ the sort $s \in S$.

The marking of an AHO-net is denoted by tuples consisting of a data element and the place it resides on. The behaviour is given by the firing of transitions, i.e. tokens are moved over the set of places.

Definition 2. *Given an AHO-net N as in Definition 1, then a marking m of N is an element of the free commutative monoid $(A \otimes P)^{\oplus}$ where $(A \otimes P) = \{(a,p)|a \in A_{type(p)}, p \in P\}$. Consequently*

$$m \in (A \otimes P)^{\oplus}$$

Definition 3. *Given an AHO-net N as in Definition 1 and a transition $t \in T$. $Var(t)$ denotes the set of variables occurring in $pre(t), post(t)$, and $cond(t)$. An assignment $asg : Var(t) \to A$ is called consistent if the equations $cond(t)$ are satisfied in A under asg.*

Transitions are enabled under a marking m for an assignment asg inducing $ASG : (T_{\text{SPEC}}(Var(t)) \times P)^{\oplus} \to (A \otimes P)$ with $ASG(term, p) = (\overline{asg}(term), p)$ if $ASG(pre(t)) \leq m$. Here $\overline{asg}$ is the extension of the assignment asg to an evaluation of terms (analogously to the first-order case in [EM85]). The follower marking m' then is constructed by

$$m' = m \ominus ASG(pre(t)) \oplus ASG(post(t))$$

The notion of processes is well-known for low level Petri nets (see e.g. [Rei85]). It represents a semantical model to study the non-sequential behaviour of Petri nets. In [EHP+02] the notion of processes based on occurrence nets is transferred to Algebraic High-Level Nets leading to the notion of high-level processes. The main difference between AHO-nets and Algebraic High-Level Nets is that we use an extension of the first-order approach to higher-order. So, the definition of higher-order processes seems to be quite natural. An example is presented in Fig. 9 in Section 5.

In our example we use AHO-nets in the area of hospital therapeutic processes. For this reason the data type part is fixed by the HASCASL-specification of workflow nets (see Fig. 5 in Section 3) leading to the notion of AHO-nets with worklow nets as tokens.

AHO-nets with graphs as tokens are more or less analogously defined to AHO-nets with workflow nets as tokens. We only have to use the HASCASL-specification GRAPH (see Fig. 1 in Section 3) instead of the HASCASL-specification WORKFLOWNETS. Furthermore, a GRAPH-algebra A implements graph objects and operations appropriate for our application domain.

5 Example: Hospital Therapeutic Processes

In this section we motivate the notions and results of this paper in terms of a practical example inspired by the case study proposal on hospital therapeutic processes in [Han97]. The case study deals with a part of business processes in a hospital, namely patient therapeutic treatments.

The idea is to model the following situation: The hospital consists of four different departments, each of which has a set of internal activities. First patients are received at the reception office. After a diagnosis is made, the specialist prescribes medication and certain general treatments, e.g. taking the blood pressure and the temperature. Then patient care plans are carried out, if demanded. Although the case study left quite some room for interpretation, it pinpoints a baseline process that concerns the receiving and curing of patients. We intend to follow the baseline process. It subsumes the coordination of patient care plans, i.e. the execution, extension and modification of patient care plans. Hence, our model is specified by AHO-nets with workflow nets as tokens.

For the modeling of patient care plans, we use more or less the specification of workflow nets (see Fig. 5 in Section 3). In our model, workflow nets determine the order in which therapeutic processes have to be performed. We consider a WORKFLOWNET-algebra $CarePlans$, such that the patient care plans $initPlan, planA$ and $planB$ depicted in Fig. 6 are elements of the corresponding algebra carrier set $Care Plans_{WFNet}$.

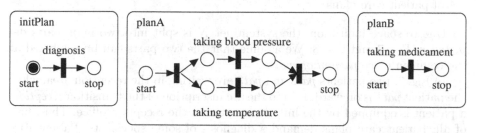

Fig. 6. Basic patient care plans

The transition *diagnosis* in the initial care plan (*initPlan* in Fig. 6) indicates, that a diagnosis of patient diseases is required. The care plan *planA* models the parallel measuring of two vital values, the blood pressure and the temperature, respectively. The care plan *planB* represents the taking of some prescribed medication.

Note, that the carrier set $Care Plans_{WFNet}$ consists of more then three elements. The WORKFLOWNET-algebra $Care Plans$ provides further operations $paste_{Care Plans}$ and $par_{Care Plans}$ to implement the sequential and parallel composition of workflow nets. Thus, we achieve further care plans $plan_1, \ldots, plan_n \in Care Plans_{WFNet}$ during the composition of the patient care plans $initNet$, $planA$ and $planB$.

To individualize patient care plans we introduce a notion of IDs. In our example there are three different patients, i.e. $\{pat1, pat2, pat3\} \in Care Plans_{ID}$.

The organizational structure of the hospital is reflected in the static system net, while the patients and their care plans are the tokens of the AHO-net N (see Definition 1 in Section 4):

$$N = (\textsc{WorkflowNet}, CarePlans, P, T, pre, post, cond, type)$$

- the specification $\textsc{WorkflowNet}$ (see Fig. 5 in Section 3),
- the $\textsc{WorkflowNet}$-algebra $Care\ Plans$ (see above),
- sets P and T of places and transitions with

$P = \{patient,\ patient\ on\ ward,\ care\ plans,\ healthy\ person\}$
$T = \{reception,\ making\ care\ plans,\ carrying\ out\ care\ plans,\ discharge\}$

such that each transition $t \in T$ models one specific department,
- pre- and post domain functions pre and $post$ assigning to each transition $t \in T$ a set of internal activities. The arc inscriptions have to ensure that patient care plans are performed in a predefined order.
- the firing condition function $cond$ assigning to each transition $t \in T$ the empty set,
- the type function $type$ assigning to each place $p \in P$ the sort of identification and workflow nets. Obviously, we have $type(patient\ on\ ward) = ID \times WFNets$ for patient inside the hospital. Furthermore $type(care\ plans) = WFNets$ indicates that we use some kind of resources for the modification of patient care plans.

Due to space limitation, the system net N is split into two main parts depicted in Fig. 7 and Fig. 8. We assume that the two parts can be combined at the common place *patient on ward* .

In Fig. 7, two patients *pat2* and *pat3* are waiting in the reception area, while the patient *pat1* is on ward. Due to the arc inscription of the transition *reception*, a patient is equipped by the initial care plan in the reception office. Thus, first of all, patient care plans demand a diagnosis of some specialists. Patient care plans need to be carried out. Notice, that workflow nets have their own firing behaviour. Thus, the arc inscription $n[t\rangle$ of the transition *carrying out care plans* (see Fig. 7) determines the treatment *diagnosis* in the patient care plan *pat1*. Subsequently, the care plan *pat1* is carried out by some nurse and the current status is updated, i.e. the follower marking is computed.

The fact is that care plans are not fixed once and for all, because they are constantly modified according to the treatments effects (e.g. the effectiveness of medication). Here, we use the care plan *planA* and the care plan *planB* to extend the specific care plan of a patient on ward. Due to the structuring technique *paste* implemented in our specification $\textsc{WorkflowNets}$ (see Fig. 5 in Section 3), workflow nets can be sequentially composed at a common place. Assume that *pat1* have to take medicaments. By firing of the transition *making care plan*, the patient care plan *pat1* and *planB* are pasted together at a common place; the output place of *pat1* and the input place of *planB* are identified. To complete the hospital therapeutic process, the discharge of patients is in some

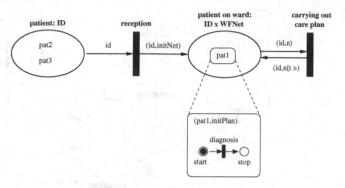

Fig. 7. Reception office and carrying out care plans

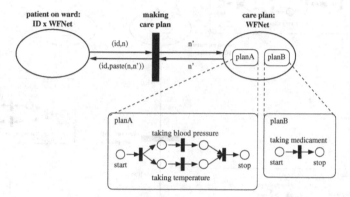

Fig. 8. Making care plans

sense the reverse activity of the reception office (see Fig. 7), i.e. patients loose their care plans, which is now completed.

To illustrate the semantics of the hospital system N in more detail one of many possible higher-order processes (see Section 4) is depicted in Fig. 9. We assume that the patient *pat*1 is on ward. First of all the diagnosis is made by some specialists. Then, we assume that the medical examination permits the treatment by *planA*. Thus, the actual care plan *plan*1 of patient *pat*1 is enlarged by the *planA*. Subsequently, the care plan is carried out by some nurse and the current status is updated. We use the operation *paste* a second time to extend the patient care plan by the *planB*. Finally the patient care plan is carried out step by step. The patient is discharged at a certain point of time. But this yields another process, which is not depicted here.

6   Conclusion

First of all, in this paper the notion of AHO-nets with graphs and Petri nets as tokens is introduced and then formalized. The main benefit of this paper is to provide a suitable specification of workflow nets in HASCASL (see Section 3). We

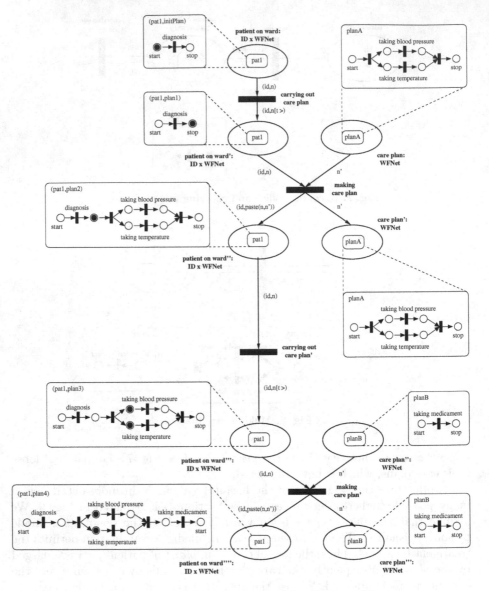

Fig. 9. A higher-order process of the AHO-net N

have introduced four operations that can be used at the transitions of higher-order nets and that suffice to generate all finite (low-level) workflow nets due to the operation of parallel composition. In the second part we model a simplified version of a hospital system showing the usefulness of our approach in the application area of business processes (see Section 5). Our model illustrates some of the aspects inherited from the expressiveness of HASCASL-specifications.

One of the open research problems with regard to our specifications is to develop further structuring techniques to change the structure of workflow nets, e.g. operations based on homomorphisms. Thus, we are able to generalize the paste operation of workflow nets to the well-known pushout construction of Petri nets (see e.g. [GHP99]).

Another promising topic is the flexible modification of workflow nets, e.g. removing specific treatments in patient care plans, using the general framework of high level replacement systems [EHKP91]. Technically, the operations are based on rules and transformations to guarantee the local modification of workflow nets. In the resulting formalism of AHO-nets a set of predefined rules is interpreted as a special kind of resources, while transformation represents certain internal activities in our model.

In this paper, patient care plans are considered by workflow nets. Indeed, patient documents are more complex in practice. A first-order specification of patient documents on a large scale can be found in [Erm96], which can be fully integrated into our model.

In the application area of agent modeling objects are marked by autonomy and mobility (see e.g. [Han00]). Here we can use AHO-nets to model the agent platform with places containing agent structures and transitions modeling the autonomous behaviour and communication of agent objects. But this is part of further research.

Acknowledgement

This work has been partially supported by the DFG project Multiple (KR 1191/5-2) and the DFG researcher group "Petrinetztechnologie".

A Set representations

If we want to iterate the constructions product, disjoint union and quotients on sets, we need to circumvent the problem that the type of the sets changes with each construction. This means that we need a specific type S which comes with the possibility to represent the result of the above mentioned constructions again in S. The specification SETCONSTRUCTIONS (Fig. 10) states the abstract requirement that such representations exists for a given type S. We also provide some sample set representation for the natural numbers. It uses Cantor's diagonalization for pairs of natural numbers, even and odd numbers as two copies of the natural numbers (for coproducts), and chooses the minimal element as representative of an equivalence class.

Given a set representation, we now can define products, coproducts, coequalizers and pushouts, while staying within the same type of sets. Note that $\_ * \_$ is now overloaded: for two given S-sets, it delivers either an $S * S$-set, as defined in the specification SET above, or an S-set, as defined here. We also could specify the mediating morphisms that exist by the respective (co)universal properties of the constructions.

spec RELATION = SETthen
    var $S: Type$
    ops $reflexive, symmetric, transitive : Pred(Set(S \times S))$
    forall $r : Set(S \times S)$
        • $reflexive\ r \Leftrightarrow \forall x : S \bullet r(x,x)$
        • $symmetric\ r \Leftrightarrow \forall x, y : S \bullet r(x,y) \Rightarrow r(y,x)$
        • $transitive\ r \Leftrightarrow \forall x, y, z : S \bullet r(x,y) \wedge r(y,z) \Rightarrow r(x,y)$
    type $PER\ S = \{r : Set(S \times S) \bullet symmetric\ r \wedge transitive\ r\}$
    op $dom : PER\ S \to S$
    forall $x : S;\ r : PER\ S$
        • $x\ isIn\ dom\ r \Leftrightarrow (x,x)isIn\ r$

spec SETCONSTRUCTIONS = MAP **and** RELATION **then**
    type S; %%*arbitrary but fixed!*
    ops $pair : S \times S \to S$; %%*products*
          $inl, inr : S \to S$; %%*coproducts*
          $coeq : PER\ S \to Map\ S$; %%*quotients*
        • *injective pair*
        • *injective inl*
        • *injective inr*
        • $(range\ inl)disjoint(range\ inr)$
    forall $r : PER\ S$
        • $ker(coeq\ r) = r$

spec NATSETCONSTRUCTIONS = MAP **and** NAT **then**
    ops $pair : Nat \times Nat \to Nat$;
          $inl, inr : Nat \to Nat$;
          $coeq : PER\ Nat \to Map\ Nat$;
          $min : Pred\ Nat \to? Nat$
    forall $r : PER\ Nat;\ m, n : Nat$
        • $pair(m,n) = ((m+n) \times (m+n+1) + 2 \times m)div\,2$
        • $inl\ m = 2 \times m$
        • $inr\ m = 2 \times m + 1$
        • $min\ p = n \Leftrightarrow (p\ n \wedge \forall m : Nat \bullet m < n \Rightarrow \neg p\ m)$
        • $coeq\ r\ n = min(\ m \bullet (m,n)isIn\ r)$

view SETCONSTRUCTIONS*to*NATSETCONSTRUCTIONS = $S \mapsto Nat$

spec ABSTRACTSETCONSTRUCTIONS[SETCONSTRUCTIONS] **given** MAP = %*def*
    ops $\_ * \_, \_coproduct\_ : Set\ S \times Set\ S \to Set\ S$;
          $pi1, pi2 : S \to? S$; %%*product prjections*
          $left, rigt : S \to? S$; %%*partial inverses of the coproduct injections*
          $factor : PER\ S \to Set\ S$; %%*quotient* : $coeq\ r :: dom\ r \longrightarrow factor\ r$
    forall $x, y : S;\ s, s1, s2, t : Set\ S;\ f, g, h : Map\ S;\ r : PER\ S$
        • $pi1(pair(x,y)) = x$
        • $pi2(pair(x,y)) = y$
        • $s1 \times s2 = image\ pair(s1 \times s2)$
        • $s1\ coproduct\ s2 = (image\ inl\ s1)union(image\ inr\ s2)$
        • $def\ left\ x \Leftrightarrow x\ isIn\ range\ inl$
        • $def\ right\ x \Leftrightarrow x\ isIn\ range\ inr$
        • $left(inl\ x) = x$
        • $right(inr\ x) = x$
        • $factor\ r = range(coeq\ r)$
        • $f :: dom\ r \longrightarrow t \wedge ker\ f \subseteq r \Rightarrow mediate\ r\ f(coeq\ x) = f\ x$

Fig. 10. Set representations: a tool for "internal" (co)limits

References

[ADCR01] G. Agha, F. De Cindio, and G. Rozenberg, editors. *Concurrent Object-Oriented Programming and Petri Nets*, volume LNCS 2001. Springer, 2001.

[BH] The bremen cofi homepage.

[CAS99] CASL – The CoFI Algebraic Specification Language – Summary, version 1.0. Documents/CASL/Summary, in [CoF], July 1999.

[CoF] CoFI. The Common Framework Initiative for algebraic specification and development, electronic archives. http://www.cofi.info.

[EHKP91] H. Ehrig, A. Habel, H.-J. Kreowski, and F. Parisi-Presicce. From graph grammars to high level replacement systems. In *Lecture Notes in Computer Science 532*, pages 269–291. Springer Verlag, 1991.

[EHP+02] E. Ehrig, K. Hoffmann, J. Padberg, P. Baldan, and R. Heckel. High Level Net Processes. volume LNCS 2300, pages 191–219. Springer, 2002.

[EM85] H. Ehrig and B. Mahr. *Fundamentals of Algebraic Specification 1: Equations and Initial Semantics*, volume 6 of *EATCS Monographs on Theoretical Computer Science*. Springer Verlag, Berlin, 1985.

[EPR94] H. Ehrig, J. Padberg, and L. Ribeiro. Algebraic High-Level Nets: Petri Nets Revisited. In *Recent Trends in Data Type Specification*, pages 188–206. Springer Verlag, 1994. Lecture Notes in Computer Science 785.

[Erm96] C. Ermel. Anforderungsanalyse eines medizinischen Informationssystems mit Algebraischen High-Level-Netzen. Technical Report 96-15, TU Berlin, 1996. (Masters Thesis TU Berlin).

[GHP99] M. Gajewsky, K. Hoffmann, and J. Padberg. Place Preserving and Transition Gluing Morphisms in Rule-Based Refinement of Place/Transition Systems. Technical Report 99-14, Technical University Berlin, 1999.

[Han97] Yanbo Han. *Software Infrastructure for Configurable Workflow System - A Model-Driven Approach Based on Higher-Order Nets and CORBA*. PhD thesis, Technical University of Berlin, 1997.

[Han00] M. Hannebauer. *Autonomous Dynamic Reconfiguration in Collaborative Problem Solving*. PhD thesis, Technical University Berlin, 2000.

[Hof00] K. Hoffmann. Runtime Modifikation between Algebraic High Level Nets and Algebraic Higher Order Nets using Folding and Unfolding Construction. In G. Hommel, editor, *Communication-Based Systems, Proceedings of the 3rd International Workshop*, pages 55–72. TU Berlin, Kluwer Academic Publishers, 2000.

[PER95] J. Padberg, H. Ehrig, and L. Ribeiro. Algebraic high-level net transformation systems. *Mathematical Structures in Computer Science*, 5:217–256, 1995.

[Rei85] W. Reisig. *Petri Nets*, volume 4 of *EATCS Monographs on Theoretical Computer Science*. Springer Verlag, 1985.

[SM02] L. Schröder and T. Mossakowski. HasCASL: Towards integrated specification and development of Haskell programs. In H. Kirchner and C. Reingeissen, editors, *Algebraic Methodology and Software Technology, 2002*, volume 2422 of *Lecture Notes in Computer Science*, pages 99–116. Springer-Verlag, 2002.

[Val98] Rüdiger Valk. Petri Nets as Token Objects: An Introduktion to Elementary Object Nets. *Proc. of the International Conference on Application and Theory of Petri Nets*, LNCS 1420:1–25, 1998.

[Val00] R. Valk. Concurrency in Communicating Object Petri Nets. In G. Agha, F. de Cindio, and G. Rozenberg, editors, *Concurrent Object-Oriented Programming and Petri Nets*, volume LNCS, pages 164–195. Springer, 2000.

The Coinductive Approach
to Verifying Cryptographic Protocols

Jesse Hughes and Martijn Warnier

Computing Science Institute, University of Nijmegen
Toernooiveld 1, 6525 ED Nijmegen, The Netherlands
{jesseh,warnier}@cs.kun.nl

Abstract. We look at a new way of specifying and verifying cryptographic protocols using the Coalgebraic Class Specification Language.
Protocols are specified into CCSL (with temporal operators for "free")
and translated by the CCSL compiler into theories for the theorem
prover PVS. Within PVS, the desired security conditions can then be
(dis)proved.
In addition, we are interested in using assumptions which are reflected
in real-life networks. However, as a result, we present only a partial
solution here. We have not proved full correctness of a protocol under
such loose restrictions. This prompts discussion of what assumptions
are acceptable in protocol verification, and when practical concerns may
outweigh theoretical motivations.

1 Introduction

Cryptographic protocols (also called "security protocols") give an abstract representation of a method for communicating securely over an open (insecure)
network in such a way that goals such as *secrecy, authentication, integrity* or
a combination of these conditions are satisfied. Verification of such protocols is
(since [2]) an important subject of research where formal methods are applied.
Their small size (usually just a few lines) and the difficulties in designing 'safe'
protocols make them very suitable for formal analysis.

Of course, one must be careful in the formalization. The small expression of
a protocol typically hides many implicit assumptions about the behavior of the
principals, freshness, etc. Formalization of a protocol in a logic exposes implicit
assumptions about the network, the behavior of participants, and so on.

The difficulty of analyzing security protocols is apparent from the history
of the so called *Needham Schroeder public key authentication* protocol from [8].
This protocol, first published in 1976 and proved to be secure in 1989 (in [2]),
contained a flaw which was finally found by Gavin Lowe (in [5]), 17 years after
its original publication.

We propose a new way of analyzing cryptographic protocols, using the theory
of coalgebras and hence "coinductive" reasoning[1] at its core. This approach is

---

[1] We use the term *coinductive* loosely here. We do not mean the principle that bisimilar
elements of a final coalgebra are equal. Rather, we mean that our distinguished
predicates are coalgebraic invariants, which are analogues to inductive predicates
for algebras.

M. Wirsing, D. Pattinson, and R. Hennicker (Eds.): WADT 2002, LNCS 2755, pp. 268–283, 2003.

closely related to Paulson's inductive approach[11]. The difference between the two is conceptual. Paulson's model is fundamentally algebraic: he models the situation by an initial algebra of finite traces, and so relies on induction as his fundamental principle. We, instead, use a coalgebraic model (of infinite traces), and reason about invariants.

As a first case study, we analyzed the *Bilateral Key Exchange protocol* (BKE) [3], which we describe in Sect. 2. We built an informal model of the protocol, first by constructing an informal model of message passing in a network setting. This model includes the presence of a powerful spy, in the Dolev-Yao tradition [4].

When modeling the protocol, we took a decidedly liberal approach. We avoided certain assumptions present in other protocol verifications, particularly those that are not realistic, i.e., do not hold in real networks in which the protocol will be implemented. In particular, it is common to require that, if **A** and **B** are using a specific protocol, then the only messages they send are messages required by the protocol itself. Let us call this the *protocol message axiom*. But, in a natural setting, it is likely that, while **A** communicates with **B** over the network, she[2] is also holding other conversations with other participants (or even with **B**). So, it does not seem a realistic assumption.

The protocol message axiom may still be reasonable, if one has an a priori argument that it is a harmless assumption. That is, perhaps there is a good argument that, if there is a successful attack on the BKE protocol, then there is a successful attack during which **A** and **B** send only protocol messages. If so, then one ought to assume the protocol message axiom, since it simplifies the analysis. However, we do not have such an a priori argument. Consequently, we chose *not* to assume the axiom, so that our resulting correctness proofs are that much more persuasive.

The result was not wholly successful. If one drops the protocol message axiom, then one must carefully choose more reasonable axioms that allow the participants to act flexibly without behaving stupidly. These axioms include that the participants do not send messages with their private keys or with received secrets and so on. As well, one needs to decide when a received message contains a secret, so that the participant does not reveal it.

However, this is the minor part of the difficulty. We found that the resulting correctness proofs become very difficult indeed, because the reasoning involved requires subtler partitioning of classes of messages. This partitioning may be introduced via appropriate fixed point constructions, and adds considerable complexity to the analysis. As well, the resulting proofs become subtle and difficult, and certainly are not amenable to automation.

In the end, we proved that, if the principals never send messages containing protocol secrets (the nonces, keys, etc., relevant to the protocol), aside from those required by the protocol, then these secrets will not become public (and, in particular, the Spy will never acquire a secret). However, we did not prove that this condition is met under reasonable assumptions about the behavior of

---

[2] It is customary in the literature on cryptographic protocols to refer to **A** and **B** as Alice and Bob.

the participants. In particular, one must show that the Spy can't compel one of the participants to reveal a secret to him (presumably by forgery). We believe that, indeed, this conclusion is provable, but found that the difficulty of the proofs was such that it became dubious that the proofs were worth the effort[3]. For purely practical concerns, perhaps the protocol message axiom ought to be adopted in formal security verifications, contrary to our approach.

This paper contributes a discussion of two orthogonal issues, then. First, we discuss coalgebraic models of security protocols. We advocate the Coalgebraic Class Specification Language, CCSL, for specifying protocols, exhibit the features of CCSL that make it attractive, and indicate how correctness proofs proceed. As a separate issue, we discuss a particular model for Bilateral Key Exchange, in which we have avoided the protocol message axiom. We explain the philosophical motivation behind this decision, and the large practical consequences as well. We offer the conflict between the philosophical and practical concerns as an open problem to be discussed.

The remainder of this paper is organized as follows: Sect. 2 gives a concrete example of a security protocol. Section 3 gives a informal overview of how we model the protocol without going into implementation details. Section 4 briefly describes the specification language CCSL [12], and how we implement the informal model using the specification language. Due to space restrictions, the technical discussions in Sects. 3, 4 and elsewhere are unfortunately curt. We hope that they are sufficient to convey the motivation and basic implementation of our work. Section 5 discusses the protocol message axiom and the effect of omitting it from our specification. Section 6 describes the relation between our approach and others and Sect. 7 ends with some conclusions.

2   An Example Protocol

As an example of a security protocol, we consider the *Bilateral Key Exchange with Public Key Protocol* (see §6.6.6. of [3]). It is a simple protocol for distributing a session key between two principals **A** and **B**, who are also authenticated to each other. The session key can be used for secure communication thereafter.

We take the protocol as given, and do not criticize or alter it in our presentation. Thus, any redundant features are left unchanged in our analysis, etc.

The usual abstract notation from the literature for such a protocol can be seen in Fig. 1.

Informally, the protocol describes the following exchange.

1. **B** sends to **A** a message containing **B**'s identity, B, and an encrypted submessage including a nonce (i.e., a random number, assumed to be not guessable), N_b, and again the text B. It is encrypted with **A**'s public key.

---

[3] We think these problems are independent of our coalgebraic basis. They would also arise in Paulson's inductive approach if one omits the consequences of the protocol message axiom.

1. $\mathbf{B} \to \mathbf{A} : B, \{N_b, B\}_{pk_a}$
2. $\mathbf{A} \to \mathbf{B} : \{Hash(N_b), N_a, A, K_{ab}\}_{pk_b}$
3. $\mathbf{B} \to \mathbf{A} : \{Hash(N_a)\}_{K_{ab}}$

Fig. 1. Bilateral Key Exchange with Public Key Protocol

2. When **A** receives the message, she decrypts it with her private key pk_a^{-1}. She responds by replying with a *Hash* of **B**'s nonce N_b, a new nonce N_a generated by **A**, the plain text message A and the session key K_{ab}. All are encrypted under **B**'s public key pk_b.
3. **B** receives and decrypts the message and sends a response to **A** containing a *Hash* of **A**'s nonce, N_a, encrypted using the session key K_{ab}.

We will use this protocol as a running example throughout this paper. We chose it for three reasons, (i) because it is simple, (ii) it is a hybrid protocol, meaning it uses both symmetric and asymmetric encryption, (iii) it has two different security goals, *authentication* and *secrecy*. These three features make the protocol (at least in our eyes) a useful example for a new verification approach.

3 Modeling

In this section we describe our model informally, giving basic concepts without technical details. We use the word "model" loosely here. Its meaning will stay informal for this section. Also, in this section, the coalgebraic features of our model are hidden. These features and a more technical description of our implementation in the Coalgebraic Class Specification Language [12, 14] will be discussed in Sect. 4.

3.1 Security Model and Related Assumptions

We make a number of general assumptions and abstractions to form our *Security Model*. These should reflect the situations in which protocols are used.

Although the protocols are called *cryptographic* protocol we will not consider attacks on the underlying cryptography. Analysis of cryptographic protocols deals with the application of cryptographic primitives, not cryptography itself. This so-called *Perfect Cryptography* assumption treats cryptographic primitives as black box processes that work as intended.

We consider a network with any number of participants and one attacker. The attacker is the most powerful one known to us; the spy from the Dolev-Yao model [4]. He is an active attacker who can intercept, redirect and alter arbitrary messages between principals. However, the spy cannot decrypt or encrypt messages with keys he does not know, so he has control over the network, but cannot defeat the encryption algorithms. The spy can also send and receive messages as a normal participant in the network.

Some of the other assumptions we use are not literally true about real-life networks and computations, but have practical justifications. In particular we use concepts such as "perfect" one way, collision free, hashes and "real" randomness of nonces.

There are a number of restrictions we omit that are common in the literature. For instance, we assume neither a fixed number of participants (as in [7]) nor a limited number of parallel protocol runs (as in [6, 7]). Moreover, we avoid the following assumption.

> **Protocol Message Axiom:** If a participant **P** distinct from the Spy sends a message m, then m is required by the protocol.

For instance, in [11], the author explicitly includes axioms of the form:

> If *evs* is a trace containing "B says to **S** the message X," then *evs* may be extended by "**S** says to B' the message X'."

Because there is no other allowance that **S** (distinct from the Spy) sends a message[4], Paulson is committed to the protocol message axiom. To the best of our knowledge, all other approaches assume a similar restriction.

We chose to avoid this axiom because it is unrealistic. In a network, the participants send many different messages, the majority of which bear no resemblance to protocol messages. We wished to be clear on what assumptions on the behavior of participants are sufficient to conclude that the protocol will reach a correct state. For this end, we chose to adopt only those assumptions which are reasonably reflected in actual networks, or have a priori arguments for their acceptance. We do not have an argument for the acceptance of the protocol message axiom (although, in the end, practical concerns form a very strong argument).

Indeed, there is a simple argument against the axiom. Suppose that the initial message $B, \{N_b, B\}_{pk_a}$ in the Bilateral Key Exchange protocol is sent from **B** to **A**. Suppose as well that, prior to **A**'s receipt of same, the Spy sends **A** a message, containing $\{N_b, B\}_{pk_a}$ and a note asking for assistance decrypting the message. If **A** complies with this request, and sends a response back to the Spy, then clearly the protocol has gone awry.

This is not to say that such behavior is reasonable on the part of **A**, but one would like protocol analyses to provide a more or less explicit description of what behavior *is* reasonable, i.e., what behavior ensures successful completion of the protocol.

Unfortunately, this aim is difficult indeed, and our present analysis did not wholly succeed in clarifying these issues. One must decide where feasibility supersede the motivations above.

3.2 Message Passing

We begin by describing a general setting in which principals pass messages in the presence of a powerful *Spy* (the Dolev-Yao spy described above). For now,

---

[4] Although, Paulson does allow accidental loss via Oops events, these accidents do not really clarify what behavior among principals ensures success.

messages can simply be seen as concatenations of *identifiers*, *nonces*, *hashes* and *keys*, all of which can be (multiple-times) encrypted. Example messages can be seen in the three steps from Fig. 1.

The model consists of states and transitions between states, where each state is intended as a temporal snapshot of the network. A state is fully characterized by the following:

- its action, that is, what has just occurred.
- the mode of the Principals (see below).
- its next state, i.e., what state follows the current state.

A state's action describes the Event that has occurred in that state. There are three Event types: (i) nothing happens, i.e., the system is idle; (ii) some Principal **P** sends a message m to another Principal **Q**; (iii) some Principal **P** receives message m. Note that the *Spy* is a Principal which can play the role of either **P** or **Q**.

A Principal's mode is fully given by (i) the public/private key pair a principal owns, (ii) his *knowledge* and (iii) the *beliefs* of the principal. Note that public and private keys are unique to each principal and that both *knowledge* and *beliefs* are state dependent, i.e., change as the system progresses (via next).

When we refer to an agent's knowledge, we mean the collection of messages with which he is familiar. A belief, for our purpose, is a statement of the form "P knows m". An agent ought to believe "P knows m" just in case he has seen evidence that m is indeed in **P**'s knowledge, but what counts as evidence depends on a specific protocol. Note: the objects of knowledge, for our purposes, are *messages*, while the objects of belief are *statements*. Knowledge does *not* consist of some variation of "true beliefs".

Given a state x, the next state (written x.next) is intended as the moment after x. Thus, our interpretation of "next" has a temporal flavor, yielding informal notions of prior, eventually, etc. It is constrained by the x.action and the mode of the principals at present. The constraints include:

- If x.action = idle, then the mode of each principal is unchanged in x.next.
- If, in x, **P** sent a message m to **Q** then
  1. The Spy learns m (see below).
  2. All other agents do not change their mode.
- If **P** receives message m, then
  1. **P** learns m.
  2. All other agents do not change their mode.

Notice that a model is a deterministic, but underspecified, transition system. Given a particular state x, there is a unique next state, x.next. Hence, the system is deterministic. However, even if one knows everything about state x, he will not know the action of x.next. Thus, it is underspecified.

Here we omit certain other constraints on our model. For instance, we require that the sender of a message already knows the message, and that sent messages are *Eventually* received[5]. Also particular protocols add additional constraints, including axioms regarding beliefs. More on this in Sect. 4.

When a principal **P** learns a message m (that is when **P**= *Spy* and m is sent, or when **P** receives m), then his knowledge increases. Clearly, after learning m, then **P** knows m. But he also knows the sub-messages of m, subject to the constraint that he only gains knowledge of an encrypted sub-message if he knows the appropriate key. As well, if m contains a key, then **P** may be able to decrypt other messages which he already knew. This complicated effect of learning is modeled by appropriate fixed points, about which more later.

3.3   Protocol

The general setting described in the previous section provides a foundation for the specification of particular protocols. Each protocol description includes also the following:

- rules governing reactions to certain messages;
- axioms governing belief updating;
- correctness conditions for the protocol.

The first item is complicated by the fact that we rejected the protocol message axiom in our model. In the presence of that axiom, one would include rules like: If a principal **Q** receives a message of the form $P, \{P, N_1\}_{pk_q}$, then he should respond with a message to **P** of the form $\{Hash(N_1), N_2, Q, K\}_{pk_p}$, where N_2 and K are fresh. Also, this second message is the only message which **Q** can send which contains N_1 (as a subterm).

However, a problem arises in that, **Q** may learn the nonce N_1 without receiving the full message $P, \{P, N_1\}_{pk_q}$. For instance, the *Spy* may intercept **P**'s message to **Q** and send simply $\{P, N_1\}_{pk_q}$. If **Q** does not treat N_1 as a secret with **P**, then the protocol will go awry. Therefore, our rule must be more restrictive. We require that, if **Q** receives a message m from which he can extract $\{P, N_1\}_{pk_q}$, then he will send an appropriate response to **P** and no other message from **Q** will contain N_1 as a subterm.

As mentioned previously, beliefs are state and protocol dependent. For the Bilateral Key Exchange Protocol beliefs are modeled as follows:

- If, in some state x, protocol message 2 is received ($\{Hash(N_2), N_3, Q, K\}_{pk_p}$) by **P** and is consistent, then **P** believes in state x.next that **Q** knows the session key K.
- If, in some state x, protocol message 3 is received ($\{Hash(N_4)\}_K$) by **Q** and is consistent, then the recipient of the message believes in state x.next that his correspondent knows the session key K.

We include these axioms so we may prove that, at the end of the protocol, both participants believe correctly that each knows the session key and hence they may communicate.

---

[5] But, in the meantime, the *Spy* can do whatever he likes.

3.4   Correctness Conditions

Correctness conditions depend on the goals of a protocol. Looking again at the *Bilateral Key Exchange* protocol we are interested in two security concepts. First, we want to be certain that only the principals running the protocol learn the session key and no one else does, i.e., that the protocol ensures *secrecy* of the session key. The other interesting property is *authentication*, in this case mutual authentication. This means that both principals become certain of the identity of their correspondent.

In the end we want to prove something like the following: Given that the principals behave in a responsible manner and given the general security model, when an agent **B** starts the *Bilateral Key Exchange* protocol with another agent **A**, both agent **A** and agent **B** will eventually know the session key and both principal **A** and **B** will believe that each other knows the session key. This session key should be a shared secret between **A** and **B**. Explicitly, our notion of correctness requires the successful (but not timely!) completion of a protocol run.

Again, we stress that, at present, we have not proved such a strong condition given our weak axioms. We have shown that, if the participants never send any message containing the nonces N_a, N_b or keys pk_a^{-1}, $pk_b^{-}1$ and K_{ab}, aside from those messages required by the protocol, then K_{ab} will never become known to the *Spy*. This is an important step toward correctness, but there is much real work to finish the proof. This work involves subtle least fixed point constructions for classes of messages, among other considerations (see Sect. 4.3 for examples).

In the next section we will describe how we implemented this model in CCSL.

4   Implementation

The Coalgebraic Class Specification Language [12,14] provides a logical framework for writing specifications involving both algebras and coalgebras. Users can freely nest algebraic and coalgebraic specifications. This allows one to use algebraic models for inductively generated data types, e.g., the *static* structure in our implementation, and coalgebras for behavioral types, e.g., the *dynamic* structure of the implementation. Indeed, we call our approach "coinductive" to reflect our reliance on coalgebraic reasoning.

Let us return to the message passing structure from Sect. 3.2. We want to explicitly show how this description involves a coalgebraic structure. An implementation for the message passing context consists of a set X together with three functions:

$$
\begin{aligned}
\text{action} &: X \longrightarrow \text{Events} \\
\text{mode} &: X \times \text{Principal} \longrightarrow \text{PrincMode} \\
\text{next} &: X \longrightarrow X
\end{aligned}
$$

This is equivalent to a set X and a single function with type

$$X \longrightarrow \text{Events} \times (\text{PrincMode})^{\text{Principal}} \times X$$

Hence we can view a model as a coalgebra, specifically a coalgebra for the functor $X \mapsto \text{Events} \times (\text{PrincMode})^{\text{Principal}} \times X$.

More generally, one can view object oriented classes as coalgebras for a suitable functor. This is the key motivation behind CCSL, where a class specification is modeled by a coalgebra. CCSL provides the formal language for describing a class specification, and a compiler for translating specifications into logical theories.

One writes a formal specification in the language CCSL and compiles it into a theory for PVS [13, 9] (or Isabelle [10]). This translation includes generating basic definitions, axioms and lemmas from the informal theory of coalgebras. For instance, CCSL creates the definitions for invariant predicate, homomorphism, etc., the proof principle of coinduction (in the case that our model is canonical), basic temporal operators for *Eventually* and *Always* as well as axioms (**S4**) and lemmas for these operators, and so on.

Our motivation for using CCSL, then, was threefold. First, a coalgebraic model seemed natural for security protocols. The dynamic features of a message passing system are easily translated to a coalgebraic structure. Second, the CCSL language is a well-developed formal setting for specifying coalgebraic models, and includes temporal operators useful for expressing correctness of security protocols. As well, CCSL supports mixing algebraic and coalgebraic specifications, which is particularly useful here. Third, the compiler creates PVS theories (in our use), so that we can exploit the speed and power of PVS in order to prove our theorems. This means that we can use a formal language built just for coalgebraic specifications to model the protocol, and use an existing theorem prover to prove our theorems. Also, the theory generated by CCSL goes well beyond what we include in the specification, since it includes the temporal axioms and coalgebraic features "for free".

The fact that, in the end, we offer only a partial proof of the correctness of the protocol does not reflect on the appropriateness of CCSL and a coalgebraic approach generally. We still firmly believe that a coalgebraic model is perfectly appropriate and natural for a dynamic system, and that in particular CCSL offers a good setting for specifications. However, we must question whether our aim of avoiding the protocol message axiom is a practical decision. We are confident that the correctness proofs can be completed even without this axiom, but the resulting arguments are very long and difficult and the extra payoff largely theoretical at present.

4.1   Specification

In implementing the model we use two key features of CCSL, (i) modularization and (ii) inheritance.

By modularization, we mean that distinct features of our specification are treated separately. For instance, the features of a principal's mode (knowledge, beliefs, public/private key pair) can be largely separated from the message passing specification, although the latter depends on the former. More explicitly, we have a class specification, PrincMode, for the principal's mode, and another specification, MsgContext, for the message passing system. The effects of learning are segregated in the former, and changes there influence the latter only indirectly (via the mode : $X \times$ Principal $\longrightarrow$ PrincMode method). This simplifies the creation of the specification, and allows one to change his implementation of knowledge-updating without explicitly changing the MsgContext specification (although, existing proofs will often need to be revisited).

The language CCSL includes an inheritance relation, so that one specification may inherit from another. This feature allows one to treat message passing generally, and inherit from this treatment for each individual protocol specification. In our work, we restrict our use of inheritance so that it preserves the models, in the sense that, given a particular protocol Prot, we have

$$\text{Model(Prot)} \subseteq \text{Model(MsgContext)}.$$

In other words, all of the restrictions (i.e., axioms) that we have placed on the general setting apply to the particular setting of a given protocol. Thus, theorems proved about the general setting are inherited by the particular protocols as well, minimizing the amount of repeat effort.

Figure 2 shows the dependencies of the different classes, where *double* arrows ($\Longrightarrow$) show the inheritance relations. The *broken* arrow ($- - \rightarrow$) shows that MsgContext depends on PrincMode. This dependency is an application of modularization in our specification.

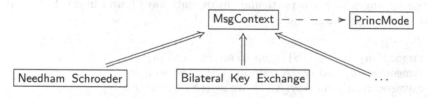

Fig. 2. Theory dependencies

4.2 Temporal Operators in CCSL

As mentioned previously, the language CCSL includes the temporal *Always* ($\square$) and *Eventually* ($\Diamond$) operators. In this section we will explain how these are defined.

Recall that the CCSL compiler defines an appropriate notion of *class invariant* for each class specification. The "always" operator is defined in terms of invariants. Specifically, $\square P$ is the largest invariant which implies P, i.e.,

$$\square P = \lambda(x : \text{Self}) : \exists(Q : \text{Self} \rightarrow \text{bool})(Q(x) \wedge \text{invariant}(Q) \wedge Q \Rightarrow P).$$

In our specification, the only method with non-constant codomain is next, which has type Self $\rightarrow$ Self. This method has a natural temporal meaning, and this imposes a purely temporal interpretation on $\Box$. Indeed, we see that $\Box P(x)$ holds iff P holds at x and every state thereafter, so $\Box$ really is the *Always* (or "henceforth") temporal operator.

As usual, $\Diamond$ is defined as $\neg\Box\neg$. Therefore, $\Diamond P(x)$ holds iff there is some state "in the future" of x in which P holds, i.e., *Eventually* P.

The compiler provides, in addition to the definitions for $\Box$ and $\Diamond$, a number of lemmas about their behavior. In particular, the compiler proves that the operators satisfy S4.

4.3   Messages and Learning

Messages are formally modeled by an *Abstract Data Type*. Explicitly, our model of messages is the least set S such that:

- For all $P \in$ Principal, $N \in$ Nonce, and $K \in$ Key, the syntactical objects princ(P), nonce(N) and key(K) are in S;
- If $m \in S$ then hash(m) is also in S;
- If $m_1 \in S$ and $m_2 \in S$ then msgpair(m_1, m_2) is also in S;
- If $m \in S$ and $K \in$ Key then symenc(K, m) and asymenc(K, m) are also in S (representing symmetric and asymmetric encryption, resp.).

In other terms, our model of messages is an initial algebra, and in particular comes with a principle of induction.

We define two closure operators[6] on $\mathcal{P}$(Message), the power set of Message. The first, $\downarrow$, (read: *downward closure*) is intended to capture extractability. That is, given $S \subseteq$ Message, a message m is in $\downarrow(S)$ just in case one can extract m from the messages in S (in particular, using only keys found in S). Explicitly, $\downarrow(S)$ is the least set satisfying

- $S \subseteq \downarrow(S)$;
- if msgpair(m_1, m_2) $\in \downarrow(S)$, then so are m_1 and m_2;
- if symenc(K, m) and key(K) are in $\downarrow(S)$, then so is m;
- if asymenc(K, m) and key(K^{-1}) are in $\downarrow(S)$, then so is m;

The upward closure, $\uparrow$, of a set S of Message is also given by a least fixed point condition. It consists of all of the messages that can be constructed from messages in S. We omit an explicit definition here, since it is analogous to $\downarrow$.

These two operators are used to axiomatize the effect of learning a message. Suppose that, in state x, an agent P knows exactly the set of messages S, and that he receives (or, if $P = Spy$, snoops) message m. Then, in state x.next, the agent should know

$$\uparrow\downarrow(S \cup \{m\}).$$

(Note: given a set S, $\downarrow\uparrow\downarrow(S) \subseteq \uparrow\downarrow(S)$, and so $\uparrow\downarrow$ is idempotent.)

The fixed point constructions here are *not* sufficient to prove correctness of a protocol in the absence of the protocol message axiom. For that, subtler constructions are required, which we omit here for concern of space.

---

[6] The closure $\downarrow$ corresponds to Paulson's **analz** operator in [11], while $\uparrow$ corresponds to **synth**.

4.4 MsgContext

We can not show the complete specification of our model in this paper. Thus, in this section we will give a CCSL specifications of the constructs explained in Sect. 3.2.

Recall that we restrict the next operator according to the event and the mode of the principals at the current state. In the general MsgContext class we have defined an axiom action that does just this.

————Ccsl————————————————————————————

$\forall\,(P : \mathsf{Principal}), \forall(Q : \mathsf{Principal}), \forall(x : \mathsf{Self}), \forall(m : \mathsf{Message}) :$
action:
  CASES x.action OF
    idle :
      $\forall\,(R : \mathsf{Principal}) :$
        x.next.princmode$(R) = x$.princmode(R)
    sent$(\mathsf{P}, \mathsf{Q}, \mathsf{m})$:
      $\forall\,(R : \mathsf{Principal}) :$
        $\neg(R = Spy) \to$
        x.next.princmode$(R) = x$.princmode$(R)\,\wedge$
        x.next.princmode$(Spy) = x$.princmode(Spy).learns$(m)\,\wedge$
        x.princmode.knows$(m)\,\wedge$
        $(\Diamond(\lambda(z : \mathsf{Self}) :$
          receivedBy?$(z$.action$)\,\wedge$
          receivedBy_rcp$(z$.action$) = Q\,\wedge$
          receivedBy_msg$(z$.action$) = m)$
        FOR$\{$next$\}(x$.next$))$
    receivedBy(P, m) :
      $\forall\,(R : \mathsf{Principal}) :$
        $\neg(R = P) \to$
        x.next.princmode$(R) = x$.princmode$(R)\,\wedge$
        x.next.princmode$(P) = x$.princmode(P).learns$(m)\,\wedge$
        $(\forall(z : \mathsf{Self}) : \Diamond(\lambda(x1 : \mathsf{Self}) : x1 = x)$ FOR $\{$next$\}(z) \to$
          $(\Diamond(\lambda(z1 : \mathsf{Self}) :$
          sent?$(z1$.action$)\,\wedge$
          sent_rcp$(z1$.action$) = P\,\wedge$
          sent_msg$(z1$.action$) = m\,\wedge$
          $\Diamond(\lambda(x1 : \mathsf{Self}) : x1 = x)$ FOR $\{$next$\}(z1)\,\wedge$
          $\Diamond(\lambda(z2 : \mathsf{Self}) : z2 = x)$ FOR $\{$next$\}(z1)$ FOR
        $\{$next$\}(z) \vee$
        $\exists(z1 : \mathsf{Self}) : (\Diamond(\lambda(x1 : \mathsf{Self}) : x1 = z)$ FOR $\{$next$\}(z1)\wedge$
        sent?$(z1$.action$)\,\wedge$
        sent_rcp$(z1$.action$) = P\,\wedge$
        sent_msg$(z1$.action$) = m\,\wedge$
        $\Diamond(\lambda(x1 : \mathsf{Self}) : x1 = x)$ FOR $\{$next$\}(z1))))$

Notice that the general structure explained in Sect. 3.2 can be seen: the three types of Events are clearly distinct. There are also some additional temporal constraints. I.e., if a message is received in state x.next, then for all states z prior to x.next, there is a state $z1$ either prior to z or after z (but prior to x.next) such that message m was sent in state $z1$.

As well, we assume that when a message m is sent, it *Eventually* ($\diamond$) will be received. At first glance, this seems to place a restriction on the ability of the *Spy*, since he cannot prevent a message from arriving. However, there is no time constraint on when the message will arrive, and so the *Spy* has as much time as necessary to interrupt the protocol. Thus, we do not feel that this assumption weakens the *Spy*.

4.5   PVS Theories and Proofs

There is a compiler for the CCSL language which can be used to translate the CCSL class specifications into theories for the PVS theorem prover [13, 9]. In PVS then, we can prove a number of lemmas and of course the correctness conditions for different protocols we are interested in.

A theorem proving environment like PVS has advantages over informal mathematical arguments. In particular, PVS acts as a proof checker as the proof is being built, so that one is more confident of the final result. As well, with user-defined strategies tailor-made for the problems at hand, the theorem prover can take over much of the tedious work. Of course, such an environment comes with the usual difficulties of any formal system. Namely, proofs can become very long and tedious and the resulting proof may not be illuminating to a reader. In the case of protocol verification, however, one is interested in a proof only for establishing the theorem at hand, and not for its explanatory power.

5   A Word about Generality

There are two main features of our work which distinguish it from other approaches. First, as we have made clear, we take a coalgebraic approach to modeling message passing in general, and security protocols in particular. Second, we have followed the principle that an assumption about our model should be justified by an argument about its reasonableness or necessity. This principle is evident in the liberal assumptions regarding what messages may be sent.

These features are orthogonal, in the sense that (i) a coalgebraic model may certainly impose stronger assumptions than we do and (ii) other methods of verification may, in principle, weaken their assumptions as well. We anticipate that other methods will encounter similar complications if they choose to drop the protocol message axiom, however. It appears that this assumption comes with a real practical payoff, although it is unclear that the resulting proofs show the correctness of a protocol in a real-life setting. Again, one must decide at what point the practical concerns ought to trump the theoretical motivations.

In the end, we have proved an important part of the correctness of the Bilateral Key Exchange protocol. Namely, we have shown that, if **A** and **B** do not send messages containing N_a, N_b or K_{ab}, aside from the required protocol messages, then these secrets will never be known to the *Spy*. Note that, with the protocol message axiom, one needs merely to show that once the initial message is sent, then eventually the second and third messages will be accepted by their recipients to conclude that the protocol is correct. Even without the protocol message axiom, the direction for the remainder of the proof is clear, but it is not easy. When one allows a large body of legal messages, then it becomes difficult to express the relationship between the secrets and the body of messages as a whole. As well, when one allows liberal behavior on the part of the two correspondents, then restrictions that ensure the correct completion of the protocol become subtle. These proofs can be completed, but not without some sweat.

6   Related Work

There are a lot of different approaches described in the literature which deal with the verification of security protocols (see e.g. [1, 2, 6, 7, 11, 15].) Generally speaking one can say that there are two kind of approaches. The ones that work (semi)automatic, these usually concentrate on some aspects of security while ignoring others, and in particular have some limitations such as the number of parallel sessions or the number of principals or protocol steps involved.

Then there are the more "total" approaches (most notably [11], see Sect. 6.1) which require a lot of user interaction but mostly do not have restrictions as the ones mentioned above. Clearly, our work must be placed in this latter setting. We want to stress that both kind of approaches are useful and are in many ways each other's complement.

6.1   Paulson's Inductive Approach

In his influential paper *"The Inductive Approach to verifying Cryptographic Protocols"* [11], Larry Paulson describes an approach for analyzing security protocols using the theorem prover Isabelle [10]. Although our approach is similar to his, there are some notable conceptual differences (aside from the practical differences in the specifications, including his assumption of the protocol message axiom).

Paulson's models are inherently algebraic, rather than coalgebraic. He considers the set of finite traces for a protocol. This set can be given by a least fixed point construction, i.e., by an initial algebra.

His basic proof principle is induction over traces of events (*evs*). To prove P always holds, he shows that

$$P[] \tag{1}$$

holds, and if

$$P(evs) \Rightarrow P(ev \# evs). \tag{2}$$

i.e., P holds for the empty trace of events [] and if P holds for some trace *evs*, P has to hold for all traces $ev \# evs$ containing one event ev more.

This is analogous to showing that P is an invariant, in the coalgebraic sense. The main theoretical difference is that we consider finite and infinite traces as models, while Paulson considers finite traces. Moreover, since Paulson relies on the inductive proof principle, he is restricted to canonical algebras as models. Because we do not assume finality of our coalgebraic model, we have a more general approach. In particular, Paulson's inductive models can be viewed as models in our coalgebraic setting (since initial algebras are also coalgebras), at least given an appropriate specification.

7   Conclusion

Our protocol verification departs from other approaches in two distinct features: its explicitly coalgebraic (and temporal) foundation, and the weakness of our axiomatization of the protocol. The former feature is an illustration of an application of abstract mathematics to verifications, making use of coinductively specified structures. The latter feature was motivated by a desire to strengthen the persuasiveness of the correctness proofs, by avoiding restrictions that did not hold in real networks. In particular, we avoided the so-called protocol message axiom.

We have argued that coalgebras are particularly useful for modeling dynamic systems. We think that the separate specification language (allowing also modularization and inheritance) as well as the higher abstraction in the proofs are other attractive features of our approach.

It would be useful to compare coinductive versus inductive reasoning on even terms, which we have not done here. One should adapt Paulson's axiomatization to a coalgebraic setting (or vice versa). With such a study, one could see the advantages and disadvantages of coalgebraic/algebraic conceptual frameworks in cryptographic protocol analysis.

We ran into considerable difficulties in proving the correctness of our sample protocol without the protocol message axiom. We have shown that, if the participants **A** and **B** transmit messages as called for by the protocol, then the Spy never learns the session key (or any of the nonces or private keys). We have not proved that the Spy can't compel **A** or **B** to send a message to him, revealing the session key or nonces. We also have not shown that beliefs which are acquired during the protocol run are true.

These partial successes emphasize the importance of the protocol message axiom, which has not been previously considered. While we have, we believe, strong theoretical motivations for omitting the protocol message axiom, we found perhaps stronger practical reasons to accept it. This negative result prompts questions about protocol correctness proofs. At what point ought practical concerns allow one to accept assumptions which are demonstrably false in natural settings, and which are not apparently "harmless"? We do not find an obvious answer to this fundamentally philosophical question.

Acknowledgments

We want to thank Bart Jacobs for suggesting the research topic and helping us on our way using CCSL, and Hendrik Tews for explaining CCSL constructs and repeatedly modifying the CCSL compiler to satisfy our whims.

The Coinductive Approach to Verifying Cryptographic Protocols

References

1. M. Abadi and A. D. Gordon. A calculus for cryptographic protocols: The spi calculus. In *Proceedings of the Fourth ACM Conference on Computer and Communications Security*, pages 36–47. ACM Press, April 1997.
2. M. Burrows, M. Abadi, and R. Needham. A logic of authentication. *Proc. Royal Soc.*, Series A, Volume 426:233–271, 1989.
3. J. Clark and J. Jacob. A Survey of Authentication Protocol Literature, version 1.0, 1997. available at URL http://www-users.cs.york.ac.uk/~jac/papers/drareview.ps.gz.
4. D. Dolev and A. Yao. On the security of public key protocols. *IEEE Transactions on Information Theory*, 29(6), 1983.
5. Gavin Lowe. An attack on the Needham-Schroeder public-key authentication protocol. *Information Processing Letters*, 56:131–133, 1995.
6. Gavin Lowe. Casper: A compiler for the analysis of security protocols. In *PCSFW: Proceedings of The 10th Computer Security Foundations Workshop*. IEEE Computer Society Press, 1997.
7. J. Millen and V. Shmatikov. Constraint solving for bounded-process cryptographic protocol analysis. In *8th ACM Conference on Computer and Communication Security*, pages 166–175. ACM SIGSAC, November 2001.
8. R.M. Needham and M.D. Schroeder. Using encryption for authentication in large networks of computers. *Communications of the ACM*, 21(12):993–999, 1978.
9. S. Owre, J.M. Rushby, N. Shankar, and F. von Henke. Formal verification for fault-tolerant architectures: Prolegomena to the design of PVS. *IEEE Trans. on Softw. Eng.*, 21(2):107–125, 1995.
10. L.C. Paulson. *Isabelle: A Generic Theorem Prover*. Number 828 in Lect. Notes Comp. Sci. Springer, Berlin, 1994.
11. L.C. Paulson. The inductive approach to verifying cryptographic protocols. *Journ. of Computer Security*, 6:85–128, 1998.
12. J. Rothe, H. Tews, and B. Jacobs. The coalgebraic class specification language CCSL. *Journal of Universal Comp. Sci.*, 7(2), 2001.
13. N. Shanker, S. Owre, J.M. Rushby, and D. Stringer-Calvert. *PVS prover guide*, 1999. Version 2.3.
14. Hendrik Tews. *Coalgebraic Methods for Object Oriented Specification*. PhD thesis, Technical University of Dresden, October 2002.
15. F. J. Thayer, J. C. Herzog, and J. D. Guttman. Strand spaces: Proving security protocols correct. *Journal of Computer Security*, 7(1), 1999.

Behavioural Equivalence and Indistinguishability in Higher-Order Typed Languages

Shin-ya Katsumata

Division of Informatics, University of Edinburgh, King's Buildings,
Edinburgh EH9 3JZ, Scotland

Abstract. We extend the study of the relationship between *behavioural equivalence* and *the indistinguishability relation*[4, 7] to the simply typed lambda calculus, where higher-order types are available. The relationship between these two notions is established in terms of *factorisability*[4]. The main technical tool of this study is *pre-logical relations[8]*, which give a precise characterisation of behavioural equivalence. We then consider a higher-order logic to reason about models of the simply typed lambda calculus, and relate the resulting standard satisfaction relation to behavioural satisfaction.

1 Introduction

This work is a contribution to the understanding of the relationship between *behavioural equivalence* and the *indistinguishability relation*. These notions arose from the study of data abstraction in the context of algebraic specifications. Behavioural equivalence identifies models which show the same behaviour for any program yielding an observable value. This formalises an intuitive equivalence between two programming environments that show the same behaviour to programmers, regardless of differences in the representation of non-observable data types. The indistinguishability relation is a partial equivalence relation which identifies values in a model that are interchangeable with each other in any program context. This provides an abstract view of the programming environment based on behaviour, rather than denotation.

These two notions are useful when reasoning about specifications, and their relationship has been studied in a series of papers beginning with [4] by Bidoit, Hennicker and Wirsing. They established the key idea of *factorisability* to relate behavioural equivalence and the indistinguishability relation. Their framework is infinitary first-order logic over Σ algebras. Hofmann and Sannella[7] extended the logic over Σ algebras to higher-order logic, which enables us to quantify over predicates and axiomatise the indistinguishability relation when the underlying signature is finite.

We further extend the target of reasoning to a language having higher-order types and functions. Higher-order functions enable us to write program-parameterised programs, and are useful in program development. Thus we are interested in reasoning about specifications in such languages.

In this paper, we take the simply typed lambda calculus as the formalisation of higher-order typed languages, and give the semantics of the lambda calculus by *typed combinatory algebras*, which subsume a wide range of semantic frameworks including Henkin models, type frames and full-type hierarchies.

M. Wirsing, D. Pattinson, and R. Hennicker (Eds.): WADT 2002, LNCS 2755, pp. 284–298, 2003.

Once we introduce higher-order types, we need to consider how to extend behavioural equivalence and the indistinguishability relation to higher-order types. In the study of the simply typed lambda calculus, there is a well-known extension method using exponential relations (the following shows the case of binary relations between two combinatory algebras $\mathcal{A}$ and $\mathcal{B}$; we can extend this to n-ary relations):

$$R^{\tau \to \tau'} = R^\tau \to R^{\tau'} = \{(f, g) \in A^{\tau \to \tau'} \times B^{\tau \to \tau'} \mid \forall(x, y) \in R^\tau . (fx, gy) \in R^{\tau'}\}$$

and the resulting relation is called a *logical relation* if it relates the interpretations in $\mathcal{A}$ and $\mathcal{B}$ of each constant. However, in this study, logical relations are not adequate for a couple of reasons:

1. Reachability at first-order types cannot be extended to higher-order types using the exponential relation. We see this by an example; let us consider the simply typed lambda calculus with zero and successor, namely z^{nat} and $s^{nat \to nat}$. We give the semantics of the lambda calculus by the full-type hierarchy. A full-type hierarchy constructed from $A^{nat} = \mathbf{N}$. We write S^τ for the set of reachable elements at type τ. We can reach any $n \in \mathbf{N}$ by the term $s^n(z)$, thus $S^{nat} = A^{nat}$. However the unary logical relation R constructed from $R^{nat} = S^{nat}$ does not give reachability correctly at higher-order types, since $R^{nat \to nat} = A^{nat \to nat}$ but clearly $S^{nat \to nat} \subset A^{nat \to nat}$.

2. Logical relations are not suitable to characterise behavioural equivalence. A restricted notion of behavioural equivalence, called closed observational equivalence, was studied in [10]. Mitchell showed *representation independence theorem*; if there exists a binary logical relation between two models such that the relation is bijective on the observable types, then these two models are closed observationally equivalent. He showed that the converse is also true when the underlying signature has only first-order constants. However this is not satisfactory for two reasons; one is the above restriction to first-order constants, and the other is that in general logical relations do not compose, despite the fact that behavioural equivalence is a transitive relation.

To solve these problems, we use *pre-logical relations*[8] by Honsell and Sannella instead of logical relations. They are a generalisation of logical relation, and have several characterisations; a relation is pre-logical iff it satisfies the basic lemma (theorem 1 below), and a pre-logical relation can be seen as a correspondence in the sense of Schoett[12] between two combinatory algebras. Roughly, a pre-logical relation is a relation satisfying $R^{\tau \to \tau'} \subseteq R^\tau \to R^{\tau'}$. Thus pre-logical relations allow flexibility at higher-order types while logical relations are determined uniquely at all types from the relations at base types. Of course logical relations are included in pre-logical relations, but also the reachability predicate and other relations are included in this class. Another advantage of pre-logical relations is that they are closed under composition, which is a desirable property for characterising behavioural equivalence.

This paper is organised as follows: section 2 introduces basic definitions of the simply typed lambda calculus, pre-logical relations and partial equivalence relations(PERs). Section 3 establishes a relation between behavioural equivalence and existence of pre-logical relations. We also introduce another model equivalence and show that it is equivalent to behavioural equivalence. In section 4 we study properties of the indistinguishability relation, which turns to be a pre-logical PER over the underlying model. In section 5 we show that behavioural equivalence is factorisable by indistinguishability. We move to higher-order logic and its semantics in section 6. We introduce two semantics; one is the standard model and the other is the relative model w.r.t. some PER. We show that the quotient model of higher-order logic by a PER and the behavioural model w.r.t. the PER are logically equivalent. We prove this by showing that they are behaviourally equivalent w.r.t. boolean observations. In section 7 we apply these results to reasoning about specifications.

2   Preliminaries

2.1   Syntax of the Simply Typed Lambda Calculus

Definition 1 (Higher-Order Signature). *Let U be a set. We define the set of types* $\mathbf{Typ}(U)$ *by BNF* $\tau ::= b \mid \tau \to \tau$ *where* $b \in U$. *A higher-order signature (or simply* signature*) is a pair of sets* (U, C) *where* U *gives the set of base types and* $C \subseteq C_0 \times$ $\mathbf{Typ}(U)$ *gives the set of typed constants for some universe* C_0 *of constant symbols. We write* c^τ *for the pair* $(c, \tau) \in C$. *We fix a higher-order signature* $\Sigma = (U, C)$. *We often write* $\mathbf{Typ}(\Sigma)$ *for* $\mathbf{Typ}(U)$.

We assume that readers are familiar with the simply typed lambda calculus. The calculus considered in this paper is the minimal fragment; it has only $\to$ types. The lambda terms are built on a countably infinite set of variables X. We define a *context* by a partial function $\Gamma : X \rightharpoonup \mathbf{Typ}(\Sigma)$. Two contexts Γ and Δ are *separated* if $\mathrm{dom}(\Gamma) \cap \mathrm{dom}(\Delta) = \emptyset$. For $T \subseteq \mathbf{Typ}(\Sigma)$, we say Γ is a T-*context* if for all $x \in \mathrm{dom}(\Gamma)$, $\Gamma(x) \in T$. We say $\Gamma \vdash M : \tau$ is a *well-formed term* if $\Gamma \vdash M : \tau$ is derived only from the inference rules of the simply typed lambda calculus.

2.2   Semantics of the Simply Typed Lambda Calculus

In this study, we take typed combinatory algebras as the basis for the semantics of the simply typed lambda calculus. The reason is twofold: one is that they are general enough to subsume other classes of models, such as Henkin models and type frames, and the other is that combinatory algebras and the notion of pre-logical relation, introduced later, are compatible. Indeed the class of combinatory algebras is closed under quotient by pre-logical PERs(proposition 2).

We write C_{SK} for the extension of a set of constants C with S, K combinators:

$$C_{SK} = C \cup \{S^{(\tau \to \tau' \to \tau'') \to (\tau \to \tau') \to \tau \to \tau''} \mid \tau, \tau', \tau'' \in \mathbf{Typ}(\Sigma)\}$$
$$\cup \{K^{\tau \to \tau' \to \tau} \mid \tau, \tau' \in \mathbf{Typ}(\Sigma)\}.$$

Definition 2 (Typed Combinatory Algebra). *A Σ-typed combinatory algebra (or sim*ply* combinatory algebra) is a tuple* $\mathcal{A} = (A, \bullet_\mathcal{A}, (-)_\mathcal{A})$ *such that:*

1. A is a $\mathbf{Typ}(\Sigma)$-indexed family of sets (called carrier sets).
2. The application operator $\bullet_{\mathcal{A}}^{\tau,\tau'}$ is a family of functions having type $A^{\tau \to \tau'} \to A^\tau \to A^{\tau'}$ for any $\tau, \tau' \in \mathbf{Typ}(\Sigma)$.
3. $c_{\mathcal{A}}^\tau \in A^\tau$ for each $c^\tau \in C_{SK}$.
4. The combinators satisfy equations $K \bullet x \bullet y = x$ and $S \bullet p \bullet q \bullet r = p \bullet r \bullet (q \bullet r)$ for any x, y, p, q, r in appropriate carrier sets (superscripts and subscripts are omitted for readability).

We write $\bullet_{\mathcal{A}}$ as a left-associative infix operator. We may omit superscripts and sub-scripts if they are obvious from the context. Script letters $(\mathcal{A}, \mathcal{B}, \cdots)$ are used to denote combinatory algebras while the carrier sets of these algebras are referred by normal letters $(A, B, \cdots)$. We write $\mathbf{CA}(\Sigma)$ for the collection of Σ-combinatory algebras.

Definition 3 (Henkin Model/Type Frame/Full Type Hierarchy). A combinatory algebra $\mathcal{A}$ is called:

- a Σ-Henkin model if extensionality holds: $\forall f, g \in A^{\tau \to \tau'} . (\forall x \in A^\tau . f \bullet_{\mathcal{A}} x = g \bullet_{\mathcal{A}} x) \Longrightarrow f = g$.
- a Σ-type frame if $A^{\tau \to \tau'} \subseteq A^\tau \to A^{\tau'}$ and $f \bullet_{\mathcal{A}} x = f(x)$.
- a Σ-full type hierarchy if $A^{\tau \to \tau'} = A^\tau \to A^{\tau'}$ and $f \bullet_{\mathcal{A}} x = f(x)$. We note that a full-type hierarchy is uniquely determined by the carrier sets for base types.

Example 1. The following is a higher-order signature $\Sigma_{set} = (U_{set}, C_{set})$ for the finite sets of natural numbers.

$$U_{set} = \{bool, nat, set\}$$
$$C_{set} = \{\text{tt}^{bool}, \text{ff}^{bool}, \text{not}^{bool \to bool}, 0^{nat}, \text{succ}^{nat \to nat}, \text{eq}^{nat \to nat \to bool},$$
$$\emptyset^{set}, \{-\}^{nat \to set}, (- \cup -)^{set \to set \to set}, \text{filter}^{(nat \to bool) \to set \to set},$$
$$\text{isempty}^{set \to bool}\}$$

The constant filter takes a predicate p and a set s and yields the set which consists of the elements in s satisfying the predicate p. The constant isempty judges whether a given set is empty or not.

We introduce two full-type hierarchies $\mathcal{A}_{set}$ and $\mathcal{B}_{set}$ over Σ_{set}. In $\mathcal{A}_{set}$, base types are interpreted as $A_{set}^{bool} = \{tt, ff\}, A_{set}^{nat} = \mathbf{N}, A_{set}^{set} = \mathcal{P}(\mathbf{N})$. We interpret filter and isempty in $\mathcal{A}_{set}$ as follows:

$$\text{filter}_{\mathcal{A}_{set}} f X = \{x \in X \mid f(x) = tt\}$$
$$\text{isempty}_{\mathcal{A}_{set}} X = tt \iff X = \emptyset$$

The interpretation of the other constants is naturally defined.

In $\mathcal{B}_{set}$, base types are interpreted as $B_{set}^{bool} = \{tt, ff\}, B_{set}^{nat} = \mathbf{N}, B_{set}^{set} = \mathbf{N} \to B_{set}^{bool}$. In this interpretation, a set $X \subseteq \mathbf{N}$ is represented by its characteristic function $\phi_X : \mathbf{N} \to B_{set}^{bool}$. We interpret filter and isempty in $\mathcal{B}_{set}$ as follows:

$$(\text{filter}_{\mathcal{B}_{set}} p f) x = tt \iff p(x) = tt \wedge f(x) = tt$$
$$\text{isempty}_{\mathcal{B}_{set}} f = tt \iff \forall x \in \mathbf{N} . f(x) = ff$$

The interpretation of the other constants is naturally defined.

An *environment* (ranged over by η, ρ) over a combinatory algebra $\mathcal{A}$ is a partial function $X \rightharpoonup \bigcup_{\tau \in \mathbf{Typ}(\Sigma)} A^\tau$. We write $\eta \in A^\Gamma$ for an environment η such that $\mathrm{dom}(\eta) = \mathrm{dom}(\Gamma)$ and $\eta(x) \in A^{\Gamma(x)}$ for all $x \in \mathrm{dom}(\eta)$.

Given a combinatory algebra $\mathcal{A}$, we can interpret well-formed lambda terms in $\mathcal{A}$ by the *meaning function* $\mathcal{A}[\![-]\!]-$, which maps a well-formed term $\Gamma \vdash M : \tau$ and an environment $\eta \in A^\Gamma$ to a value in A^τ. The meaning function is defined by induction on the derivation of well-formed lambda terms, and uses a trick of compiling lambda abstraction using S and K in a combinatory algebra when interpreting $\lambda x^\tau . M$. For details, see [2, 11].

Proposition 1. (Semantic Substitution Lemma) *Let Γ and Δ be separated contexts, $\Gamma, x : \tau \vdash M : \tau'$ and $\Delta \vdash N : \tau$ be well-formed terms, $\rho \in A^\Gamma$ and $\eta \in A^\Delta$. Then $\mathcal{A}[\![M]\!]\rho\{x \mapsto \mathcal{A}[\![N]\!]\eta\} = \mathcal{A}[\![M[N/x]]\!]\rho \cup \eta$.*

Definition 4 (Σ-homomorphism). *A Σ-homomorphism is a $\mathbf{Typ}(\Sigma)$-indexed family of functions $\{h^\tau : A^\tau \rightarrow B^\tau\}_{\tau \in \mathbf{Typ}(\Sigma)}$ such that for all $c^\tau \in C_{SK}$, $h^\tau(c_\mathcal{A}^\tau) = c_\mathcal{B}^\tau$ and for all $\tau, \tau' \in \mathbf{Typ}(\Sigma), x \in A^{\tau \rightarrow \tau'}$ and $y \in A^\tau$, $h^{\tau'}(x \bullet_\mathcal{A} y) = h^{\tau \rightarrow \tau'}(x) \bullet_\mathcal{B} h^\tau(y)$. A Σ-isomorphism is a Σ-homomorphism h such that h^τ is bijective on each $\tau \in \mathbf{Typ}(\Sigma)$. We write $\mathcal{A} \cong \mathcal{B}$ if there exists a Σ-isomorphism between $\mathcal{A}$ and $\mathcal{B}$.*

2.3 Pre-logical Relations

First, some definitions. A *relation* between $\mathcal{A}$ and $\mathcal{B}$ (written $R \subseteq \mathcal{A} \times \mathcal{B}$) is a $\mathbf{Typ}(\Sigma)$-indexed family of sets R satisfying $R^\tau \subseteq A^\tau \times B^\tau$ for all $\tau \in \mathbf{Typ}(\Sigma)$. We write $(\eta, \eta') \in R^\Gamma$ if $\eta \in A^\Gamma, \eta' \in B^\Gamma$ and for all $x \in \mathrm{dom}(\Gamma)$, $(\eta(x), \eta'(x)) \in R^{\Gamma(x)}$. The composition of relations $R \subseteq \mathcal{A} \times \mathcal{B}$ and $R' \subseteq \mathcal{B} \times \mathcal{C}$ is defined by type-wise composition of R and R'. The *exponential relation* of R^τ and $R^{\tau'}$ is defined by

$$R^\tau \rightarrow R^{\tau'} = \{(f, g) \in A^{\tau \rightarrow \tau'} \times B^{\tau \rightarrow \tau'} \mid \forall (x, y) \in R^\tau . (f \bullet_\mathcal{A} x, g \bullet_\mathcal{B} y) \in R^{\tau'}\}.$$

Pre-logical relations were proposed by Honsell and Sannella[8], and are a generalised notion of logical relations. In this paper, we adopt the following definition of pre-logical relations[1].

Definition 5 (Pre-logical Relations[8]). *A relation $R \subseteq \mathcal{A} \times \mathcal{B}$ is pre-logical if*

1. *$R^{\tau \rightarrow \tau'} \subseteq R^\tau \rightarrow R^{\tau'}$, or equivalently for all $(f, g) \in R^{\tau \rightarrow \tau'}$ and $(x, y) \in R^\tau$, the pair $(f \bullet_\mathcal{A} x, g \bullet_\mathcal{B} y)$ is included in $R^{\tau'}$, and*
2. *for all $c^\tau \in C_{SK}$, the pair $(c_\mathcal{A}^\tau, c_\mathcal{B}^\tau)$ is included in R^τ.*

We contrast the above to the definition of logical relations. A logical relation is a type-indexed family of binary relations R satisfying $R^{\tau \rightarrow \tau'} = R^\tau \rightarrow R^{\tau'}$ and for each $c^\tau \in C, (c_\mathcal{A}^\tau, c_\mathcal{B}^\tau) \in R^\tau$. Thus when we give a logical relation, we perform the following

---

[1] Originally pre-logical relations were defined over lambda applicative structures, which is a general class of set-theoretic models of the simply typed lambda calculus. In the case that the underlying models are combinatory algebras, the definition coincides with definition 5 as observed in [8].

steps: we first give a relation R on base types, then extend it to higher-order types using the above scheme and check whether the interpretations in $\mathcal{A}$ and $\mathcal{B}$ of each of the constants are related by R. In contrast, the definition of pre-logical relations lacks right-to-left inclusion in the above scheme[2]. Thus it allows flexibility of choice of relations at higher-order types. We note that logical relations are also pre-logical relations, since the above scheme implies that the relation R relates S, K combinators at all types.

In [8] various characterisation of pre-logical relations are studied. One notable characterisation is via the basic lemma.

Theorem 1 (Basic Lemma for Pre-logical Relations[8]). *Let* $R \subseteq \mathcal{A} \times \mathcal{B}$. *Then* $\forall (\eta, \eta') \in R^\Gamma .(\mathcal{A}[\![M]\!]\eta, \mathcal{B}[\![M]\!]\eta') \in R^\tau$ *holds for any well-formed term* $\Gamma \vdash M : \tau$ *if and only if* R *is a pre-logical relation.*

Another notable property of pre-logical relations is that the composition of two pre-logical relations is again a pre-logical relation.

Theorem 2 (Composability of Pre-logical Relations[8]). *Let* $R \subseteq \mathcal{A} \times \mathcal{B}$ *and* $R' \subseteq \mathcal{B} \times \mathcal{C}$ *be pre-logical relations. Then* $R \circ R' \subseteq \mathcal{A} \times \mathcal{C}$ *is a pre-logical relation.*

2.4 Partial Equivalence Relations

Recall that a *PER* (ranged over by E) over $\mathcal{A}$ is a relation $E \subseteq \mathcal{A} \times \mathcal{A}$ such that for all $\tau \in \mathbf{Typ}(\Sigma)$, E^τ is symmetric and transitive. We write the domain of E^τ by $|E^\tau| = \{x \in A^\tau \mid (x, x) \in E^\tau\}$. Then E^τ is just an equivalence relation over $|E^\tau|$, so we write $[x]$ for the equivalence class of $x \in |E^\tau|$ by E^τ and A/E^τ for the quotient $|E^\tau|/E^\tau$.

When a PER $E \subseteq \mathcal{A} \times \mathcal{A}$ is pre-logical (or logical), we call E a *pre-logical (or logical) PER*. The quotient of a combinatory algebra by a pre-logical PER is again a combinatory algebra.

Proposition 2 ([8]). *Let* E *be a pre-logical PER over* $\mathcal{A}$.

1. *The tuple* $(A/E, \star, [(-)_\mathcal{A}])$ *where* $[x] \star [y] = [x \bullet_\mathcal{A} y]$ *is a* Σ-*combinatory algebra. We call this the* quotient *of* $\mathcal{A}$ *by* E, *and write it by* $\mathcal{A}/E$.
2. *Let* $\Gamma \vdash M : \tau$ *be a well-formed term and* $\eta \in A/E^\Gamma$. *Then* $\mathcal{A}/E[\![M]\!]\eta = [\mathcal{A}[\![M]\!]\rho]$ *where* $\rho \in A^\Gamma$ *and* $\rho(x) \in \eta(x)$ *for all* $x \in \mathrm{dom}(\Gamma)$.

Definition 6 (Projection). *We define the* projection relation $\Pi(E) \subseteq \mathcal{A}/E \times \mathcal{A}$ *as the following* $\mathbf{Typ}(\Sigma)$-*indexed family of binary relations:*

$$\Pi(E)^\tau = \{([e], e) \in A/E^\tau \times A^\tau \mid e \in |E^\tau|\}.$$

Lemma 1. *The projection* $\Pi(E)$ *is a pre-logical relation.*

Proof. Clearly $\Pi(E)$ relates all constants in C_{SK}. From the definition of pre-logical PERs, E is closed under the application operator. Therefore $\Pi(E)$ is so as well.   □

---

[2] Indeed, the reverse direction is required to hold only for lambda-definable elements because of the presence of the combinators in the set of constants.

3 Behavioural Equivalence and Pre-logical Relations

Behavioural equivalence identifies two models showing the same behaviour in response to all observations. Each observation compares the values of two terms of observable types, whose values are directly accessible to programmers. This definition of observation formalises the use of experiments to detect the difference of behaviour of visible data types between two models. Thus, intuitively speaking, if two models are behaviourally equivalent, they provide the same programming environment to programmers, even though they may have different implementations of invisible data types.

We establish a link between behavioural equivalence and pre-logical relations. The result is a natural extension of [8] to allow free variables of observable types, and an extension of [12] to handle higher-order types.

The definition of behavioural equivalence is adapted from [7]. There are other possibilities for the treatment of free variables, but we do not discuss them. For detail, see [7]. We fix a set $OBS \subseteq \mathbf{Typ}(\Sigma)$ called the *observable types*. We first introduce an auxiliary notion of OBS-surjective environment.

Definition 7. *An OBS-surjective environment between $\mathcal{A}$ and $\mathcal{B}$ is a tuple (Γ, ρ, ρ') where Γ is an OBS-context and $\rho \in A^\Gamma$ and $\rho' \in B^\Gamma$ are environments such that* $\mathrm{im}(\rho) = \bigcup_{\tau \in OBS} A^\tau$ *and* $\mathrm{im}(\rho') = \bigcup_{\tau \in OBS} B^\tau$.
We say $\mathcal{A}$ is OBS-countable if the set $\bigcup_{\tau \in OBS} A^\tau$ is countable.

We note that if there exists an OBS-surjective environment between $\mathcal{A}$ and $\mathcal{B}$, then $\mathcal{A}$ and $\mathcal{B}$ are OBS-countable. This is due to the cardinality of the set of variables.

Definition 8 (Behavioural Equivalence). *We say $\mathcal{A}$ and $\mathcal{B}$ are behaviourally equivalent w.r.t. OBS (written $\mathcal{A} \equiv_{OBS} \mathcal{B}$) if there exists an OBS-surjective environment (Γ, ρ, ρ') between $\mathcal{A}$ and $\mathcal{B}$ such that for any $\tau \in OBS$ and well-formed terms $\Gamma \vdash M, N : \tau$, we have $\mathcal{A}[\![M]\!]\rho = \mathcal{A}[\![N]\!]\rho \iff \mathcal{B}[\![M]\!]\rho' = \mathcal{B}[\![N]\!]\rho'$.*

We also give another formalisation of behavioural equivalence. We first introduce a program equivalence in a model, then we say two models are behaviourally equivalent if the program equivalence in both models coincides.

Definition 9. *1. Let Γ be an OBS-context, $\tau \in OBS$ and $\Gamma \vdash M, N : \tau$ be well-formed terms. We write $\mathcal{A} \models \Gamma \vdash M \sim N : \tau$ if for all $\eta \in A^\Gamma$, we have $\mathcal{A}[\![M]\!]\eta = \mathcal{A}[\![N]\!]\eta$.*
2. We write $\mathcal{A} \sim_{OBS} \mathcal{B}$ if $A^\tau \cong B^\tau$ for any $\tau \in OBS$, and for any OBS-context Γ, $\tau \in OBS$ and well-formed terms $\Gamma \vdash M, N : \tau$, we have $\mathcal{A} \models \Gamma \vdash M \sim N : \tau \iff \mathcal{B} \models \Gamma \vdash M \sim N : \tau$.

We introduce *observational pre-logical relations* to characterise behavioural equivalence (c.f. Schoett's correspondence[12]).

Definition 10 (Observational Pre-logical Relations). *An observational pre-logical relation $R \subseteq \mathcal{A} \times \mathcal{B}$ w.r.t. OBS is a pre-logical relation such that for all $\tau \in OBS$, $R^\tau \subseteq A^\tau \times B^\tau$ is a bijection.*

Proposition 3. *Let $R \subseteq \mathcal{A} \times \mathcal{B}$ and $S \subseteq \mathcal{B} \times \mathcal{C}$ be observational pre-logical relations w.r.t. OBS. Then $R \circ S$ is an observational pre-logical relation w.r.t. OBS.*

The following theorem characterises behavioural equivalence in terms of observational pre-logical relations. Simultaneously, it shows that the two formalisations of behavioural equivalence coincide[3].

Theorem 3. *The following are equivalent[4]:*

1. $\mathcal{A} \equiv_{OBS} \mathcal{B}$.
2. $\mathcal{A}$ and $\mathcal{B}$ are OBS-countable and $\mathcal{A} \sim_{OBS} \mathcal{B}$.
3. $\mathcal{A}$ and $\mathcal{B}$ are OBS-countable and there is an observational pre-logical relation $R \subseteq \mathcal{A} \times \mathcal{B}$ w.r.t. OBS.

Proof. (Sketch) (1 $\Longrightarrow$ 2) The assumption implies that there exists an OBS-surjective environment (Γ, ρ, ρ'). From this, $\mathcal{A}$ and $\mathcal{B}$ are OBS-countable. We then define for each $\tau \in OBS$, $h^\tau : A^\tau \to B^\tau$ by $h^\tau(a) = \rho'(x_a)$, where $x_a \in \text{dom}(\Gamma)$ is a variable such that $\rho(x_a) = a$. We can show h^τ is well-defined and gives an isomorphism. Next we show $\mathcal{B} \models \Delta \vdash M \sim N : \tau$ implies $\mathcal{A} \models \Delta \vdash M \sim N : \tau$; the reverse direction is by symmetry. Let $\eta' \in B^\Delta$. From the definition of OBS-surjective environment, for all $x \in \text{dom}(\Delta)$, there exists $y_x \in \text{dom}(\Gamma)$ such that $\eta'(x) = \rho'(y_x)$. Then we define an environment $\eta \in A^\Delta$ by $\eta(x) = \rho(y_x)$ and a variable renaming σ by $\sigma(x) = y_x$. Now we have $\mathcal{A}[\![M\sigma]\!]\rho = \mathcal{A}[\![M]\!]\eta = \mathcal{A}[\![N]\!]\eta = \mathcal{A}[\![N\sigma]\!]\rho$. This implies $\mathcal{B}[\![M]\!]\eta' = \mathcal{B}[\![M\sigma]\!]\rho' = \mathcal{B}[\![N\sigma]\!]\rho' = \mathcal{B}[\![N]\!]\eta'$ from $\mathcal{A} \equiv_{OBS} \mathcal{B}$.

(2 $\Longrightarrow$ 3) From $\mathcal{A} \sim_{OBS} \mathcal{B}$, we can choose bijections R_0^τ for each $\tau \in OBS$ satisfying $(\mathcal{A}[\![M]\!], \mathcal{B}[\![M]\!]) \in R_0^\tau$ for all $\emptyset \vdash M : \tau$. Then it is easy to see that the following relation $R \subseteq \mathcal{A} \times \mathcal{B}$ is an pre-logical relation:

$$R^\tau = \{(\mathcal{A}[\![M]\!]\rho, \mathcal{B}[\![M]\!]\rho') \mid \Gamma \text{ is an } OBS\text{-context} \wedge \Gamma \vdash M : \tau \wedge (\rho, \rho') \in R_0^\Gamma\}$$

where R^τ is clearly a bijection for each $\tau \in OBS$.

(3 $\Longrightarrow$ 1) Since $\mathcal{A}$ and $\mathcal{B}$ are OBS-countable, for each $\tau \in OBS$ and each pair $(e, f) \in R^\tau$, it is possible to assign a distinct variable $x_{e,f}^\tau$. Then we define an OBS-surjective environment (Γ, ρ, ρ') by $\Gamma(x_{e,f}^\tau) = \tau$, $\rho(x_{e,f}^\tau) = e$ and $\rho'(x_{e,f}^\tau) = f$. The goal $\mathcal{A} \equiv_{OBS} \mathcal{B}$ follows from lemma 1. $\qquad\square$

Example 2. We construct a logical relation $R_{set} \subseteq \mathcal{A}_{set} \times \mathcal{B}_{set}$ from the following relations at base types:

$$R^{bool} = \text{Id}^{bool}, \quad R^{nat} = \text{Id}^{nat},$$
$$R^{set} = \{(X, \phi) \in \mathcal{P}(\mathbf{N}) \times (\mathbf{N} \to B^{bool}) \mid \forall x . x \in X \iff \phi(x) = tt\}$$

We can easily show that R relates the interpretation of all constants, and by definition, it is bijective on $\{bool, nat\}$. Therefore we have $\mathcal{A}_{set} \equiv_{\{bool,nat\}} \mathcal{B}_{set}$.

Example 3. In [9], the notion of constructive data refinement is formalised in terms of the existence of a pre-logical relation. They demonstrate that an implementation δ of

---

[3] The proof of theorem 3 does not rely on particular properties of combinatory algebras. Thus we can expect that it holds over lambda applicative structures.

[4] In fact 2 $\iff$ 3 still holds when dropping the condition that $\mathcal{A}$ and $\mathcal{B}$ are OBS-countable.

real number computation in the programming language PCF forms a data refinement in their sense; for any model $\mathcal{B}$ of PCF, there exists a model $\mathcal{A}$ of real number computation such that $\mathcal{A}$ and $\mathcal{B}|_\delta$ (the δ-reduct of $\mathcal{B}$) are closed observationally equivalent w.r.t. *bool*. To show this, they give an actual construction of $\mathcal{A}$ and $R \subseteq \mathcal{A} \times \mathcal{B}|_\delta$ from any PCF model $\mathcal{B}$, where R is a pre-logical relation but not a logical relation. For details, see [9]. We believe that we can replace closed observational equivalence with behavioural equivalence and we can still construct a model $\mathcal{A}$ such that $\mathcal{A} \equiv_{\{bool\}} \mathcal{B}|_\delta$.

4 Indistinguishability Relations

We introduce an equivalence of values called indistinguishability based on their behaviour rather than their denotation. We regard two values in a model as "behaviourally" indistinguishable if they are interchangeable in any program. This is shown by performing a set of experiments; we fit one value into a program yielding a visible result, and see whether any difference is detected when we exchange the one with the other. If two values pass the above experiment over all possible programs, then we say that they are indistinguishable. This identification of values is more suitable to provide an abstract aspect of specifications.

There are several ways to formalise the above idea. In this paper we adopt the same definition of indistinguishability as [7] for combinatory algebras.

Definition 11 (Reachable). *Let $\tau \in \mathbf{Typ}(\Sigma)$. A value $v \in A^\tau$ is OBS-reachable if there exists an OBS-context Γ, a well-formed term $\Gamma \vdash M : \tau$ and $\rho \in A^\Gamma$ such that $v = \mathcal{A}[\![M]\!]\rho$.*

Definition 12 (Indistinguishability Relation). *Let $\tau \in \mathbf{Typ}(\Sigma)$. We say two values $v, w \in A^\tau$ are indistinguishable (written $v \approx_\mathcal{A}^\tau w$) if they are OBS-reachable and for any OBS-context Γ, $\tau' \in OBS$, $\rho \in A^\Gamma$ and well-formed term $\Gamma, x : \tau \vdash M : \tau'$, we have $\mathcal{A}[\![M]\!]\rho\{x \mapsto v\} = \mathcal{A}[\![M]\!]\rho\{x \mapsto w\}$.*

The indistinguishability relation is defined on each combinatory algebra. Thus $\approx$ gives rise to a family of PERs indexed by $\mathbf{CA}(\Sigma)$. The results in this section are proved for only one combinatory algebra, but readers may regard them as statements for the family of indistinguishability PERs.

In $\mathcal{A}_{set}$, $x \approx_{\mathcal{A}_{set}}^{set} y$ implies $x = y$. We note that $\approx_{\mathcal{A}_{set}}^{set}$ is a partial equivalence relation but not a total one since infinite sets of natural numbers are not OBS-reachable.

Theorem 4. *The indistinguishability relation $\approx_\mathcal{A}$ is a pre-logical PER such that $\approx_\mathcal{A}^\tau = \mathrm{Id}_{A^\tau}$ for all $\tau \in OBS$.*

Proof. (Sketch) It is easy to see that $\approx_\mathcal{A}$ is a PER which relates all constants and is Id^τ for all $\tau \in OBS$. We show it is closed under application. We assume $e \approx_\mathcal{A}^{\tau \to \tau'} f$ and $x \approx_\mathcal{A}^\tau y$. Let Γ be an OBS-context, $\tau'' \in OBS$, $\Gamma, z : \tau' \vdash M : \tau''$ be a well-formed term and $\rho \in A^\Gamma$. Since e is OBS-reachable, we can write $e = \mathcal{A}[\![E]\!]\eta$ with a well-formed term E and an environment such that $\mathrm{dom}(\rho) \cap \mathrm{dom}(\eta) = \emptyset$. Then from proposition 1 and $x \approx_\mathcal{A}^\tau y$, we have $\mathcal{A}[\![M]\!]\rho\{z \mapsto e \bullet x\} = \mathcal{A}[\![M[Ew/z]]\!](\rho \cup \eta)\{w \mapsto x\} = \mathcal{A}[\![M]\!]\rho\{z \mapsto e \bullet y\}$. We can similarly swap e and f. Thus we obtain $e \bullet x \approx_\mathcal{A}^{\tau'} f \bullet y$. $\square$

By analogy with the terminology of denotational semantics, a Σ-combinatory algebra is *fully abstract* when the indistinguishability relation on $\mathcal{A}$ and the set-theoretic equality coincide (the proof is omitted).

Theorem 5. *The quotient model $\mathcal{A}/\approx_A$ is $\approx$-fully abstract, i.e. for all $\tau \in \mathbf{Typ}(\Sigma)$ and $a, b \in (A/\approx_A)^\tau$, $a = b$ iff $a \approx_{A/\approx_A}^\tau b$.*

5 Factorisability

We have seen two approaches to obtain abstract models of specifications; behavioural equivalence on the one hand and indistinguishability relation on the other hand. Both of them naturally arise from the motivation of reasoning about specifications from a behavioural point of view. Thus we are interested in considering their relationship. The key idea is the notion of factorisability[4].

Definition 13 (Factorisability). *Let E be a $\mathbf{CA}(\Sigma)$-indexed family of PERs and $\equiv$ be an equivalence relation over $\mathbf{CA}(\Sigma)$. Then*

  − *$\equiv$ is* left-factorisable *by E if for all $\mathcal{A}, \mathcal{B} \in \mathbf{CA}(\Sigma)$, $\mathcal{A}/E_A \cong \mathcal{B}/E_B \Longrightarrow \mathcal{A} \equiv \mathcal{B}$.*
  − *$\equiv$ is* right-factorisable *by E if for all $\mathcal{A}, \mathcal{B} \in \mathbf{CA}(\Sigma)$, $\mathcal{A} \equiv \mathcal{B} \Longrightarrow \mathcal{A}/E_A \cong \mathcal{B}/E_B$.*

We say $\equiv$ is factorisable *by E if both of the above hold.*

In this section we show that behavioural equivalence is factorisable by the indistinguishability relation. First we prove left-factorisabilty.

Theorem 6 (Left-Factorisability). $\mathcal{A}/\approx_A \cong \mathcal{B}/\approx_B \Longrightarrow \mathcal{A} \equiv_{OBS} \mathcal{B}$.

Proof. $\mathcal{A}/\approx_A \cong \mathcal{B}/\approx_B$ implies $\mathcal{A}/\approx_A \equiv_{OBS} \mathcal{B}/\approx_B$. From theorem 4, we have $\mathcal{A}/\approx_A \equiv_{OBS} \mathcal{A}$ and $\mathcal{B}/\approx_B \equiv_{OBS} \mathcal{B}$. Thus $\mathcal{A} \equiv_{OBS} \mathcal{B}$ by transitivity. □

In [7], Hofmann and Sannella represented the indistinguishability relation and the "experiments" for behavioural equivalence in a higher-order logic, then showed that the satisfiability of the experiments coincide in each model when quotients of two models are isomorphic. However this approach seems not to work in this paper, since their method depends on the finiteness of specifications to represent the indistinguishability relation, while combinatory algebras have a countably infinite number of types and S, K-combinators.

The proof of right-factorisability is essentially the same as the one in [7].

Theorem 7 (Right-Factorisability). $\mathcal{A} \equiv_{OBS} \mathcal{B} \Longrightarrow \mathcal{A}/\approx_A \cong \mathcal{B}/\approx_B$.

Proof. (Sketch) From $\mathcal{A} \equiv_{OBS} \mathcal{B}$, there is an observational pre-logical relation $R \subseteq \mathcal{A} \times \mathcal{B}$ w.r.t. OBS. Now we define a relation $h \subseteq \mathcal{A}/\approx_A \times \mathcal{B}/\approx_B$:

$$h^\tau = \{(\mathcal{A}/\approx_A[\![M]\!]\rho, \mathcal{B}/\approx_B[\![M]\!]\rho') \mid \Gamma \text{ is an } OBS\text{-context} \wedge \Gamma \vdash M : \tau \wedge (\rho, \rho') \in R^\Gamma\}$$

We can show that h gives a partial injection in both directions. Moreover, all elements in $\mathcal{A}/\approx_A$ and $\mathcal{B}/\approx_B$ are OBS-reachable. Therefore h^τ is total and surjective, i.e. is bijective for each $\tau \in \mathbf{Typ}(\Sigma)$. It is easy to see that h is a Σ-isomorphism. □

6 Higher-Order Logic to Reason about Higher-Typed Languages

We consider a higher-order logic to reason about specifications in the higher-order typed languages. We introduce two models of the higher-order logic; one is the standard model, which equates two programs when they have the same denotations, and the other is the behavioural model, which equates two programs when they have the same behaviour. The latter model is useful when we reason about specifications based on behaviour of programs. We then relate standard satisfaction and behavioural satisfaction.

6.1 Syntax

The higher-order logic considered in this section is designed to reason about combinatory algebras over a signature Σ. Thus the logic has constants for the application operator $\bullet$ and S, K-combinators corresponding to those in combinatory algebras.

The syntax of the higher-order logic can be formalised in the framework of the simply typed lambda calculus—it is just a lambda calculus over a certain signature (which is an extension of Σ) providing a type of propositions and constants for logical connectives.

Although we can reuse definitions, syntax and terminology of the simply typed lambda calculus, we re-define them for higher-order logic to make it clear which calculus we are talking about. We use a different function type symbol $\Rightarrow$ instead of $\rightarrow$, and write $\mathbf{Typ}^{\Rightarrow}(U)$ for the set defined by BNF $\phi ::= b \mid \phi \Rightarrow \phi$ where $b \in U$.

Definition 14 (Higher-Order Logic). *The syntax of higher-order logic over Σ is given by the syntax of the simply typed lambda calculus over the signature $\Sigma_{HOL} = (U_{HOL}, C_{HOL})$ defined by:*

$$U_{HOL} = \{\Omega\} \cup \mathbf{Typ}(\Sigma)$$
$$C_{HOL} = \{\supset^{\Omega \Rightarrow \Omega \Rightarrow \Omega}\} \cup \{=^{\phi \Rightarrow \phi \Rightarrow \Omega} \mid \phi \in \mathbf{Typ}^{\Rightarrow}(U_{HOL})\}$$
$$\cup \{\bullet^{\tau \rightarrow \tau' \Rightarrow \tau \Rightarrow \tau'} \mid \tau, \tau' \in \mathbf{Typ}(\Sigma)\} \cup C_{SK}$$

We may omit types of constants in superscripts if they are obvious from the context. The constants $\supset, =$ and $\bullet$ are used as infix operators.

We call types of Σ_{HOL} *formula types* (ranged over by ϕ) and terms of Σ_{HOL} *formulas* (ranged over by F). In this logic, a lambda term M is represented by a formula $M^{\bullet}_{CL}$, which is a combinatorial representation of M by constants in C_{SK} and $\bullet^{\tau \rightarrow \tau' \Rightarrow \tau \Rightarrow \tau'}$.

Lambda abstraction plus logical constants $\supset$ and $=$ are powerful enough to derive other familiar logical constants such as $\mathrm{tt}, \mathrm{ff}, \neg, \wedge, \vee, \neq$ and quantifiers $\forall x : \phi . F$ and $\exists x : \phi . F$. See [1] for the definition and the axioms for the logical constants.

Example 4. The higher-order logic considered in this section is dedicated to reasoning about the combinatory algebras providing the semantics of the lambda calculus. Thus the higher-order logic has the axiom schema for S and K combinators (see definition 2). We may need to add extra axioms depending on the properties of the underlying combinatory algebra. If one assumes that it is a Henkin model, one adds the axiom scheme of extensionality: $\forall x, y : \tau \rightarrow \tau' . (\forall z : \tau . x \bullet z = y \bullet z) \supset x = y$.

In the higher-order logic over Σ_{set}, we would like to assume the induction principle for type nat:

$$\forall p : nat \Rightarrow \Omega \,.\, p\,0 \supset (\forall x : nat \,.\, p\,x \supset p(\text{succ} \bullet x)) \supset \forall x : nat \,.\, p\,x.$$

We can also specify the behaviour of constants, like

isempty $\bullet\, \emptyset = tt$

$\forall p, q : nat \rightarrow bool, s : set \,.\, \text{filter} \bullet p \bullet (\text{filter} \bullet q \bullet s) = \text{filter} \bullet q \bullet (\text{filter} \bullet p \bullet s)$

Proof systems for the higher-order logic and its soundness and completeness are not covered in this paper. For details see [6].

6.2   Standard Satisfaction and Behavioural Satisfaction

We apply the semantic framework of the simply typed lambda calculus to give semantics to the higher-order logic. A model of the higher-order logic, say $\mathcal{M}$, is built on top of a combinatory algebra $\mathcal{A}$. Constant symbols for combinatory algebras such as S, K and $\bullet$ are interpreted by the corresponding elements $S_{\mathcal{A}}, K_{\mathcal{A}}$ and $\bullet_{\mathcal{A}}$. Do not confuse the underlying combinatory algebra $\mathcal{A}$ and the model $\mathcal{M}$ of the higher-order logic.

We introduce two models. One is the *standard model* which interprets Ω as the two-point set $\mathbf{2} = \{tt, ff\}$, the function type $\Rightarrow$ as the set-theoretic function space, $\supset$ as the (curried form of) boolean implication and $=$ as the (curried form of) characteristic function of set-theoretic equality.

Definition 15 (Standard Model). *The* standard model $\mathcal{L}_{\mathcal{A}}$ *of the higher-order logic over $\mathcal{A}$ is a Σ_{HOL}-full type hierarchy over $L_{\mathcal{A}}^{\Omega} = \mathbf{2}$ and $L_{\mathcal{A}}^{\tau} = A^{\tau}$ together with the following interpretation of the logical constants:*

$$\supset_{\mathcal{L}_{\mathcal{A}}}^{\Omega \Rightarrow \Omega \Rightarrow \Omega} (x)(y) = tt \iff (x = tt \implies y = tt)$$
$$=_{\mathcal{L}_{\mathcal{A}}}^{\phi \Rightarrow \phi \Rightarrow \Omega} (x)(y) = tt \iff x = y$$
$$\bullet_{\mathcal{L}_{\mathcal{A}}}^{\tau \rightarrow \tau' \Rightarrow \tau \Rightarrow \tau'} (x)(y) = x \bullet_{\mathcal{A}} y$$
$$c_{\mathcal{L}_{\mathcal{A}}}^{\tau} = c_{\mathcal{A}}^{\tau} \quad (c^{\tau} \in C_{SK}).$$

We say that a closed formula $F : \Omega$ is satisfied (*written $\mathcal{A} \models F$*) *if $\mathcal{L}_{\mathcal{A}}[\![F]\!] = tt$.*

The other model is the *behavioural model* w.r.t. a pre-logical PER E over $\mathcal{A}$. The standard model is not appropriate when we want to reason about specifications up to their behaviour rather than their denotation. This is because the equality may distinguish two different denotations even though they have the same behaviour. The behavioural model solves this problem by interpreting each predicate type ϕ as $|\overline{E}^{\phi}|$ where $\overline{E}$ is the extension of E to all predicate types using the exponential relation, and the equality as the equivalence relation $\overline{E}$ over $|\overline{E}|$. In particular, E is often taken as the indistinguishability relation over $\mathcal{A}$ (see theorem 4).

Definition 16 (Behavioural Model). *Given a pre-logical PER E over $\mathcal{A}$, we define a PER $\overline{E}$ over carrier sets $L_{\mathcal{A}}$ as follows:*

$$\overline{E}^{\Omega} = \text{Id}_2, \ \overline{E}^{\tau} = E^{\tau}, \ \overline{E}^{\phi \Rightarrow \phi'} = \overline{E}^{\phi} \rightarrow \overline{E}^{\phi'}$$

Then the behavioural model $\mathcal{L}_{\mathcal{A}}^E$ of the higher-order logic over $\mathcal{A}$ with respect to E is given by a Σ_{HOL}-type frame $(L_{\mathcal{A}}^E, (-)_{\mathcal{L}_{\mathcal{A}}^E})$ where $(L_{\mathcal{A}}^E)^\phi = |\overline{E}^\phi|$ and $(-)_{\mathcal{L}_{\mathcal{A}}^E}$ gives the interpretation of the logical constants as follows:

$$\supset_{\mathcal{L}_{\mathcal{A}}^E}^{\Omega \Rightarrow \Omega \Rightarrow \Omega} (x)(y) = tt \iff (x = tt \implies y = tt)$$

$$=_{\mathcal{L}_{\mathcal{A}}^E}^{\phi \Rightarrow \phi \Rightarrow \Omega} (x)(y) = tt \iff (x, y) \in \overline{E}^\phi$$

$$\bullet_{\mathcal{L}_{\mathcal{A}}^E}^{\tau \to \tau' \Rightarrow \tau \Rightarrow \tau'} (x)(y) = x \bullet_{\mathcal{A}} y$$

$$c_{\mathcal{L}_{\mathcal{A}}^E}^\tau = c_{\mathcal{A}}^\tau \quad (c^\tau \in C_{SK}).$$

We say that a closed formula $F : \Omega$ is satisfied w.r.t. E (written $\mathcal{A} \models^E F$) if $\mathcal{L}_{\mathcal{A}}^E[\![F]\!] = tt$.

We show that the standard model over $\mathcal{A}/E$ and the behavioural model over $\mathcal{A}$ w.r.t. E satisfy the same formula (c.f. theorem 3.35 in [7]). We notice that this is implied by $\mathcal{L}_{\mathcal{A}/E} \equiv_{\{\Omega\}} \mathcal{L}_{\mathcal{A}}^E$, since from $\mathcal{L}_{\mathcal{A}/E} \equiv_{\{\Omega\}} \mathcal{L}_{\mathcal{A}}^E$, for all closed formula F, we have $\mathcal{L}_{\mathcal{A}/E}[\![F]\!] = \mathcal{L}_{\mathcal{A}/E}[\![tt]\!] = tt \iff \mathcal{L}_{\mathcal{A}}^E[\![F]\!] = \mathcal{L}_{\mathcal{A}}^E[\![tt]\!] = tt$.

Theorem 8. $\mathcal{L}_{\mathcal{A}/E} \equiv_{\{\Omega\}} \mathcal{L}_{\mathcal{A}}^E$.

Proof. (Sketch) We construct an observational pre-logical relation $R \subseteq \mathcal{L}_{\mathcal{A}/E} \times \mathcal{L}_{\mathcal{A}}^E$ w.r.t. $\{\Omega\}$ by theorem 3. First we show that there is a family of isomorphisms $h^\phi : L_{\mathcal{A}/E}^\phi \cong L_{\mathcal{A}}^\phi/\overline{E}^\phi$ for the carrier sets of $\mathcal{L}_{\mathcal{A}/E}$. Let $I^\phi \subseteq L_{\mathcal{A}}^\phi \times |\overline{E}^\phi|$ be the inclusion relation. Then the pre-logical relation in question is given by the composition relation $h^\phi \circ \Pi(\overline{E})^\phi \circ I^\phi \subseteq L_{\mathcal{A}/E}^\phi \times |\overline{E}^\phi|$.   □

Corollary 1. For all closed formula $F : \Omega$, $\mathcal{A}/E \models F$ iff $\mathcal{A} \models^E F$.

7   Reasoning about Specifications

We revisit the model theory of behavioural and abstractor specification studied in [4]. Behavioural equivalence $\equiv_{OBS}$ is factorisable by the indistinguishability relation $\approx$ (theorem 6), and for any $\mathcal{A} \in \mathbf{CA}(\Sigma)$, $\mathcal{A}/\approx$ is fully abstract (theorem 5). The latter implies that $\approx_{\mathcal{A}}$ is a regular relation[5]. One important consequence from this setting is the following relationship between behavioural and abstractor specification. Due to space limitations, we only state the theorem without giving the definitions of symbols. For details, see [4].

Theorem 9 (Bidoit et al. [4]). Let $SP = (\Sigma, \Phi)$ be a specification, where Φ is a set of formulas in the higher-order logic over Σ. Then we have:

$$\mathbf{Mod}(\text{behaviour } SP \text{ w.r.t. } \approx) = \mathbf{Abs}^{\equiv_{OBS}}(\mathbf{FA}^\approx(\mathbf{Mod}(SP)))$$

$$\mathbf{Mod}(\text{abstract } SP \text{ w.r.t. } \equiv_{OBS}) = \mathbf{Mod}(\text{behaviour } SP/\approx \text{ w.r.t. } \approx)$$

$$\mathbf{Th}^\approx(\mathbf{Mod}(\text{behaviour } SP \text{ w.r.t. } \approx)) = \mathbf{Th}(\mathbf{FA}^\approx(\mathbf{Mod}(SP)))$$

$$\mathbf{Th}^\approx(\mathbf{Mod}(\text{abstract } SP \text{ w.r.t. } \equiv_{OBS})) = \mathbf{Th}(\mathbf{Mod}(SP/\approx))$$

Proof. See theorem 5.16, 6.8 and 7.4 in [4].   □

---

[5] A $\mathbf{CA}(\Sigma)$-indexed family of PERs E is regular if $\mathcal{A}/E_{\mathcal{A}}$ is fully abstract (see [4]).

Related Work

The work by Bidoit, Hennicker and Wirsing[4] established the key idea of factorisability to relate behavioural equivalence and the indistinguishability relation, and they used this to reason about the semantics of behavioural and abstractor specifications. Subsequent work by Bidoit and Hennicker[3] discussed a proof method for showing behavioural equivalence in first order logic, and considered finitary axiomatisation of behavioural equality. The above work is extended by Hofmann and Sannella[8] to higher-order logic. This work is an extension of their work from first-order Σ-algebras to combinatory algebras. Bidoit and Tarlecki[5] gave a categorical generalisation of [4]. See the end of section 5 for comments on the relationship to the present paper.

Our characterisation of behavioural equivalence is related to Mitchell's representation independence theorem[10]. Honsell and Sannella[8] removed the restriction on the constants by using pre-logical relations instead of logical relations. This paper shows a similar result about behavioural equivalence, which subsumes closed observational equivalence.

In [5], Bidoit and Tarlecki give a relationship between behavioural satisfaction, behavioural equivalence, indistinguishability and correspondences in an abstract setting using concrete categories (a faithful functor to the category of (type-indexed) sets). By instantiating their concrete categories with various real examples, such as the category of multi-sorted algebras and regular algebras, we can derive suitable notions of behavioural equivalence, indistinguishability, etc. and theorems on them.

We can instantiate their abstract framework with the category of Σ-combinatory algebras $\mathbf{CA}(\Sigma)$, and obtain various results on behavioural equivalence and indistinguishability. Pre-logical relations correspond to spans (moreover correspondences), and pre-logical PERs correspond to partial congruences in their terminology. Category $\mathbf{CA}(\Sigma)$ satisfies certain properties[6], thus we can obtain a theorem characterising behavioural equivalence via correspondences (see theorem 28 of [5]).

Their definitions of indistinguishability and behavioural equivalence are abstract: they define the indistinguishability relation $\approx_{\mathcal{A}}$ as the largest congruence over the full subobject $\langle|\mathcal{A}|_{OBS}\rangle_{\mathcal{A}}$ of $\mathcal{A}$. Then behavioural equivalence is defined by $\mathcal{A} \equiv_{OBS} \mathcal{B}$ iff $\mathcal{A}/\approx_{\mathcal{A}} \cong \mathcal{B}/\approx_{\mathcal{B}}$. In contrast, in this paper we give an explicit definition of behavioural equivalence and indistinguishability, and show the relationship between them in an elementary way.

8 Conclusion

We have extended the study of the relationship between *behavioural equivalence* and *indistinguishability* [4, 7] to the simply typed lambda calculus, where higher-order types are available. We characterised behavioural equivalence between two combinatory algebras by pre-logical relations, and showed that behavioural equivalence is factorised by indistinguishability. We also showed that standard satisfaction over $\mathcal{A}/E$ is equivalent to behavioural satisfaction w.r.t. a PER E over $\mathcal{A}$.

---

[6] Category $\mathbf{CA}(\Sigma)$ admits renaming and has full subobjects and surjective quotients. All full subobjects are compatible with $\mathbf{CA}(\Sigma)$-morphisms. Pullbacks preserve surjective quotients, and quotients are fully compatible with subobjects in $\mathbf{CA}(\Sigma)$.

It is interesting to restrict the class of models from combinatory algebras to Henkin models, where the extensionality axiom holds. This changes the properties of the class of models; in particular it is not closed under quotient by pre-logical PERs. It will be interesting to see how behavioural equivalence and the indistinguishability relation are characterised in the class of Henkin models.

Acknowledgements

I am grateful to Donald Sannella for his continuous encouragement to this work, and to John Longley for helpful discussions. This work has been partly supported by an LFCS studentship.

References

1. P. Andrews. *An introduction to mathematical logic and type theory: to truth through proof.* Academic Press, 1986.
2. H. Barendregt. *The Lambda Calculus-Its Syntax and Semantics.* North Holland, 1984.
3. M. Bidoit and R. Hennicker. Behavioural theories and the proof of behavioural properties. *Theoretical Computer Science*, 165(1):3–55, 1996.
4. M. Bidoit, R. Hennicker, and M. Wirsing. Behavioural and abstractor specifications. *Science of Computer Programming*, 25(2–3):149–186, 1995.
5. M. Bidoit and A. Tarlecki. Behavioural satisfaction and equivalence in concrete model categories. In *Proc. 21st Int. Coll. on Trees in Algebra and Programming (CAAP '96)*, volume 1059 of *LNCS*, pages 241–256. Springer, 1996.
6. L. Henkin. Completeness in the theory of types. *Journal of Symbolic Logic*, 15:81–91, 1950.
7. M. Hofmann and D. Sannella. On behavioural satisfaction and behavioural abstraction in higher-order logic. *Theoretical Computer Science*, 167(1–2):3–45, 1996.
8. F. Honsell and D. Sannella. Pre-logical relations. In *Proc. Computer Science Logic*, volume 1683 of *LNCS*, pages 546–561. Springer, 1999. An extended version is in *Information and Computation* 178:23–43,2002.
9. Furio Honsell, John Longley, Donald Sannella, and Andrzej Tarlecki. Constructive data refinement in typed lambda calculus. In *Proc. FoSSACS*, volume 1784 of *LNCS*, pages 149–164. Springer, 2000.
10. J. Mitchell. On the equivalence of data representations. In V. Lifschitz, editor, *Artificial Intelligence and Mathematical Theory of Computation: Papers in Honor of John McCarthy*, pages 305–330. Academic Press, San Diego, 1991.
11. J. Mitchell. *Foundations for Programming Languages.* MIT Press, 1996.
12. O. Schoett. Behavioural correctness of data representations. *Science of Computer Programming*, 14:43–57, 1990.

Approach-Independent Structuring Concepts for Rule-Based Systems

Hans-Jörg Kreowski and Sabine Kuske

Universität Bremen, Fachbereich 3
Postfach 33 04 40
D-28334 Bremen, Germany
{kreo,kuske}@informatik.uni-bremen.de

Abstract. In this paper, we propose new structuring concepts for rule-based systems that are independent of the type of rules and of the type of configurations to which rules are applied. Hence the concepts are applicable in various rule-based approaches allowing one to build up large systems from small components in a systematic way.

1 Introduction

In many areas of computer science (like term rewriting, theorem proving, logic programming, functional programming, knowledge representation, algebraic specification, graph transformation, theory of formal languages, etc.), data-processing systems of various kinds are frequently modeled by means of rules. The basic features are always alike. There are configurations (like strings, terms, trees, graphs, sets, etc.) to represent the states of systems, and there are rules which can be applied to configurations yielding configurations. In this way, each rule provides a binary relation on configurations, and an operational system semantics is obtained by sequential composition of these relations – usually obeying some kind of control condition in addition. But a common understanding of the structuring of rule-based systems seems to be missing. For example, the area of algebraic specification is rich in structuring concepts (see e.g., [EM85, EM90, AKK99]), but their semantics is usually based on the notion of an algebra which is not available in other rule-based approaches. In functional and logic programming, rules can be grouped together under the heading of a function or a predicate, which is not possible in other approaches.

In this paper, we propose the notions of transformation units and transformation modules as structuring principles that can be used within many rule-based approaches. A transformation unit encapsulates a set of rules accompanied by specifications of initial and terminal configurations as well as a control condition. Moreover, it may import other transformation units for structuring purposes. The semantics of a transformation unit is given as a binary relation between initial and terminal configurations by interleaving rule applications and calls of imported units according to the control condition.

A collection of transformation units which are closed under import form a network of transformation units. The semantics of such a network is given by the

M. Wirsing, D. Pattinson, and R. Hennicker (Eds.): WADT 2002, LNCS 2755, pp. 299–311, 2003.

iteration of the interleaving semantics. This allows to broadcast the interleaving effects at the nodes along the network struture to other nodes. Under suitable circumstances, the iterated interleaving semantics is a fixed point. A network becomes a transformation module if it is provided with an import interface and an export interface.

The introduced concepts of transformation units and transformation modules generalize our respective notions for graph transformation approaches (see [KK96, KKS97, KK99a, KK99b, DKKK00, HHKK00]).

The running example, illustrating the new structuring concepts, is taken from the area of algebraic specification and term rewriting and specifies a balancedness test for binary trees as a transformation module.

2   Transformation Units

In this section, we introduce the notion of a transformation unit as a means to decompose a system with a large set of rules into a family of subcomponents which use each other. The notion is independent of a particular rule-based framework. This is achieved by assuming an underlying rule base which comprises a domain of configurations, a class of rules, and a rule application operator describing how a rule is applied to a configuration. Moreover, there are configuration class expressions to describe initial and terminal configurations, and control conditions to specify rule application strategies. A rule base provides the syntactic and semantic components of which transformation units and their behaviour are put together. A transformation unit encapsulates a set of rules together with a control condition and descriptions of initial and terminal configurations. For structuring purpose, it has a set of identifiers which refer to imported entities. Semantically, a transformation unit specifies a binary relation between its initial and terminal configurations. The relation is obtained operationally by interleaving rule applications and calls of imported items according to the control condition. It is well-defined whenever some choice of binary relations on configurations is fixed for the import identifiers. In this sense, the interleaving semantics of transformation units is generic, and many choices of the imported parameters are possible. The typical case that the identifiers refer to other transformation units is discussed in Section 3.

2.1   Syntax and Semantics

1. A *rule base* $\mathcal{B}$ consists of a class $\mathcal{K}$ of *configurations*, a class $\mathcal{R}$ of *rules*, a *rule application operator* $\Rightarrow$ yielding a binary relation $\Rightarrow_r \subseteq \mathcal{K} \times \mathcal{K}$ for each $r \in \mathcal{R}$, a class $\mathcal{E}$ of *configuration class expressions* where each $e \in \mathcal{E}$ specifies a subclass $SEM_E(e) \subseteq \mathcal{K}$, and a class $\mathcal{C}$ of *control conditions* where each $c \in \mathcal{C}$ specifies a binary relation $SEM_E(c) \subseteq \mathcal{K} \times \mathcal{K}$. In the latter two cases, the semantics depends on the *environment* E, i.e. a choice of a binary relation $E(id) \subseteq \mathcal{K} \times \mathcal{K}$ for each $id \in ID$ where ID is a given set of *identifiers*.

2. A *transformation unit* over a rule base $\mathcal{B}$ is a system $tu = (I, U, R, C, T)$ where I and T are configuration class expressions specifiying *initial* and *terminal configurations*, U is a finite set of *import identifiers*, R is a set of rules, and C is a control condition. Moreover, each rule $r \in R$ has got an identifier $id(r) \in ID$ with $id(r) \notin U$.

3. Let *SEM* be a mapping associating a binary relation on configurations, $SEM(id) \subseteq \mathcal{K} \times \mathcal{K}$, to each $id \in U$. Then a transformation unit $tu = (I, U, R, C, T)$ defines a binary relation $INTER_{SEM}(tu) \subseteq \mathcal{K} \times \mathcal{K}$ such that $(con, con') \in INTER_{SEM}(tu)$ if and only if

   - $con \in SEM_{E(tu,SEM)}(I)$, $con' \in SEM_{E(tu,SEM)}(T)$,
   - there is a sequence of configurations $con_0, \ldots, con_m$ such that $con = con_0$, $con' = con_m$, and, for $i = 1, \ldots, m$,
     - $(con_{i-1}, con_i) \in SEM(id)$ for some $id \in U$, or
     - $con_{i-1} \Rightarrow_r con_i$ for some $r \in R$,
   - $(con, con') \in SEM_{E(tu,SEM)}(C)$

   where the environment $E(tu, SEM)$ is given by $E(tu, SEM)(id) = SEM(id)$ for $id \in U$, $E(tu, SEM)(id) = \Rightarrow_r$ if $id = id(r)$ for some $r \in R$, and $E(tu, SEM)(id) = \emptyset$ otherwise.

   Sequences of rule applications and calls of imported items (as considered above) are called *interleaving sequences*, and the resulting semantic relation $INTER_{SEM}(tu)$ is called *interleaving semantics* of tu (with respect to *SEM*).

A rule base provides the most elementary syntactic and semantic prerequisites of a rule-based specification language. As in nearly all rule-based settings, a rule defines a binary relation on the domain of configurations to which it can be applied. In addition, we assume features to specify subdomains and subrelations. They may use identifiers to refer to rules and imported units. Therefore, the semantics of configuration class expressions and control conditions depends on the semantics of the named entities. This is reflected by arbitrary environments on the level of the rule base while the interleaving semantics of a transformation unit depends on special environments where only the import identifiers are still variably interpreted. The identifiers of rules are fixed according to the rule application operator, and all other identifiers are interpreted by the empty relation such that no interference can happen.

In this paper, we consider regular expressions over *ID* as only control conditions. Identifiers are atomic conditions allowing to apply a rule or to call an imported entity. And if e and e' are control conditions, $e\,;\,e'$, $e\,|\,e'$, and e^* are also control conditions allowing to require certain sequences, alternatives, and iterations, respectively. Every environment E is recursively extended to all regular expressions by sequential composition, union, and Kleene-star closure of binary relations: $SEM_E(e\,;\,e') = SEM_E(e) \circ SEM_E(e')$, $SEM_E(e\,|\,e') = SEM_E(e) \cup SEM_E(e')$, and $SEM_E(e^*) = (SEM_E(e))^*$ using initially $SEM_E(id) = E(id)$ for $id \in ID$. In this way, regular expressions control the order in which rules are applied and imported units are called. For a finite set of identifiers, there are regular expressions that admit any order. We are using this case as a default control condition, denoted by *OK*.

As a typical kind of configuration class expressions, we use $T \subseteq ID$ specifying the set $RED_E(T)$ of reduced normal forms with respect to the environment E: $con \in RED_E(T)$ if and only if there are no $t \in T$ and $con' \in \mathcal{K}$ with $(con, con') \in E(t)$. The set T is also denoted by $red(T)$ to stress that reduced normal forms are specified. If each $t \in T$ is the identifier of a rule $r(t)$ and $E(t) = \Rightarrow_{r(t)}$, then $RED_E(T)$ contains all reduced normal forms with repect to a given set of rules. Such reduced normal forms are often used as terminal configurations in the area of term rewriting. We are using the symbol all as default expression with $SEM_E(all) = \mathcal{K}$ for all environments E. It may be considered as a special case of reduced normal forms because $RED_E(\emptyset) = \mathcal{K}$.

Given a transformation unit $tu = (I, U, R, C, T)$ with $U = \emptyset$ or with an environment E such that $E(id) = \emptyset$ for all $id \in U$, interleaving sequences are nothing else than ordinary derivation sequences composed of rule applications only. In other words, the notion of transformation units generalizes all kinds of rule-based systems which are given by sets of rules. It allows one to use other semantic relations besides the local rules to perform computations. There are several possibilities to interpret the import part U of a transformation unit tu:

- The elements of U may be identifiers of transformation units that are known in the context, for example, as entries of a library of transformation units stored for reuse. If such imported units specify a unique semantic relation, the interleaving semantics of tu is also uniquely determined.
- The same applies in a heterogeneous setting where the import identifiers refer to some known entities (specifications, programs, etc.) that are interpreted by binary relations whatsoever.
- If an import identifier does not refer to a known item, it may be specified as tu itself. This leads to the notion of a network of transformation units which use each other. This case is considered in more detail in the Section 3.

2.2   Examples

Chomsky grammars. Given an alphabet Σ, the set Σ^* of all strings over Σ can serve as a class of configurations, and the set $\Sigma^* \times \Sigma^*$ of all pairs of strings as a class of rules. Then a rule $r = (u, v) \in \Sigma^* \times \Sigma^*$ can be applied to $w \in \Sigma^*$ yielding $w' \in \Sigma^*$ if there are $x, y \in \Sigma^*$ with $w = xuy$ and $w' = xvy$. In other words, $w \Rightarrow_r w'$ is the usual notion of a direct derivation of semi-Thue systems and Chomsky grammars. Moreover, one may use $S \in \Sigma$ and $T \subseteq \Sigma$ as string class expressions with $SEM(S) = \{S\}$ and $SEM(T) = T^*$ and OK as only control condition which accepts everything, i.e., $SEM_E(OK) = \Sigma^* \times \Sigma^*$ for all environments E.

A transfomation unit over this rule base of the form $G = (S, \emptyset, P, OK, T)$ is just a Chomsky grammar. The interleaving sequences are ordinary derivations, and the elements of the interleaving semantics have the form (S, w) with $S \Rightarrow_P^* w$ and $w \in T^*$. In other words, the projection to the second component yields the language generated by G. If a grammar is very large, it may be helpful to define it in a structured way by breaking up the set of rules into small units.

Term rewriting. Let $\Sigma = (S, OP)$ be a many-sorted *signature* where S is a set of *sorts* and OP an $S^* \times S$-sorted set of *operators*. Let X be an S-sorted set of *variables*. Then the S-sorted set of *terms* over Σ with variables in X, $T_\Sigma(X)$, provides a class of configurations. The set of all pairs $T_\Sigma(X) \times T_\Sigma(X)$ may be used as a class of rules. Such a rule $r = (L, R)$ – also written $L \to R$ – is applied to a term in the usual sense of term rewriting: Let t_0 be a context term with a single occurrence of a variable, and let $ass: X \to T_\Sigma(X)$ be an assignment of terms to variables, then $t_0[L[ass]] \Rightarrow_r t_0[R[ass]]$ where $t_0[t]$ denotes the substitution of the single variable of t_0 by t, and $\bar{t}[ass]$ is obtained from $\bar{t}$ by substituting all variables according to the assignment.

Alternatively, one may use only ground terms as configurations, i.e. T_Σ instead of $T_\Sigma(X)$. In this case, a rule application requires an assignment of the form $ass: X \to T_\Sigma$.

Together with the control condition OK and reduced normal forms, term rewriting provides a rule base. A transformation unit over it of the form $spec = (all, \emptyset, R, OK, red(id(R)))$, where $id(R)$ contains the identifiers of the rules of R, is a term rewrite system in the ordinary sense (see, e.g., [HO80, Kir99]).

Often, the aim of a term rewrite system is to evaluate terms of a particular form like $op(t_1, \ldots, t_n)$ where $op \in OP$ and $t_1, \ldots, t_n$ are constructor terms. To be able to specify such initial terms, we consider pairs t_0 *on* Σ_0 with $t_0 \in T_\Sigma(X)$ and $\Sigma_0 \subseteq \Sigma$ as term class expressions. The semantics of t_0 *on* Σ_0 is given by all substitutions with Σ_0-terms, i.e. $SEM(t_0 \text{ } on \text{ } \Sigma_0) = \{t_0[ass] \mid ass: X \to T_{\Sigma_0}\}$. This is independent of the environment.

As an explicit example of a term rewrite system in form of transformation units, we specify a predicate on binary trees that tests whether the input tree is totally balanced or not. The idea is to compute the height of the tree and its "balance" which is the height up to which the tree is totally balanced and then to check the equality of the two values. To illustrate the role and effect of the import component, height, balance, and the equality test on natural numbers are imported.

is-balanced

  initial: *is B balanced on* **bintree**$_0$
  uses: **height, balance, nat**
  rules: *(test) is B balanced* $\to$ *height*$(B) \equiv$ *balance*(B)
  terminal: *red*(*test*, **height, balance, nat**)

where *test* is the name of the rule and B is a variable of sort **bintree**. As generally introduced above, the underlying rule base is given by the signature

bintree = **alphabet** + **nat** +
  sorts: *bintree*
  opns: *leaf*: $A \to$ *bintree*
            $(-, -, -)$: *bintree A bintree* $\to$ *bintree*
            height: *bintree* $\to$ *nat*
            balance: *bintree* $\to$ *nat*
            is $-$ *balanced*: *bintree* $\to$ *bool*

where **alphabet** provides the sort A and **nat** the usual arithmetic on natural numbers including minimum min, maximum max, and an equality test $\equiv$. The subsignature of **bintree** given by **alphabet**, the sort $bintree$ and the two operations $leaf$ and $(-,-,-)$ is denoted by **bintree$_0$**.

Starting an interleaving sequence with the initial term *is b balanced* where b is some **bintree$_0$**-term, like for example $((leaf(e), a, leaf(t)), p, leaf(s))$, one can apply the rule and gets $height(b) \equiv balance(b)$. This is already the result if the imported relations are empty. If **height** and **balance** refer to relations that allow to replace subterms of the form $height(b)$ and $balance(b)$ by **nat**-terms, then we can call these relations one after the other obtaining $m \equiv balance(b)$ (or $height(b) \equiv n$) and $m \equiv n$. If finally **nat** provides a relation that evaluates $m \equiv n$ and yields some truth value, then this can be called. To end up with a reduced normal form, the four steps are also necessary. Whether the result is correct, depends on the correctness of the imported relations, which may be specified analogously.

height
    initial: $height(B)$ *on* **bintree$_0$**
    uses: **nat**
    rules: (h_1) $height(a) \rightarrow 0$
              (h_2) $height((L, a, R)) \rightarrow 1 + max(height(L), height(R))$
    terminal: $red(h_1, h_2, \mathbf{nat})$

balance
    initial: $balance(B)$ *on* **bintree$_0$**
    uses: **nat**
    rules: (b_1) $balance(a) \rightarrow 0$
              (b_2) $balance((L, a, R)) \rightarrow 1 + min(balance(L), balance(R))$
    terminal: $red(b_1, b_2, \mathbf{nat})$

With respect to the input identifier **nat**, we assume that the corresponding relation is available in the context, as a predefined transformation unit, for example.

Graph transformation. The area of graph transformation provides a great variety of rule-based approaches (see e.g., [Roz97, EEKR99, EKMR99]) because there are many types of graphs and many ways to apply rules to them. As we have introduced originally the concepts of transformation units for arbitrary graph transformation approaches, which are special cases of rule bases, our considerations in [KK96, KKS97, KK99a] provide many examples in the more general setting of this paper, too, including, for instance, a structured specification of the Floyd-Warshall algorithm for shortest paths and a graph colouring algorithm in the style of constraint programming.

3   Transformation Modules

The basic idea of transformation units is to specify a semantic relation by the interleaving of local rules and imported relations. If these relations are known in the context, the interleaving semantics describes the situation properly. But what about the case that the import identifiers refer to other transformation units which are not predefined, but are defined at the same time or level as the actual transformation unit? What happens in particular if there is a cycle in the import structure of a collection of transformation units? To handle such cases, we introduce networks of transformation units and aggregate and iterate the interleaving semantics over the networks. This allows to start even with empty relations and to broadcast the computational effects at the nodes of the net along the net structure.

While a transformation unit specifies a single binary relation on configurations, a network provides a family of such relations, one for each node. Some of these relations may be of particular interest while others are of an auxiliary nature. Hence it is meaningful to distinguish between units of interest and auxiliary units by means of an export interface which consists of a subset of nodes of the network. Moreover, some of the nodes of the network may refer to transformation units which are already known in the context and are reused by the network. This can be described by an import interface. A network of transformation units together with import and export interfaces is a transformation module. Semantically, it specifies a family of relations on configurations, one relation for each export node. This is obtained by the restriction of the iterated interleaving semantics of the network to the export interface. The resulting relations depend on the semantics of the import interface.

It is actually quite easy to extend the notion of the interleaving semantics to a network of transformation units: Assuming that each node has got a semantic relation, one can just construct the interleaving semantics for each node using the semantic relations of the imported units for interleaving. And because this yields a semantic relation for each node, the process can be repeated as often as one likes. Unfortunately, it is not always adequate to use the interleaving semantics of an imported node to build up the interleaving semantics of the importing node. In the second example of Section 2.2, **height** specifies a relation with entries of the form $(height(b), n)$ where b is some **bintree$_0$**-term and n some **nat**-term. But in the evaluation of *is b balanced* $\Rightarrow$ *height(b)* $\equiv$ *balance(b)*, *height(b)* occurs in a larger context so that the **height**-relation cannot be called directly. It must be modified and adapted before it can be applied. To overcome such problems, we allow reconstructions of relations before they are called in the interleaving semantics of a network.

3.1   Networks of Transformation Units

A *network of transformation units* is a system N consisting of a set $V \subseteq ID$ and a mapping tu that assigns a transformation unit $tu(id) = (I_{id}, U_{id}, R_{id}, C_{id}, T_{id})$ to each $id \in V$ such that $U_{id} \subseteq V$, i.e. each transformation unit of N imports transformation units of N.

N can be seen as a directed and node-labelled graph with V as the set of nodes, tu as the node labelling and a set of edges containing a pair $(id, id') \in V \times V$ if and only if $id' \in U_{id}$.

If all local transformation units of N have got semantic relations, i.e. there is a mapping $SEM: V \to 2^{\mathcal{K} \times \mathcal{K}}$ [1], the interleaving semantics $INTER_{SEM|U_{id}}(tu(id))$ is defined for each $id \in V$ where $SEM|U_{id}$ is the restriction of SEM to U_{id}. In this way, the interleaving semantics is extended to the *interleaving operator* $INTER_0$ with $INTER_0(SEM): V \to 2^{\mathcal{K} \times \mathcal{K}}$ given by

$$INTER_0(SEM)(id) = INTER_{SEM|U_{id}}(tu(id))$$

for $id \in V$.

Alternatively, one can consider an *aggregating interleaving operator* $INTER_1$ that keeps the given semantic relations and adds the interleaving effect. This means $INTER_1(SEM): V \to 2^{\mathcal{K} \times \mathcal{K}}$ is defined for $id \in V$ by

$$INTER_1(SEM)(id) = SEM(id) \cup INTER_0(SEM)(id).$$

Moreover, if $F: 2^{\mathcal{K} \times \mathcal{K}} \to 2^{\mathcal{K} \times \mathcal{K}}$ is some construction on binary relations of configurations, both interleaving operators can be modified in such a way that the relations $F(SEM(id))$ for $id \in V$ are used in the interleaving instead of $SEM(id)$. Formally, $INTER_i^F$ for $i = 0, 1$ is defined for SEM and $id \in V$ by $INTER_i^F(SEM)(id) = INTER_i(F \circ SEM)(id)$ where $F \circ SEM$ is the sequential composition of SEM followed by F.

Each of the variants of the interleaving operator is a function mapping a mapping from V to $2^{\mathcal{K} \times \mathcal{K}}$ into a mapping from V to $2^{\mathcal{K} \times \mathcal{K}}$ and, therefore, can be iterated ad infinitum. Explicitly, let $INTER$ be some interleaving operator and let $ITERATE_0(N): V \to 2^{\mathcal{K} \times \mathcal{K}}$ be some mapping associating an initial relation to each node of N. Then we get a sequence of such mappings $ITERATE_i(N): V \to 2^{\mathcal{K} \times \mathcal{K}}$ defined inductively by $ITERATE_{i+1}(N) = INTER(ITERATE_i(N))$. Its union $ITERATE(N): V \to 2^{\mathcal{K} \times \mathcal{K}}$ given by

$$ITERATE(N)(id) = \bigcup_{i \in \mathbb{N}} ITERATE_i(N)(id)$$

for each $id \in V$ is called *iterated interleaving semantics*.

The interleaving operators import semantic relations and interleave them with the relations of the local rules. The resulting relations depend on the import so that the iterations of the interleaving operator may produce new results. The following second example shows that this can go on ad infinitum.

3.2   Examples

Term rewriting continued. The transformation units **is-balanced**, **height**, **balance**, and **nat** in Section 2.2 form a network.

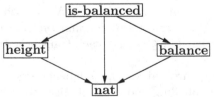

---

[1] $2^{\mathcal{K} \times \mathcal{K}}$ denotes the set of all binary relations on configurations

To discuss its iterated interleaving semantics, let us consider $INTER_1^{CC}$ as underlying interleaving operator where CC is the context closure of pairs of terms, i.e. $CC(P)$ for $P \subseteq T_\Sigma(X) \times T_\Sigma(X)$ contains all pairs $(t_0[L], t_0[R])$ for $(L, R) \in P$ and all context terms t_0. We may assume that **nat** is predefined with the ordinary evaluation of **nat**-terms as semantic relation. Choosing this relation as $ITERATE_0(\textbf{nat})$, it does not change in further steps of iteration provided that **nat** is properly specified. If the initial semantics of the other three nodes is empty, $ITERATE_1$ contains everything that is computed by rule applications and calls of **nat**. For **height** and **balance**, this is already the final semantics because there are no further interleaving sequences than those in the first step. In particular, $ITERATE_2(\textbf{is-balanced})$ makes use of it such that all proper computations take place in this step. No further interleaving sequences can appear in further steps.

It can be shown by induction on the length of interleaving sequences that **height** computes the length of the longest paths in a binary tree and **balance** the height of the largest totally balanced subtree. Hence, a binary tree is totally balanced if and only if both values coincide. Exactly this is tested by **is-balanced**.

Binary trees as a recursive domain. To illustrate the effect of cyclic import, we consider a transformation unit that imports itself.

> **tree**
> > initial: $\emptyset$
> > uses: **tree**
> > rules: *leaf, make*
> > conds: *leaf* | (**tree** ; *make*)
> > terminal: *all*

To keep things simple, the underlying rule base is tailored just for this example. The configurations are sets of binary trees. There are two rules, *leaf* and *make*. If applied to a set B of binary trees, they yield $leaf(B) = A$ and $make(B) = B \cup \{(L, a, R) \mid L, R, \in B, a \in A\}$ where A is some label alphabet. There are two domain expressions, $\emptyset$ and *all* the meanings of which are independent of the environment: $SEM(\emptyset) = \{\emptyset\}$, and $SEM(all)$ contains all sets of binary trees.

Accordingly, each interleaving sequence starts with the empty set of binary trees. The control condition requires that either *leaf* is applied or **tree** is called and then *make* applied. In the first case, the result in always A independent of the actual semantics of **tree**. This means that in particular $(\emptyset, A) \in ITERATE_i(\textbf{tree})$ for $i > 0$ even if we start with $ITERATE_0(\textbf{tree}) = \emptyset$.

The second case does not apply in the first step because **tree** refers to the empty relation. Hence, $ITERATE_1(\textbf{tree}) = \{(\emptyset, A)\}$. And in the second step, there is another interleaving sequence: $\emptyset, A, make(A)$, such that $ITERATE_2(\textbf{tree}) = \{(\emptyset, A), (\emptyset, make(A))\}$. In each further step, we can add another application of *make* such that $ITERATE_i(\textbf{tree}) = \{(\emptyset, make^j(A)) \mid 0 \leq j \leq i\}$.

Moreover, it is easy to see that $make^i(A)$ contains all binary trees up to the height i. In other words, **tree** specifies a generation process for binary trees in a way similar to the domain equation $TREE = A + TREE \cdot A \cdot TREE$.

3.3   Some Properties

Given a network of transformation units $N = (V, tu)$, let us consider the interleaving operator $INTER_1^F$ for some construction $F: 2^{\mathcal{K} \times \mathcal{K}} \to 2^{\mathcal{K} \times \mathcal{K}}$ with $rel \subseteq F(rel)$ for all $rel \subseteq \mathcal{K} \times \mathcal{K}$. This means that F extends every given relation as the context closure in our term rewrite example. Obviously, the identity on $2^{\mathcal{K} \times \mathcal{K}}$ is such an extending construction such that the interleaving operator $INTER_1$ is a special case.

Because F extends relations, $INTER_1^F$ extends every family of relations $SEM: V \to 2^{\mathcal{K} \times \mathcal{K}}$: $SEM \subseteq F \circ SEM \subseteq INTER_1(F \circ SEM) = INTER_1^F(SEM)$. In particular, the following holds: $ITERATE_i(N) \subseteq INTER_1^F(ITERATE_i(N)) \subseteq ITERATE_{i+1}(N)$ for $i \in \mathbb{N}$, and $ITERATE(N) \subseteq INTER_1^F(ITERATE(N))$.

This means that the iteration of the interleaving semantics yields an infinite increasing chain and the iterated interleaving semantics is extended by the interleaving operator $INTER_1^F$.

If the network is acyclic, the nodes can be numbered by levels inductively. Nodes the transformation units of which have an empty import component get level 0. And if all import identifiers of some node $id \in V$ have got a level and m is the maximum, then id has level $m + 1$. By induction on the levels, it is easy to show that the iteration of the interleaving semantics becomes stable. More precisely, the following holds for each $id \in V$ with level m:

$$ITERATE_{m+1}(N)(id) = ITERATE_{m+k}(N)(id) \text{ for all } k \in \mathbb{N}.$$

In contrast to that, the iteration may increase forever in networks with cyclic import as the last example above shows. But if the modifier F behaves nicely, the increase of the process of iterated interleaving stops for $ITERATE(N)$. More precisely, let F be monotonic, i.e., $F(rel) \subseteq F(rel')$ for all $rel \subseteq rel'$, and continuous, i.e., $F(\cup_{i \in \mathbb{N}} rel_i) = \cup_{i \in \mathbb{N}} F(rel_i)$ for all increasing chains of relations $rel_0 \subseteq rel_1 \subseteq \cdots$. Then the iterated interleaving semantics is a fixed point of the interleaving operator:

$$ITERATE(N) = INTER_1^F(ITERATE(N)).$$

The inclusion from left to right is shown above. The inclusion from right to left can be seen as follows:

$$INTER_1^F(ITERATE(N)) =_{(1)} INTER_1(F \circ ITERATE(N))$$
$$=_{(2)} INTER_1(F \circ \bigcup_{i \in \mathbb{N}} ITERATE_i(N))$$
$$=_{(3)} INTER_1(\bigcup_{i \in \mathbb{N}} F \circ ITERATE_i(N))$$
$$\subseteq_{(4)} \bigcup_{i \in \mathbb{N}} INTER_1(F \circ ITERATE_i(N))$$
$$=_{(5)} \bigcup_{i \in \mathbb{N}} INTER_1^F(ITERATE_i(N))$$
$$=_{(6)} \bigcup_{i \in \mathbb{N}} ITERATE_{i+1}(N) \subseteq_{(7)} ITERATE(N)$$

The equalities $1, 2, 5$, and 6 and the inclusion 7 hold by definition. The equality 3 follows from the continuity of F. Finally, the inclusion 4 follows from the fact that each interleaving sequence using relations of $\cup_{i \in \mathbb{N}} F \circ ITERATE_i(N)$ is also one using only relations of $F \circ ITERATE_k(N)$ for some $k \in \mathbb{N}$.

3.4 Transformation Modules

If one looks at the term rewriting example of Section 3.2 the nodes of a network of transformation units play different roles. The node **nat** refers to a standard data type and should be available in the context. Hence it may belong to the import interface. The node **is-balanced** and maybe also the node **height** are units of interest while the node **balance** is rather of an auxiliary nature useful for internal purposes, but not meaningful for the public. Therefore, it should not appear in the export interface. This consideration leads to the following notion.

A transformation module is a system $MOD = (IMPORT, BODY, EXPORT)$ where $BODY$ is a network of transformation units with the node set V_{BODY}, and $IMPORT$ and $EXPORT$ are subsets of V_{BODY}.

The semantics of a module is given by the semantics of its body restricted to the export where the semantics of the import can be chosen arbitrarily. Formally, let $SEM \colon IMPORT \to 2^{\mathcal{K} \times \mathcal{K}}$ be a mapping, and let $INTER$ be some interleaving operator. Then $SEM_{MOD} \colon EXPORT \to 2^{\mathcal{K} \times \mathcal{K}}$ is defined for all $id \in EXPORT$ by

$$SEM_{MOD}(id) = ITERATE(BODY)(id)$$

where $ITERATE_0(BODY)$ is given by $ITERATE_0(BODY)(id) = SEM(id)$ for $id \in IMPORT$ and $ITERATE_0(BODY)(id) = \emptyset$ otherwise.

4 Conclusion

In this paper, we have made an attempt at providing structuring principles for rule-based systems that are independent of a particular rule-based framework. The proposal has been guided by our earlier investigations of structuring concepts for graph transformation systems. But the presented notions are different and more general in several significant respects:

- Most obviously, we are no longer assuming graphs as underlying structures, but arbitrary configurations.
- The semantics of configuration class expressions may depend on the environment. This allows one, for example, to introduce reduced normal forms with respect to imported relations.
- In a network and hence in a transformation module, the transformation units define each other by calling each other within interleaving sequences. But instead of calling just the interleaving semantics, a preprocessing may take place in advance to adapt the interleaving semantics in some proper way. As an example, it may be closed under context to make it locally applicable.

- Moreover, the interleaving operator is provided with a cumulative effect explicitly by keeping the input relations and adding the outcome of interleaving to them.

On the syntactic level of transformation units and transformation modules, the proposed structuring concepts are of a simple nature: Transformation units can be imported and exported by a kind of call-by-reference. On the semantic level, the structure is reflected by the interleaving semantics that computes a binary relation on certain types of configurations for each transformation unit in a network. In suitable circumstances, the iterated interleaving semantics is a fixed point of the interleaving operator such that further investigations can be undertaken on a sound mathematical fundament.

The interleaving semantics is based on the sequential composition of relations such that transformation units and modules may be suitable for the modeling of sequential systems. Another direction of future research will be the generalization of the semantics in such a way that the modeling of parallel, concurrent, and distributed systems is covered. A first step in this direction has been done in [KK02] where graph transformation units are provided with a distributed semantics.

References

[AKK99]   Egidio Astesiano, Hans-Jörg Kreowski, and Bernd Krieg-Brückner. *Algebraic Foundations of Systems Specification*. IFIP State-of-the-Art Reports. Springer Verlag, 1999.

[DKKK00]   Frank Drewes, Peter Knirsch, Hans-Jörg Kreowski, and Sabine Kuske. Graph transformation modules and their composition. In *Proc. Applications of Graph Transformations with Industrial Relevance*, volume 1779 of *Lecture Notes in Computer Science*, pages 15–30, 2000.

[EEKR99]   Hartmut Ehrig, Gregor Engels, Hans-Jörg Kreowski, and Grzegorz Rozenberg, editors. *Handbook of Graph Grammars and Computing by Graph Transformation, Vol. 2: Applications, Languages and Tools*. World Scientific, Singapore, 1999.

[EKMR99]   Hartmut Ehrig, Hans-Jörg Kreowski, Ugo Montanari, and Grzegorz Rozenberg, editors. *Handbook of Graph Grammars and Computing by Graph Transformation, Vol. 3: Concurrency, Parallelism, and Distribution*. World Scientific, Singapore, 1999.

[EM85]   Hartmut Ehrig and Bernd Mahr. *Fundamentals of Algebraic Specification 1: Equations and Initial Semantics*. EATCS Monographs on Theoretical Computer Science. Springer Verlag, 1985.

[EM90]   Hartmut Ehrig and Bernd Mahr. *Fundamentals of Algebraic Specification 2: Module Specifications and Constraints*. EATCS Monographs on Theoretical Computer Science. Springer Verlag, 1990.

[HHKK00]   Reiko Heckel, Berthold Hoffmann, Peter Knirsch, and Sabine Kuske. Simple modules for GRACE. In Hartmut Ehrig, Gregor Engels, Hans-Jörg Kreowski, and Grzegorz Rozenberg, editors, *Proc. Theory and Application of Graph Transformations*, volume 1764 of *Lecture Notes in Computer Science*, pages 383–395, 2000.

[HO80]   G. Huet and D. Oppen. Equations and rewrite rules: A survey. In R.V. Book, editor, *Formal Language Theory: Perspectives and Open Problems*, pages 349–405. Academic Press, 1980.

[Kir99]   Hélène Kirchner. Term rewriting. In *Algebraic Foundations of Systems Specification* [AKK99], pages 273–320.

[KK96]   Hans-Jörg Kreowski and Sabine Kuske. On the interleaving semantics of transformation units — a step into GRACE. In Janice E. Cuny, Hartmut Ehrig, Gregor Engels, and Grzegorz Rozenberg, editors, *Proc. Graph Grammars and Their Application to Computer Science*, volume 1073 of *Lecture Notes in Computer Science*, pages 89–108, 1996.

[KK99a]   Hans-Jörg Kreowski and Sabine Kuske. Graph transformation units and modules. In Ehrig et al. [EEKR99], pages 607–638.

[KK99b]   Hans-Jörg Kreowski and Sabine Kuske. Graph transformation units with interleaving semantics. *Formal Aspects of Computing*, 11(6):690–723, 1999.

[KK02]   Peter Knirsch and Sabine Kuske. Distributed graph transformation units. In Andrea Corradini, Hartmut Ehrig, Hans-Jörg Kreowski, and Grzegorz Rozenberg, editors, *Proc. First International Conference on Graph Transformation (ICGT)*, volume 2505 of *Lecture Notes in Computer Science*, pages 207–222, 2002.

[KKS97]   Hans-Jörg Kreowski, Sabine Kuske, and Andy Schürr. Nested graph transformation units. *International Journal on Software Engineering and Knowledge Engineering*, 7(4):479–502, 1997.

[Roz97]   Grzegorz Rozenberg, editor. *Handbook of Graph Grammars and Computing by Graph Transformation, Vol. 1: Foundations*. World Scientific, Singapore, 1997.

Notions of Behaviour and Reachable-Part
and Their Institutions

Alexander Kurz

Department of Mathematics and Computer Science, University of Leicester
kurz@mcs.le.ac.uk

Abstract. Notions of observability and reachability induce a relation
of indistinguishability on the models of specifications. We show how to
obtain, in a systematic way, from a given institution, an institution that
respects this indistinguishability relation. Moreover, observability and
reachability are treated in formally dual way.

Introduction

This paper is concerned with combining notions of observability/reachability
with institutions. Notions of observability and reachability play a major role in
specification and verification. From the point of view of the user or specifier, such
notions introduce an equivalence relation of indistinguishability on the models
(implementations) of specifications: the user/specifier wants to reason only up
to this equivalence.

Institutions [10] are logics which allow to reason about models for different
signatures which is essential to achieve modularity. The central question of the
paper is, therefore:

> Given a notion of observability or reachability and our favourite
> institution, is there a systematic way to build an institution re-
> specting the corresponding relation of indistinguishability?

The aim of this work is twofold. First, to provide one possible answer to the
question raised above. Second, following the ideas in [6], to show that observability and reachability can be treated in a formally dual way.

We proceed in three steps and establish four conditions which are sufficient for
a positive answer to our question, leading to the notion of black-box institution.
Section 1 formalises observability and reachability by taking a black-box point
of view. In case of observability, we assume that for each model M, there is a
black-box view BM, also called the behaviour of M, together with a quotient
$M \to BM$. Dually, in case of reachability, we assume that for each model M,
there is a black-box view RM, also called the reachable-part of M, together with
an embedding $RM \to M$. The corresponding definitions introduce Conditions
(1) and (2).

Section 2 extends the considerations from categories to indexed categories,
that is, functors $Mod : \mathbf{Sig}^{op} \to \mathbf{CAT}$. For this, we have to add Condition (3)

M. Wirsing, D. Pattinson, and R. Hennicker (Eds.): WADT 2002, LNCS 2755, pp. 312–327, 2003.

stating that taking behaviours/reachable-parts commutes with reindexing (taking reducts).

Section 3 shows that we can extend our favourite 'standard' logic to a logic respecting the appropriate indistinguishability relation by requiring Condition (4) which states that a model satisfies a formula if the black-box view standardly satisfies the formula. This leads to the notion of black-box institutions, answering the question above in a positive way.

Section 4 shows that our framework applies indeed to our motivating examples of observational logic [13], constructor-based logic [5] and COL [4].

We conclude with a discussion of related work. The technical notions are recalled in an appendix reviewing categorical duality, indexed categories and institutions, and fibred categories.

Acknowledgements.   I gratefully acknowledge valuable remarks of Michel Bidoit, Rolf Hennicker, and the anonymous referees.

1   Notions of Behaviour and Reachable-Part

Our first task is to formalise observability and reachability in a way that is suitable for the concerns of the paper. To establish a formal duality we use the language of category theory which is briefly reviewed in appendix A.

1.1   Observability

Many notions of behaviour found in the literature can be described abstractly by two simple properties.

Definition B 1 *Let C be a category. Given an operation B on the objects of C and a family of arrows $\eta = (\eta_M : M \to BM)_{M \in C}$, we call (B, η) a* **notion of behaviour** *for C iff*

$$\text{all } \eta_M \text{ are epi} \tag{1}$$

$$\forall M, N \in C \text{ there is } (\cdot)^{\sharp} : C(M, BN) \to C(BM, BN) \text{ such that } f^{\sharp} \circ \eta_M = f \tag{2}$$

Note that the operation $(\cdot)^{\sharp}$ is uniquely determined due to (1). In case that C is a concrete category we call two elements of M **behaviourally equivalent**, and write $x \simeq_M y$, iff $\eta_M(x) = \eta_M(y)$. Intuitively, (1) expresses that every model M has a quotient BM which we call the **behaviour** of M; (2) is pictured as the left-hand diagram below

$$
\begin{array}{ccc}
BM & \xrightarrow{f^{\sharp}} & BN \\
\eta_M \uparrow & \nearrow f & \\
M & &
\end{array}
\qquad\qquad
\begin{array}{ccc}
BM & \xrightarrow{(\eta_N \circ g)^{\sharp}} & BN \\
\eta_M \uparrow & & \uparrow \eta_N \\
M & \xrightarrow{g} & N
\end{array}
$$

and expresses that *morphisms in $\mathcal{C}$ preserve behavioural equivalence*. Indeed, for $g : M \to N$ in $\mathcal{C}$, we have, as indicated in the right-hand diagram above, $x \simeq_M y \Leftrightarrow \eta_M(x) = \eta_M(y) \Rightarrow (\eta_N \circ g)^\sharp \circ \eta_M(x) = (\eta_N \circ g)^\sharp \circ \eta_M(y) \Leftrightarrow \eta_N(g(x)) = \eta_N(g(y)) \Leftrightarrow g(x) \simeq_N g(y)$.

Here are some straightforward examples, more will follow later.

Example 1. 1. A well-known example is provided by the category of deterministic automata (without initial states, see [20] for details), BM being the minimal realisation of an automaton M and $\eta_M : M \to BM$ mapping a state in M to the state in BM that accepts the same language.

2. In the category of labelled transition systems, a notion of behaviour is given by the quotient BM of a transition system M wrt the largest bisimulation.

3. More generally, subsuming the above, for any set-endofunctor T, the category $\mathsf{Coalg}(T)$ of T-coalgebras [21] has a canonical notion of behaviour. For any T-coalgebra M, BM is given by the image factorisation $M \to BM \to Z$ of the unique arrow $M \to Z$ into the final T-coalgebra Z [1].

4. Consider the signature for stacks over elements E given by $new : 1 \to X$, $push : E \times X \to X$, $pop : X \to X + 1$, $top : X \to E + 1$. Define a relation of observational (or behavioural) equivalence by $x \simeq y$ iff $top(pop^n(x)) = top(pop^n(y))$ for all $n \in \mathbb{N}$. Let $\mathcal{C}$ be the category consisting of the structures $(X, top, pop, push, new)$ for which $\simeq$ is a congruence[2]. Define BM to be the structure identifying all observationally equal states. Note that the array-pointer implementation and the standard model have the same behaviour.

Note that the last example is different from its predecessors in that $\mathcal{C}$ there has no final object due to the liberty in implementing *new* and *push*.

A notion of behaviour on $\mathcal{C}$ gives rise to two related categories, the category $\mathcal{C}^B$ consisting of all behaviours and the category $\mathcal{C}_B$ consisting of all models but seen from the point of view of their behaviours.

Definition B 2 *Consider a notion of behaviour B on $\mathcal{C}$.*

– $\mathcal{C}^B$ *is the full subcategory of $\mathcal{C}$ consisting of all objects M with $M \cong BM$.*
– $\mathcal{C}_B$ *has the same objects as $\mathcal{C}$ and morphisms $\mathcal{C}_B(M, N) = \mathcal{C}(BM, BN)$.*

Intuitively, $\mathcal{C}$ consists of all possible realizations of a specification whereas $\mathcal{C}^B$ only contains the black box views. $\mathcal{C}_B$ combines both aspects. The models are the same as in $\mathcal{C}$ but the morphisms incorporate the black box view, $\mathcal{C}_B(M, N) = \mathcal{C}(BM, BN)$.

Example 2 (Observational Logic). Our motivating example is observational logic [13]. In observational logic, each signature Σ determines a set of state sorts and a

---

[1] The final coalgebra always exists [1] although its carrier may be a proper class. Avoiding the final coalgebra, BM can be defined as the quotient wrt the largest behavioural equivalence or as the co-intersection of all quotients of M.

[2] That $\simeq$ is a congruence means that using all operations does not allow us to distinguish more states than using only *top* and *pop*.

notion of observational equality for each state sort. The class $\mathcal{C}$ of models, called observational algebras, of a signature allows to take for any algebra $A \in \mathcal{C}$ the quotient BA of A wrt the observational equality induced by Sig. The important point to note about the category $\mathcal{C}_B$ is that its morphisms coincide with the observational morphisms of [13]. Since, on state sorts, a morphism maps a state to an equivalence class of states, these morphisms can be described as relations as in [13].

The reader not familiar with monads [19] can skip the next proposition and continue with its corollary and the following example. The proposition characterises notions of behaviour as *monads whose unit is componentwise epi* and shows that these monads are idempotent. In our context, idempotence means idempotence of taking behaviours, ie, $BBM \cong BM$.

Proposition B 3 *There is an bijection between notions of behaviour and monads whose unit is componentwise epi. Moreover these monads are idempotent.*

Proof. Given a notion of behaviour (B, η) with lifting $(\cdot)^\sharp$, we show that $(B, \eta, (\cdot)^\sharp)$ is a Kleisli triple[3]. By definition of a Kleisli triple one has to check for all $g : M \to BN, f : L \to BM$ the laws (i) $f = f^\sharp \circ \eta_M$, (ii) $(\eta_M)^\sharp = \mathrm{id}_{BM}$, (iii) $(g^\sharp \circ f)^\sharp = g^\sharp \circ f^\sharp$. This is straight forward. Conversely, it is immediate that every Kleisli-triple $(B, \eta, (\cdot)^\sharp)$ (and hence monad) with η_M epi gives rise to a notion of behaviour (B, η). Finally, to see that these monads are idempotent, that is, that the multiplication μ_M is iso, apply Condition (2) to $f = \mathrm{id}_{BM}$ and use Condition (1). (Details of the proof can be found in [16], Lemma 2.2).

The fact that B is an idempotent monad determines the structure described in the following corollary.

Corollary B 4 *Let (B, η) be a notion of behaviour for $\mathcal{C}$. There are functors*

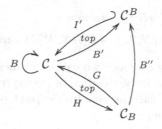

where B', B'', and G map an object to its behaviour, I' is the inclusion of behaviours, and H is the identity on objects, all satisfying $I'B' = B = GH$, $B''H = B'$, $I'B'' = G$. Moreover, η is a natural transformation, behaviour is

---

[3] There is a well-known bijection between Kleisli triples and monads. The monad (B, η, μ) associated to the Kleisli triple $(B, \eta, (\cdot)^\sharp)$ is given as follows: B is defined on arrows as $Bf = (\eta_N \circ f)^\sharp$ for $f : M \to N$ and $\mu_M = (\mathrm{id}_{BM})^\sharp$ for $M \in \mathcal{C}$. Conversely, given a monad (B, η, μ), the Kleisli triple is $(B, \eta, (\cdot)^\sharp)$ with $f^\sharp = \mu_M \circ Bf$ for $f : M \to BN$.

left adjoint to inclusion $(B' \dashv I')$, B'' *is an equivalence of categories, and* C^B *is a full reflective subcategory of* C.

Proof. The category of algebras for the monad is (isomorphic to) C^B since the multiplication is iso. The Kleisli-category is (isomorphic to) C_B. It now follows from B being a monad: functoriality, the equations, the adjunctions, naturality of η, and B'' full and faithful. Since the multiplication is iso, every object in C^B is isomorphic to an object in the image of B'', hence B'' is an equivalence. C^B is a full reflective subcategory due to the bijection between full reflective subcategories and idempotent monads. (See [7], Vol.2, Proposition 4.1.6 and Proposition 4.2.3 for proofs of the facts we used).

Example 3.   1. In the case of deterministic automata, C^B is (equivalent to) the category of languages[4], B' can be understood as mapping an automaton to the language it accepts and I' as mapping a language to the minimal realising automaton. The insight that behaviour is left-adjoint to minimal realisation, $B' \dashv I'$, is due to Goguen [11].
   2. Continuing Example 2, the 'observational black-box functor' of observational logic appears here as $C_B \xrightarrow{B''} C^B \hookrightarrow \mathsf{Alg}(\Sigma)$. It is full and faithful since B'' is full and faithful. Moreover, the category C^B of fully abstract algebras is a full reflective subcategory of the category C of observational algebras.

1.2   Reachability

(R, ε) is a notion of reachable-part for C iff $(R^{\mathrm{op}}, \varepsilon^{\mathrm{op}})$ is a notion of behaviour for C^{op} [5]. In detail:

Definition R 1 *Let C be a category. Given an operation R on the objects of C and a family of arrows $\varepsilon = (\varepsilon_M : RM \to M)_{M \in C}$, we call (R, ε) a **notion of reachable-part** for C iff*

$$\text{all } \varepsilon_M \text{ are mono}$$

$$\forall N, M \in C \text{ there is } (\cdot)^\sharp : C(RN, M) \to C(RN, RM) \text{ such that } \varepsilon_M \circ f^\sharp = f$$

In case that C is a concrete category we call an element x of M **reachable** iff x is in the image of ε_M. Intuitively, the dual of (1) expresses that every model has a submodel RM which we call the **reachable-part** of M. The dual of (2), see the left-hand diagram below,

---

[4] More precisely, given an alphabet Σ, objects are subsets $L \subseteq \Sigma^*$ ordered by inclusion.

[5] The $(\cdot)^{\mathrm{op}}$-notation is explained in appendix A.

expresses that morphisms in $\mathcal{C}$ preserve reachability. Indeed, with the notation of the right-hand diagram above, an element y in N is reachable iff there is y' in RN with $\varepsilon_N(y') = y$. Now, for $g : N \to M$ it holds that $g(y) = g \circ \varepsilon_N(y') = \varepsilon_M((g \circ \varepsilon_N)^\sharp(y'))$, that is, $g(y)$ is reachable.

Here is a typical example of reachability:

Example 4. 1. In the category of deterministic automata (with initial state), there is an operation R mapping an automaton M to the automaton RM which contains only states reachable from the initial state.
  2. More generally, given a signature for algebras, there is the operation R mapping an algebra M to the algebra RM which contains only elements which are denoted by a term.

Dualising Definition B 2, we obtain categories $\mathcal{C}^R$ and $\mathcal{C}_R$. The objects of $\mathcal{C}^R$ are called **reachable**, an interpretation of the morphisms of $\mathcal{C}_R$ is given in the remark following the definition.

Definition R 2 *Consider a notion of reachable-part R on $\mathcal{C}$.*

 – $\mathcal{C}^R$ *is the full subcategory of $\mathcal{C}$ consisting of all objects M with $M \cong RM$.*
 – $\mathcal{C}_R$ *has the same objects as $\mathcal{C}$ and morphisms $\mathcal{C}_R(M, N) = \mathcal{C}(RM, RN)$.*

Remark 1. In categorical language a partial map $X \to Y$ with domain D can be described, see the left-hand diagram,

as a mono $D \to X$ and a map $D \to Y$. Because a morphism $RN \to RM$ of $\mathcal{C}_R$ can be considered as a morphism $RN \to M$ [6], the right-hand diagram shows that morphisms in $\mathcal{C}_R$ are partial morphisms defined on reachable-parts.

Example 5 (Constructor-Based Logic). Our motivating example is constructor-based logic [6]. In constructor-based logic, each signature Σ determines a set of *constrained sorts* along with a set of *constructor* operations (defining how to construct elements of constrained sorts). The class $\mathcal{C}$ of models, called constructor-based algebras, of a signature allows to take for any $A \in \mathcal{C}$ the reachable-part (called generated part in [5]) RA of A [7]. The constructor-based morphisms of [6] coincide with the morphisms in $\mathcal{C}_R$. Their distinguishing feature is that they are, on constrained sorts, partial functions defined only on the reachable-part.

---

[6] It follows from Corollary R 4 below (or directly from the dual of Conditions (1) and (2)) that $\mathcal{C}(RN, RM) \to \mathcal{C}(RN, M)$, $f \mapsto \varepsilon_M \circ f$, is a bijection, the inverse given by $(\cdot)^\sharp$.

[7] Compared to earlier treatments of reachability, the distinguishing feature of constructor-based logic is that models are not required to be reachable, operations only have to preserve reachability.

It follows from the duality principle and Proposition B 3 that notions of reachable-part coincide with comonads having a counit that is componentwise mono and that these comonads are, moreover, idempotent. Idempotence means here idempotence of taking reachable-parts, ie, $RRM \cong RM$.

Proposition R 3 *There is a bijection between notions of reachable-part and comonads having a counit that is componentwise mono. Moreover, this comonads are idempotent.*

Corollary R 4 *Let (R, ε) be a notion of behaviour for $\mathcal{C}$. There are functors*

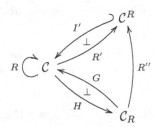

where R', R'', and G map an object to its reachable-part, I' is the inclusion of reachable-parts, and H is the identity on objects, all satisfying $I'R' = R = GH$, $R''H = R'$, $I'R'' = G$. Moreover, ε is a natural transformation, restriction to reachable-parts is right adjoint to inclusion $(I' \dashv R')$, R'' is an equivalence of categories, and $\mathcal{C}^R$ is a full coreflective subcategory of $\mathcal{C}$.

2   Notions of Behaviour and Reachable-Part for Indexed Categories

In a second step we extend our notions of behaviour/reachable-part from categories to indexed categories.

2.1   Observability

We require a notion of behaviour for each index and that reindexing preserves behaviours.

Definition B 5 (Notion of Behaviour for Indexed Categories) *Given a functor $Mod : \mathrm{Sig}^{\mathrm{op}} \to \mathsf{CAT}$ we call a family $\mathsf{B} = (B_\Sigma, \eta_\Sigma)_{\Sigma \in \mathrm{Sig}}$ a notion of behaviour for Mod iff each (B_Σ, η_Σ) is a notion of behaviour for $Mod(\Sigma)$ and for all arrows $\sigma : \Sigma \to \Sigma'$ and all $M' \in Mod(\Sigma')$ an isomorphism making the diagram*

$$B(Mod(\sigma)(M')) \xrightarrow{\ \cong\ } Mod(\sigma)(B(M')) \qquad (3)$$

$$\eta_{Mod(\sigma)(M')} \searrow \qquad \swarrow Mod(\sigma)(\eta_{M'})$$

$$Mod(\sigma)(M')$$

commute. We write B *for* $B_{\Sigma'}, B_{\Sigma}$, *respectively, the subscripts being clear from the context.*

Intuitively, (3) states that $Mod(\sigma)$ preserves behaviours. The habit of avoiding the subscripts of $B_{\Sigma'}, B_{\Sigma}$ comes from an equivalent formulation of the above definition in terms of fibred categories (aka fibrations). The reader not familiar with fibred categories may continue with Corollary B 8.

The following definition generalises the characterisation of notions of behaviour given by Proposition B 3 from categories to fibred categories.

Definition B 6 (Notion of Behaviour for Fibrations) *Let* $p : C \to X$ *be a fibration. A fibred monad* (B, η, μ) *on* p *is called a notion of behaviour for* p *iff all* η_M, $M \in C$, *are epi.*

Remark 2. The concept of a monad with componentwise epi unit makes sense for any 2-category. Definition B 6 formulates the instance in the 2-category of fibrations over X whereas the previous section dealt with the instance in the 2-category of categories.

One advantage of Definition B 6 is that monads on fibrations are easier to define than monads on indexed categories and hence the analogue of Corollary B 4 comes more natural in terms of fibrations. But before going into this let us first sketch the equivalence of the two definitions.

Proposition B 7 *Definition B 5 and Definition B 6 are equivalent.*

Proof. The proof is a corollary to the equivalence of (non-split) indexed categories and cloven fibrations, see eg [12]. We sketch the argument for our special case. Let $Mod : \mathsf{Sig}^{op} \to \mathsf{CAT}$ be a functor and $p : C \to \mathsf{Sig}$ the corresponding fibration obtained via the Grothendieck-construction. We give a bijection between notions of behaviour for Mod and p. First, given (B, η, μ) on p we define B_{Σ} and η_{Σ} as the restriction of B and η to the fibre over Σ. Condition (3) is satisfied since B preserves cartesian arrows. Second, given $(B_{\Sigma}, \eta_{\Sigma})$, we can extend B to all of C as follows. Every morphism $f : M \to N$ in C over $\sigma : \Sigma \to \Sigma'$ can be factored as $\bar{\sigma} \circ f^{|}$ where $\bar{\sigma}$ is cartesian over σ and $f^{|}$ is vertical. Let $Bf = \bar{\bar{\sigma}} \circ \hat{\sigma}_N \circ Bf^{|}$ where $\bar{\bar{\sigma}}$ is cartesian over σ in BN and $\hat{\sigma}_N$ is the (unique) iso given by Condition (3). Thanks to Condition (1), the isomorphism $\hat{\sigma}_N$ given by Condition (3) is natural in N and satisfies $\hat{\sigma}_N^{id} = \mathrm{id}_{BN}$ and $\widehat{\sigma \circ \varrho}_N = Mod(\varrho)(\hat{\sigma}_N) \circ \hat{\varrho}_{Mod(\sigma)(N)}$. Chasing the obvious big diagram now shows that B is indeed a functor. B is fibrewise a monad due to Proposition B 3 and preserves cartesian liftings by construction. $\qquad\qquad\Box$

A pleasant consequence of the proposition is that we now obtain a version of Corollary B 4 for indexed categories. Indeed, let $B = (B_{\Sigma}, \eta_{\Sigma})_{\Sigma \in S_{ig}}$ be a notion of behaviour for $Mod : \mathsf{Sig}^{op} \to \mathsf{CAT}$. By Proposition B 7, we can extend B to a monad B on the total category C of the fibration p corresponding to Mod. Now we can apply Corollary B 4 to C and B.

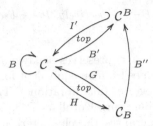

Moreover, we know (see [16], Proposition 4.18) that $\mathcal{C}^B$ and $\mathcal{C}_B$ are fibred over Sig, that is, there are fibrations p^B and p_B making the diagram above into a diagram in the (2-)category of fibrations over Sig. Translating the diagram back to indexed categories, it follows that the constructions of Corollary B 4 extend from categories to indexed categories.

For future reference we record (recall Definition B 2 for the operations $(\cdot)^{B_\Sigma}$ and $(\cdot)_{B_\Sigma}$)

Corollary B 8 *Let* $\mathsf{B} = (B_\Sigma, \eta_\Sigma)_{\Sigma \in \mathsf{Sig}}$ *be a notion of behaviour for* Mod : $\mathsf{Sig}^{\mathrm{op}} \to \mathsf{CAT}$. *Then there are functors* Mod^{B} : $\mathsf{Sig}^{\mathrm{op}} \to \mathsf{CAT}$, $\Sigma \mapsto Mod(\Sigma)^{B_\Sigma}$ *and* Mod_{B} : $\mathsf{Sig}^{\mathrm{op}} \to \mathsf{CAT}$, $\Sigma \mapsto Mod(\Sigma)_{B_\Sigma}$.

2.2   Reachability

$(R_\Sigma, \varepsilon_\Sigma)_{\Sigma \in \mathsf{Sig}}$ is a notion of reachable-part for the indexed category Mod : $\mathsf{Sig}^{\mathrm{op}} \to \mathsf{CAT}$ iff $(R_\Sigma{}^{\mathrm{op}}, \varepsilon_\Sigma{}^{\mathrm{op}})$ is a notion of behaviour for the dual indexed category Mod^{co} [8]. In detail:

Definition R 5 *Given a functor* Mod : $\mathsf{Sig}^{\mathrm{op}} \to \mathsf{CAT}$ *we call a family* $\mathsf{R} = (R_\Sigma, \varepsilon_\Sigma)_{\Sigma \in \mathsf{Sig}}$ *a* **notion of reachable-part** *for* Mod *iff each* $(R_\Sigma, \varepsilon_\Sigma)$ *is a notion of reachable-part for* $Mod(\Sigma)$ *and for all arrows* $\sigma : \Sigma \to \Sigma'$ *and all* $M' \in Mod(\Sigma')$ *an isomorphism making the diagram*

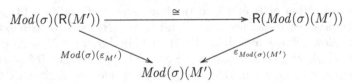

$$Mod(\sigma)(\mathsf{R}(M')) \xrightarrow{\ \cong\ } \mathsf{R}(Mod(\sigma)(M'))$$

with $Mod(\sigma)(\varepsilon_{M'})$ and $\varepsilon_{Mod(\sigma)(M')}$ to $Mod(\sigma)(M')$

commute. We write R *for* $R_{\Sigma'}, R_\Sigma$, *respectively, the subscripts being clear from the context.*

Intuitively, the dual of (3) states that $Mod(\sigma)$ preserves reachability.

A notion of reachable-part on a cofibration q is dual to a notion of behaviour on the fibration q^{op}, in detail:

Definition R 6 *Let* $q : \mathcal{C} \to \mathcal{X}$ *be a cofibration. A cofibred comonad* (R, ε, δ) *on* q *is called a notion of reachable-part for* q *iff all* ε_M, $M \in \mathcal{C}$, *are mono.*

---

[8] The definition of $(\cdot)^{\mathrm{co}}$ is recalled in appendix B.

Proposition R 7 *Definition R 5 and Definition R 6 are equivalent.*

Proof. Let Mod : $\mathsf{Sig}^{\mathrm{op}} \to \mathsf{CAT}$ be a functor and $q : \mathcal{C} \to \mathsf{Sig}$ the corresponding co-fibration obtained via the co-Grothendieck-construction. Apply Proposition B 7 to Mod^{co} and q^{op}.

Corollary R 8 *Let* $\mathsf{R} = (R_{\Sigma}, \varepsilon_{\Sigma})_{\Sigma \in \mathsf{Sig}}$ *be a notion of reachable-part for Mod :* $\mathsf{Sig}^{\mathrm{op}} \to \mathsf{CAT}$. *Then there are functors* Mod^{R} : $\mathsf{Sig}^{\mathrm{op}} \to \mathsf{CAT}$, $\Sigma \mapsto Mod(\Sigma)^{R_{\Sigma}}$ *and* Mod_{R} : $\mathsf{Sig}^{\mathrm{op}} \to \mathsf{CAT}$, $\Sigma \mapsto Mod(\Sigma)_{R_{\Sigma}}$.

3 Institutions for Observability and Reachability

In this section, we show how to obtain institutions whose satisfaction relation respects a given notion of behaviour or reachable-part. In order to treat both cases simultaneously, we call the behaviour/reachable-part of a model the black-box view of the model.

We assume that we are given a 'standard' logic as eg first-order logic which we use to reason about black-box views and that we want to extend this logic to all models in a way such that formulas of the logic can not distinguish between a model and its black box view.

Definition 9 *Let* $\mathcal{C}$ *be a class (of models),* V *an operation on* $\mathcal{C}$ *(with* $\mathsf{V}M$ *called the black-box view of* $M \in \mathcal{C}$*),* $\mathcal{L}$ *a class (of formulas) and* $\models, \mathrel{\|\!\!\!=} \subseteq \mathcal{C} \times \mathcal{L}$ *two relations. Then* $\mathrel{\|\!\!\!=}$ *is called the* **black-box view** *relation induced by* $(\mathsf{V}, \models)$ *iff*

$$M \mathrel{\|\!\!\!=} \varphi \iff \mathsf{V}M \models \varphi \tag{4}$$

We can now define the notion of a black-box institution. Recall that a notion of behaviour (reachable-part) V for an indexed category Mod : $\mathsf{Sig}^{\mathrm{op}} \to \mathsf{CAT}$ is given by functors V_{Σ} on $Mod(\Sigma)$ satisfying (the dual of) Conditions (1)–(3). According to Definition B(R) 2 and Corollary B(R) 8, V and Mod give rise to the indexed categories Mod^{V} and Mod_{V}, Mod^{V} being the category of the black-box views.

Definition 10 *Let* V *be a notion of behaviour (reachable-part) for Mod and* $(Mod^{\mathsf{V}}, Sen, \models)$ *be an institution and* $\mathrel{\|\!\!\!=}$ *be the black-box view relation induced by* $(\mathsf{V}, \models)$. *Then* $(Mod, Sen, \mathrel{\|\!\!\!=})$ *and* $(Mod_{\mathsf{V}}, Sen, \mathrel{\|\!\!\!=})$ *are the* **black-box institutions** *induced by* $(\mathsf{V}, \models)$.

The definition is justified by the following theorem (we assume that institutions have satisfaction relations that do not distinguish isomorphic models).

Theorem 1. *If* V *is a notion of behaviour or reachable-part for Mod then* Mod^{V} *and* Mod_{V} *are functors. If, moreover,* $(Mod^{\mathsf{V}}, Sen, \models)$ *is an institution then* $(Mod, Sen, \mathrel{\|\!\!\!=})$ *and* $(Mod_{\mathsf{V}}, Sen, \mathrel{\|\!\!\!=})$ *are institutions.*

Proof. The first part of the theorem is Corollary B 8. For the second part we have to show that for $(Mod, Sen, \models\!\!\!|)$ and $(Mod_V, Sen, \models\!\!\!|)$ the satisfaction condition of institutions holds. The proof being the same in both cases, we pick the first. Let $\sigma : \Sigma \to \Sigma'$ be a signature morphism, $\varphi \in Sen(\Sigma)$, $M' \in Mod(\Sigma')$. We have

$$Mod(\sigma)(M') \models\!\!\!| \varphi \Leftrightarrow V(Mod(\sigma)(M')) \models \varphi \Leftrightarrow Mod(\sigma)(VM') \models \varphi \Leftrightarrow$$
$$VM' \models Sen(\sigma)(\varphi) \Leftrightarrow M' \models\!\!\!| Sen(\sigma)(\varphi).$$

The equivalences are due to, respectively, Condition (4), Condition (3), the satisfaction condition for $\models$, and Condition (4). $\square$

In the same way, we obtain theorems for combining notions of behaviour and reachable-part. For example, consider a situation where we start with some indexed category of models *Mod* and assume that we have a notion of reachable-part R for *Mod* and then a notion of behaviour on reachable algebras. The black-box view of a model M is then given by BRM and all black-box views are assembled in the indexed category $(Mod^R)^B$. As before, we suppose that we have a standard logic $\models$ on black-box views that we want to transfer to all models, organised in the indexed categories *Mod* or $(Mod_R)_B$.

Theorem 2. *If R is a notion of reachable-part for Mod and B a notion of behaviour for* Mod^R *then* $(Mod^R)^B$ *and* $(Mod_R)_B$ *are functors. If, moreover,* $((Mod^R)^B, Sen, \models\!\!\!|)$ *is an institution and* $\models\!\!\!|$ *is the black-box view relation induced by* $(BR, \models)$*, then* $(Mod, Sen, \models\!\!\!|)$ *and* $((Mod_R)_B, Sen, \models\!\!\!|)$ *are institutions.*

Proof. For the first part, note that a notion of behaviour B for Mod^R induces a notion of behaviour for the equivalent category Mod_R, justifying to consider $(Mod_R)_B$. The second part is shown as for the theorem above.

We have shown how to reduce the proof of institutions to the verification of Conditions (1)–(4). In all applications we will consider in the next section the verification of (1) and (2) is straightforward. Similarly, $\models\!\!\!|$ is usually designed to satisfy (4). So the point is to carefully choose Sig in order to guarantee (3).

4   Applications

We sketch how our framework applies to the examples that motivated our presentation, namely observational logic, constructor-based logic, and COL.

Observational Logic. Observational logic was introduced in [13]. Our discussion refers to the presentation in [5]. Conditions (1) and (2) are satisfied, because the class of models, called observational algebras, is defined in such a way that the quotient wrt observational equality is again a model (cf. Definition 2.6 in [5]). Technically this is expressed by the requirement that observational equality is a congruence ([5], Definition 2.6) which in turn formalises the intuition that non-observer operations are not allowed to contribute to observations. Condition (3) is satisfied because the category Sig has been defined in a way such that reindexing preserves behaviours ([5], Corollary 2.19). Condition (3) can be understood as requiring that a signature morphism $\sigma : \Sigma \to \Sigma'$ from an 'old' signature Σ to a 'new' signature Σ' may not introduce new observations on old sorts (formalised as [5], Definition 2.14). Condition (4) appears as Theorem 2.12 in [5].

Constructor-Based Logic. Constructor-based logic was introduced in [6]. Our discussion refers to the presentation in [5]. Conditions (1) and (2) are satisfied, because the class of models, called constructor-based algebras, is defined in such a way that the sub*sets* of elements which are generated by constructors form a sub*algebra* ([5], Definition 3.6 and 3.7). This formalises the intuition that the use of non-constructors cannot lead to elements which are not reachable by constructors. Condition (3) is [5], Corollary 3.19, and a consequence of the definition of Sig which can be phrased as 'new constructors may not generate new elements on old sorts'. Condition (4) is Theorem 3.12 in [5].

Constructor-Based Observational Logic (COL). COL is introduced in [4]. COL consists of an integration of the two concepts above. Let us denote by Mod : $\mathsf{Sig}^{\mathrm{op}} \to \mathsf{CAT}$ the indexed category consisting of all COL-algebras with standard algebra morphisms. Constructors and observers determine a notion of reachable-part R. For any COL-algebra A, $\mathsf{R}A$ consists of all elements which are observably equal to an element generated by constructors. This yields an functor Mod_{R}. (Taking into account only constructible experiments,) observers then determine a notion of behaviour B on Mod_{R}, yielding the indexed category $(Mod_{\mathrm{R}})_{\mathrm{B}}$ of COL-algebras with COL-morphisms. That COL now forms an institution follows from Theorem 2 by establishing Conditions (3) and (4).

5   Conclusion

The aim of this paper was to give a semantic analysis of how to obtain institutions respecting notions of observability and reachability. We formalised observability and reachability as *notions of behaviour and reachable-part* and showed how they give rise to *black-box institutions*. We believe that the simplicity of Conditions (1)–(4) defining black-box institutions, together with the duality inherent in our approach, provide a nice and unifying picture for several known examples of institutions.

Nevertheless, it would certainly be interesting to find more instances of our framework. One possibility might be to look at generalisations. For example, our approach relied on the notion of an idempotent monad which is slightly more general than a notion of behaviour since the unit is not required to be componentwise epi. Moreover, most results would also hold for monads in general.

Institutions and Indexed Categories.   On the one hand, we followed the tradition to say that, in an institution, Mod : $\mathsf{Sig}^{\mathrm{op}} \to \mathsf{CAT}$ is a functor (= split indexed category) and not a pseudo-functor (= indexed category). On the other hand, in Definition B 5 it seemed more natural to consider isomorphisms in (3) rather than identity. But this means that B is a functor in the category of (non-split) indexed categories and not in the category of split indexed categories. It therefore seems natural to allow pseudo-functors in the definition of an institution[9].

Induction and Coinduction. It is well-known that notions of behaviour and reachable-part give rise to coinduction and induction proof principles. On our level of abstraction, however, we cannot deal with these more concrete[10] issues. On a more concrete level this has been done eg in [5].

Related Work. The duality of observability and reachability goes back to Kalman [15] in the context of linear systems in control theory. His results were generalised by Arbib and Manes [2, 3] to, essentially, algebras for an endofunctor on an arbitrary category. In [6] we took up their idea of duality of observability and reachability and applied it to specification formalisms. This paper continues [6] by showing that also the institution proofs for the observability and reachability case can be obtained from each other by using a formal duality principle.

The idea to use Conditions (1)–(4) to prove a general theorem about the existence of institutions appears in [16] where it was extracted from the proof of the observational logic institution in [13]. Compared to [16], which contains most of the technicalities used here, we have simplified matters by insisting on Condition (1) instead of working with idempotent monads in general and by moving the fibrations more into the background. More importantly, we show here that this approach fits nicely with the duality of observability and reachability and can also be applied to institutions which integrate both observability and reachability.

Let us also note that our approach has been designed for cases where we start from a standard logic which is an institution but does not respect the black-box view. A different approach is to start with a logic which respects the black-box view and directly build an institution. This approach is natural, for instance, if modal logic is used as a specification language and the black-box view is given by bisimulation, for examples see [9, 8, 17].

References

1. P. Aczel and N. Mendler. A final coalgebra theorem. In D. H. Pitt et al, editor, *Category Theory and Computer Science*, volume 389 of *LNCS*, pages 357–365. Springer, 1989.
2. M.A. Arbib and E.G. Manes. Foundations of system theory: decomposable systems. *Automatica*, 10:285–302, 1974.

---

[9] In the three examples of Section 4, the isomorphism of Condition (3) is an identity.

[10] We can take *concrete* here in the technical sense as referring to categories having a forgetful functor to sets: induction is about sub*sets* being closed under operations and coinduction is about *elements* being bisimilar.

3. M.A. Arbib and E.G. Manes. Adjoint machines, state-behaviour machines, and duality. *Journ. of Pure and Applied Algebra*, 6:313–344, 1975.
4. M. Bidoit and R. Hennicker. On the integration of observability and reachability concepts. In M. Nielsen and U. Engberg, editors, *Proceedings of FoSSaCS 2002*, volume 2303 of *LNCS*, pages 21–36. Springer, 2002.
5. M. Bidoit, R. Hennicker, and A. Kurz. Observational logic, constructor-based logic, and their duality. *Theoretical Computer Science*. To appear. Extended version of [6]. Available as CWI Technical Report SEN-R0223, 2002.
6. M. Bidoit, R. Hennicker, and A. Kurz. On the duality between observability and reachability. In F. Honsell and M. Miculan, editors, *Foundations of Software Science and Computation Structures (FOSSACS'01)*, volume 2030 of *LNCS*, pages 72–87, 2001.
7. Francis Borceux. *Handbook of Categorical Algebra*. Cambridge University Press, 1994.
8. Corina Cîrstea. Institutionalising many-sorted coalgebraic modal logic. In L.S. Moss, editor, *Coalgebraic Methods in Computer Science (CMCS'02)*, volume 65.1 of *ENTCS*, 2002.
9. Corina Cîrstea. On specification logics for algebra-coalgebra structures: Reconciling reachability and observability. In M. Nielsen and U. Engberg, editors, *Proceedings of FoSSaCS 2002*, volume 2303 of *LNCS*, pages 82–97. Springer, 2002.
10. J. Goguen and R. Burstall. Institutions: abstract model theory for specification and programming. *Journal of the ACM*, 39 (1):95–146, 1992.
11. Joseph Goguen. Realisation is universal. *Mathematical System Theory*, 6(4):359–374, 1973.
12. John W. Gray. *Formal category theory: adjointness for 2-categories.*, volume 391 of *Lecture Notes in Mathematics*. Springer, 1974.
13. R. Hennicker and M. Bidoit. Observational logic. In Armando Haeberer, editor, *Algebraic Methodology and Software Technology (AMAST'98)*, volume 1548 of *LNCS*. Springer, 1999.
14. Bart Jacobs. *Categorical Logic and Type Theory*. Elsevier, 1998.
15. R. E. Kalman, P. L. Falb, and M. A. Arbib. *Topics in Mathematical System Theory*. McGraw-Hill, 1969.
16. A. Kurz and R. Hennicker. On institutions for modular coalgebraic specifications. *Theoretical Computer Science*, 280:69–103, 2002.
17. A. Kurz and D. Pattinson. Coalgebras and modal logics for parameterised endofunctors. Technical Report SEN-R0040, CWI, 2000.
18. A. Kurz and D. Pattinson. Notes on coalgebras, co-fibrations and concurrency. In H. Reichel, editor, *Coalgebraic Methods in Computer Science (CMCS'00)*, volume 33 of *ENTCS*, pages 199–233, 2000.
19. Saunders Mac Lane. *Category Theory for the Working Mathematician*. Springer, 1971.
20. J.J.M.M. Rutten. Automata and coinduction - an exercise in coalgebra. In D. Sangiorigi and R. de Simone, editors, *Proceedings of CONCUR '98*, volume 1466 of *LNCS*. Springer, 1998.
21. J.J.M.M. Rutten. Universal coalgebra: A theory of systems. *Theoretical Computer Science*, 249:3–80, 2000.
22. Thomas Streicher. Fibred categories á la J. Bénabou. `http://www.mathematik.tu-darmstadt.de/~streicher/FIBR/fib.ps.gz`, 1999.
23. Andrzej Tarlecki. Institutions: An abstract framework for formal specifications. In E. Astesiano, H.-J. Kreowski, and B. Krieg-Brückner, editors, *Algebraic Foundations of Systems Specification*, chapter 4. Springer, 1999.

A   Categorical Duality

We briefly review categorical duality, for background see [19]. A category $\mathcal{C}$ consists of a class of objects, also denoted by $\mathcal{C}$, and for all $A, B \in \mathcal{C}$ of a set of arrows (or morphisms) $\mathcal{C}(A, B)$. The *dual* (or opposite) category $\mathcal{C}^{\mathrm{op}}$ has the same objects and arrows $\mathcal{C}^{\mathrm{op}}(A, B) = \mathcal{C}(B, A)$. We write A^{op} and f^{op} for $A \in \mathcal{C}$ and $f \in \mathcal{C}(B, A)$ to indicate when we think of A as an object in $\mathcal{C}^{\mathrm{op}}$ and of f as an arrow in $\mathcal{C}^{\mathrm{op}}(A, B)$. Duality can now be formalised as follows. Let P be a property of objects or arrows in $\mathcal{C}$. We then say that:

> An object A (arrow f, respectively) in $\mathcal{C}$ has property co-P
> iff A^{op} (f^{op}, respectively) has property P.

For example, a morphism $f \in \mathcal{C}(A, B)$ is co-epi (usually called mono) iff f^{op} is epi.

The duality principle extends to functors and natural transformations. The dual of a functor $F : \mathcal{C} \to \mathcal{D}$ is the functor $F^{\mathrm{op}} : \mathcal{C}^{\mathrm{op}} \to \mathcal{D}^{\mathrm{op}}$ which acts on objects and morphisms as F does. The dual of a natural transformation $\eta = (\eta_A)$ is $\eta^{\mathrm{op}} = (\eta_A{}^{\mathrm{op}})$. For instance, for an endofunctor F, the category of F-coalgebras is dual to the category of F^{op}-algebras; a functor F is left adjoint to G iff F^{op} is right adjoint to G^{op}; and (T, η, ε) is a comonad on $\mathcal{C}$ iff $(T^{\mathrm{op}}, \eta^{\mathrm{op}}, \varepsilon^{\mathrm{op}})$ is a monad on $\mathcal{C}^{\mathrm{op}}$.

B   Indexed Categories and Institutions

For our purposes, unless stated otherwise, an *indexed category* over a (large) category $\mathcal{I}$ is a functor $H : \mathcal{I}^{\mathrm{op}} \to \mathsf{CAT}$ to the (superlarge) category CAT of all large categories.

An institution $(Mod, Sen, \models)$ consists of an indexed category Mod over a category of *signatures* Sig, a functor $Sen : \mathsf{Sig} \to \mathsf{Set}$, and for each signature $\Sigma \in \mathsf{Sig}$ a *satisfaction relation* $\models_\Sigma \subseteq Mod(\Sigma) \times Sen(\Sigma)$ such that for all $\sigma : \Sigma \to \Sigma'$, all $\varphi \in Sen(\Sigma)$, and all $M' \in Mod(\Sigma')$

$$Mod(\sigma)(M') \models_\Sigma \varphi \quad \Leftrightarrow \quad M' \models_{\Sigma'} Sen(\sigma)(\varphi).$$

This condition is called the *satisfaction condition* of institutions. Moreover, we require that $M_1 \models_\Sigma \varphi \Leftrightarrow M_2 \models_\Sigma \varphi$ whenever M_1, M_2 are isomorphic. We usually write $M \models \varphi$ instead of $M \models_\Sigma \varphi$ because the signature Σ can always be inferred from the signature of the model M. For more on institutions see [23].

Dual Notions. The *dual of an indexed category* $H : \mathcal{I}^{\mathrm{op}} \to \mathsf{CAT}$ is the indexed category $H^{\mathrm{co}} : \mathcal{I}^{\mathrm{op}} \to \mathsf{CAT}$ given by $H^{\mathrm{co}}(u) = (Hu)^{\mathrm{op}} : (HJ)^{\mathrm{op}} \to (HI)^{\mathrm{op}}$ for $u : I \to J$. Note that the dual H^{co} of H as an indexed category is different from the dual H^{op} of H as a functor.

C   Fibrations

We briefly recall the basic notions of fibred category. This material is only needed for Definition B 6 and Proposition B 7. For more on fibred categories see [14, 7, 22].

A functor $H : \mathcal{I}^{\mathrm{op}} \to \mathsf{CAT}$ is called an indexed category, since for each 'index' $I \in \mathcal{I}$ there is a category $H(I)$ and for each arrow $u : I \to J$ in $\mathcal{I}$ there is a functor $H(u) : H(J) \to H(I)$ 'reindexing' objects in $H(J)$ along u.

Alternatively, one can use a functor $p : \mathcal{A} \to \mathcal{I}$ to describe how objects of $\mathcal{A}$ are indexed over objects of $\mathcal{I}$. Consider an arrow $u : I \to J$ in $\mathcal{I}$. We say that A is over I [f is over u] if $p(A) = I$ [$p(f) = u$] and that f is vertical if $p(f)$ is the identity. $\mathcal{A}$ is called the total category and the fibre over I is the category consisting of the objects over I and the vertical arrows between them.

Using this terminology, we define a fibration over $\mathcal{I}$ to be a functor $p : \mathcal{A} \to \mathcal{I}$ such that for all $u : I \to J$ in $\mathcal{I}$ and all A over J there is an object $u^*(A)$ and a so-called cartesian arrow $\bar{u} : u^*(A) \to A$ over u, that is, for all f over u there is a unique vertical g such that $\bar{u} \circ g = f$. Moreover, cartesian arrows are required to be closed under composition.

A fibration is called a split fibration if it comes equipped with a choice $u^*(A) \to A$ of cartesian arrow for each u and A and, moreover, $(id_I)^*$ is the identity functor and $(u \circ v)^* = v^* \circ u^*$.

There is an equivalence between functors $\mathcal{I}^{\mathrm{op}} \to \mathsf{CAT}$ and split fibrations over $\mathcal{I}$. Given a split fibration $p : \mathcal{A} \to \mathcal{I}$, define $H(I)$ to be the fibre over I and, for $u : I \to J$, let $H(u) = u^*$. Given a functor $H : \mathcal{I}^{\mathrm{op}} \to \mathsf{CAT}$ the corresponding fibration $p : \mathcal{A} \to \mathcal{I}$ is obtained via the Grothendieck construction as follows. The objects of $\mathcal{A}$ are pairs $(I, A), I \in \mathcal{I}, A \in H(I)$. Morphisms of $\mathcal{A}$ are pairs $(u, f) : (I, A) \to (J, B)$ with $u : I \to J \in \mathcal{I}$ and $f : A \to (Hu)(B)$. $p : \mathcal{A} \to \mathcal{I}$ is the first projection.

A fibred functor $F : p \to p'$ between fibrations $p : \mathcal{A} \to \mathcal{I}$, $p' : \mathcal{A}' \to \mathcal{I}$ is a functor $F : \mathcal{A} \to \mathcal{A}'$ such that $p'F = p$ and F preserves cartesian liftings. A fibred natural transformation between fibred functors is a natural transformation which has vertical components.

Dual Notions. The functor $q : \mathcal{A} \to \mathcal{I}$ is a (split) cofibration (over I) iff $q^{\mathrm{op}} : \mathcal{A}^{\mathrm{op}} \to \mathcal{I}^{\mathrm{op}}$ is a (split) fibration. There is an equivalence between indexed categories over $\mathcal{I}$ and split cofibrations over $\mathcal{I}^{\mathrm{op}}$. Given a functor $H : \mathcal{I}^{\mathrm{op}} \to \mathsf{CAT}$ the corresponding cofibration $q : \mathcal{A} \to \mathcal{I}^{\mathrm{op}}$ is obtained via the co-Grothendieck construction as follows. The objects of $\mathcal{A}$ are pairs $(I, A), I \in \mathcal{I}, A \in H(I)$. The morphisms of $\mathcal{A}$ are pairs $(u, f) : (J, B) \to (I, A)$ with $u : J \to I \in \mathcal{I}^{\mathrm{op}}$ and $f : (Hu)(B) \to A$.

Note that, given an indexed category H, the corresponding fibration p and cofibration q have identical fibres but the non-vertical arrows are generally different. The cofibration obtained from H^{co} is dual to the fibration obtained from H.

For examples of institutions which are more naturally associated with cofibrations than fibrations, so called co-institutions, see [18].

Combining Specification Formalisms in the 'General Logic' of Multialgebras

Yngve Lamo[1] and Michał Walicki[2]

[1] Bergen University College, Bergen, Norway
yla@hib.no
[2] Dept. of Informatics, University of Bergen, Bergen, Norway
michal@ii.uib.no

Abstract. We recall basic facts about the institution of multialgebras, $\mathcal{MA}$, and introduce a new, quantifier-free reasoning system for deriving consequences of multialgebraic specifications. We then show how $\mathcal{MA}$ can be used for combining specifications developed in other algebraic frameworks. We spell out the definitions of embeddings of institution of partial algebras, $\mathcal{PA}$, and membership algebras, $\mathcal{MEMB}$ into $\mathcal{MA}$. We also show an alternative relation, namely, institution transformation of $\mathcal{PA}$ into $\mathcal{MA}$ and discuss its role as compared to the embedding.

1 Introduction

Multialgebras provide a powerful algebraic framework for specification – primarily, but not exclusively, of nondeterministic behavior [5, 21, 20]. A nondeterministic operation returns the set of all possible outcomes, so it is interpreted as a function from the carrier to the powerset of the carrier. We follow, and in few places generalize, the definitions of multialgebras from earlier works like [20, 19]. We summarize earlier results on multialgebras in section 2 leading to the fact that they form an exact institution $\mathcal{MA}$, [15]. This result underlies the study of parameterized specifications and specification of parameterized data types, reported in [8, 9] – it will be used as the current paper addresses the related issue of composing specifications. Section 2 ends with the presentation of a new, quantifier-free, sound and strongly complete logic for multialgebraic specifications. We thus obtain a general logic of multialgebras (in the sense of [11]). The central focus of the paper, however, is neither nondeterminism nor reasoning but, instead, the possibilities offered by $\mathcal{MA}$ to combine different algebraic specification formalisms in one framework (where the reasoning system can, of course, faciliate proving consequences of the combined specifications).

Section 3 presents embeddings of institutions $\mathcal{PA}$ of partial algebras and $\mathcal{MEMB}$ of membership algebras into $\mathcal{MA}$. We do it by means of an example in which a $\mathcal{PA}$-specification **Set** and a $\mathcal{MEMB}$-specification **Nat** are embedded into $\mathcal{MA}$, the former is augmented with nondeterministic choice $\sqcup$, and then extended to a parameterized specification $\sqcup$**Set[El]**; finally, the imported **Nat** specification is passed as an actual parameter. We thus illustrate the potential

M. Wirsing, D. Pattinson, and R. Hennicker (Eds.): WADT 2002, LNCS 2755, pp. 328–342, 2003.
© Springer-Verlag Berlin Heidelberg 2003

of combining specifications from various algebraic formalisms in a unified framework. We also indicate in the concluding section 4 the possibility of extending the model class of $\mathcal{PA}$ (and $\mathcal{MEMB}$) specifications with nondeterminism using institution transformations, [10], and refer to related work [7, 6] on how this can be utilized for a smooth introduction of flexible error recovery strategies into $\mathcal{PA}$-specifications.

For the reason of space limitations we have to assume the reader to be familiar not only with the general background on algebraic specifications and category theory, but also with the institutions and their mappings. We use the definitions and notation from [11].

2   The Institution of Multialgebras

We summarize the relevant notions about multialgebras (see e.g., [21]; [6] contains the proofs concerning the institution of multialgebras). The algebraic signature, $\Sigma = (\mathbf{S}, \Omega)$, and terms over Σ with variables from a set X, $\mathcal{T}_{\Sigma,X}$, are defined in the usual way. A multialgebra for a signature Σ is an algebra where operations may be set-valued.

Definition 1. *A multialgebra A for $\Sigma = (\mathbf{S}, \Omega)$ is given by:*

- *a set s^A, the carrier set, for each sort symbol $s \in \mathbf{S}$*
- *a subset $c^A \in \mathcal{P}(s^A)$, for each constant, $c :\to s$*
- *an operation $\omega^A : s_1{}^A \times \cdots \times s_k{}^A \to \mathcal{P}(s^A)$ for each symbol $\omega : s_1 \times \cdots \times s_k \to s \in \Omega$, where $\mathcal{P}(s^A)$ denotes the power set of s^A. Composition of operations is defined by pointwise extension, i.e., $f^A(g^A(x)) = \bigcup_{y \in g^A(x)} f^A(y)$.*

The disjoint union of the carrier set(s) of a multialgebra A is denoted by $|A|$. One sometimes demands that constants and operations are total [21, 19], i.e. never return the empty set – we will not make this assumption. An operation is total if it returns nonempty result set for all arguments (a partial operation returns empty result set for some arguments). An operation returning not more than one value for any argument is deterministic (a nondeterministic operation returns more than one value for some arguments). So an operation that is total and deterministic is a function.

We generalize earlier work by allowing not only the result sets of operations but also the carrier sets to be empty. Note that for a constant $c \in \Omega$, c^A denotes a (sub)set of the carrier s^A. Thus constants can be used for unary predicates (as will be done in 3.3, when we relate membership algebras with multialgebras).

We use weak homomorphisms of multialgebras (see [18] for alternatives).

Definition 2. *Given two Σ-multialgebras A and B, a (weak) homomorphism $h : A \to B$ is a set of functions $h_s : s^A \to s^B$ for each sort $s \in \mathbf{S}$, such that:*

- *$h_s(c^A) \subseteq c^B$, for each constant $c :\to s$*
- *$h_s(\omega^A(a_1, \ldots, a_n)) \subseteq \omega^B(h_{s_1}(a_1), \ldots, h_{s_n}(a_n))$, for each operation $\omega : s_1 \times \cdots \times s_n \to s \in \Omega$ and for all $a_i \in s_i^A$.*

The two definitions give us, for a signature Σ, the category of Σ-multialgebras, $\mathbf{MAlg}_\Sigma$, with Σ-multialgebras as objects and weak homomorphisms as arrows.

Multialgebraic specifications are written using the atomic predicates of "deterministic" equality and set inclusion. Since even atomic multialgebraic theories do not, in general, possess initial models, there seems to be little reason to restrict the formulae to Horn clauses and so we allow general sequents.

Definition 3. *Given a signature Σ and a set of variables X, the set $\mathcal{F}_{\Sigma,X}$ of well formed formulae is given by:*

1. *Atoms: if $t, t' \in \mathcal{T}_{\Sigma,X}$ then:*
   - *$t \doteq t' \in \mathcal{F}_{\Sigma,X}$ (equality) – t and t' denote the same one-element set.*
   - *$t \prec t' \in \mathcal{F}_{\Sigma,X}$ (inclusion) – the set t is included in the set t'.*
2. *$\Gamma \to \Delta \in \mathcal{F}_{\Sigma,X}$, when Γ, Δ are finite (possibly empty) sets of atoms.*

Occasionally, we may write $\neg\phi$, where ϕ is an atom, for $\phi \to$. Notice that we do not (have to) attach the variable contexts to formulae – empty carriers are treated differently, mainly by a slight modification of the notion of assignment.

Definition 4. *Given a set of variables X and a multialgebra A, an assignment α is a function:*

$$\alpha : X \to |A| \uplus \{\emptyset\}, \text{ where } \alpha(x_s) = \emptyset \iff s^A = \emptyset$$

The subscript at x_s indicates the sort of the variable, so that x_s is assigned $\emptyset$ iff the sort s^A is empty. Otherwise variables are assigned individual elements – not sets thereof! (For convenience, we identify 1-element sets and individual elements.) Unlike the standard definitions, our definition guarantees the existence of an assignment even if the carrier $|A|$ is empty. Then any term with variables from empty sort will be empty, since operations in multialgebra applied to the empty set yield the empty set.

In the standard way, an assignment α to A induces a unique interpretation of any term t in A, denoted $\alpha(t)$. Now, satisfaction of formulae in A is defined first relatively to an assignment, written $\langle A, \alpha \rangle$.

Definition 5. *The satisfaction relation $\models$ in a multialgebra A is defined by:*

1. *$\langle A, \alpha \rangle \models t \prec t'$ iff $\alpha(t) \subseteq \alpha(t')$;*
2. *$\langle A, \alpha \rangle \models t \doteq t'$ iff $\alpha(t) = \{e\} = \alpha(t')$, for some $e \in |A|$;*
3. *$\langle A, \alpha \rangle \models \Gamma \to \Delta$ iff $(\exists \delta \in \Delta : \langle A, \alpha \rangle \models \delta \vee \exists \gamma \in \Gamma : \langle A, \alpha \rangle \not\models \gamma)$;*
4. *$A \models \phi$ iff $\forall \alpha : Var(\phi) \to A : \langle A, \alpha \rangle \models \phi$, where $Var(\phi)$ denotes all variables in ϕ*

Putting all the definitions together, we obtain the institution $\mathcal{MA}$, [6]. It has the following property, which is used in defining the semantics of parameterized specifications:

Proposition 1. *The model functor $\mathrm{Mod}_{\mathcal{MA}} : \mathbf{Sign}^{op} \to \mathbf{Cat}$ is finitely continuous.*

Since **Sign** is (finitely) cocomplete, we obtain that $\mathcal{MA}$ is an exact institution, [15]. Consequently, it satisfies the amalgamation lemma (for its formulation and proof the reader is referred to [17], where exact institutions are called institutions with composable signatures).

2.1   The Gentzen Proof System

We give a quantifier free Gentzen style proof system $\mathcal{G}$ for multialgebraic specifications. First a notational convention.

Remark 1. According to point 2 in Definition 5, an equality may hold only if the carrier (of the respective sort) is non-empty. Given an algebra A, we have that:

$$A \models x_s \doteq x_s \iff s^A \neq \emptyset \text{ and } A \models \neg(x_s \doteq x_s) \iff s^A = \emptyset$$

Also, if $s^A = \emptyset$, we have for any terms t_s, t'_s: $A \models t_s \prec t'_s$ and $A \not\models \neg(t_s \prec t'_s)$

We introduce logical symbols abbreviating the formulae stating that a carrier is empty or not.

Definition 6. *We define the symbols* $\mathcal{E}_s \equiv \neg(x_s \doteq x_s)$, *for any* $x_s \in X_s$, *and* $\neg\mathcal{E}_s \equiv x_s \doteq x_s$, *for any* $x_s \in X_s$. *By Remark 1, for any algebra* A:

$$A \models \mathcal{E}_s \iff s^A = \emptyset \text{ and } A \models \neg\mathcal{E}_s \iff s^A \neq \emptyset$$

The axioms and rules of the system $\mathcal{G}$ are given below. $\mathcal{G}$ allows us to derive sequents from a set Θ of sequents. (We also allow $\mathcal{E}_s$ as additional ground atomic formulae.) For deriving only multialgebraic tautologies, the system $\mathcal{G}^-$ containing only the logical axioms, the replacement and the expansion rules will suffice. For strong completeness, however, we need also the additional specific cut rules originating from [16]. Notice that these are much more tractable than the general cut, since they give a specific prescription as to what cut-formulae can be used in a bottom-up proof construction. These rules allow us to dispense with the quantifiers when proving consequences of axiomatic theories. (In [2], proving ϕ from a specification Θ required translation of the whole into a first-order formula, corresponding to $\forall(\Theta) \rightarrow \phi$. Besides the translation work, it required the use of quantifiers and limited the result to finite Θ only.) On the other hand, Definition 4 of assignment and Remark 1 explain that quantifiers can be avoided also when dealing with empty carriers. Except for the presence of the atomic predicate $\doteq$, these are the most significant improvements with respect to the closely related system from [2]. For the details concerning the system below, the reader is referred to [6].

Axioms

$$\Gamma \rightarrow x \prec x, \Delta \ : x \in X \qquad\qquad \Gamma, \gamma \rightarrow \gamma, \Delta \qquad\qquad \Gamma, \mathcal{E}_s \rightarrow t_s \prec t'_s, \Delta$$

Replacement rules

$$\frac{\Gamma, x \prec t \rightarrow \Delta, x \prec t'}{\Gamma \rightarrow \Delta, t \prec t'}$$
$t \notin X$, and $x \in X$ is fresh

$$\frac{\Gamma \rightarrow \Delta, x \prec t \mid \Gamma, x \prec t' \rightarrow \Delta}{\Gamma, t \prec t' \rightarrow \Delta}$$
$t \notin X$ and $x \in X$ arbitrary

$$\frac{\Gamma \rightarrow \Delta, y \prec t \mid \Gamma \rightarrow \Delta, x \prec f(\ldots, y, \ldots)}{\Gamma \rightarrow \Delta, x \prec f(\ldots, t, \ldots)}$$
where $y \in X$ arbitrary and $t \notin X$

$$\frac{\Gamma, y \prec t, x \prec f(\ldots, y, \ldots) \rightarrow \Delta}{\Gamma, x \prec f(\ldots, t, \ldots) \rightarrow \Delta}$$
where $y \in X$ is fresh and $t \notin X$

$$\frac{\Gamma \to \Delta, t \doteq x \mid \Gamma \to \Delta, t' \doteq x}{\Gamma \to \Delta, t \doteq t'}$$
$$t, t' \notin X \text{ and } x \in X \text{ arbitrary}$$

$$\frac{\Gamma, t_s \doteq x_s, t'_s \doteq x_s \to \Delta}{\Gamma, t_s \doteq t'_s \to \Delta}$$
$$t_s, t'_s \notin X \text{ and } x_s \in X \text{ is fresh}$$

$$\frac{\Gamma \to \Delta, t_s \prec x_s \mid \Gamma \to \Delta, x_s \prec t_s \mid \Gamma, \mathcal{E}_s \to \Delta}{\Gamma \to \Delta, t_s \doteq x_s}$$
$$\text{where } x_s \in X \text{ and } t_s \neq x_s$$

$$\frac{\Gamma, t_s \prec x_s, x_s \prec t_s \to \Delta, \mathcal{E}_s}{\Gamma, t_s \doteq x_s \to \Delta}$$
$$\text{where } x_s \in X \text{ and } t_s \neq x_s$$

$$\frac{\Gamma \to \Delta, t_s \prec x_s \mid \Gamma \to \Delta, x_s \prec t_s \mid \Gamma, \mathcal{E}_s \to \Delta}{\Gamma \to \Delta, x_s \doteq t_s}$$
$$\text{where } x_s \in X \text{ and } t_s \neq x_s$$

$$\frac{\Gamma, t_s \prec x_s, x_s \prec t_s \to \Delta, \mathcal{E}_s}{\Gamma, x_s \doteq t_s \to \Delta}$$
$$\text{where } x_s \in X \text{ and } t_s \neq x_s$$

Expansion rules

$$\frac{\Gamma, y \prec f(\bar{z}) \to \Delta}{\Gamma, y \prec x, x \prec f(\bar{z}) \to \Delta}$$
$$(\text{sound for arbitrary } x \in X)$$

$$\frac{\Gamma, x \prec y \to \Delta}{\Gamma, y \prec x \to \Delta}$$

$$\frac{\Gamma, z \prec f(\dots, y, \dots) \to \Delta}{\Gamma, y \prec x, z \prec f(\dots, x, \dots) \to \Delta}$$
$$(\text{sound for arbitrary } x \in X)$$

$$\frac{\Gamma, y \prec z \to \Delta}{\Gamma, y \prec x, x \prec z \to \Delta}$$

$$\frac{\Gamma, \mathcal{E}_s, x_{s'} \prec f(\dots, y_s, \dots), \mathcal{E}_{s'} \to \Delta}{\Gamma, \mathcal{E}_s, x_{s'} \prec f(\dots, y_s, \dots) \to \Delta}$$

Specific cut rules (relative to the non-logical axioms Θ)

$$\frac{\Gamma \to \Delta, \gamma'_1 \mid \dots \mid \Gamma \to \Delta, \gamma'_n \mid \Gamma, \delta'_1 \to \Delta \mid \dots \mid \Gamma, \delta'_m \to \Delta}{\Gamma \to \Delta}$$
for each axiom $\gamma_1, \dots, \gamma_n \to \delta_1, \dots, \delta_m \in \Theta$, with arbitrary renaming $'$ of variables

Writing $\mathcal{G}^-$ for the above system $\mathcal{G}$ without the specific cut rules, we have:

Proposition 2. *The systems are sound and complete – for every sequent ϕ:*

- $\vdash_{\mathcal{G}^-} \phi \iff \models \phi$
- *For every set of axioms* $\Theta : \Theta \vdash_{\mathcal{G}} \phi \iff \Theta \models \phi$.

3   Other Specification Frameworks in $\mathcal{MA}$

In this section we show the embeddings of two institutions – $\mathcal{PA}$ of partial algebras and $\mathcal{MEMB}$ of membership algebras – into $\mathcal{MA}$. By means of an example, we illustrate the power of $\mathcal{MA}$ for reuse and combination of specifications written in other algebraic formalisms. We start by taking a $\mathcal{PA}$-specification of sets and embed it to $\mathcal{MA}$ in 3.1; in 3.2 we extend the resulting specification of sets with a nondeterministic choice operation; in 3.3 we embed a $\mathcal{MEMB}$-specification of natural numbers to $\mathcal{MA}$; in 3.4 we extract a (sub)specification of elements from the specification of sets with nondeterministic choice, obtaining a parameterized specification, and we use the result of the embedding of naturals from $\mathcal{MEMB}$ as an actual parameter. Thus, we have combined $\mathcal{PA}$ and $\mathcal{MEMB}$ specifications in $\mathcal{MA}$, moreover we have extended the specifications with nondeterminism. Outline of the following example:

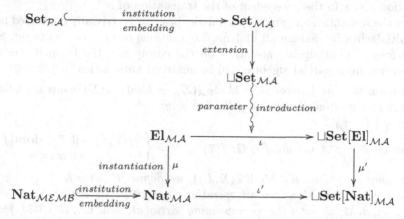

3.5 shows another way of including $\mathcal{PA}$ in $\mathcal{MA}$ and compares it to the embedding.

3.1 Partial Algebras and Multialgebras

We recall the basics of the institution of partial algebras, $\mathcal{PA}$ (e.g., [3, 4, 13]).

Definition 7. *A partial algebra A for a signature $\Sigma = (\mathbf{S}, \Omega)$ is given by:*

- *A set s^A for each sort $s \in \mathbf{S}$*
- *A partial function $\omega^A : s_1{}^A \times \cdots \times s_n{}^A \to s^A$, for each $\omega : s_1 \times \cdots \times s_n \to s \in \Omega$*

Given two partial algebras A and B, a (weak) homomorphism $h : A \to B$ is a set of total functions $h_s : s^A \to s^B$, for each sort $s \in \mathbf{S}$ such that:

- *$h_s(\omega^A(x_1, \ldots, x_n)) = \omega^B(h_{s_1}(x_1), \ldots, h_{s_n}(x_n))$ for each $\omega : s_1 \times \cdots \times s_n \to s \in \Omega$ and arguments $x_1, \ldots, x_n$, whenever $\omega^A(x_1, \ldots, x_n)$ is defined.*

A weak homomorphism h can be equivalently described as an ordinary homomorphism such that for each operation $\omega \in \Omega : h(\mathbf{dom}(\omega^A)) \subseteq \mathbf{dom}(\omega^B)$, where $\mathbf{dom}$ identifies ω's definition domain in a given algebra A.

Definition 8. *Formulae are universally quantified Horn clauses over existential equations, $t \overset{e}{=} t'$. Satisfaction by a partial algebra A is defined as follows (let $\alpha : X \to |A|$ range over standard assignments (total functions)):*

- *$\langle A, \alpha \rangle \models (X; t \overset{e}{=} t')$ iff $\alpha(t)$ and $\alpha(t')$ are defined and $\alpha(t) = \alpha(t')$*
- *$\langle A, \alpha \rangle \models (X; a_1, \ldots, a_n \to a)$ iff $\exists i : 1 \leq i \leq n : \langle A, \alpha \rangle \not\models a_i$ or $\langle A, \alpha \rangle \models a$*
- *$A \models (X; \phi)$ iff $\forall \alpha : X \to |A| : \langle A, \alpha \rangle \models (X; \phi)$*

Embedding $\mathcal{PA}$ into $\mathcal{MA}$. To relate multialgebras with partial algebras we observe the similarities between the element equality $\doteq$ and the existential equality $\overset{e}{=}$. In a multialgebra, the equality does not hold if one side is interpreted by a set with cardinality greater than one or by the empty set. The existential equality does not hold in a partial algebra if one side of the equality sign is undefined. These similarities suggest the straightforward translation: replace $\overset{e}{=}$ by $\doteq$. Each x in the variable context X in $(X; \phi)$, will give rise to an extra condition $x \doteq x$ in the antecedent of the translation of ϕ [1].

On the semantic side, any Σ-partial algebra can be trivially converted into a Σ-multialgebra by making all undefined operations return the empty set. Since operations in multialgebra are strict on the empty set, the implicit strictness assumption from partial algebras will be enforced automatically.

Definition 9. *The functor* $\beta^- : \mathsf{Mod}_{\mathcal{PA}}(\Sigma) \to \mathsf{Mod}_{\mathcal{MA}}(\Sigma)$ *maps a partial algebra A to a multialgebra in the following way:*

- $|\beta^-(A)| = |A|$
- *for all $\overline{x} \in |\beta^-(A)|$ and $f \in \Omega$:* $f(\overline{x})^{\beta^-(A)} = \begin{cases} \{f(\overline{x})^A\} & \text{–if } \overline{x} \in \mathsf{dom}(f^A) \\ \emptyset & \text{–otherwise} \end{cases}$

For a homomorphism: $h \in \mathsf{Mod}(\Psi(\Sigma, \Gamma))$, we define $\beta^-(h) = h$.
For a multialgebra M where all operations are either deterministic or return empty set, $\beta(M)$ denotes the corresponding partial algebra, i.e., $\beta^-(\beta(M)) = M$.

Saying that a multialgebra and a partial algebra are "essentially the same", we will mean that they are obtained from each other by means of $\beta(\_)$, resp. $\beta^-(\_)$.

The embedding of $\mathcal{PA}$ into $\mathcal{MA}$ is now obtained by augmenting the partial algebra specification with additional axioms forcing all operations to return either a unique element or the empty set. This is the underlying model in partial algebras which in the generalized context of $\mathcal{MA}$ need explicit axioms. For an operation $f(\overline{x})$, the axiom forcing it to be empty or deterministic is of the form $y \doteq y, y \prec f(\overline{x}) \to f(\overline{x}) \doteq f(\overline{x})$, where y is a fresh variable. The example below shows the embedding of the $\mathcal{PA}$-specification of sets (it is the constant $\{\}$ which can be partial in this example).

$\mathsf{Set}_{\mathcal{PA}} =$		$\mathsf{Set}_{\mathcal{MA}} =$	
$\mathbf{S} : Set, El$		$\mathbf{S} : Set, El$	
$\Omega : \{\} :$	$\to Set$	$\Omega : \{\} :$	$\to Set$
$\circ : El \times Set \to Set$		$\circ : El \times Set \to Set$	
$\Theta : 1.\ \{x, y, S\};$		$\Theta : 1.\ x \doteq x, y \doteq y, S \doteq S$	
$x \circ (y \circ S) \overset{e}{=} y \circ (x \circ S)$		$\to x \circ (y \circ S) \doteq y \circ (x \circ S)$	
$2.\ \{x, S\};$		$2.\ x \doteq x, S \doteq S$	
$x \circ (x \circ S) \overset{e}{=} x \circ S$		$\to x \circ (x \circ S) \doteq x \circ S$	
		$3.\ y \doteq y, y \prec x \circ S \to x \circ S \doteq x \circ S$	
		$4.\ y \doteq y, y \prec \{\} \to \{\} \doteq \{\}$	

---

[1] For the use with the logic $\mathcal{G}$, we would instead add, to the consequent of the translated ϕ, the disjunct $\mathcal{E}_s$ for each sort s for which there is a variable $x_s \in X$. I.e., $(\{x_s\}, \Gamma \to \alpha)$ would become $\Gamma \to \alpha, \mathcal{E}_s$, rather than $x_s \doteq x_s, \Gamma \to \alpha$. But we retain this later notation to emphasize that all translated formulae remain Horn clauses.

The following lemma shows that the axioms of the above form (e.g., 3., 4.) are sufficient – multialgebras satisfying such axioms are "essentially" partial algebras.

Lemma 1. *Let $SP = (S, \Omega, \Theta)$ be a specification in $\mathcal{MA}$ such that, for each operation $f : \overline{s} \to s \in \Omega : SP \models y \doteq y, y \prec f(\overline{x}) \to f(\overline{x}) \doteq f(\overline{x})$, where y is distinct from all $\overline{x}$. Then in any $M \in \mathrm{Mod}_{\mathcal{MA}}(SP)$ we have that for all $\overline{x} \in |M| : f(\overline{x})^M = \emptyset$ or $f(\overline{x})^M$ is deterministic.*

Proof. Let $M \in \mathrm{Mod}_{\mathcal{MA}}(SP)$ and $\alpha : \{y, \overline{x}\} \to M$ be an assignment. Two cases:

1. $\langle M, \alpha \rangle \not\models y \doteq y$: then the carrier set of s is empty so $\alpha(f(\overline{x}))^M = \emptyset$.
2. $\langle M, \alpha \rangle \models y \doteq y$: if $\langle M, \alpha \rangle \not\models f(\overline{x}) \doteq f(\overline{x})$ then, since $\langle M, \alpha \rangle \models y \doteq y, y \prec f(\overline{x}) \to f(\overline{x}) \doteq f(\overline{x})$, we must have that $\langle M, \alpha \rangle \not\models y \prec f(\overline{x})$, i.e., $\alpha(f(\overline{x}))^M = \emptyset$.

$(\alpha(f(\overline{x}))^M$ may be $\emptyset$ also when some of $\overline{x}$ range over empty sorts, but this case is covered by point 2. Thus we do not need additional conditions $x \doteq x$ for $x \in \overline{x}$.)

$\square$

The lemma and the above discussion lead to the main fact:

Proposition 3. *There is an embedding (Ψ, α, β) of institutions from $\mathcal{PA}$ to $\mathcal{MA}$; the embedding is a simple map.*

- *The functor $\Psi : \mathrm{Sign}_{\mathcal{PA}} \to \mathbf{Th}_{0}\mathcal{MA}$ is given by: $\Psi(S, \Omega) = ((S, \Omega), \Theta_\Sigma)$, where $\Theta_\Sigma = \{y \doteq y, y \prec f(\overline{x}) \to f(\overline{x}) \doteq f(\overline{x}) : f \in \Omega\}$.*
  For morphisms $\Psi(\mu_S, \mu_\Omega)$ is the identity.
- *The natural transformation $\alpha : \mathrm{Sen}_{\mathcal{PA}} \to \Psi; \mathrm{Sen}_{\mathcal{MA}}$ is given by:*
  - *$\alpha(t \overset{e}{=} t') \equiv t \doteq t'$ – auxiliary definition for atoms*
  - *$\alpha(\{x_1 ... x_k\}; a_1 ... a_n \to a) \equiv x_1 \doteq x_1 ... x_k \doteq x_k, \alpha(a_1) ... \alpha(a_n) \to \alpha(a)$*
  Ψ is extended to a functor $\Psi : \mathbf{Th}_{0\mathcal{PA}} \to \mathbf{Th}_{0\mathcal{MA}}$ by: $\Psi(\Sigma, \Theta) = (\Sigma, \Theta_\Sigma \cup \alpha_\Sigma(\Theta))$.
- *The components of the natural transformation $\beta : \Psi^{op}; \mathrm{Mod}_{\mathcal{MA}} \to \mathrm{Mod}_{\mathcal{PA}}$ are β's from Definition 9, i.e.:*
  - *$|\beta_\Sigma(M')| = |M'|$*
  - *$f(x_1, \ldots, x_n)^{\beta_\Sigma(M')} = \begin{cases} undefined \text{ if } f(x_1, \ldots, x_n)^{M'} = \emptyset \\ x \text{ such that } f(x_1, \ldots, x_n)^{M'} = \{x\} \text{ otherwise} \end{cases}$*
  This is a well defined partial algebra by lemma 1. For a homomorphism: $h \in \mathrm{Mod}_{\mathcal{MA}}(\Psi(\Sigma, \Theta))$, we define $\beta_\Sigma(h) = h$.

We also have an immediate consequence of the above construction:

Proposition 4. *For a $\mathcal{PA}$ theory (Σ, Θ), the functor $\beta_{(\Sigma, \Theta)}$ is an equivalence (in fact, an isomorphism) of categories $\mathrm{Mod}_{\mathcal{MA}}(\Psi(\Sigma, \Theta))$ and $\mathrm{Mod}_{\mathcal{PA}}(\Sigma, \Theta)$.*

Proof. The inverse functor $\beta^{-}_{(\Sigma, \Theta)}$ sends a partial algebra $P \in \mathrm{Mod}_{\mathcal{PA}}(\Sigma, \Theta)$ onto a multialgebra $M' \in \mathrm{Mod}_{\mathcal{MA}}(\Psi(\Sigma, \Theta))$ such that $\beta(M') = P$, i.e., it is β^{-} from Definition 9. One verifies easily the isomorphism condition. $\square$

As Mossakowski showed in [13], $\mathcal{PA}$ allows to specify exactly the finitely locally presentable categories [1], i.e. we have identified the sub-institution of $\mathcal{MA}$ allowing to specify these classes of models. Given a partial algebra specification

SP, we call $\Psi(SP)$ a multialgebra specification of *partial form* and denote this sub-institution $\mathcal{MAPA}$. As the finitely locally presentable categories have initial models, we have also obtained a sub-institution of $\mathcal{MA}$ admitting such models.

3.2   Extending Specifications with Nondeterminism

In some cases it may be desirable to extend a specification with nondeterministic operations. A strategy for doing this is to embed the specification into $\mathcal{MA}$ and then adding the nondeterministic operations.

The obtained specification **Set** is extended with nondeterministic choice $\sqcup$.

$\sqcup\mathbf{Set}_{\mathcal{MA}} =$
$\mathbf{S} : Set, El$
$\Omega : \{\} : \qquad\qquad \to Set$
$\qquad \circ : El \times Set \to Set$
$\qquad \sqcup : \qquad Set \to El$
$\Theta : 1.\ x \doteq x, y \doteq y, S \doteq S \to x \circ (y \circ S) \doteq y \circ (x \circ S)$
$\quad\ 2.\qquad x \doteq x, S \doteq S \to x \circ (x \circ S) \doteq x \circ S$
$\quad\ 3.\qquad y \doteq y, y \prec x \circ S \to x \circ S \doteq x \circ S$
$\quad\ 4.\qquad y \doteq y, y \prec \{\} \to \{\} \doteq \{\}$
$\quad\ 5.\qquad z \prec \sqcup(x \circ S) \to z \doteq x, z \prec \sqcup(S)$

$\boxed{\begin{array}{l} El^A = \{a, b, c\} \\ Set^A = \mathcal{P}^{fin}(|El|^A) \\ \{\}^A = \{\emptyset\} \\ x \circ^A S = \{x\} \cup S \\ \sqcup^A S = S \end{array}}$

A possible model A, shown on the right, has total and deterministic operations $\{\}$ and $\circ$, while $\sqcup(\{\})$ is undefined and $\sqcup(S)$ for non-empty S is nondeterministic[2].

3.3   Membership Algebras and Multialgebras

We now repeat the steps made for $\mathcal{PA}$ for the institution $\mathcal{MEMB}$ of membership algebras, [12]. We will use notation corresponding to the rest of the paper, which slightly differs from that used in [12].

Definition 10. *A (membership) signature Σ is a quadruple $\Sigma = (\mathbf{S}, \Omega, \Pi, \pi)$, where $(\mathbf{S}, \Omega)$ is a standard signature, Π is a set of sub-sort predicate names, and π is a function $\pi : \Pi \to \mathbf{S}$.*

The function π labels sub-sort predicate symbols by sort symbols – its intention is to identify a predicate p with $\pi(p) = s$ as a sub-sort of sort s.

Definition 11. *A signature morphism between two membership signatures $\Sigma = (\mathbf{S}, \Omega, \Pi, \pi)$ and $\Sigma' = (\mathbf{S}', \Omega', \Pi', \pi')$ is a triple $\mu = (\mu_\Pi, \mu_\mathbf{S}, \mu_\Omega)$, where $\mu_\Pi : \Pi \to \Pi'$ and $\mu_\mathbf{S} : \mathbf{S} \to \mathbf{S}'$ are functions such that the following diagram*

---

[2] One should be wary of the possible confusion in this example: the elements of the sort of defined sets Set obtain in A the interpretation as the actual, semantic sets of elements, except that $\{\}^A$ is interpreted as a well defined element $\{\emptyset\}$. The latter must be distinguished from the $\emptyset$ returned by the undefined operation $\sqcup(\{\})^A$.

$$\begin{array}{ccc} \Pi & \xrightarrow{\mu_\Pi} & \Pi' \\ {\scriptstyle \pi}\downarrow & & \downarrow{\scriptstyle \pi'} \\ \mathbf{S} & \xrightarrow{\mu_\mathbf{S}} & \mathbf{S'} \end{array}$$

commutes, and there exists an ω' for each $\omega \in \Omega$ such that: $\mu_\Omega(\omega : \overline{s} \to s) = \omega' : \mu_\mathbf{S}(\overline{s}) \to \mu_\mathbf{S}(s)$.

Definition 12. *A membership algebra for a signature* $(\mathbf{S}, \Omega, \Pi, \pi)$ *is given by:*

- *a many sorted* $(\mathbf{S}, \Omega)$ *algebra* A, *together with*
- *an assignment of a subset* $p^A \subseteq \pi(p)^A$ *for each predicate name* $p \in \Pi$.

A membership algebra homomorphism $h : A \to B$ *is an ordinary homomorphism which, in addition, satisfies:* $h_{\pi(p)}(p^A) \subseteq p^B$.

Definition 13. *The axioms used for specifying classes of membership algebras are universally quantified Horn clauses over atomic formulae:*

- *equations,* $t = t'$,
- *membership assertions of the form* $t : p$, *for* $p \in \Pi$ *and* $t \in (\mathcal{T}_{\Sigma,X})_{\pi(p)}$.

The assignments to variables are as usual, and the satisfaction is defined as follows:

Definition 14. *Let* $\alpha : X \to |A|$ *range over standard assignments.*

1. $\langle A, \alpha \rangle \models (X; t = t')$ *iff* $\alpha(t) = \alpha(t')$
2. $\langle A, \alpha \rangle \models (X; t : p)$ *iff* $\alpha(t) \in p^A$
3. $\langle A, \alpha \rangle \models (X; a_1, \ldots, a_n \to a)$ *iff* $\exists a_i : \langle A, \alpha \rangle \not\models (X; a_i)$ *or* $\langle A, \alpha \rangle \models (X; a)$
4. $A \models (X; \varphi)$ *iff* $\forall \alpha : X \to |A| : \langle A, \alpha \rangle \models (X; \varphi)$

It is shown in [12] that the above definitions yield a liberal institution of membership algebras $\mathcal{MEMB}$. Moreover there exists an embedding of institutions both ways between $\mathcal{MEMB}$ and the institution of many sorted Horn logic with predicates and equalities, i.e. these two institutions can be viewed as sub-logics of each other.

Embedding $\mathcal{MEMB}$ into $\mathcal{MA}$. is based on the fact that nondeterministic constants play the same role as unary predicates. Hence the membership relation $t : p$ is naturally translated as $t \prec p$. Making, in addition, all operations deterministic, one obtains the straightforward translation of $\mathcal{MEMB}$ specifications into $\mathcal{MA}$, as shown in the example below:

$\text{Nat}_{\mathcal{MEMB}} =$
$\mathbf{S} : Nat$
$\Pi :\ \pi(nat) = Nat$
$\qquad \pi(pos) = Nat$
$\Omega :\ zero : \qquad \to Nat$
$\qquad suc : Nat \to Nat$
$\qquad pred : Nat \to Nat$
$\Theta :$ 1. $\emptyset;$ $\qquad\qquad zero : nat$
$\quad$ 2. $\{x\};\ x : pos \to x : nat$
$\quad$ 3. $\{x\};\ x : nat \to pred(suc(x)) = x$
$\quad$ 4. $\{x\};\ x : pos \to pred(x) : nat$
$\quad$ 5. $\{x\};\ x : nat \to suc(x) : pos$

$\text{Nat}_{\mathcal{MA}} =$
$\mathbf{S} : Nat$
$\Omega :\ nat : \qquad \to Nat$
$\qquad pos : \qquad \to Nat$
$\qquad zero : \qquad \to Nat$
$\qquad suc : Nat \to Nat$
$\qquad pred : Nat \to Nat$
$\Theta :$ 1. $\qquad\qquad\quad zero \prec nat$
$\quad$ 2. $x \doteq x, x \prec pos \to x \prec nat$
$\quad$ 3. $x \doteq x, x \prec nat \to pred(suc(x)) \doteq x$
$\quad$ 4. $x \doteq x, x \prec pos \to pred(x) \prec nat$
$\quad$ 5. $x \doteq x, x \prec nat \to suc(x) \prec pos$
$\quad$ 6. $\qquad\qquad\quad zero \doteq zero$
$\quad$ 7. $\qquad\qquad\quad suc(x) \doteq suc(x)$
$\quad$ 8. $\qquad\qquad\quad pred(x) \doteq pred(x)$

Note that axioms 2., 3. and 4. could be written equivalently in $\mathcal{MA}$ as $pos \prec nat$; $pred(pos) \prec nat$; and $suc(nat) \prec pos$, i.e., using the syntax (and the intended, if not the formal, meaning) of unified algebras [14].

Proposition 5. *There is an embedding* (Φ, α, β) *of institutions from* $\mathcal{MEMB}$ *to* $\mathcal{MA}$; *the embedding is a simple map of institutions.*

- *The functor* $\Phi : \mathbf{Sign}_{\mathcal{MEMB}} \to \mathbf{Th}_{\mathcal{MA}}$ *is given by:*
  $\Phi(\mathbf{S}, \Omega, \Pi)$ *is the theory* $(\mathbf{S}, \Omega \uplus \Pi', \Theta_\Sigma)$, *where:*
  - $\Pi' = \{p :\to \pi(p) : p \in \Pi\}$ – *a new constant of sort* $\pi(p)$ *for each* $p \in \Pi$.
  - $\Theta_\Sigma = \{\omega(\overline{x}) \doteq \omega(\overline{x}) : \omega \in \Omega\}$ – *determinacy axiom for each operation.*
  $\Phi(\mu_S, \mu_\Omega, \mu_\Pi)$ *is the signature morphism* $(\mu_S, \mu_\Omega \uplus \mu_\Pi)$
- *The natural transformation* $\alpha : \mathsf{Sen}_{\mathcal{MEMB}} \to \Phi; \mathsf{Sen}_{\mathcal{MA}}$ *is given by:*
  - $\alpha(t : c) \equiv t \prec c$, *for each atom* $t : c$
  - $\alpha(t = t') \equiv t \doteq t'$, *for each atom* $t = t'$
  - $\alpha(\{x_1...x_k\}; a_1...a_n \to a) \equiv x_1 \doteq x_1...x_k \doteq x_k, \alpha(a_1)...\alpha(a_n) \to \alpha(a)$
  Φ *is extended to* $\Phi : \mathbf{Th}_{0\mathcal{MEMB}} \to \mathbf{Th}_{0\mathcal{MA}}$ *by:* $\Phi(\Sigma, \Theta) = \Phi(\Sigma) \cup \alpha_\Sigma(\Theta)).$
- *The natural transformation* $\beta : \Phi^{op}; \mathsf{Mod}_{\mathcal{MA}} \to \mathsf{Mod}_{\mathcal{MEMB}}$ *is essentially the identity on models and homomorphisms. For any* $M' \in \mathsf{Mod}_{\mathcal{MA}}(\Phi(\Sigma, \Theta))$
  - $|\beta(M')| = |M'|$
  - $f(x_1, \dots, x_n)^{\beta(M')} = x$ *such that* $f(x_1, \dots, x_n)^{M'} = \{x\}$
  - $p^{\beta(M')} = p^{M'}$
  For a homomorphism $h : M' \to B' \in \mathsf{Mod}_{\mathcal{MA}}(\Phi(\Sigma, \Theta))$, *we let* $\beta(h) = h$.

Proposition 6. *The functor* $\beta_{(\Sigma,\Theta)}$ *is an equivalence (in fact, an isomorphism) of categories* $\mathsf{Mod}_{\mathcal{MA}}(\Phi(\Sigma, \Theta))$ *and* $\mathsf{Mod}_{\mathcal{MA}}(\Sigma, \Theta)$ *for every* $\mathcal{MEMB}$ *theory* (Σ, Θ).

Proof. The inverse functor $\beta^{-1}_{(\Sigma,\Theta)}$ sends a membership algebra $M \in \mathsf{Mod}(\Sigma, \Theta)$ onto a multialgebra $M' \in \mathsf{Mod}(\Phi(\Sigma, \Theta))$ such that $\beta(M') = M$ (i.e., $|M'| = |M|$, $f^{M'}(\overline{x}) = \{f^M(\overline{x})\}$ and for $p \in \Pi : p^{M'} = p^M$.) One verifies easily the isomorphism condition. $\qquad\square$

3.4   Parameterized Specifications in $\mathcal{MA}$

Parameterized specifications and, in particular, specifications of parameterized data types in $\mathcal{MA}$, based on a generalization of the traditional persistency requirement, has been described in [8] and is further studied in [9]. In [9] we also address the issue of refinement of specifications based on introduction of additional structure into the specified programs. In the current context, we only claim that extracting a parameter from a flat specification can be seen as a refinement. We thus view now the specification $\sqcup\mathbf{Set}_{\mathcal{MA}}$ from 3.2 as a parameterized specification

$$\iota : \mathbf{El}_{\mathcal{MA}} \to \sqcup\mathbf{Set}[\mathbf{El}]_{\mathcal{MA}} \tag{1}$$

where the parameter specification $\mathbf{El}_{\mathcal{MA}}$ is simply the sort El.

Let $\mu : \mathbf{El}_{\mathcal{MA}} \to \mathbf{Nat}_{\mathcal{MA}}$ be the obvious specification morphism. In the standard way, we obtain the result of this parameter instantiation, $\sqcup\mathbf{Set}[\mathbf{Nat}]_{\mathcal{MA}}$, as the pushout in the category $\mathbf{Th}_{\mathcal{MA}}$ of multialgebraic specifications:

$$
\begin{array}{ccc}
\mathbf{El}_{\mathcal{MA}} & \xrightarrow{\;\;\iota\;\;} & \sqcup\mathbf{Set}[\mathbf{El}]_{\mathcal{MA}} \\
\mu\downarrow & & \downarrow\mu' \\
\mathbf{Nat}_{\mathcal{MA}} & \xrightarrow{\;\;\iota'\;\;} & \sqcup\mathbf{Set}[\mathbf{Nat}]_{\mathcal{MA}}
\end{array}
$$

Proposition 1 ensures the existence of well-defined semantics for such parameter instantiation in $\mathcal{MA}$. For the parameterized specification (1), one can even find a persistent functor $F : \mathsf{Mod}_{\mathcal{MA}}(\mathbf{El}) \to \mathsf{Mod}_{\mathcal{MA}}(\sqcup\mathbf{Set}[\mathbf{El}])$:

$$
\begin{array}{ll}
El^{\mathrm{F}(A)} = El^A & x \circ^{\mathrm{F}(A)} S = \{x\} \cup S \\
Set^{\mathrm{F}(A)} = \mathcal{P}^{fin}(El^A) & \sqcup^{\mathrm{F}(A)} S = S
\end{array}
$$

The model in 3.2 was obtained by applying this F to the algebra with $|El| = \{a, b, c\}$. Using Proposition 1, the persistent functor for the instantiated specification $\iota' : \mathbf{Nat}_{\mathcal{MA}} \to \sqcup\mathbf{Set}[\mathbf{Nat}]_{\mathcal{MA}}$ can be obtained from F by amalgamation.

3.5   Institution Transformation of $\mathcal{PA}$ into $\mathcal{MA}$

$\mathcal{MA}$ offers wider possibilities of reuse and combination than embeddings exemplified so far. Instead of embedding other institutions, we can transform them into $\mathcal{MA}$. The transformation (proposed in [10] as one of the primitive operations for constructing various mappings of institutions) amounts, roughly, to extending the model class – for $\mathcal{PA}$ it will be like the embedding from proposition 3 but *without* adding the determinacy axioms:

$$\text{Set}_{\mathcal{PA}} =$$
$$\mathbf{S} : Set, El$$
$$\Omega : \{\} : \qquad\qquad \to Set$$
$$\qquad \circ : El \times Set \to Set$$
$$\Theta : 1. \quad \{x, y, S\};$$
$$\qquad x \circ (y \circ S) \overset{e}{=} y \circ (x \circ S)$$
$$\quad 2. \quad \{x, S\};$$
$$\qquad x \circ (x \circ S) \overset{e}{=} x \circ S$$

$$\text{Set}'_{\mathcal{MA}} =$$
$$\mathbf{S} : Set, El$$
$$\Omega : \{\} : \qquad\qquad \to Set$$
$$\qquad \circ : El \times Set \to Set$$
$$\Theta : 1. \quad x \doteq x, y \doteq y, S \doteq S$$
$$\qquad \to \; x \circ (y \circ S) \doteq y \circ (x \circ S)$$
$$\quad 2. \quad x \doteq x, S \doteq S$$
$$\qquad \to \; x \circ (x \circ S) \doteq x \circ S$$

Embedding into $\mathcal{MA}$ allows one to augment existing operations with new, possibly nondeterministic ones, as was done in 3.2. Transforming, on the other hand, allows to extend the model class so that some old operations may acquire nondeterministic behavior. The resulting specification $\text{Set}'_{\mathcal{MA}}$ above has a larger model class than the transformed $\text{Set}_{\mathcal{PA}}$ – not only multialgebras which are "essentially" partial algebras, where $\{\}$ is undefined or deterministic, but also ones where it is nondeterministic.

Proposition 7. *There is an institution transformation* (Ψ, α, β) *of* $\mathcal{PA}$ *to* $\mathcal{MA}$.

- *The functor* $\Psi : \mathbf{Sign}_{\mathcal{PA}} \to \mathbf{Sign}_{\mathcal{MA}}$ *is identity on objects and morphisms.*
- *The natural transformation* $\alpha : \text{Sen}_{\mathcal{PA}} \to \Psi; \text{Sen}_{\mathcal{MA}}$ *is as in Proposition 3:*
  - $\alpha(t \overset{e}{=} t') \equiv t \doteq t'$ *for each atom* $t \overset{e}{=} t'$, *and*
  - $\alpha(\{x_1, ..., x_k\}; a_1, ..., a_n \to a) \equiv x_1 \doteq x_1, ..., x_k \doteq x_k, \alpha(a_1), ..., \alpha(a_n) \to \alpha(a)$
- *The components of the natural transformation* $\beta^- : Mod_{\mathcal{PA}} \to \Psi^{op}; Mod_{\mathcal{MA}}$ *are* β^- *'s from Definition 9, i.e.:*
  - $|\beta_\Sigma^-(P)| = |P|$
  - $f(x_1, ..., x_n)^{\beta_\Sigma^-(P)} = \begin{cases} \emptyset \text{ if } (x_1, ..., x_n) \notin \mathbf{dom}(f^P) \\ \{f(x_1, ..., x_n)^P\} \text{ otherwise} \end{cases}$

For a homomorphism $h \in Mod_{\mathcal{PA}}$, *we let* $\beta_\Sigma^-(h) = h$.

In [7, 6] this transformation was used for a smooth introduction of error recovery into specifications built in $\mathcal{PA}$. Only operations which in the original $\mathcal{PA}$-specifications might have been partial can now become nondeterministic and this was used to model partiality – interpreting undefinedness as indeterminacy of the possible result. As the nondeterminism can be narrowed down during the development process, this provided simple means for introduction of error elements or other recovery strategies, including even exceptions. In this case, we would claim, $\mathcal{MA}$ offers means for a unification of the two extremes in specification of partiality: the abstractness of $\mathcal{PA}$, due to the assumption of strictness, needed in the initial stages of development and, on the other hand, the possibility for detailed error treatment offered by total algebra approaches and useful in the final stages of low-level design.

4   Conclusion

Although the parameterization mechanism applied in 3.4 is quite standard, there are two points to observe. For the first, it is used in the new context of $\mathcal{MA}$, for which the relevant properties, in particular exactness, were verified. More importantly, the example from section 3 showed how to reuse (by embedding or transformation) and combine specifications developed in *different* frameworks. One should emphasize that this reuse is based on a rather straightforward translation of the theories which, for the most, amounts to axiomatizing the additional properties of various frameworks in order to restrict the generality of the multialgebraic semantics to the appropriate subset of models. As multialgebras extend in a conservative fashion many traditional algebraic frameworks, there is no need of *coding* or other complicated representation of the imported theories. Partial and membership algebras were chosen as an example, and a trivial embedding can be given of total algebras, but we expect that most algebraic specification formalisms can be fitted – and that also means: combined – within $\mathcal{MA}$. Except embedding, we have also shown another way of including institutions into $\mathcal{MA}$, exemplified by transformation of $\mathcal{PA}$. With reference to earlier work [7, 6], we suggested that this particular transformation offers a powerful way of combining the advantages of partial algebra and total algebra approaches to partiality, providing additional means for specifying error recovery strategies like exceptions. Details of embeddings, as well as other ways of relating/including various formalisms in $\mathcal{MA}$ remain to be investigated.

The paper presented also a new reasoning system for multialgebras. As compared to earlier ones it combines various aspects: it allows empty results as well as empty carriers, it addresses both inclusion and deterministic equality predicates, it is strongly complete and yet quantifier free. Each of the earlier systems, e.g. [5, 2, 20, 19], was weaker than the present one at least at one of the above points. In [6] a related (and logically equivalent) Rasiowa-Sikorski style system is presented, which is amenable to implementation. With respect to the combination of various specifications, it might be worthwhile to implement such a system in order to experiment with the common logical space for deriving consequences of the combination of specifications originating from different frameworks.

Besides defining explicitly relations to other specification frameworks and developing methodologies for combination of different specifications, the problem which still remains open concerns finding the greatest possible (if such exists) liberal subinstitution of $\mathcal{MA}$ – preferably, in form of verifiable restrictions on the syntax of specifications.

References

1. Jiří Adámek and Jiří Rosický. *Locally Presentable and Accessible Categories*. Cambridge University Press, 1994.
2. Marcin Białasik and Beata Konikowska. Reasoning with first-order nondeterministic specifications. *Acta Informatica*, 36:357–403, 1999.
3. Peter Burmeister. Partial algebra - an introductory survey. *Algebra Universalis*, 15:306–358, 1982.
4. Maura Cerioli, Till Mossakowski, and Horst Reichel. From total equational to partial first order. In Egidio Astesiano, Hans-Jörg Kreowski, and Bernd Krieg-Brückner, editors, *on Algebraic Foundations of Systems Specification*, chapter 3. Springer, 1999.

5. Heinrich Hussmann. *Nondeterminism in Algebraic Specifications and Algebraic Programs.* Birkhäuser, 1993.
6. Yngve Lamo. *Institution of multialgebras as a general framework for algebric specification.* PhD thesis, University of Bergen, Department of Informatics, 2002.
7. Yngve Lamo and Michał Walicki. Modeling partiality by nondeterminism. In N. Callaos, J. M. Pineda, and M. Sanchez, editors, *Proceedings of SCI/ISAS 2001,* volume I. Orlando, FL, 2001.
8. Yngve Lamo and Michał Walicki. Specification of parameterized data types: Persistency revisited. *Nordic Journal of Computing,* 8:278–303, 2001.
9. Yngve Lamo and Michał Walicki. Composition and refinement of specifications and parameterised data types. In John Derrick, Eerke Boiten, Jim Woodcock, and Joakim von Wright, editors, *Proceedings of the REFINE 2002 workshop,* volume 70/3. Elsevier Science Publishers, 2002. URL: http://www.elsevier.com/gej-ng/31/29/23/125/48/show/Products/notes/index.htt#001.
10. Alfio Martini and Uwe Wolter. A systematic study of mappings between institutions. In Francesco Parisi Presicce, editor, *Recent Trends in Algebraic Development Techniques,* volume 1376 of *Lecture Notes in Computer Science,* pages 300–315. Springer, 1998.
11. José Meseguer. General logics. In H.-D. Ebbinghaus et al., editors, *Logic colloquium'87,* pages 275–329. Elsevier Science Publisher B.V.(North-Holland), 1989.
12. José Meseguer. Membership algebra as a logical framework for equational specification. In Francesco Parisi Presicce, editor, *Recent Trends in Algebraic Development Techniques,* volume 1376 of *Lecture Notes in Computer Science,* pages 18–61. Springer, 1998.
13. Till Mossakowski. Equivalences among various logical frameworks of partial algebras. In H. K. Büning, editor, *Computer Science Logic,* volume 1092 of *Lecture Notes in Computer Science,* pages 403–433. Springer, 1996.
14. Peter Mosses. Unified algebras and institutions. In *Proceedings of LICS'89, 4-th Annual Symposium,* 1989.
15. Fernando Orejas. Structuring and modularity. In Egidio Astesiano, Hans-Jörg Kreowski, and Bernd Krieg-Brückner, editors, *on Algebraic Foundations of Systems Specification,* chapter 6. Springer, 1999.
16. Aida Pliuškievičienė. Specialization of the use of axioms for deduction search in axiomatic theories with equality. *J. Soviet Math.,* 1, 1973.
17. Andrzej Tarlecki. Institutions: An abstract framework for formal specifications. In Egidio Astesiano, Hans-Jörg Kreowski, and Bernd Krieg-Brückner, editors, *on Algebraic Foundations of Systems Specification,* chapter 4. Springer, 1999.
18. Michał Walicki, Adis Hodzic, and Sigurd Meldal. Compositional homomorphisms of relational structures (modeled as multialgebras). In *Proceedings of 13e-th International Symposium on Fundamentals of Computing Theory,* Lecture Notes in Computer Science. Springer, 2001.
19. Michał Walicki and Sigurd Meldal. A complete calculus for the multialgebraic and functional semantics of nondeterminism. *ACM TOPLAS,* 17(2), March 1995.
20. Michał Walicki and Sigurd Meldal. Multialgebras, power algebras and complete calculi of identities and inclusions. In *Recent Trends in Algebraic Specification Techniques,* volume 906 of *LNCS.* Springer, 1995.
21. Michał Walicki and Sigurd Meldal. Algebraic approaches to nondeterminism-an overview. *ACM Computing Surveys,* 29, 1997.

On How Distribution and Mobility Interfere with Coordination

Antónia Lopes[1] and José Luiz Fiadeiro[2]

[1] Department of Informatics, Faculty of Sciences, University of Lisbon
Campo Grande, 1749-016 Lisboa, Portugal
mal@di.fc.ul.pt

[2] Department of Mathematics and Computer Science, University of Leicester
University Road, Leicester LE1 7RH, UK
jose@fiadeiro.org

Abstract. With the advent of the e-Economy, system architectures need to take into account that Distribution and Mobility define a dimension that is not orthogonal to the traditional two – coordination and computation. In this paper, we address the way that this additional dimension interferes with coordination. First, we address the effectiveness of connectors in place in a system when the properties of the media through which its components can be effectively coordinated are taken into account. Then, we consider the modelling of coordination styles that are location-dependent or even involve the management of the location of the coordinated parties.

1 Introduction

Architecture-based approaches to the design and construction of software systems focus on the gross modularisation of systems as collections of interacting components. At the architecture level, the aspects related to distribution, such as the way the system is supposed to be mapped into a network, are not usually addressed. This conforms with traditional forms of distributed systems that consider that the environment where a system executes — the physical nodes and links — is statically configured, the distribution of the system over these nodes is also static and that all forms of communication and access to resources can be provided (directly or indirectly) by the operating system or the middleware.

With the advent of Mobile Computing in wide and ad-hoc networks, new forms of distributed systems come into play. In mobile computing systems, components are entitled to move across a network that is not necessarily statically determined. The network itself may be constituted by mobile nodes without a fixed infrastructure and, hence, their connectivity may change over time. In this situation, it is no longer reasonable to abstract away from component location and the properties of the physical distribution topology of locations and communication infrastructure. This is particularly true at the architectural level. For instance, in architectures that are structured in

M. Wirsing, D. Pattinson, and R. Hennicker (Eds.): WADT 2002, LNCS 2755, pp. 343–358, 2003.

terms of components and connectors [12], it is no longer possible to assume that the coordination mechanisms put in place through connectors can be made effective across the physical links that connect the hosts of the components in the underlying network. Another reason to address the distribution and mobility dimension of systems at an architectural level is the fact that this third dimension (together with computation and coordination) is an additional source of complexity in system development and, hence, should be addressed at the highest possible level of abstraction. Therefore, concepts and techniques are needed to support the description of the aspects related to distribution and mobility at the architectural level of software design.

As a first step towards the more ambitious goal of having an architectural approach to distribution and mobility, we have been investigating the addition of this new dimension to the architectural framework that we developed in the past [8]. In our first step in this direction [11], an extension of CommUnity was proposed in order to support the description of the distribution and mobility dimension of systems. This extension was developed having in mind the goal of being able to represent distribution and mobility explicitly in architectures. In spite of its apparent simplicity, this goal comprises many different aspects that result from the different ways computation, coordination and mobility can interfere. For instance, in what concerns the interference between computation and mobility, we have already shown that patterns of distribution and mobility of components, or groups of components, can be explicitly represented in architectural descriptions through what we have called *distribution connectors*, similarly to the way *coordination connectors* represent component interactions.

In this paper we will explore the interference between coordination and mobility, in particular the emergence of several new abstractions for program interaction such as transient interaction (e.g., transient variable sharing) [13] and code mobility (e.g., remote evaluation and mobile agents) [5]. We will show that these coordination constructs, that facilitate component interactions in mobile systems, can be modelled as coordination connectors, and used in systems architectures together with the connectors that model traditional communication primitives such as asynchronous communication and remote procedure call. We shall also address the impact on system architecture of the properties of the locations in which components perform their computations and the properties of the media through which their interconnections can be effectively coordinated.

The paper is organised as follows. Section 2 briefly introduces the extension we have proposed for CommUnity. It encompasses an extension of the program design language in order to support the design of components that have location-dependent patterns of computing, namely mobile components, an extension of the primitive mechanisms that support program interaction, and the definition of program composition in this context. Then, in section 3, we shall illustrate how coordination mechanisms put in place through coordination connectors can become ineffective and need to be replaced with ones that are compatible with the connection topology among the components available at physical level. Then, section 4 shows how some of the new forms of coordination raised by mobility can be expressed as coordination connectors and used in systems architectures. Section 5 closes the paper with some conclusions.

2   The CommUnity Framework

CommUnity, introduced in [6], is a parallel program design language that is similar to Unity [4] in its computational model but adopts a different coordination model. CommUnity relies on the sharing (synchronisation) of actions and exchange of data through input and output channels and, moreover, it requires interactions between components to be made explicit. In CommUnity, the separation between "computation" and "coordination" is taken to an extreme in the sense that the definition of the individual components of a system is completely separated from the interconnections through which these components interact.

Syntax

CommUnity was recently extended in order to support the design of the distribution and mobility dimension of systems [11]. It adopts an explicit representation of the space within which movement takes place, but no specific notion of space is assumed. This is achieved by considering that "space" is constituted by the set of possible values of a special data type Loc included in a fixed data type specification over which components are designed. The data sort Loc models the positions of the space in a way that is considered to be adequate for the particular application domain in which the system is or will be embedded. The only requirement that we make is for a special location $-\perp-$ to be distinguished (its role will be discussed further below).

In order to model systems that are location-aware, we make explicit how system "constituents" (output and private channels, actions, or any group of these) are mapped to the positions of the space statically determined by Loc. This is achieved by associating each "constituent" of a system with a location variable. Mobility is then associated with the change of value of location variables.

A CommUnity design is defined in terms of input and output location variables (resp. denoted by II, IO), input, ouput and private channels (resp. denoted by I, O and V) and a set of action names (Γ).

Channels. Private channels model internal communication. Input channels are used for reading data from the environment of the component and the component has no control on the values that are made available there. Output channels allow the environment to read data produced by the component. Each channel v is typed with a sort $sort(v)$. We shall use X to denote $I \cup O \cup V$.

Location Variables. Location variables (locations, for short) in a component design can be declared as *input* or *output* in the same way as channels but are all typed with sort Loc. Input locations are read from the environment and cannot be modified by the component. Hence, if $l \in II$, the movement of any constituent located at l is under the control of the environment. Output locations can only be modified locally but can be read by the environment. Hence, if $l \in IO$, the movement of any constituent located at l is under the control of the component.

Each local channel x of a design is associated with a location l. We make this assignment explicit by writing $x@l$. The value of l indicates the current position of the space where the values of x are made available. A modification in the value of l entails the movement of x as well as of the other channels and actions located at that location variable.

Input channels are located at a special output location λ whose value is invariant and given by $\perp$. The intuition is that this location variable is a non-commitment to any particular location. The idea is that input channels will be assigned a location when connected with a specific output channel of some other component of the system. Every channel x is associated with a set of locations $\Lambda(x)$ which is $\{\lambda\}$ if x is an input channel and is $\{l,\lambda\}$ in the case of an output channel $x@l$. We shall use L to denote the pointed set of locations $(lI\cup lO)_\lambda$ and $local(X,L)$ to denote the union $V\cup O\cup lO$ of *local* channels and locations.

Actions. Actions can be declared either as *private* or *shared*. Private actions represent internal computations in the sense that their execution is uniquely under the control of the component. Shared actions represent possible interactions between the component and the environment, meaning that their execution is also under the control of the environment. As we will see, actions provide points of *rendez-vous* at which components can synchronise.

Each action name g is associated with a set $\Lambda(g)$ of locations including λ, meaning that the execution of action g is distributed over those locations. In other words, the execution of g consists of the synchronous execution of a guarded command in each of these locations. Guarded commands are associated with located actions, i.e. pairs $g@l$ s.t. $l\in\Lambda(g)$, as follows:

$$g@l[D(g@l): [G(g@l)\rightarrow R(g@l)]$$

- $D(g@l)$ is a subset of $local(X,L)$ consisting of the local channels into which executions of the action can place values and of the locations to which the action can inflict movement. This is what is sometimes called the *write frame* of $g@l$. For simplicity, we will often omit the explicit reference to the write frame when $D(g@l)$ can be inferred from the assignments. Given a local channel or location v, we will also denote by $D(v)$ the set of located actions $g@l$ such that $v\in D(g@l)$. The fact that the special location λ is invariant is ensured by the condition $D(\lambda)=\varnothing$. We denote by $F(g@l)$ the *frame* of $g@l$, i.e., the channels and locations that are in $D(g@l)$ or used in $G(g@l)$.

- $G(g@l)$ is the guard condition.

- $R(g@l)$ is a conditional multiple assignment on the local channels and locations declared in $D(g@l)$. When the write frame $D(g@l)$ is empty, $R(g@l)$ is denoted by *skip*.

When a design does not explicitly refer the guarded command associated with $g@\lambda$, this is because it is the empty command *[true $\rightarrow$ skip]*. Notice also that every standard CommUnity design (location-unaware) defines trivially a distributed design: the one that has all its actions and channels located at λ.

Variations in the context of execution of a mobile system are not limited to the locations of its components and respective hosts. It is important that other observables, s.a. network bandwidth, battery power or the communication range, can be used at the programming level. In CommUnity, we have only a construct *inrange:Loc->bool* that allows a program to observe if a given position of the space is in its communication range.

Semantics

We assumed that the distribution space is constituted by the set of possible values of a given data sort Loc. We consider that, once we fix an algebra U for the data types, namely a domain U_{Loc} for Loc, the relevant properties of the mobility space are captured by two binary relations over U_{Loc}:

- A relation bt s.t. $n \, bt \, m$ means that n and m are positions in the space "in touch" with each other. Interactions among components can only take place when they are located in positions that are "in touch" with one another. Because the special location variable λ intends to be a position to locate entities that can communicate with any other entity in a location-transparent manner, we require that the value of λ is always set at configuration time as being $\perp_U$ and, furthermore, $\perp_U \, bt \, m$, for every $m \in U_{Loc}$.

- A relation $reach$ s.t. $n \, reach \, m$ means that position n is reachable from m. Permission to move a component or a group of components is conceded when the new position "is reachable" from the current one.

In general, the topology of locations is dynamic and, hence, the operational semantics for a program is given in terms of an infinite sequence of relations $(bt_i, reach_i)_{i \in N}$. At each execution step, one of the actions that can be executed is chosen and executed. The conditions under which a distributed action g can be executed at time i are the following:

1. *for every* $l_1, l_2 \in \Lambda(g)$, $[l_1]^i \, bt_i \, [l_2]^i$: the execution of g involves the synchronisation of its local actions and, hence, their locations have to be in touch.

2. *for every* $l \in \Lambda(g)$, $g@l$ *can be executed, i.e.*,
   i. *for every* $x \in F(g@l)$, $[l]^i \, bt_i \, [\Lambda(x)]^i$: the execution of $g@l$ requires that every channel in the frame of $g@l$ can be accessed and, hence, l has to be in touch with their locations.
   ii. *for every location* $l_1 \in D(g@l)$ *and* $m \in [R(g)]^i(l_1)$, $m \, reach_i \, [l_1]^i$: if a location l_1 can be effected by the execution of $g@l$, then every possible new value of l_1 must be a position reachable from the current one.
   iii. *the local guard* $G(g@l)$ *evaluates to true*

where $[e]^i$ denotes the value of the expression e at time i. In the case of expressions e involving $inrange(exp)$, the value of $[e]^i$ also depends on the location l where the expression is being evaluated and is defined by $[l]^i \, bt \, [exp]^i$.

Given this, when, in an execution step, one of the actions whose enabling condition holds of the current state is selected, its assignments are executed atomically as a transaction. Furthermore, it is guaranteed that private actions that are infinitely often enabled are selected infinitely often.

Interaction and Composition

The primitive mechanisms that support component interaction in CommUnity are the synchronisation of actions and the interconnection of input channels of a component with output channels of other components. The extension of the language with a distribution dimension also entails the definition of mechanisms that support the interaction of designs at the level of their locations. Such mechanisms are essentially the interconnection of input locations of a component with output locations of other components.

In CommUnity, interactions between components have to be made explicit and external to the components by providing the name bindings that express input/output communication and action synchronisation. This externalisation of interactions can be expressed via the following notion of morphism:

Given designs P_i with channels X_i, actions Γ_i and locations L_i, a morphism $\sigma: P_1 \rightarrow P_2$ consists of a total function $\sigma_{ch}: X_1 \rightarrow X_2$, a partial mapping $\sigma_{ac}: \Gamma_2 \rightarrow \Gamma_1$ and a total function $\sigma_{lc}: L_1 \rightarrow L_2$ that preserves the pointed element (λ), satisfying:

1. *for every $x \in X_1$ and $l \in L_1$:*

    a. $sort_2(\sigma_{ch}(x)) = sort_1(x)$ **b.** *if $x \in out(X_1)$ $(prv(X_1))$ then $\sigma_{ch}(x) \in out(X_2)$ $(prv(X_2))$*   **c.** *if $x \in in(X_1)$ then $\sigma_{ch}(x) \in out(X_2) \cup in(X_2)$*   **d.** *if $l \in outloc(L_1)$ then $\sigma_{lc}(l) \in outloc(L_2)$*   **e.** $\sigma_{lc}(\Lambda_1(x)) \subseteq \Lambda_2(\sigma_{ch}(x))$

2. *for every $g \in \Gamma_2$ s.t. $\sigma_{ac}(g)$ is defined:*

    a. *if $g \in sh(\Gamma_2)$ $(prv(\Gamma_2))$ then $\sigma_{ac}(g) \in sh(\Gamma_1)$ $(prv(\Gamma_1))$*
    b. $\sigma_{lc}(\Lambda_1(\sigma_{ac}(g))) \subseteq \Lambda_2(g)$

3. *for every $x \in local(X_1, L_1)$ and $g@l_2 \in D_2(\sigma(x))$:*

    a. $\sigma_{ac}(g)$ *is defined and* **b.** $\sigma_{ac}(g)@l_1 \in D_1(x)$ *for some $l_1 \in \sigma_{lc}^{-1}(l_2) \cap \Lambda_1(\sigma_{ac}(g))$*

4. *for every $g \in \Gamma_2$ s.t. $\sigma_{ac}(g)$ is defined and $l \in \Lambda_1(\sigma_{ac}(g))$:*

    a. $\sigma(D_1(\sigma_{ac}(g)@l)) \subseteq D_2(g@\sigma_{lc}(l))$
    b. $\Phi \vDash R_2(g@\sigma_{lc}(l)) \supset \underline{\sigma}(R_1(\sigma_{ac}(g)@l))$ **c.** $\Phi \vDash G_2(g@\sigma_{lc}(l)) \supset \underline{\sigma}(G_1(\sigma_{ac}(g)@l)$

*where $\vDash$ means validity in the first-order sense taken over the axiomatisation Φ of the underlying data types (which includes the location space). Designs and morphisms constitute a category **DSGN**.*

This notion of morphism extends what in the literature on parallel program design is known as superposition [4,9,10] by taking the distribution aspects into account. A morphism $\sigma: P_1 \rightarrow P_2$ identifies a way in which P_1 is "augmented" to become P_2 so that it can be considered as having been obtained from P_2 through the superposition of additional behaviour, namely the interconnection of one or more components. In other words, σ identifies P_1 as a component of P_2.

The map σ_{ch} identifies for every channel of the component the corresponding channel of the system. The first group of constraints establish that sorts and types of channels have to be preserved but input channels of a component may become output channels of the system.

The partial mapping σ_{ac} identifies the action of the component that is involved in each action of the system, if ever. This mapping is partial and contravariant to account for the fact that, on the one hand, superposition may unfold actions of the original program and, on the other hand, new actions may be added. Condition 2.a states that the type of actions is preserved.

The map σ_{lc} identifies for every location variable of the component the corresponding location variable of the system. As for channels, output locations are mapped into output locations but input locations of a component may become output locations of the system as the result of interconnecting an input location of a component with an output location of another component. Conditions 1.e and 2.b state that the locations of channels and actions are preserved.

Conditions 3.a, 3.b and 4.a require that change within a component be completely encapsulated in the structure of actions defined for the component, namely they ensure that the actions of a component leave the local channels and locations of the other components out of their scope. The last two conditions require that the computations performed by the system reflect the interconnections established between its components. Condition 4.b reflects the fact that the effects of the actions of the components can only be preserved or made more deterministic in the system and condition 4.c allows the guard of the action to be strengthened but not weakened. Strengthening of the guard reflects the fact that all the components that participate in the execution of a joint action have to give their permission for the action to occur.

In the next sections several examples illustrate the way these morphisms can be used for establishing interconnections and the way diagrams in the category *DSGN* define systems configurations. The semantics of a system configuration is given by a categorical construction: the colimit of the underlying diagram. Taking the colimit of a diagram collapses the configuration into an object by internalising all the interconnections and distribution aspects, thus delivering a design for the system as a whole. Colimits in CommUnity capture a generalised notion of parallel composition in which the designer makes explicit what interconnections are used between components:

- Channels and locations involved in each I/O-communication established by the configuration are amalgamated.
- Every set $\{g_1,...,g_n\}$ of actions that are synchronised is represented by a single action $g_1\|...\|g_n$ whose occurrence captures the joint execution of the actions in the set. The transformations performed by the joint action are distributed over the locations of the synchronised actions. Each located action $g_1\|...\|g_n@l$ is specified by the conjunction of the specifications of the local effects of each of the synchronised actions g_i that is distributed over l, and the guards of joint actions are also obtained through the conjunction of the guards specified by the components.

3   Effectiveness of Coordination Connectors

In *structural models* of Software Architecture, i.e., models that share the view that system architectures are structured in terms of components and coordination connectors, the properties of the physical distribution topology of locations and communication links are not usually taken into account. It is assumed that the physical links ("wires") that enable communication between hosts in the underlying communication network are fixed and statically determined. Consequently, these models rely on the fact that the coordination mechanisms put in place through connectors can be made effective across the wires that link the components' hosts.

When mobility comes to play, for instance in contexts where physical mobility of computation hosts, such as laptops, exists, the wired or wireless physical links that enable communication between hosts can change. Furthermore, the ability of a component to communicate with others is influenced by its location because of the barriers to communication that are erected in communication networks by system managers. Finally, the communication network is also subject to frequent changes.

In this section, we use an example of a client-server system to show that when the distribution aspects of systems are addressed at the architectural level, it is necessary to take into account the properties of the network communication infrastructure in order to understand whether the coordination connectors in place are effective or have to be replaced with ones that are compatible with the topology of distribution available at physical level.

For developing the example, we use the CommUnity architectural framework. In particular, we shall made use of distribution connectors that were proposed in [11], through which the mechanisms that define the distribution topology of systems can be externalised and explicitly represented in system architectures.

We consider a very simple client-server system. As usual in this kind of systems, the server exports a service that the client may request at some point in its execution. The service is the calculation of $f(v,x)$ for a certain function f, where v is a resource local to the server and x is a value given by the client.

In CommUnity, clients and servers can be modelled as follows.

```
design server is                   design client is
in   x: T                          in   res: T
out  r: T                          out  val: T
prv  v: T', lx: T, s: [0..2]       prv  rq, inp: bool
do   req: [s=0→lx:=x∥s:=1]         do   req: [¬rq∧ inp→rq:=true∥
[]   serv:[s=1→r:=f(v,lx)∥s:=2]                       inp:=false]
[]   ret: [s=2→s:=0]               []   read: [rq → rq:=false]
[]   chg: [true→v:∈T']             [] prv prod:[¬inp→inp:=true∥val:∈T]
```

In *server,* it is modelled that a server repeatedly accepts requests, executes the service and returns the result. Moreover, its resource v may be updated at any time through the execution of action *chg*. In *client,* the typical behaviour of a client is modelled, ignoring for the moment the details of its internal computation: First, it produces the data needed for the service, then it requests the service and, finally, it reads the result.

It remains to describe the way the client and the server interact. There are several possibilities; we start by considering the exchange of messages through synchronous communication. In other words, the client and the server must synchronise first for the transmission of the request's data and then again for the transmission of the result.

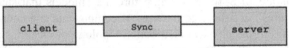

In order to achieve this form of interaction, we have to make explicit the synchronisation of the actions *req* of *client* and *server,* the synchronisation of actions *read* of *client* and *ret* of *server,* and the I/O interconnection of the channels used for the transmission of the service's parameter and result. Such interconnections can be described by the following diagram in the category ***DSGN,***

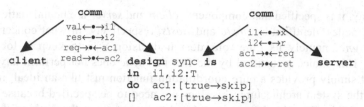

where *comm* consists of two input variables to model the medium through which data is to be transmitted between the client and the server, and two shared actions for the two components to synchronise in order to transmit the data. Because names in CommUnity are local, the identities of the shared input variables and the shared actions in *comm* are not relevant: they are just placeholders for the projections to define the relevant bindings. Hence, we normally do not bother to give them explicit names, and represent them through the symbol •.

This architectural description of the client-server system abstracts away how the system is supposed to be distributed on the nodes of a network (every design considered so far is location-unaware) and takes for granted that the synchronous communication between the client and the server can be made effective. This can be easily confirmed by analysing the design of the system as a whole, obtained by taking the colimit of its configuration diagram. Such design has, for instance, the joint action

req|req: [¬rq∧inp∧s=0 →rq:=true‖inp:=false‖1x:=val‖s:=1]

that models the synchronous execution of actions *req* of *client* and *server*. This action is enabled provided that enabling conditions of both actions evaluate to true. Nothing else is required, even though the execution of *req* by *server* involves the reading of *val*, a resource local to the client.

Let us now suppose that the server exists at a fixed location — *hosts*, which is a node in a subnet protected by a firewall. That is to say, there exists a *filter* node and every external communication with *hosts* has to pass through it. If the *client* exists in a node *hostc* that is not in this subnet, then it is not realistic to expect that the coordination of the client and the server put in place through the connector *Sync* is effective. Such distribution aspects of the client-server system can be explicitly modelled as follows,

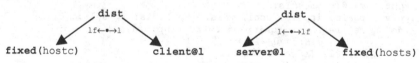

where

- *fixed(hostc)* and *fixed(hosts)* are instances of a parameterised design *fixed(v:Loc)* that consists of an output location named *lf* that is constrained to be initialised at configuration time (i.e. when this component is included in the configuration of the system being built) with a value denoted by *v*.

- *client@l* (resp., *server@l*) denotes the design *client* (resp., *server*) augmented with an input location variable *l*, where every action and local channel of *client* (resp., *server*) is located.

In this way, it is specified that components *client* and *server* are bound statically to the network nodes identified by *hostc* and *hosts*, respectively. In what concerns the connector *Sync*, namely its glue, our design decision was to keep it location-transparent. This choice is justified by the fact that *sync* does not perform any computation but simply provides a pure coordination function just like an ideal, neutral "cable". In the system architecture, this does not need to be specified because every constituent of a design is by default considered to be located at the distinguished location variable λ.

It is assumed that *hosts, hostc* and *filter* are constants of type *Loc*. This data type in this case has no other requirements that to have these constants and the axioms *hosts≠hostc, hosts≠filter, hostc≠filter*, that ensure that the three hosts are actually different.

By adopting a simple box-and-line notation, the architecture of the system can be represented as follows.

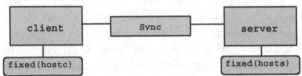

Notice that, in this architecture, we achieved the envisaged separation between the individual software components of the system, the coordination connector through which they interact, and the mechanisms that are responsible for the distribution topology of the system.

A design of the system as a whole, where all interconnections and distribution aspects are internalised (obtained by taking the colimit of the underlying categorical diagram) is given by *client-sync-server*. This design exposes clearly the system dependence on the properties of the communication infrastructure that is available at physical levels, namely the links that connect *hostc* and *hosts* in the underlying network.

```
design client-sync-server is
outloc lc, ls
inv    lc=chost ∧ ls=shost
out    val@lc, res@ls: T
prv    rq@lc, inp@lc: bool, v@ls: T', lx@ls: T, s@ls: [0..2]
do     req|req@lc:[¬rq∧inp →rq:=true||inp:=false]
             @ls:[s=0 →lx:=val||s:=1]
[]     serv@ls:[s=1→res:=f(v,lx)||s:=2]
[]     read|ret@lc:[rq→rq:=false]
[]             @ls:[s=2→s:=0]
[] prv prod@lc:[¬inp →inp:=true||val:∈T]
[]     chg@ls:[true→v:∈T']
```

In the situation described previously, *hosts* is in a subnet protected by a firewall and every external communication to *hosts* has to pass through *filter* node. This means that the current relation *be in touch* has the following property:

*for every n∈ U_{Loc}, if (hosts $_U$ **bt** n) then n=⊥$_U$ or n∈ FW or n=filter$_U$*

where *FW⊆U_{Loc}* represents the set of nodes of the subnet protected by the firewall. Given that *hostc$_U$∉ FW*, we have that *¬(hosts$_U$**bt** hostc$_U$)* and, hence, we may conclude that, in *client-sync-server*, once the client produces the data for the service, the issue of the request gets blocked and the system enters a deadlock. This clearly confirms that the connector in place in the system is currently not effective and that a connector compatible with the current distribution topology has to be used instead. Moreover, it points out the importance of being able to make systems evolve through dynamic reconfiguration in response to changes in the topology of the network.

4 New Patterns of Coordination

In the context of the separation of computation and coordination supported by architecture-based approaches, the distributed and mobile dimension of systems is also reflected on the kind of the coordination constructs that are appropriate and should be available for component interconnection.

On the one hand, the advent of mobility calls for new abstractions for component interaction that facilitate systems design. The fact that components may move across a network of locations demands the adoption of coordination styles that are location-dependent. *Transient interaction*, for instance, is a form of interaction that is conditional on the relative positions of components. More precisely, interaction is limited to the situations in which the components are in the communication range of each other.

On the other hand, the opportunities made available by technological developments have been explored and have led to new programming paradigms and new conceptual mechanisms for structuring systems. For instance, mobile code paradigms such as *Remote Evaluation* and *Mobile Agents* became very popular in the design of distributed applications. As mentioned in [14], coordination no longer involves just communication, it may also involve the management of the locations of the coordinated parties.

Our aim in this section is to show, by means of examples, that at an architectural level of design, such new abstractions for coordination can be formally specified as connectors and used in systems architectures together with the connectors that model traditional coordination primitives. Furthermore, we shall also show how in situations in which coordination encompasses both communication and relocation, the design of these aspects can be carried out independently.

When Coordination Is Location-dependent

Transient variable sharing is a context-dependent pattern of interaction proposed in [13] as a variant of traditional variable sharing, which is suitable for mobile computing systems that are subject to frequent disconnections. Roughly speaking, it is based on variable sharing limited to the situations in which the components are in the communication range of each other.

In CommUnity, read-only sharing of variables is supported through I/O interconnection of variables. For instance, the interconnection of an input variable y of design R with the output variable x of design W establishes that the value of y is read from x.

If, for instance, W is a mobile component, the reading of variable x is conditioned by the "connectivity" between W and R. The semantics defined for CommUnity designs ensures that, when W is out of the range of communication of R, every action of R that needs to access the value of y gets blocked.

In read-only transient sharing, even while W is out of the range of communication of R the value of y can be accessed. Of course, in this situation there is no way to continue to guarantee that the value of y is given by x. Instead, the value of y is given by the value of x on disconnection.

This form of transient sharing can be modelled through the binary coordination connector *TranSh* with roles *writer* and *reader* and the glue *transh*.

```
design writer is              design reader is
inloc lw                      inloc lr
out    x@lw: T                 in     y: T
do     wr@lw:[true→ x:∈T]      do     read@lr:[true→skip]
```

The roles define the behaviour required of the components to which the connector can be applied. For a *writer,* we require an action that models every kind of possible operation on x. For a *reader,* we require an action that models the access to the input variable y. This is because it is essential to know in which location this action is executed.

The glue ensures that updates to x are propagated to y whenever the *reader* and the *writer* are in contact with each other. Whenever the communication between the two components is possible, *transh* prevents the *writer* from writing x before the previous change of x has been propagated to y. In the other situations, lr is not in the range of lw and, hence, y remains with the value of x at disconnection time. On re-connection, the value of x is sooner or later propagated to y. This is achieved through the execution of the action *auto* that is private to *transh* and, hence, subject to fairness requirements.

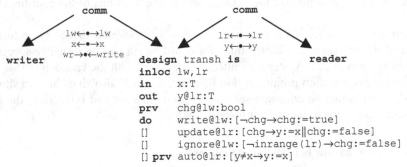

```
                 comm                      comm
          lw←─•─→lw              lr←─•─→lr
          x←─•─→x                y←─•─→y
          wr→•←─write    design transh is                    reader
writer                   inloc lw,lr
                         in     x:T
                         out    y@lr:T
                         prv    chg@lw:bool
                         do     write@lw:[¬chg→chg:=true]
                         []     update@lr:[chg→y:=x‖chg:=false]
                         []     ignore@lw:[¬inrange(lr)→chg:=false]
                         [] prv auto@lr:[y≠x→y:=x]
```

Synchronous execution of actions is another form of communication available in many models of distributed systems that has also inspired a transient counterpart [13]. A simple form of transient synchronisation of an action a with an action b consists of requiring the two actions to be executed synchronously whenever they are located at connected hosts. When this is not the case, the two actions can be executed independently.

This form of transient synchronisation can be represented by a binary connector with two identical roles — *comp*. This design simply identifies the action subject to the synchronisation. The glue of the connector ensures the synchronous execution of the two actions whenever they are located in locations in contact with each other and allows their independent execution in the other situations.

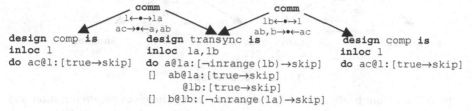

```
design comp is            design transync is          design comp is
inloc l                   inloc la,lb                 inloc l
do ac@l:[true→skip]       do a@la:[¬inrange(lb)→skip]  do ac@l:[true→skip]
                          []  ab@la:[true→skip]
                              @lb:[true→skip]
                          []  b@lb:[¬inrange(la)→skip]
```

When Coordination requires Movement

The fact that the location of components becomes a first-class design element also leads to new coordination patterns that explore the relocation of components, namely *Remote Evaluation (RE)* [15]. Like transient sharing and transient synchronisation, *RE* is a variant of a traditional style of interaction — *Client-Server*.

Let us consider again the client-server system described in section 3. Recall that the server offers a fixed service — the calculation of a certain function *f*, which depends on some resource *v* that is local to the server. The server holds the know-how needed to use the resource (the code that implements function *f*) as well as the resource involved in the service (*v*). In *RE*, the server offers its resources but it is the client that holds the code that describes how to perform the service. The client needs to transmit the data and must provide the server with the code that implements the service.

In this situation, the server simply has to be ready to cooperate with the client in the execution of the service. It receives the request and then it allows the use of its local resource, which can be locally changed at any time through the execution of action *chg*. We further consider that the server is bound to the same location for it whole life (see design *re-server*).

```
design re-server is           design knowhow is
outloc l                      inloc l
out     v@l: T'                in   v: T', val:T
prv     s@l: [0..1]            out  res@l: T
do      req@l: [s=0→s:=1]      prv  lval@l:T, t@l: [0..3]
[]      serv@l: [s=1→s:=0]     do   get_val@l: [t=0→lval:=val‖t:=1]
[]      chg@l: [true→v:∈T']    []   req@l: [t=1→t:=2]
                              []   serv@l: [t=2→res:=f(v,lval)‖t:=3]
                              []   send_res@l: [t=3→t:=0]
```

The client in this case is slightly more complex because a part of it concerns the know-how needed to use the server's resource and has to be relocated in order to achieve local interaction with the server. The simpler way of describing *re-client* is in terms of a configuration involving a traditional client bound to a fixed location and a design *knowhow* that encapsulates the portion of *re-client* that has to be moved. As modelled above, *knowhow* repeatedly gets the data needed for the execution of the service, requests the use of the server resource, uses the resource for the calculation of *f* and returns the result. The configuration of *re-client* (see below) establishes that *knowhow* and *client* exchange messages (service's data and result) through synchronous communication.

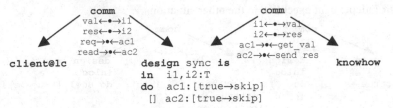

```
                    comm                           comm
                 val←─•→i1                      i1←─•→val
                 res←─•→i2                      i2←─•→res
                 req→•←ac1                      ac1→•←get_val
                 read→•←ac2                     ac2→•←send_res
   client@lc              design sync is                    knowhow
                          in   i1,i2:T
                          do   ac1:[true→skip]
                          []   ac2:[true→skip]
```

In *RE*, the coordination of *re-client* and *re-server* involves communication and code migration. It is necessary to describe the movement of the subcomponent *knowhow* of *re-client* that makes local interaction with the server possible and, once there, the kind of communication it is used. These two dimensions of coordination can be addressed and described separately.

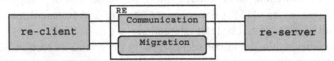

```
                          ┌──────────────────────┐
                          │ RE                    │
                          │ ┌──────────────────┐  │
   ┌────────────┐         │ │ Communication    │  │         ┌────────────┐
   │ re-client  │─────────│ └──────────────────┘  │─────────│ re-server  │
   └────────────┘         │ ┌──────────────────┐  │         └────────────┘
                          │ │ Migration        │  │
                          │ └──────────────────┘  │
                          └──────────────────────┘
```

In what concerns communication, *RE* does not entail a particular form of communication. Message passing through synchronous communication or remote procedure call are, for instance, possible choices. Hence, we use the connector *Communication*, which models a generic bi-directional communication protocol in the sense that the glue of this connector leaves completely unspecified the way in which messages are processed and transmitted. In this way, *RE* can be regarded as a parameterised entity that takes the communication protocol as a parameter (similar to higher-order connectors defined in [12]).

The pattern of migration of *RE* can be defined separately by the connector *Migration* below.

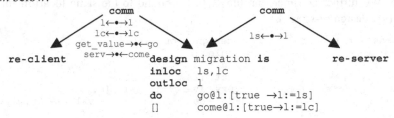

```
                      comm                         comm
                   l←─•→l                        ls←─•→l
                   lc←─•→lc
                   get_value→•←go
   re-client       serv→•←come design migration is          re-server
                               inloc   ls,lc
                               outloc  l
                               do      go@l:[true →l:=ls]
                               []      come@l:[true→l:=lc]
```

The glue carries out the relocation of *knowhow* in the server location as soon as it gets the data needed for the service. Once *knowhow* has terminated the use of the server resources, it returns to the client location. Then, the client may collect the result. Notice that this corresponds to the synchronous execution of actions *send_res* at *knowhow* and *read* at *client*. At this point, these two actions are co-located (i.e., *lc* and *l* have the same value) and hence the location-guards of *send_res* and *read* are both true.

Finally, by putting together the glues of *Migration* and *Communication* (without any kind of interaction) as well as the underlying role connections, we obtain a description of *RE* as a coordination connector.

5   Concluding Remarks

In this paper, we have addressed the interference of distribution and mobility with coordination in the context of system architectures that are structured in terms of components and connectors. By adopting an architectural framework that was extended in order to support the description of the distribution aspects of systems, we have shown how the effectiveness of connectors in place in a system may depend on the locations of the components that are being coordinated as well as on the properties of the wires through which the coordination has to take place. In this situation, it is essential that when connectors in place become ineffective, they can be replaced by different ones. Support for describing this kind of evolution can be conceived in terms of a dynamic reconfiguration language equipped with a set of observables including, for instance, the component locations and the network topology (many other relevant observables are identified in [3]). We plan to extend the reconfiguration language developed for CommUnity architectures [17] in order to make systems evolve in reaction to these new sources of change.

Furthermore, we have focused on new styles of coordination that rely on and use the distribution/mobility dimension of systems. We have considered situations in which the coordination patterns depend on the location of the coordinated parties or even determine their relocation. We have shown that these new forms of coordination, like the traditional ones, do not have to be programmed in the code of components. They can be externalised as first class entities and superposed over other components to regulate their behaviour or their location without intruding in the way they have been implemented. In fact, components will not even know that they are being regulated in the sense that the coordination is performed externally.

The examples we have analyzed also suggest that in the situations in which communication, distribution and mobility, can be regarded as different dimensions of the coordination pattern, these dimensions can be addressed separately and then composed. This separation is important not only because it makes it easier to describe complex interaction patterns and promotes reuse, but also addresses the evolutionary dimension of systems.

Not every language supports the separation of concerns just described. CommUnity was developed precisely to illustrate how the separation between computation and coordination can be supported by Formal Description Techniques. Moreover, the extension with distribution and mobility (in contrast with a former extension presented in [16]) was developed having in mind the support required for the externalisation of the mechanisms that are responsible for managing the distribution topology of systems. In [7], we have provided a mathematical characterisation of the language features that the separation of computation and coordination requires. Future work is going on to establish the corresponding extension taking into account distribution dimension.

Acknowledgements. This work was partially supported through the IST-2001-32747 Project *AGILE – Architectures for Mobility*.

References

1. R.Allen and D.Garlan, "A Formal Basis for Architectural Connectors", *ACM TOSEM*, 6(3), 1997, 213-249.
2. L.Bass, P.Clements and R.Kasman, *Software Architecture in Practice*, Addison Wesley 1998.
3. L.Cardelli and A.Gordon, "Mobile Ambients", in Nivat (ed), *FoSSACs'98*, LNCS 1378, 140-155, Springer-Verlag, 1998.
4. K.Chandy and J.Misra, *Parallel Program Design – A Foundation*, Addison-Wesley 1988.
5. G.Cugola, C.Ghezzi, G.P.Picco and G.Vigna, "Analyzing Mobile Code Languages", J.Vitek and C.Tschudin, eds., *Mobile Object Systems*, LNCS 1222, Springer-Verlag, 1997.
6. J.L.Fiadeiro and T.Maibaum, "Categorical Semantics of Parallel Program Design", *Science of Computer Programming* 28, 111-138, 1997.
7. J.L.Fiadeiro and A.Lopes, "Algebraic Semantics of Coordination, or what is in a signature?", in A. Haeber (ed), *Proc. International Conference on Algebraic Methodology and Software Technology (AMAST'98)*, LNCS 1548, Springer-Verlag 1999.
8. J.L.Fiadeiro, A.Lopes and M.Wermelinger, "A Mathematical Semantics for Architectural Connectors", to appear in the Lecture notes of the Summer School on Generic Programming, LNCS, Springer-Verlag, 2003.
9. N.Francez and I.Forman, *Interacting Processes,* Addison-Wesley 1996.
10. S.Katz, "A Superimposition Control Construct for Distributed Systems", *ACM TOPLAS* 15(2), 1993, 337-356.
11. A.Lopes, J.L.Fiadeiro and M.Wermelinger, "Architectural Primitives for Distribution and Mobility", *SIGSOFT 2002/FSE-10*, 41-50, ACM Press, 2002.
12. A.Lopes, J.L.Fiadeiro and M.Wermelinger, "A Compositional Approach to Connectors Construction", *Recent Trends in Algebraic Development Techniques – 14th Workshop on ADTs*, LNCS 2267, 201-220, Springer-Verlag, 2001.
13. G.-C.Roman, P.J.McCann and J.Y.Plun, "Mobile UNITY: reasoning and specification in mobile computing", *ACM TOSEM*, 6(3), 250-282, 1997.
14. G.-C.Roman, A.L.Murphy and G.P.Picco, "Coordination and Mobility", in A.Omicini et al (eds), Coordination of Internet designs: Models, Techniques, and Applications, 253-273, Springer-Verlag, 2001.
15. J.Stamos and D.Gifford, "Remote Evaluation", *ACM TOPLAS* 12 (4),537-565, 1990.]
16. M.Wermelinger and J.L.Fiadeiro, "Connectors for Mobile Programs", *IEEE TOSE* 24(5), 331-341, 1998.
17. M.Wermelinger, A.Lopes and J.Fiadeiro, "A Graph Based Architectural (Re)configuration Language", *Proc. 8th ESEC/9th FSE*, 21-32, ACM Press, 2001.

Foundations of Heterogeneous Specification

Till Mossakowski

BISS, Dept. of Computer Science, University of Bremen

Abstract. We provide a semantic basis for heterogeneous specifications that not only involve different logics, but also different kinds of translations between these. We show that Grothendieck institutions based on spans of (co)morphisms can serve as a unifying framework providing a simple but powerful semantics for heterogeneous specification.

1 Introduction

For the specification of large software systems, heterogeneous multi-logic specifications are needed, since complex problems have different aspects that are best specified in different logics. A combination of all the used logics would become too complex in many cases. Moreover, using heterogeneous specifications, different approaches being developed at different sites can be related, i.e. there is a formal interoperability among languages and tools. In many cases, specialized languages and tools have their strengths in particular aspects. Using heterogeneous specification, these strengths can be combined with comparably small effort.

The most prominent approach to heterogeneous specification is CafeOBJ with its cube of eight logics and twelve projections (formalized as *institution morphisms*) among them [13], and having a semantics based on Diaconescu's notion of Grothendieck institution [12]. However, this approach has a limitation: only one type of translation between institution is used, namely institution morphisms. Tarlecki [47] is more general, he introduces a whole bunch of heterogeneous constructs for different kinds of translations. However, only one kind of translation can be used at a time. The goal of the present work is to overcome these limitations while simultaneously staying as simple as possible.

2 Institutions and Their (Co)Morphisms

Following [21], we formalize logics as institutions.

Definition 1. *An* institution $I = (\mathbf{Sign}^I, \mathbf{Sen}^I, \mathbf{Mod}^I, \models^I)$ *consists of*

- *a category $\mathbf{Sign}^I$ of signatures,*
- *a functor $\mathbf{Sen}^I \colon \mathbf{Sign}^I \longrightarrow \mathbf{Set}$ giving, for each signature Σ, the set of sentences $\mathbf{Sen}^I(\Sigma)$, and for each signature morphism $\sigma \colon \Sigma \longrightarrow \Sigma'$, the sentence translation map $\mathbf{Sen}^I(\sigma) \colon \mathbf{Sen}^I(\Sigma) \longrightarrow \mathbf{Sen}^I(\Sigma')$, where often $\mathbf{Sen}^I(\sigma)(\varphi)$ is written as $\sigma(\varphi)$,*

M. Wirsing, D. Pattinson, and R. Hennicker (Eds.): WADT 2002, LNCS 2755, pp. 359–375, 2003.
© Springer-Verlag Berlin Heidelberg 2003

– a functor $\mathbf{Mod}^I \colon (\mathbf{Sign}^I)^{op} \longrightarrow \mathcal{CAT}^1$ giving, for each signature Σ, the category of models $\mathbf{Mod}^I(\Sigma)$, and for each signature morphism $\sigma \colon \Sigma \longrightarrow \Sigma'$, the reduct functor $\mathbf{Mod}^I(\sigma) \colon \mathbf{Mod}^I(\Sigma') \longrightarrow \mathbf{Mod}^I(\Sigma)$, where often $\mathbf{Mod}^I(\sigma)$ (M') is written as $M'|_\sigma$ (the σ-reduct of M'),
– a satisfaction relation $\models^I_\Sigma \subseteq |\mathbf{Mod}^I(\Sigma)| \times \mathbf{Sen}^I(\Sigma)$ for each $\Sigma \in \mathbf{Sign}^I$,

such that for each $\sigma \colon \Sigma \longrightarrow \Sigma'$ in $\mathbf{Sign}^I$ the following satisfaction condition holds:

$$M' \models^I_{\Sigma'} \sigma(\varphi) \Leftrightarrow M'|_\sigma \models^I_\Sigma \varphi$$

for each $M' \in \mathbf{Mod}^I(\Sigma')$ and $\varphi \in \mathbf{Sen}^I(\Sigma)$.   □

The notion of institutions gains much of its importance by the fact that several languages for modularizing specifications have been developed in a completely institution independent way [42, 16, 14, 22, 15, 34], one of which also has been extended to the heterogeneous case [47]. Most of their constructs can be translated into the formalism of *development graphs* introduced below, which hence can be seen as a core formalism for structured and heterogeneous theorem proving. For the language CASL, such a translation has been laid out explicitly in [4].

Definition 2. *Given an arbitrary but fixed institution I, a development graph over I is an acyclic directed graph $\mathcal{S} = \langle \mathcal{N}, \mathcal{L} \rangle$.*

$\mathcal{N}$ is a set of nodes. Each node $N \in \mathcal{N}$ is a tuple (Σ^N, Γ^N) such that $\Sigma^N \in$ $\mathbf{Sign}^I$ is a signature and $\Gamma^N \subseteq \mathbf{Sen}^I(\Sigma^N)$ is the set of local axioms of N.

$\mathcal{L}$ is a set of directed links, so-called definition links, between elements of $\mathcal{N}$. Each definition link from a node M to a node N is either

– *global (denoted $M \xrightarrow{\sigma} N$), annotated with a signature morphism σ :* $\Sigma^M \to \Sigma^N \in \mathbf{Sign}^I$, *or*
– *hiding (denoted $M \xrightarrow[h]{\sigma} N$), annotated with a signature morphism σ :* $\Sigma^N \to \Sigma^M \in \mathbf{Sign}^I$ *going against the direction of the link. Typically, σ will be an inclusion, and the symbols of Σ^M not in Σ^N will be hidden.*

What is the meaning of such development graphs? Development graphs without hiding have a theory-level semantics, see [4]. For development graphs with hiding, a model-level semantics seems to be more appropriate:

Definition 3. *Given a node $N \in \mathcal{N}$, its associated class $\mathbf{Mod}_\mathcal{S}(N)^2$ of models (or N-models for short) consists of those Σ^N-models n for which*

– *n satisfies the local axioms Γ^N,*
– *for each $K \xrightarrow{\sigma} N \in \mathcal{S}$, $n|_\sigma$ is a K-model, and*
– *for each $K \xrightarrow[h]{\sigma} N \in \mathcal{S}$, n has a σ-expansion k (i.e. $k|_\sigma = n$) which is a K-model.*

---

[1] $\mathcal{CAT}$ be the (quasi-)category of categories and functors.
[2] $\mathbf{Mod}_\mathcal{S}$ is not to be confused with the model functor $\mathbf{Mod}$ of the institution.

Complementary to definition and hiding links, which *define* the theories of related nodes, we introduce the notion of a *theorem link* with the help of which we are able to *postulate* relations between different theories. Global theorem links[3] (denoted by $N - \overset{\sigma}{-} \blacktriangleright M$, where $\sigma\colon \Sigma^N \longrightarrow \Sigma^M$) are the central data structure to represent proof obligations arising in formal developments.

Definition 4. *Let S be a development graph. S implies a global theorem link* $N - \overset{\sigma}{-} \blacktriangleright M$ *(denoted $S \models N - \overset{\sigma}{-} \blacktriangleright M$), iff for all $m \in \mathbf{Mod}_S(M)$, $m|_\sigma \in \mathbf{Mod}_S(N)$.*

We now come to the task of relating different institutions. Institution *morphisms* [21] relate two given institutions. A typical situation is that an institution morphism expresses the fact that a "larger" institution *is built upon* a "smaller" institution by *projecting* the "larger" institution onto the "smaller" one.

Given institutions I and J, an *institution morphism* [21] $\mu = (\Phi, \alpha, \beta)\colon I \longrightarrow J$ consists of

- a functor $\Phi\colon \mathbf{Sign}^I \longrightarrow \mathbf{Sign}^J$,
- a natural transformation $\alpha\colon \mathbf{Sen}^J \circ \Phi \longrightarrow \mathbf{Sen}^I$ and
- a natural transformation $\beta\colon \mathbf{Mod}^I \longrightarrow \mathbf{Mod}^J \circ \Phi^{op}$,

such that the following *satisfaction condition* is satisfied for all $\Sigma \in \mathbf{Sign}^I$, $M \in \mathbf{Mod}^I(\Sigma)$ and $\varphi' \in \mathbf{Sen}^J(\Phi(\Sigma))$:

$$M \models_\Sigma^I \alpha_\Sigma(\varphi') \Leftrightarrow \beta_\Sigma(M) \models_{\Phi(\Sigma)}^J \varphi'$$

The notion of institution morphism can be varied in several ways by changing the directions of the arrows or even, in the case of semi-morphisms, omitting the arrows [20, 46]:

	morphism	comorphism	
Sign	$\longrightarrow$	$\longrightarrow$	**Sign**′
Sen	$\longleftarrow$	$\longrightarrow$	**Sen**′ $\circ\, \Phi$
Mod	$\longrightarrow$	$\longleftarrow$	**Mod**′ $\circ\, \Phi$

	forward morphism	forward comorphism	
Sign	$\longrightarrow$	$\longrightarrow$	**Sign**′
Sen	$\longrightarrow$	$\longleftarrow$	**Sen**′ $\circ\, \Phi$
Mod	$\longrightarrow$	$\longleftarrow$	**Mod**′ $\circ\, \Phi$

	semi morphism	semi comorphism	
Sign	$\longrightarrow$	$\longrightarrow$	**Sign**′
Sen			**Sen**′ $\circ\, \Phi$
Mod	$\longrightarrow$	$\longleftarrow$	**Mod**′ $\circ\, \Phi$

The respective satisfaction conditions are quite obvious (note that for semi-(co)morphisms, none is required).

---

[3] There are also local and hiding theorem links, which are omitted here for simplicity.

Finally, each type of morphism also comes in a *simple theoroidal* variant [20], meaning that signatures may be mapped to theories. Following [26], the category **Th** of theories has as objects theories, i.e. signatures plus sets of axioms. A theory morphism is a signature morphism mapping axioms to logical consequences. Let $Sig: \textbf{Th} \longrightarrow \textbf{Sign}$ be the functor forgetting axioms, and $\iota: \textbf{Sign} \longrightarrow \textbf{Th}$ denote the obvious inclusion, which is a right inverse to Sig.

Again following [26], a theoroidal comorphism $\mu = (\Phi, \alpha, \beta): I \longrightarrow J$ is said to be a *subinstitution comorphism* (and I is said to be a subinstitution of J) if Φ is an embedding of categories, α is a pointwise injection, and β is a natural isomorphism.

In the literature, a whole bunch of different types of translations has been used. The following table partitions them by some informal classification scheme (a "th" stands for the simple theoroidal case, "semi" denotes semi-morphisms, and an "x" stand for folklore knowledge or trivialities):

	(semi) morphism	(theoroidal) comorphism	(theoroidal) forward morphism	forward comorphism
Inclusion	[41]	[2, 26]	x	x
Coding	$([41, 1])^4$	th [3, 8, 9, 24] [25–28, 31, 33, 45]	th [5, 48], $([39, 40, 44])^5$	
Projection	[2, 41, 13]			
Feature interaction	[30]	th		
Implementation	semi [43, 46, 41]	x	[48]	

3   Heterogeneous Specification

One typical scenario (cf. e.g. [19, 18]) of heterogeneity arises in the specification of reactive systems: some equational or first-order logic is used to specify the data (here, lists over arbitrary elements), some process algebra (here, CSP) is used to describe the system (here, a buffer implemented as a list), and some temporal logic is used to state fairness or eventuality properties that go beyond the expressiveness of the process algebra (here, we express the fairness property that the buffer cannot read infinitely often without writing). A corresponding heterogeneous specification (using the structuring constructs of CASL) is given in Fig. 1, the corresponding development graph in Fig. 2. Here, $(List, Ax)$ is a specification of lists, Buf is the buffer process, $List'$ is the signature resulting from the translation to $CFOL^=$-LTL, and $Fair$ is the fairness axiom.

Actually, one should add that the process Buf does not meet the fairness constraint, since it can read infinitely often without ever writing. However, a simplistic buffer such as

---

[4] It is not entirely clear whether these should be really called encodings, since —unlike the other codings in this row— it is not clear that they are suitable for re-use of theorem provers.

[5] Salibra and Scollo introduce a relaxed kind of forward morphism mapping models to sets of models.

logic CSP-CFOL$^=$
spec BUFFER $=$
    data LIST
    channels $read, write : Elem$
    process let $Buf(l : List[Elem]) =$
                     $read?x \rightarrow Buf(cons(x, nil))$
                  $\Box$ **if** $l = nil$ **then** $STOP$
                      else $write!last(l) \rightarrow Buf(rest(l))$
        in $Buf(nil)$
    with logic $\rightarrow$ CFOL$^=$-LTL
    then %implies
    $\forall x : ds \, . \, in\_any\_case(x, always \; eventually \; label\_cond(y \, . \, fst(y) = write))$
       %%   Roughly corresponds to $AGF \; fst(label) = write$ in CTL*
end

Fig. 1. A sample heterogeneous specification.

$$(List, Ax) \in CFOL^= \xrightarrow{\quad pr \quad} (List, \{Buf\}) \in \text{CSP-}CFOL^=$$

$$\downarrow toLTL$$

$$(List', \{Fair\}) \in CFOL^=\text{-LTL}$$

Fig. 2. An informal sample heterogeneous development graph.

$$Copy = read?x \rightarrow write!x \rightarrow Copy$$

satisfies the fairness constraint, and so does a buffer using bounded lists.

We now briefly introduce several institutions involved:

$FOL^=$ is *many-sorted first-order logic with equality.* Signatures consist of a set of sorts, a set of function symbols and a set of predicate symbols (each symbol coming with a string of argument sorts and, for function symbols, a result sort). Signature morphisms map the three components in a compatible way. Models are first order structures, and sentences are the usual first-order sentences built from equations, predicate applications and logical connectives and quantifiers $\forall, \exists$. Sentence translations and model reducts are quite straightforward. Satisfaction is defined inductively in the usual way. A detailed description of this institution can be found in [21].

$CFOL^=$ adds new sentences, namely *sort generation generation constraints*, to $FOL^=$. These express that a particular set of sorts is generated by terms built from a particular set of operations (and possibly variables valued with values from other sorts). This allows specifying inductive datatypes like lists. Details can be found in the semantics of CASL [11, 31]. Actually, $CFOL^=$ is a subinstitution of CASL. CASL additionally admits the use of subsorts and partiality, but we omit these here for simplicity.

CSP-$CFOL^=$ (actually a subinstitution of CSP-CASL [36, 37]) combines $CFOL^=$ with the process algebra CSP. Signatures and signature morphisms are those from $CFOL^=$, but restricted to signature morphisms that are injective on sorts.

There are several notions of model for CSP-$CFOL^=$. We here choose the most informative one, based on labeled transition systems (LTS) [38]. Thus, models are $CFOL^=$-models, possibly equipped with an LTS being labeled in the disjoint union of all carriers. On the $CFOL^=$-part, reducts are as in $CFOL^=$. If a model is not equipped with an LTS, neither is its reduct. If a model is equipped with an LTS, and the LTS is labeled only with labels from carriers of the $CFOL^=$-reduct, it is quite straightforward to construe the LTS as an LTS for the $CFOL^=$-reduct (injectivity of signature morphisms on sorts ensures that carrier sets are not doubled). Otherwise, the LTS is deleted.

Sentences are either $CFOL^=$ sentences, or CSP process terms [23, 38] involving $CFOL^=$-terms in place of alphabet letters. Sentence translation is straightforward.

Satisfaction for $CFOL^=$ sentences is as in $CFOL^=$. A CSP process term P is evaluated using the $CFOL^=$-part a model M, leading to an LTS L with labels in the disjoint union of all carriers. Now M satisfies P iff M is equipped with L. Details can be found in [36, 37].

$CFOL^=$-LTL (actually a subinstitution of CASL-LTL [35]) combines $CFOL^=$ with the computation tree logic CTL* [17][6].

Signatures are $CFOL^=$-signatures with

- a distinguished set DS of *dynamic sorts*,
- an injective assignment of *label sorts Label_ds* (outside DS) to dynamic sorts ds, such that
- there exists a transition predicate $\_ \xrightarrow{\;\;} \_ : ds \times Label\_ds \times ds$ for each dynamic sort ds.

Signature morphisms are $CFOL^=$-signature morphisms preserving the extra structure on the nose (the latter is also called the *dynamic part* of the signature).

Models and model reducts are inherited from $CFOL^=$. The presence of the transition predicates means that each dynamic sort is interpreted as an LTS. Sentences are either first-order sentences, or CTL* formulae anchored by the elements of dynamic sorts. Satisfaction is that of CTL*. Details can be found in [35].

We will use the notation $\mathbf{Sign}^{CFOL^=}$ etc. to denote the individual components of these institutions.

Among these institutions, we now introduce some morphisms and comorphisms (cf. Fig. 3):

---

[6] Actually, the "LTL" in CASL-LTL is a bit misleading. It does not stand for linear temporal logic, as one might expect, but for labeled transition logic.

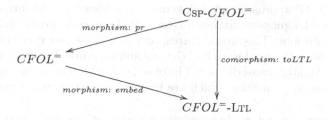

Fig. 3. A (non-commutative) diagram of institutions an (co)morphisms.

- A morphism pr: CSP-$CFOL^=$ $\longrightarrow$ $CFOL^=$. At the signature and sentence level, it is the obvious inclusion. For models, just the LTS (if present) is forgotten.
- A theoroidal semi-comorphism $toLTL$: CSP-$CFOL^=$ $\longrightarrow$ $CFOL^=$-LTL. A signature is extended by a dynamic sort ds, a label sort $Label\_ds$ and a transition predicate, and injection operations inj: $s \longrightarrow Label_{ds}$ for each sort s. These are axiomatized to be injective and to jointly generate $Label\_ds$. Signature morphisms are just extended to map the extra structure on the nose in the obvious way. (Actually, the translation of signature morphisms is only defined for identity signature morphisms; hence we have a *partial* semi-comorphism only. We will not formalize this here, since Section 6 offers a simple solution.)
  A $CFOL^=$-LTL-model is translated to a CSP-$CFOL^=$-model by forgetting the interpretation of the dynamic part to get a $CFOL^=$-model, and equipping it with the LTS determined by the interpretation of the transition relation.
- A comorphism $embed$: $CFOL^=$ $\longrightarrow$ $CFOL^=$-LTL. This is just the obvious subinstitution comorphism.

This shows that practical examples may involve several types of translation between institutions. One might argue that one could try to modify the above introduced institutions in such a way that only one type of translation would be needed. However, the institutions are taken from the literature[7], and it would require research effort (and in some cases tool development) in order to change them.

4 The Bi-Grothendieck Institution

How can we give a precise meaning to the development graph involving several kinds of translations between institutions in Fig. 2?

---

[7] In the case of CSP-$CFOL^-$, the present author has worked on the formalization as an institution [37], but not on the design [36]. There seems to be no alternative formulation of CSP-$CFOL^=$ as an institution simplifying the above picture. Moreover, the problems to be solved when formalizing the quite popular language LOTOS [7] as an institution should be similar to those of CSP-$CFOL^=$.

Tarlecki's [47] approach to heterogeneous specification is to introduce a new heterogeneous language construct for each of the various kinds of translations between institutions. This would correspond to adding new types of definition, hiding and theorem links for each of the translation kinds. However, it is unclear how several kinds of translations will interact, e.g. with respect to amalgamation and interpolation properties, which are important for structured proof systems [6, 32].

Even if one does not use heterogeneous specifications *simultaneously* involving different kinds of translations, a unifying framework is highly desirable in order not to have to switch to a different tool when considering a heterogeneous specification with a different type of translation.

A good way to deal with these problems is to flatten the graph of institutions and translations, as it is done with Diaconescu's Grothendieck institution [12]. We here recall the Grothendieck institution for the comorphism-based case [29]:

Definition 5. *An* indexed coinstitution *is a functor* $\mathcal{I}\colon Ind^{op} \longrightarrow$ **CoIns** *into the category* **CoIns** *of institutions and institution comorphisms*[8]. *A* discrete indexed coinstitution *is one with Ind discrete.* □

The basic idea of the Grothendieck institution is that all signatures of all institutions are put side by side, and a signature morphism in this large realm of signatures consists of an intra-institution signature morphism plus an inter-institution translation (along some institution comorphism). The other components are then defined in a straightforward way.

Definition 6. *Given an* indexed coinstitution $\mathcal{I}\colon Ind^{op} \longrightarrow$ **CoIns**, *define the* Grothendieck institution $\mathcal{I}^{\#}$ *as follows:*

- *signatures in $\mathcal{I}^{\#}$ are pairs (Σ, i), where $i \in |Ind|$ and Σ a signature in the institution $\mathcal{I}(i)$,*
- *signature morphisms $(\sigma, e)\colon (\Sigma_1, i) \longrightarrow (\Sigma_2, j)$ consist of a morphism $e\colon j \longrightarrow i \in Ind$ and a signature morphism $\sigma\colon \Phi^{\mathcal{I}(e)}(\Sigma_1) \longrightarrow \Sigma_2$ (here, $\mathcal{I}(e)\colon \mathcal{I}(i) \longrightarrow \mathcal{I}(j)$ is the institution comorphism corresponding to the arrow $e\colon j \longrightarrow i$ in the indexed coinstitution, and $\Phi^{\mathcal{I}(e)}$ is its signature translation component),*
- *the (Σ, i)-sentences are the Σ-sentences in $\mathcal{I}(i)$, and sentence translation along (σ, e) is the composition of sentence translation along σ with sentence translation along $\mathcal{I}(e)$,*
- *the (Σ, i)-models are the Σ-models in $\mathcal{I}(i)$, and model reduction along (σ, e) is the composition of model translation along $\mathcal{I}(e)$ with model reduction along σ, and*
- *satisfaction w.r.t. (Σ, i) is satisfaction w.r.t. Σ in $\mathcal{I}(i)$.* □

---

[8] Indeed, the name is justified by the fact that the category of institutions and institution comorphisms is isomorphic to the category of coinstitutions and coinstitution morphisms. A coinstitution is an institution with model translations covariant to signature morphisms, while sentence translations are contravariant.

A Grothendieck institution for an indexed institution $\mathcal{I}: Ind^{op} \longrightarrow \mathbf{Ins}$ can be defined similarly[12], it will also be denoted by $\mathcal{I}^{\#}$. By contrast, the Grothendieck construction does not obviously generalize to diagrams consisting of forward or semi (co)morphisms, because of the lacking (contravariance between model and) sentence translation. Let us therefore for a moment concentrate on morphisms and comorphisms only.

We consider heterogeneous specification over a set of institutions, a set of morphisms and a set of comorphisms. This is formalized as an indexed institution $\mathcal{I}_m$ (collecting the morphisms) together with an indexed coinstitution $\mathcal{I}_c$ (collecting the comorphisms), both with the same underlying set of institutions, regarded as a discrete indexed institution $\mathcal{I}_0$.

Definition 7. *Let* $(\mathcal{I}_m, \mathcal{I}_c, \mathcal{I}_0)$, *with* $\mathcal{I}_m$ *an indexed institution,* $\mathcal{I}_c$ *an indexed coinstitution and* $\mathcal{I}_0$ *a discrete indexed institution be given, such that* $|Ind_m| = |Ind_c| = |Ind_0|$, *and* $\mathcal{I}_m$, $\mathcal{I}_c$ *and* $\mathcal{I}_0$ *agree on these.*

Then we form the Grothendieck institutions $\mathcal{I}_0^{\#}$, $\mathcal{I}_m^{\#}$ *and* $\mathcal{I}_c^{\#}$. *Since* $\mathcal{I}_0^{\#}$ *obviously is included in* $\mathcal{I}_m^{\#}$ *and* $\mathcal{I}_c^{\#}$ *via a (co)morphism, we can take the pushout*

$$
\begin{array}{ccc}
\mathcal{I}_0^{\#} & \longrightarrow & \mathcal{I}_m^{\#} \\
\downarrow & & \downarrow \\
\mathcal{I}_c^{\#} & \dashrightarrow & \mathcal{J}
\end{array}
$$

in the category of institutions and institution morphisms (or comorphisms, this would make no difference here). The pushout in either category exists by results of [20]. By abuse of notation, we will denote the pushout $\mathcal{J}$ *by* $(\mathcal{I}_m, \mathcal{I}_c)^{\#}$. *It will be called the* Bi-Grothendieck institution.

The heterogeneous development graph in Fig. 2 can now formally be understood as a development graph over the Bi-Grothendieck institution.

5 Inducibility

The Bi-Grothendieck institution is quite complex, and it is not immediately clear how to obtain e.g. proof support for it. It is therefore tempting to try to reduce the complexity of this construction by mapping morphisms to comorphisms or vice versa. This can be done by weakening the adjunction between morphisms and comorphisms introduced in [2]:

Definition 8. *Given an institution comorphism* $\rho = (\Phi, \alpha, \beta): I \longrightarrow J$, *a functor* $\Psi: \mathbf{Sign}^J \longrightarrow \mathbf{Sign}^I$ *and a natural transformation* $\varepsilon: \Phi \circ \Psi \longrightarrow Id$, *we say that* ρ ε-*induces the institution morphism* $\mu = (\Psi, \bar{\alpha}, \bar{\beta}): J \longrightarrow I$ *given by*

$$
\bar{\alpha} = (\mathbf{Sen}^J \cdot \varepsilon) \circ (\alpha \cdot \Psi)
$$
$$
\bar{\beta} = (\beta \cdot \Psi^{op}) \circ (\mathbf{Mod}^J \cdot \varepsilon^{op})
$$

A morphism that is ε-*induced by some comorphism is called* inducible.

Dually, given an institution morphism $\mu = (\Psi, \bar{\alpha}, \bar{\beta}): J \longrightarrow I$, a functor $\Phi: \mathbf{Sign}^I \longrightarrow \mathbf{Sign}^J$ and a natural transformation $\eta: Id \longrightarrow \Psi \circ \Phi$, we say that μ η-induces the institution comorphism $\rho = (\Phi, \alpha, \beta): I \longrightarrow J$ given by

$$\alpha = (\bar{\alpha} \cdot \Phi) \circ (\mathbf{Sen}^I \cdot \eta)$$
$$\beta = (\mathbf{Mod}^I \cdot \eta^{op}) \circ (\bar{\alpha} \cdot \Phi)$$

A comorphism that is η-induced by some morphism is called inducible. *Moreover, it is straightforward to extend inducibility to the simple theoroidal case. Here,* $\mathbf{Sign}^I \xrightarrow{\Phi} \mathbf{Sign}^J$ *has to be replaced with* $\mathbf{Sign} \xrightarrow{\Phi} \mathbf{Th}^J \xrightarrow{Sig} \mathbf{Sign}^J$, *leading to the equations*

$$\alpha = (\bar{\alpha} \cdot Sig \cdot \Phi) \circ (\mathbf{Sen}^I \cdot \eta)$$
$$\beta = (\mathbf{Mod}^I \cdot \eta^{op}) \circ (\bar{\beta} \cdot Sig \cdot \Phi)$$

Furthermore, inducibility also extend to semi-(co)morphisms.   □

With this, we can easily obtain the desired reduction:

Theorem 9. *Let $(\mathcal{I}_m, \mathcal{I}_c, \mathcal{I}_0)$ as in Definition 7 be given.*

If each morphism in $\mathcal{I}_m$ is ε-induced by some comorphism in $\mathcal{I}_c$, then there is a retraction of $(\mathcal{I}_m, \mathcal{I}_c)^\#$ onto $\mathcal{I}_c^\#$.

Dually, if each comorphism in $\mathcal{I}_c$ is η-induced by some morphism in $\mathcal{I}_m$, then there is a retraction of $(\mathcal{I}_m, \mathcal{I}_c)^\#$ onto $\mathcal{I}_m^\#$.

Proof. Consider the pushout construction in Definition 7. Clearly, $\mathcal{I}_m^\#$, $\mathcal{I}_c^\#$ and $\mathcal{I}_0^\#$ all have the same object class. Moreover, since $\mathcal{I}_0$ is discrete, the signature morphisms in $\mathcal{I}_0^\#$ are basically those of the individual institutions. With this, it is easy to see that the signature morphisms in $(\mathcal{I}_m, \mathcal{I}_c)^\#$ are paths of morphisms coming from $\mathcal{I}_m^\#$ and $\mathcal{I}_c^\#$ in an alternating way.

The retraction of $(\mathcal{I}_m, \mathcal{I}_c)^\#$ onto $\mathcal{I}_c^\#$ (having the obvious inclusion as right inverse) is therefore given by the identity for the objects, while for a path of alternating morphisms, each morphism

$$(\Sigma_1 \xrightarrow{\sigma} \Psi^d(\Sigma_2) \, , \, i \xrightarrow{d} j \,): (\Sigma_1, i) \longrightarrow (\Sigma_2, j)$$

from $\mathcal{I}_m^\#$ is replaced with

$$(\Phi^e(\Sigma_1) \xrightarrow{\Phi^e(\sigma)} \Phi^e(\Psi^d(\Sigma_2)) \xrightarrow{\varepsilon_{\Sigma_2}} \Sigma_2 \, , \, j \xrightarrow{e} i \,),$$

where e is the index of the comorphism inducing $\mathcal{I}_m(d)$, and ε the corresponding natural transformation. Since all the resulting morphisms live in $\mathcal{I}_c^\#$, they can be composed to a single morphism.

The other statement follows by a dual argument.   □

However, unfortunately there are practically relevant situations where this is not applicable.

Proposition 10. *Neither the morphism pr:* CSP-$CFOL^= \longrightarrow CFOL^=$ *nor the theoroidal semi-comorphism toLTL:* CSP-$CFOL^= \longrightarrow CFOL^=$-LTL *is inducible.*

Proof. Assume that $pr = (\Psi, \bar{\alpha}, \bar{\beta})$: CSP-$CFOL^= \longrightarrow CFOL^=$ is ε-induced by a comorphism $\rho = (\Phi, \alpha, \beta)$: $CFOL^= \longrightarrow$ CSP-$CFOL^=$. Let Σ_1 consist of a sort s and Σ_2 of sorts t and u (both seen as CSP-$CFOL^=$-signatures). Let $\sigma \colon \Psi(\Sigma_2) \longrightarrow \Psi(\Sigma_1)$ map both t and u to s (recall that Ψ is just an inclusion). Now $\bar{\beta}_{\Sigma_2}$ just forgets the optional LTS component, and hence is surjective. Since $\bar{\beta}_{\Sigma_2} = \beta_{\Psi(\Sigma_2)} \circ \varepsilon_{\Sigma_2}$, $\beta_{\Psi(\Sigma_2)}$ is surjective as well. Since all signature morphisms in CSP-$CFOL^=$ are injective (and carriers are assumed to be non-empty), the corresponding reduct functors are easily seen to be surjective. Hence, the lower right path in the naturality diagram for β

$$
\begin{array}{ccc}
\mathbf{Mod}^{\text{CSP-}CFOL^=}(\Phi(\Psi(\Sigma_1))) & \xrightarrow{\ \beta_{\Psi(\Sigma_1)}\ } & \mathbf{Mod}^{CFOL^=}(\Psi(\Sigma_1)) \\[2pt]
\Big\downarrow{\scriptstyle -|_{\Phi(\sigma)}} & & \Big\downarrow{\scriptstyle -|_{\sigma}} \\[2pt]
\mathbf{Mod}^{\text{CSP-}CFOL^=}(\Phi(\Psi(\Sigma_2))) & \xrightarrow{\ \beta_{\Psi(\Sigma_2)}\ } & \mathbf{Mod}^{CFOL^=}(\Psi(\Sigma_2))
\end{array}
$$

is surjective as well. Hence, also the upper left path must be surjective, and hence its second component $-|_\sigma$. But $-|_\sigma$ just doubles the carrier set, and this is clearly not surjective.

The semi-comorphism $toLTL$ is more precisely defined on the subinstitution CSP-$CFOL^=$-d consisting of identity signature morphisms only, i.e. $toLTL = (\Phi, \beta)$: CSP-$CFOL^=$-$d \longrightarrow CFOL^=$-LTL. Assume that it is η-induced by a semimorphism $\mu = (\Psi, \bar{\beta})$: $CFOL^=$-LTL $\longrightarrow$ CSP-$CFOL^=$-d. Since all signature morphisms in CSP-$CFOL^=$-d are identities, η is the identity as well. Hence, $\bar{\beta} \cdot \Phi = \beta$, and one easily obtains a contradiction to the $\bar{\beta}$-naturality diagram for a signature morphism $\sigma \colon \Phi(\Sigma_1) \longrightarrow \Phi(\Sigma_2)$ in $CFOL^=$-LTL. This proof relies on the severe restriction of the CSP-$CFOL^=$-d signature morphisms; however, also a proof not exploiting this is possible. □

6   Spans of Comorphisms

The method of the previous section to use inducibility to reduce the complexity of heterogeneous specifications involving different kinds of translations between institutions works for some cases, but the counterexamples of Proposition 10 have shown that the method is not general enough.

A more general idea is to express all the different kinds of translations as *spans* of morphisms or of comorphisms. The question is now whether to work with morphisms or with comorphisms. Indeed, comorphisms interact with amalgamation properties in a much simpler way than morphisms do, see Propositions 3.5 and 3.6 of [29]. Amalgamation properties are important in many respects, e.g. for heterogeneous theorem proving [29]. Therefore, we work with

spans of comorphisms. Nevertheless, the results presented below easily dualize
to spans of morphisms.

Each institution morphism $\mu\colon I \longrightarrow J = I \xleftarrow{\quad\alpha\quad} \begin{smallmatrix}\xrightarrow{\;\Psi\;}\\[-2pt]\\[-2pt]\xrightarrow{\;\beta\;}\end{smallmatrix} J$ can be translated

into a span $I \xleftarrow{\;\mu^{+}\;} J \circ \Psi \xrightarrow{\;\mu^{-}\;} J$ of institution comorphisms as follows:

$$
\begin{array}{ccccc}
\mathbf{Sign}^I & \xleftarrow{\quad id \quad} & \mathbf{Sign}^I & \xrightarrow{\quad \Psi \quad} & \mathbf{Sign}^J \\[4pt]
\mathbf{Sen}^I & \xleftarrow{\quad \alpha \quad} & \mathbf{Sen}^J \circ \Psi & \xrightarrow{\quad id \quad} & \mathbf{Sen}^J \circ \Psi \\[4pt]
\mathbf{Mod}^I & \xrightarrow{\quad \beta \quad} & \mathbf{Mod}^J \circ \Psi & \xleftarrow{\quad id \quad} & \mathbf{Mod}^J \circ \Psi
\end{array}
$$

Here, the "middle" institution $J \circ \Psi$ is the institution with signature category
inherited from I, but sentences and models inherited from J via Ψ.

This span construction can also be lifted to the indexed level: Given an
indexed institution $\mathcal{I}_m$ and an indexed coinstitution $\mathcal{I}_c$ (both over the same
set of institutions in the sense of Definition 7), form an indexed coinstitution
$Span(\mathcal{I}_m, \mathcal{I}_c)$ as follows: the index category is obtained by freely adding pairs of
formal morphisms to the index category of $\mathcal{I}_c$; more precisely, one pair (obtained
in a straightforward way) for each span of comorphisms corresponding to a
morphism in $\mathcal{I}_m$.

Unfortunately, we cannot expect that Theorem 9 carries over to the present
situation. But we have some weaker property that still is sufficiently strong for
practical needs:

Theorem 11. *Given an indexed institution $\mathcal{I}_m$ and an indexed coinstitution*
$\mathcal{I}_c$ (both over the same set of institutions in the sense of Definition 7), then
each development graph over the Bi-Grothendieck institution $(\mathcal{I}_m, \mathcal{I}_c)^{\#}$ can be
translated into a development graph over the Grothendieck institution over the
span-based indexed coinstitution $Span(\mathcal{I}_m, \mathcal{I}_c)^{\#}$, such that model categories are
preserved.

Proof. As in the proof of Theorem 9, we rely on the fact that signature morphisms in the Bi-Grothendieck institution $(\mathcal{I}_m, \mathcal{I}_c)^{\#}$ are paths of morphisms
coming from $\mathcal{I}_m^{\#}$ and $\mathcal{I}_c^{\#}$ in an alternating way. A global definition link therefore
has the form

$$
M \xrightarrow{\langle (\sigma_1, e_1), (\sigma_2, d_2), \dots, (\sigma_n, e_n) \rangle} N \,,
$$

where the d_i are from $\mathcal{I}_m$ and the e_i are from $\mathcal{I}_c$ (with (σ_1, e_1), (σ_n, e_n) possibly
not present). The definition link now is replaced by a sequence of definition and
hiding links:

$$
M \xrightarrow{(\sigma_1, e_1)} M_1 \xrightarrow{(\sigma_2, id)} M_2 \xrightarrow[h]{(id, d_2^-)} M_3 \xrightarrow{(id, d_2^+)} M_4 \longrightarrow \cdots \xrightarrow{(\sigma_n, e_n)} N
$$

Here, d_2^- and d_2^+ are the indices for the span of comorphisms associated to
$\mathcal{I}_m(d_2)$, and $M_1, \dots, M_4$ are new nodes with appropriate signatures and no local

axioms. Of course, the path could also start and/or end with a d_i instead of an e_i, but this won't affect the general construction: each path element containing a d_i leads to a sequence of a definition link, a hiding link, and again a definition link, while path elements containing an e_i are just kept. The construction for theorem links is entirely analogous, except that only the last arrow in the sequence has to be a theorem link — the other ones must be definition links.

Let us now come to hiding links. The construction is very similar, so we restrict ourselves to the replacement of individual path elements of form (σ_i, d_i). Such an element leads to

$$\cdots \longrightarrow M_n \xrightarrow[h]{(id,d_i^+)} M_{n+1} \xrightarrow{(id,d_i^-)} M_{n+2} \xrightarrow[h]{(\sigma_i,id)} M_{n+3} \longrightarrow \cdots$$

In comparison to the construction above, here the arrows are reversed, and definition and hiding links interchanged.

It is straightforward to see that the model class is left unchanged by these translations.

$\square$

Consider now a semi morphism $I \xrightarrow[\beta]{\Psi} J$. It can be translated into a span of comorphisms

$$
\begin{array}{ccccc}
\mathbf{Sign} & \xleftarrow{\;id\;} & \mathbf{Sign}^I & \xrightarrow{\;\Psi\;} & \mathbf{Sign}^J \\
\mathbf{Sen}^I & \xleftarrow{\;incl\;} & \emptyset & \xrightarrow{\;incl\;} & \mathbf{Sen}^J \circ \Psi \\
\mathbf{Mod}^I & \xrightarrow{\;\beta\;} & \mathbf{Mod}^J \circ \Psi & \xleftarrow{\;id\;} & \mathbf{Mod}^J \circ \Psi
\end{array}
$$

while a semi-comorphism $I \xrightarrow[\beta]{\Phi} J$ is translated into a span of comorphisms

$$
\begin{array}{ccccc}
\mathbf{Sign} & \xleftarrow{\;id\;} & \mathbf{Sign}^I & \xrightarrow{\;\Phi\;} & \mathbf{Sign}^J \\
\mathbf{Sen}^I & \xleftarrow{\;incl\;} & \emptyset & \xrightarrow{\;incl\;} & \mathbf{Sen}^J \circ \Phi \\
\mathbf{Mod}^I & \xrightarrow{\;id\;} & \mathbf{Mod}^I & \xleftarrow{\;\beta\;} & \mathbf{Mod}^J \circ \Phi
\end{array}
$$

With this, we also can give a semantics to definition, hiding and theorem links involving semi-(co)morphisms. Partial (semi-)comorphisms and a restricted class of forward (co)morphisms can be covered as well.

Example 12. Extend the institutions and (co)morphisms introduced in Section 3 by the following ones:

- $CFOL^=$-inj is the restriction of $CFOL^=$ to signature that are injective on sorts.

- CSP-$CFOL^=$-d-$nosen$ is the restriction of CSP-$CFOL^=$ to the empty set of sentences, for each signature, and to identity signature morphisms.
- The comorphism $pr^-: CFOL^=$-$inj \longrightarrow CFOL^=$ is just the obvious subinstitution inclusion.
- The comorphism $pr^+: CFOL^=$-$inj \longrightarrow$ CSP-$CFOL^=$ behaves very similar to pr: At the signature level, it is the identity, at the sentence level, it is the obvious inclusion. For models, just the LTS (if present) is forgotten.

When applying the construction of Theorem 11 to (a formalized variant of) the development graph given in Fig. 2, we arrive at the development graph shown in Fig. 4 (for simplicity, we index the involved institutions and comorphisms by themselves here).

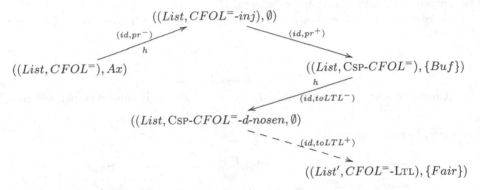

Fig. 4. The sample heterogeneous development graph with spans of comorphisms.

7   Conclusion

We have presented an example heterogeneous specification involving different kinds of translations between the involved institution. We have discussed several ways of giving a semantics to such specifications. The most promising way has turned out to use indexed coinstitutions and their Grothendieck construction as a semantical foundation for heterogeneous specification, and express other types of translation by spans of comorphisms. Of course, the dual view (followed by CafeOBJ), which takes morphism-based Grothendieck institutions as foundation [12], also can be combined with the span approach. However, as shown in [29], the comorphism-based Grothendieck construction interacts more nicely with amalgamation. Hence, we stick to comorphisms here.

Our approach also implies that we can exploit techniques such as heterogeneous borrowing [30] and the heterogeneous proof calculus [29], which are based on comorphisms only, in a much wider context. Tool support for heterogeneous specifications and development graphs is under development in form of the heterogeneous tool set (hets). The latter provides an abstract programming interface for the implementable part of institutions and comorphisms. This will serve as a basis for implementing heterogeneous analysis and proof tools that are based on corresponding tools for the individual logics.

Acknowledgments

Thanks to Andrzej Tarlecki, Joseph Goguen, Grigore Rosu, Serge Autexier and Dieter Hutter for useful cooperation and discussions, to Răzvan Diaconescu for inventing Grothendieck institutions, and to the anonymous referees for valuable criticism and suggestions. I would also like to thank the members of the Common Framework Initiative (CoFI).

This work has been supported by the *Deutsche Forschungsgemeinschaft* under Grant KR 1191/5-1.

References

1. S. Alagi. Institutions: integrating objects, XML and databases. *Information and Software Technology*, 44:207–216, 2002.
2. M. Arrais and J. L. Fiadeiro. Unifying theories in different institutions. In M. Haveraaen, O. Owe, and O.-J. Dahl, editors, *Recent Trends in Data Type Specifications. 11th WADT*, LNCS 1130, pages 81–101. Springer Verlag, 1996.
3. E. Astesiano and M. Cerioli. Relationships between logical frameworks. In M. Bidoit and C. Choppy, editors, *Proc. 8th ADT workshop*, LNCS 655, pages 126–143. Springer Verlag, 1992.
4. S. Autexier, D. Hutter, H. Mantel, and A. Schairer. Towards an evolutionary formal software-development using CASL. In C. Choppy and D. Bert, editors, *Recent Trends in Algebraic Development Techniques, 14th International Workshop, WADT'99, Bonas, France*, LNCS 1827, pages 73–88. Springer-Verlag, 2000.
5. M. Bidoit and R. Hennicker. Using an institution encoding for proving consequences of structured COL-specifications. Talk at the WADT 2002, Frauenchiemsee.
6. T. Borzyszkowski. Logical systems for structured specifications. *Theoretical Computer Science*, 286:197–245, 2002.
7. E. Brinksma, editor. *Information processing systems — open systems interconnection. LOTOS: a formal description technique based on the temporal ordering of observational behaviour*. 1988. International Standard ISO 8807.
8. M. Cerioli. *Relationships between Logical Formalisms*. PhD thesis, TD-4/93, Università di Pisa-Genova-Udine, 1993.
9. M. Cerioli and J. Meseguer. May I borrow your logic? (transporting logical structures along maps). *Theoretical Computer Science*, 173:311–347, 1997.
10. CoFI. The Common Framework Initiative for algebraic specification and development, electronic archives. Notes and Documents accessible from http://www.cofi.info.
11. CoFI Semantics Task Group. CASL – The CoFI Algebraic Specification Language – Semantics. Note S-9 (Documents/CASL/Semantics, version 0.96), in [10], July 1999.
12. R. Diaconescu. Grothendieck institutions. *Applied categorical structures*, 10:383–402, 2002.

13. R. Diaconescu and K. Futatsugi. Logical foundations of CafeOBJ. *Theoretical computer science*, 285:289–318, 2002.
14. R. Diaconescu, J. Goguen, and P. Stefaneas. Logical support for modularisation. In G. Huet and G. Plotkin, editors, *Proceedings of a Workshop on Logical Frameworks*, 1991.
15. F. Durán and J. Meseguer. Structured theories and institutions. In M. Hofmann, G. Rosolini, and D. Pavlovic, editors, *CTCS '99 Conference on Category Theory and Computer Science*, ENTCS 29. 1999.
16. H. Ehrig and B. Mahr. *Fundamentals of Algebraic Specification 2*. Springer Verlag, Heidelberg, 1990.
17. E. A. Emerson. Temporal and Modal Logic. In J. van Leeuwen, editor, *Handbook of Theoretical Computer Science, Volume B*. Elsevier / MIT Press, 1990.
18. B. Ghribi and L. Logrippo. A validation environment for LOTOS. In *Protocol Specification, Testing and Verification*, pages 93–108, 1993.
19. P. Gibson, B. Mermet, and D. Mery. Feature interactions: A mixed semantic model approach. In *IWFM*, 1997.
20. J. Goguen and G. Rosu. Institution morphisms. *Formal aspects of computing*, 13:274–307, 2002.
21. J. A. Goguen and R. M. Burstall. Institutions: Abstract model theory for specification and programming. *Journal of the Association for Computing Machinery*, 39:95–146, 1992. Predecessor in: LNCS 164, 221–256, 1984.
22. J. A. Goguen and W. Tracz. An implementation-oriented semantics for module composition. In G. T. Leavens and M. Sitaraman, editors, *Foundations of Component-Based Systems*, chapter 11, pages 231–263. Cambridge University Press, New York, NY, 2000.
23. C. A. R. Hoare. *Communicating Sequential Processes*. Prentice-Hall, 1985.
24. H.-J. Kreowski and T. Mossakowski. Equivalence and difference of institutions: Simulating Horn clause logic with based algebras. *Mathematical Structures in Computer Science*, 5:189–215, 1995.
25. N. Martí-Oliet and J. Meseguer. From abstract data types to logical frameworks. LNCS 906, pages 48–80. Springer, 1995.
26. J. Meseguer. General logics. In *Logic Colloquium 87*, pages 275–329. North Holland, 1989.
27. J. Meseguer. Conditional rewriting as a unified model of concurrency. *Theoretical Computer Science*, 96(1):73–156, 1992.
28. T. Mossakowski. Equivalences among various logical frameworks of partial algebras. In H. K. Büning, editor, *Computer Science Logic. 9th Workshop, CSL'95.*, LNCS 1092, pages 403–433. Springer Verlag, 1996.
29. T. Mossakowski. Comorphism-based Grothendieck logics. In K. Diks and W. Rytter, editors, *Mathematical foundations of computer science*, LNCS 2420, pages 593–604. Springer, 2002.
30. T. Mossakowski. Heterogeneous development graphs and heterogeneous borrowing. In M. Nielsen and U. Engberg, editors, *Foundations of Software Science and Computation Structures*, LNCS 2303, pages 326–341. Springer-Verlag, 2002.
31. T. Mossakowski. Relating CASL with other specification languages: the institution level. *Theoretical Computer Science*, 286:367–475, 2002.
32. T. Mossakowski, S. Autexier, and D. Hutter. Extending development graphs with hiding. In H. Hußmann, editor, *Fundamental Approaches to Software Engineering*, LNCS 2029, pages 269–283. Springer-Verlag, 2001.

33. T. Mossakowski, Kolyang, and B. Krieg-Brückner. Static semantic analysis and theorem proving for CASL. In F. Parisi Presicce, editor, *Recent trends in algebraic development techniques. Proc. 12th International Workshop*, LNCS 1376, pages 333–348. Springer, 1998.

34. P. D. Mosses. CoFI: The Common Framework Initiative for Algebraic Specification and Development. In *TAPSOFT '97, Proc. Intl. Symp. on Theory and Practice of Software Development*, LNCS 1214, pages 115–137. Springer-Verlag, 1997.

35. G. Reggio, E. Astesiano, and C. Choppy. CASL-LTL - a CASL extension for dynamic reactive systems - summary. Technical Report of DISI - Università di Genova, DISI-TR-99-34, Italy, 2000.

36. M. Roggenbach. CSP-CASL – a new integration of process algebra and algebraic specification. Manuscript, Bremen, submitted for publication.

37. M. Roggenbach and T. Mossakowski. The CSP-CASL institution and its relation to temporal logic. Manuscript, University of Bremen.

38. A. Roscoe. *The Theory and Practice of Concurrency*. Prentice-Hall, 1997.

39. A. Salibra and G. Scollo. A soft stairway to institutions. In M. Bidoit and C. Choppy, editors, *Proc. 8th ADT workshop*, LNCS 655, pages 310–329. Springer Verlag, 1992.

40. A. Salibra and G. Scollo. Interpolation and compactness in categories of pre-institutions. *Mathematical Structures in Computer Science*, 6(3):261–286, June 1996.

41. D. Sannella and A. Tarlecki. *Working with multiple logical systems, In: Foundations of Algebraic Specifications and Formal Program Development*, chapter 10. Cambridge University Press, to appear. See http://zls.mimuw.edu.pl/~tarlecki/book/index.html.

42. D. Sannella and A. Tarlecki. Specifications in an arbitrary institution. *Information and Computation*, 76:165–210, 1988.

43. D. Sannella and A. Tarlecki. Toward formal development of programs from algebraic specifications: implementations revisited. *Acta Inf.*, 25:233–281, 1988.

44. G. Scollo. *On the engineering of logics*. PhD thesis, 1993. University of Twente, Enschede.

45. A. Tarlecki. Institution representation. draft note, 1987.

46. A. Tarlecki. Moving between logical systems. In M. Haveraaen, O. Owe, and O.-J. Dahl, editors, *Recent Trends in Data Type Specifications. 11th Workshop on Specification of Abstract Data Types*, LNCS 1130, pages 478–502. Springer Verlag, 1996.

47. A. Tarlecki. Towards heterogeneous specifications. In D. Gabbay and M. d. Rijke, editors, *Frontiers of Combining Systems 2, 1998*, Studies in Logic and Computation, pages 337–360. Research Studies Press, 2000.

48. U. Wolter, K. Didrich, F. Cornelius, M. Klar, R. Wessäly, and H. Ehrig. How to cope with the spectrum of SPECTRUM. In M. Broy and S. Jähnichen, editors, *KORSO: Methods, Languages and Tools for the Construction of Correct Software*, LNCS 1009, pages 173–189. Springer-Verlag, New York, NY, 1995.

Algebraic-Coalgebraic Specification in CoCasl

Till Mossakowski[1], Horst Reichel[2], Markus Roggenbach[1], and Lutz Schröder[1]

[1] BISS, Dept. of Computer Science, University of Bremen
[2] Institute for Theoretical Computer Science, Technical University of Dresden

Abstract. We introduce CoCasl as a simple coalgebraic extension of the algebraic specification language Casl. CoCasl allows the nested combination of algebraic datatypes and coalgebraic process types. We show that the well-known coalgebraic modal logic can be expressed in CoCasl. We present sufficient criteria for the existence of cofree models, also for several variants of nested cofree and free specifications. Moreover, we describe an extension of the existing proof support for Casl (in the shape of an encoding into higher-order logic) to CoCasl.

In recent years, coalgebra has emerged as a convenient and suitably general way of modeling the reactive behaviour of systems [26]. While algebraic specification deals with *inductive datatypes* generated by constructors, coalgebraic specification deals with *coinductive process types* that are observable by selectors. An important role is played here by *final coalgebras*, which are complete sets of possibly infinite behaviours, such as streams or even the real numbers.

For algebraic specification, the Common Algebraic Specification Language Casl [20] has been designed as a unifying standard, while for the much younger field of coalgebraic specification there is still a divergence of notions and notations. The idea pursued here is to obtain a fruitful synergy by extending Casl with coalgebraic constructs that dualize (in the sense of e.g. [5]) the algebraic constructs already present in Casl.

In more detail, CoCasl provides a basic co-type construct, cogeneratedness constraints, and structured cofree specifications; moreover, coalgebraic modal logic is introduced as syntactical sugar. Co-types serve to describe reactive processes, equipped with observer operations whose role is dual to that of the constructors of a datatype. Cotypes can be qualified as being cogenerated or cofree, respectively, thus imposing full abstractness and realization of all observable behaviours, respectively. The most powerful construct are cofree specifications, which allow specifying final models of arbitrary specifications. Of course, this raises the question for what kinds of specifications such final models actually exist. We provide a sufficient existence condition which covers specifications that employ initially specified datatypes in observer functions and restrict behaviours by modal formulas. For such cases, we also lay out how the existing proof support for Casl, realized by means of an encoding into Isabelle/HOL [16], can be extended to CoCasl. In summary, CoCasl is a syntactically and semantically simple extension of Casl that allows a straightforward treatment of reactive behaviour.

M. Wirsing, D. Pattinson, and R. Hennicker (Eds.): WADT 2002, LNCS 2755, pp. 376–392, 2003.

1   CASL

The specification language CASL (*Common Algebraic Specification Language*) has been designed by CoFI, the international *Common Framework Initiative for Algebraic Specification and Development* [20]. Its features include first-order logic, partial functions, subsorts, sort generation constraints, and structured and architectural specifications. For the language definition and a full formal semantics cf. [20]. An important point here is that the semantics of structured and architectural specifications is independent of the logic employed for basic specifications, so that the language is easily adapted to the extension of the logic envisaged here. That said, CoCASL does introduce one additional structuring concept, namely, cofree specifications.

We now briefly recall the many-sorted CASL institution; subsorting is then defined on top of this. Full details can be found in [20, 17]. A CASL signature consists of a set of sorts, two sets of (total and partial) function symbols, and a set of predicate symbols (each symbol coming with a string of argument sorts and, for function symbols, a result sort). Signature morphisms map the four components in a compatible way. Models are many-sorted partial first order structures. Homomorphisms are so-called weak homomorphisms. That is, they are total as functions, and they preserve (but not necessarily reflect) the definedness of partial functions and the satisfaction of predicates.

Over such a signature, sentences are built from atomic sentence using the usual features of first order logic. Here, an atomic sentence is either a definedness assertion, a strong equation, an existence equation, or a predicate application; see [20] for details. There is an additional type of sentence that goes beyond first-order logic: a *sort generation constraint* states that a given set of sorts is generated by a given set of functions, i.e. that all the values of the generated sorts are reachable by some term in the function symbols, possibly containing variables of other sorts.

The CASL language is defined on top of this institution, offering a richer and more convenient syntax than the plain CASL institution. For instance, it provides powerful constructs for defining datatypes, which are briefly recalled below, in direct comparison to the corresponding CoCASL constructs.

2   Type and Cotype Definitions

The basic CASL construct for type definitions is the **types** construct. It declares constructors and, optionally, selectors for a datatype (or several datatypes at once); both constructors and selectors may be partial. Such a type declaration is expanded into the declaration of the constructor and selector operations and axioms relating the selectors and constructors. Nothing else is said about the type; thus, there may not only be 'junk' and 'confusion', but there may also be rather arbitrary behaviour of the selectors outside the range of the corresponding constructors.

In CoCASL, this construct is complemented by the **cotypes** construct. In its full form, the syntax of this construct is identical to the **types** construct; e.g., one may write

cotype *Process* ::= *cont*(*hd1* :? *Elem*; *next* :? *Process*)
             | *fork*(*hd2* :? *Elem*; *left* :? *Process*; *right* :? *Process*)

thus determining constructors and selectors as for types. However, for cotypes, the constructors are optional and the selectors are mandatory. Moreover, the **cotype** construct introduces a number of additional axioms concerning the domains of definition of the selectors, besides the axioms relating constructors with their selectors as for types:

- the domains of two selectors in the same alternative are the same,
- the domains of two selectors in different alternatives are disjoint, and
- the domains of all selectors of a given sort are jointly exhaustive.

Thus, the alternatives in a cotype are to be understood as parts of a disjoint sum.

Definition 1. A cotype in CoCASL is given by the local environment sorts and the family of observers

$$CT = (S, (obs_{i,j,k} : T_i \to T_{i,j,k})_{i=1...n, j=1...m_i, k=1...r_{i,j}}).$$

Here, S is a set of sorts (the local environment sorts, also called observable sorts), $T_1 \ldots T_n$ are the newly declared process types (or non-observable sorts) in the cotype (which possibly involve mutual recursion, and $obs_{i,j,k}$ is the k-th observer of the j-th alternative in the **cotype** definition of T_i. $T_{i,j,k}$ is the result sort of the observer; it may be either one of the T_i or one of the local environment sorts in S. The *signature* $Sig(CT)$ of a cotype CT consists of the local environment sorts S, the cotype sorts $T_1 \ldots T_n$, and the profiles of the observers; the *theory* $Th(CT)$ also adds the above listed axioms.

Cotypes correspond directly to coalgebras:

Proposition 1. *To a given* CoCASL **cotype** *definition* CT, *one can associate a functor* $F : \mathbf{Set}^n \to \mathbf{Set}^n$ *such that the category of partial* $Th(CT)$-*algebras is isomorphic to the category of* F-*coalgebras. In particular, this implies that all homomorphisms between partial* $Th(CT)$-*algebras are closed* [6], *i.e. not only preserve, but also reflect definedness.*

3   Generation and Cogeneration Constraints

In order to exclude 'junk' from models of datatypes, CASL provides generatedness constraints that essentially introduce (higher order) implicit induction axioms. Dually, CoCASL introduces *cogeneratedness constraints* that amount to an implicit coinduction axiom and thus restrict the models of the datatype to fully abstract ones. This means that equality is the largest congruence w.r.t. the

---

spec STREAM1 [**sort** *Elem*] =
   cogenerated cotype
        Stream ::= *cons*(*hd* : *Elem*; *tl* : *Stream*)
end

Fig. 1. Cogenerated specification of bit streams in CoCasl.

introduced sorts, operations and predicates (excluding the constructors). In the example in Fig. 1, the STREAM-models are (up to isomorphism) the *tl*-closed subsets of E^ω, where E is the interpretation of the sort *Elem*. A more complex example is the specification of CCS [18]. States are generated by the CCS syntax, but they are identified if they are bisimilar w.r.t. the ternary transition relation. This can be expressed in CoCasl by stating that states are cogenerated w.r.t. the transition relation.

Formally, a *cogeneration constraint* over a signature Σ is a subsignature fragment (i.e. a tuple of component-wise subsets that need not by itself form a complete signature) $\bar{\Sigma} = (\bar{S}, \bar{TF}, \bar{PF}, \bar{P})$ of Σ. In the above example, the cogeneration constraint is $(\{Elem\}, \{hd, tl\}, \emptyset, \emptyset)$.

A Σ-cogeneration constraint $\bar{\Sigma} \subseteq \Sigma$ is satisfied in a Σ-model M if each equivalence relation on M that

- is the equality relation on sorts in $\bar{\Sigma}$, and
- is a closed congruence for the operations and predicates in $\bar{\Sigma}$

is the equality relation. (Recall from [6] that a congruence on a partial algebra is closed if domains of partial functions and predicates are closed under the congruence.)

Note that selectors of cotypes, which play the role of observers, are always unary. However, like the **generated { ... }** construct in CASL, the **cogenerated { ... }** construct allows the inclusion of arbitrary signature items in the cogeneratedness constraint, so that observers of arbitrary arity are also possible. In particular, observers may have additional observable arguments (cf. the example in Fig. 6 below) as well as several non-observable arguments.

In duality to generated types in CASL, the construct **cogenerated cotype ...** abbreviates **cogenerated {cotype ... }**. No such abbreviation is provided for **cogenerated {type ... }**, the use of which is in fact expressly discouraged. A particularly discouraging example for the use of types where cotypes are expected is given in 6.

4   Free Types and Cofree Cotypes

CASL allows the exclusion not only of 'junk' in datatypes, but also of 'confusion', i.e. of equalities between different constructor terms. To this end, it provides the (basic) **free types** construct. Free datatypes carry implicit axioms that state,

---

spec STREAM2 [**sort** *Elem*] =
    cofree cotype
          Stream ::= (*hd* : *Elem*; *tl* : *Stream*)
end

Fig. 2. Cofree specification of bit streams in COCASL.

beside term-generatedness, the injectivity of the constructors and the disjointness of their images. The most immediate effect of these axioms is that *recursive definitions on a free datatype are conservative*. The elements of a free datatype can be thought of as being the (finite) constructor terms, i.e. in a suitable sense finite trees.

In COCASL, we provide, dually, a **cofree cotypes** construct that specifies the absolutely final coalgebra of infinite behaviour trees (see Example 6 on why there is no **cofree types** construct). More concretely, this means that, in addition to cogeneratedness, there is also a principle stating that there are enough behaviours, namely all infinite trees [2] (with branching as specified by the selectors). In contrast to its dual (no confusion among constructors), the latter principle cannot be expressed in first-order logic; however, a second-order specification is possible (see below). In the example in Fig. 2, the STREAM2-models are isomorphic to E^ω, where E is the interpretation of the sort *Elem*.

Definition 2. Given a set of sorts S, an *S-colouring* is just an S-sorted family of sets (of colours).

We are now ready to dualize the important algebraic concept of term algebra.

Definition 3. Given a cotype

$$CT = (S, (obs_{i,j,k} : T_i \to T_{i,j,k})_{i=1...n, j=1...m_i, k=1...r_{i,j}})$$

and an S-colouring C, the *behaviour algebra* $Beh_{CT}(C)$ is defined to be the following $Sig(CT)$-algebra:

- the carriers for observable sorts (i.e. in S) are those determined by C;
- the carriers for a non-observable sort T_{i_0} consist of all infinite trees of the following form:
  - each inner node is labelled with a pair (T_i, j), where T_i is a non-observable sort and $j \in \{1 \ldots m_i\}$ selects an alternative out of those for T_i;
  - the root is labelled with (T_{i_0}, j_0) for some j_0;
  - each leaf is labelled with an observable sort $s \in S$ and some colour from C_s;
  - each non-leaf node with label (T_i, j) has one child for each of the observers $obs_{i,j,k}$ ($k = 1 \ldots r_{i,j}$). The child node is labelled with the result sort of the observer.

– an observer operation $obs_{i_0,j,k}$ is defined for a tree with root (T_{i_0}, j_0) if and only if $j = j_0$, and in this case, it just selects the child tree corresponding to the observer.

Proposition 4. *Given a cotype*
$$CT = (S, (obs_{i,j,k} : T_i \rightarrow T_{i,j,k})_{i=1\ldots n, j=1\ldots m_i, k=1\ldots r_{i,j}})$$
and an S-colouring C, the behaviour algebra $Beh_{CT}(C)$ is final in the following sense: for any $Sig(CT)$-algebra A equipped with a C-colouring h (that is, a family of maps $(h_s : A_s \rightarrow C_s)_{s \in S}$), we can extend h in a unique way to a $Sig(CT)$-homomorphism
$$h^\natural : A \rightarrow Beh_{CT}(C).$$

Proof. Using the characterization of Prop. 1, the result follows from the general construction of final coalgebras over the category of $\{T_1, \ldots, T_n\}$-sorted sets (this generalizes the well-known result for **Set** [2]). Intuitively, $h^\natural$ constructs the behaviour of an element, which is the infinite tree given by all possible observations that can be made successively applying the observers until a value of observable sort (i.e. in S) is reached. □

Given a signature Σ, we formally add cofreeness constraints of form $cofree(CT)$, where
$$CT = (S, (obs_{i,j,k} : T_i \rightarrow T_{i,j,k})_{i=1\ldots n, j=1\ldots m_i, k=1\ldots r_{i,j}})$$
is a cotype with $Sig(CT) \subseteq \Sigma$, as Σ-sentences to our logic. A cofreeness constraint $cofree(CT)$ holds in a Σ-algebra A, if the reduct of A to $Sig(CT)$ is isomorphic to the behaviour algebra $Beh_{CT}(C)$ over the set of colours C with $C_s := A_s$ for $s \in S$.

Note that this implies the satisfaction of the cogeneratedness constraint $(S, \{sel_{i,j,k}|sel_{i,j,k} \text{ total}\}, \{sel_{i,j,k}|sel_{i,j,k} \text{ partial}\}, \emptyset)$, i.e. each cofree cotype is also cogenerated. The converse does not hold, i.e. a cogenerated cotype need not be cofree. However, cogenerated *cotypes* still behave quite nicely (in contrast to arbitrary cogenerated *types*): the elements of carriers of the non-observable sorts (i.e. those outside S) are completely determined by their *behaviours*. Thus, the elements can be identified with their behaviours, and up to isomorphism, we have a subalgebra of the cofree cotype. Hence, cofreeness essentially adds the requirement that each possible behaviour is actually represented by an element.

Note that an equivalent description of the behaviour algebra can be given in terms of contexts [11], using the "magic formula":
$$(\Pi_{v \in \bar{S}}[Ctx^{\Sigma}_{(\bar{S}, (\bar{F}_s)_{s \in S \setminus \bar{S}})}(C)[z_s]_v \rightarrow C_v])_{s \notin \bar{S}}$$
where $Ctx^{\Sigma}_{(\bar{S}, (\bar{F}_s)_{s \in S \setminus \bar{S}})}(C)[z_s]$ is the set of all terms consisting of constants in C, observer operations and a single occurrence of a variable z_s of non-observable sort s.

The main benefit of cofree cotypes (in comparison to cogenerated cotypes) is the principle

corecursive definitions in cofree cotypes are always conservative

Algebra	Coalgebra
type = (partial) algebra	**cotype** = coalgebra
constructor	selector
generation	observability
generated type	**cogenerated (co)type**
= no junk	= full abstractness
= induction principle	= coinduction principle
no confusion	all possible behaviours
free type	**cofree cotype**
= absolutely initial datatype	= absolutely final process type
= no junk + no confusion	= full abstractness + all possible behaviours
free { … } = initial datatype	**cofree { … }** = final process type

Fig. 3. Summary of dualities between CASL and COCASL.

meaning that any given model of a cofree cotype can (even uniquely) be extended to a model of a corecursive definition over this cotype.

Note that if we want to get an institution, in order to be able to translate the various constraints along signature morphisms in a way that the satisfaction condition for institutions is fulfilled, one has to equip the constraints with an additional signature morphism, as in [20, 17].

Fig. 3 contains a summary of the CASL concepts and their COCASL dualizations (including structured free and cofree covered in the next section).

5   Structured Free and Cofree Specifications

Besides institution-specific language constructs, CASL also provides institution-independent structuring constructs. In particular, CASL provides the structured **free** construct that restricts the model class to *initial* or *free* models. That is, if Sp_1 is a specification with signature Σ_1, then the models of Sp_1 **then free** $\{Sp_2\}$ are those models M of Sp_1 **then** Sp_2 that are free over $M|_{\Sigma_1}$ w.r.t. the reduct functor $\_|_{\Sigma_1}$. This allows for the specification of datatypes that are generated freely w.r.t. given axioms, as, for example, in the specification of finite sets over an element sort shown in Figure 4.

The **cofree { … }** construct dualizes the **free { … }** construct by restricting the model class of a specification to the cofree ones. This generalizes **cofree types** to the case of non-unary functions (e.g. as in Figure 5) and the presence of axioms (e.g. as in Figure 7 below).

More precisely, the semantics of **cofree** is defined as follows:

Definition 5. If Sp_1 is a specification with signature Σ_1 then the models of Sp_1 **then cofree** $\{Sp_2\}$ are those models M of Sp_1 **then** Sp_2 that are *fibre-final* over $M|_{\Sigma_1}$ w.r.t. the reduct functor $\_|_{\Sigma_1}$. Here, fibre-finality means that M is the final object in the fibre over $M|_{\Sigma_1}$. The fibre over $M|_{\Sigma_1}$ is the full subcategory of $Mod(Sp_2)$ consisting of those models whose Σ_1-reduct is $M|_{\Sigma_1}$.

```
spec GENERATEFINITESET [sort Elem] =
free
  {   type FinSet[Elem] ::= {}
                           | {__}(Elem)
                           | __∪__(FinSet[Elem]; FinSet[Elem])
      op  __∪__ : FinSet[Elem] × FinSet[Elem] → FinSet[Elem],
               assoc, comm, idem, unit {}
  }
end
```

Fig. 4. Specification of finite sets over an arbitrary element sort in CASL.

```
spec FUNCTIONTYPE =
  sorts  A, B
  then  cofree {
      sort Fun[A, B]
      op eval : Fun[A, B] × A → B
  }
end
```

Fig. 5. Cofree specification of function types.

This definition deviates somewhat from the semantics of **free** in that the latter postulates *freeness* rather than fibre-initiality. (Actually, it might be worthwhile to redefine the CASL semantics for free specifications in terms of fibre-initial models.) We will see shortly that the more liberal semantics for **cofree** is essential in cases where sorts from the local environment occur as argument sorts of selectors. Call a sort from the local environment an *output sort* if it occurs only as a result type of selectors. In the cases of interest, a more general co-universal property concerning, in the notation of the above definition, morphisms of Σ_1-models that are the identity on all sorts except possibly the output sorts follows from fibre-finality.

We shall see below (Theorem 11) that the **cofree cotypes** construct is equivalent to **cofree { cotypes ... }**. By contrast, the use of **cofree { types ... }** should be avoided:

Example 6. The specification

free type *Bool* ::= *false* | *true* **then**
cofree {type *T* ::= *c1*(*s1* :?*Bool*) | *c2*(*s2* :?*Bool*) **}**,

is inconsistent. Indeed, by applying the uniqueness part of finality to a model of the unrestricted type where T has an element on which both selectors are undefined (this is allowed for types but not for cotypes), one obtains that any model of the cofree type would be a singleton; however, singleton models fail to satisfy the finality property e.g. for the model of the unrestricted type where T is $Bool \times Bool$ and the selectors are the projections.

As an example for the significance of the relaxation of the cofreeness condition, consider the specification of Moore automata as given in Figure 6. Here, the observer *next* depends not only on the state, but additionally on an input letter.

```
spec MOORE =
   sorts  In, Out
   then  cofree {
         sort State
         ops  next  :  In × State → State
              observe  :  State → Out
   }
end
```

Fig. 6. Cofree specification of Moore automata.

In the standard theory of coalgebra, *next* would become a higher-order operation $next : State \to State^{In}$, and the cofree coalgebra indeed yields the final automaton showing all possible behaviours - but only for a *fixed* carrier for In (the inputs). The carrier for Out is also regarded as fixed; however, one can show that the co-universal property holds also for morphisms that act non-trivially on Out. If the semantics of **cofree** required actual cofreeness, i.e. a couniversal property also for morphisms that act non-trivially on In, the specification would be inconsistent!

```
spec BITSTREAM3 =
   free  type Bit ::= 0 | 1
   then  cofree {
         cotype BitStream ::= (hd : Bit; tl : BitStream)
         ∀s : BitStream
         •  hd(s) = 0 ∧ hd(tl(s)) = 0 ⇒ hd(tl(tl(s))) = 1 }
end
```

Fig. 7. Structured cofree specification of bit streams in CoCASL.

Let us now come to a further modification of the stream example. If the axiom were omitted in the specification in Figure 7, the model class would be the same as that in Figure 1, instantiated to the case of bits as elements. *With* the axiom, the streams are restricted to those where two 0's are always followed by a 1. Again, this is unique up to isomorphism.

It is straightforward to specify iterated free/cofree constructions, similarly as in [23]. Consider e.g. the specification of lists of streams of trees in Figure 8. Alternatively, one could have used structured free and cofree constructs as well:

$$SP \text{ then free } \{SP_1\} \text{ then cofree } \{SP_2\} \text{ then free } \ldots$$

Note that also in the latter case, there won't be any **free** *within* a **cofree** or vice versa.

```
spec LISTSTREAMTREE [sort Elem] =
   free type
         Tree ::= EmtpyTree| Tree(left :? Tree; elem :? Elem; : Elem; right :? Tree)
   cofree cotype
         Stream ::= (hd : Tree; tl : Stream)
   free type
         List ::= Nil| Cons(head :? Stream; tail :? List)
end
```

Fig. 8. Nested free and cofree (co)types.

An example for **free** *within* **cofree** is shown in Figure 9. Here, the inner **free** has to be a structured one, since sets cannot be specified as **free type** directly. Alternatively, sets may be specified using a **generated type** together with a first-order extensionality axiom. We have preferred the former variant over the latter one in order to be able to apply Theorem 10 below.

```
spec NONDETERMINISTICAUTOMATA =
   sort  In
   then  cofree {
      sort State
      then  free {
         type  Set ::= {} | {_}(State) | _ ∪ _(Set; Set)
         op  _ ∪ _ : Set × Set → Set,
                   assoc, comm, idem, unit {}   }
      then op next : In × State → Set   }
end
```

Fig. 9. A free type *within* a cofree type.

6   Modal Logic

Given a cotype that defines non-observable sorts with selectors acting as observers, we can define a coalgebraic modal logic in the style of [14]. A non-observable sort corresponds to a set of possible worlds, and leaving this set implicit is the main idea of modal logic. A selector t with *non-observable* result sort leads to modalities $[t]$, $<t>$, $[t*]$, $<t*>$ (all-next, some-next, always, eventually), while a selector with observable result sort leads to a flexible constant (i.e. a constant that depends on the respective world) that can be used to make observations (by equating it to a term of observable sort). The modal logic then consists of such equations as atomic sentences, which may be combined using the modalities above and the usual propositional connectives, as well as quantification over variables of observable sorts. Using this logic, we can write, in the example of Figure 7,

$$hd = 0 \land <tl>\, hd = 0 \Rightarrow <tl><tl>\, hd = 1$$

as syntactic sugar for

$$hd(s) = 0 \land hd(tl(s)) = 0 \Rightarrow hd(tl(tl(s))) = 1$$

Each modal formula ϕ has a *sort*, and is implicitly universally quantified over a variable of this sort. The sort of ϕ is determined by the non-observable argument in observers used in ϕ. In particular, a modal formula is well-formed only in case of correct sorting. One may switch to a different sort (i.e. a different state space) using the modalities, but only in a well-sorted way. If necessary (due to overloading), observers have to be provided with explicit types. The 'iterative' modalities $[t*]$ and $<t*>$ are meaningful only for observers that remain within the same non-observable sort.

Moreover, we provide a *global diamond* ⟨global⟩ as suggested in [15], where ⟨global⟩ϕ expresses the fact that ϕ holds in some state of the system. For reasons laid out in [15], the global diamond is restricted to positive occurrences. As explained in [15], the global diamond is, in terms of expressivity, equivalent to *modal rules* which state implications between *validities* of modal formulas. For full details of the modal logic see [19].

The modal logic allows expressing safety or fairness properties. For example, the model of the specification BITSTREAM4 of Figure 10 consists, up to isomorphism, of those bitstreams that will always eventually output a 1. Here, the 'always' stems from the fact that the modal formula is, on the outside, implicitly quantified over all states, i.e. over all elements of type *BitStream*.

Remark 7. The modal μ-calculus [13], which provides a syntax for least and greatest fixed points of recursive modal predicate definitions, is expressible using free and cofree specifications: μ is expressible with free recursively defined predicates, while ν is expressible with cofree recursively defined predicates, and nesting of μ and ν corresponds to nesting of **free** and **cofree**. It is an open point of discussion whether future versions of COCASL should include the μ-calculus, or whether the existing modal operators and the explicit coding of μ-formulas suffice for practical purposes.

```
spec BitStream4 =
  free type Bit ::= 0 | 1
  then cofree {
      type BitStream ::= (hd : Bit; tl : BitStream)
      •   <tl*> hd = 1
      }
end
```

Fig. 10. Specification of a fairness property.

7   Existence of Cofree Models

The theory of algebraic specification and institutions provides us with a very general characterization of the existence of free models [27]: free models exist for specifications with universally quantified Horn axioms. (Part of) a dual result has been obtained in [15]. In summary, results from [26, 14, 15] guarantee that cofree coalgebras over **Set** exist for bounded functors and modal axioms or, more generally, axioms that are stable under coproducts and quotients. This has to be generalized slightly in order to cope with specifications with several non-observable sorts, i.e. for coalgebras over $\mathbf{Set}^n$.

Even more generally, we have

Proposition 8. *Let* **C** *be a category equipped with a factorization system* $(\mathbf{E}, \mathcal{M})$ *for (large) sinks [1], and let* $\Sigma : \mathbf{C} \to \mathbf{C}$ *be a functor that preserves* $\mathcal{M}$. *Then* $(\mathbf{E}, \mathcal{M})$ *lifts to a factorization structure to the category* $\mathbf{CoAlg}(\Sigma)$ *of* Σ-*coalgebras.*

Let **B** *be a subcategory of* $\mathbf{CoAlg}(\Sigma)$ *that is closed under* **E**-*sinks, and let* $\mathbf{CoAlg}(\Sigma)$ *have a final coalgebra. Then* **B** *has a fully abstract final coalgebra in the sense of [15].*

Proof. The lifting statement is clear. The given condition on **B** is equivalent to **B** being $\mathcal{M}$-coreflective [1]. Then the $\mathcal{M}$-coreflection of the final Σ-coalgebra is a (fully abstract) final coalgebra in **B**.

The condition on Σ is always almost satisfied for the factorization structure (jointly surjective, injective) on $\mathbf{Set}^n$, since injective maps in $\mathbf{Set}^n$ are sections provided that their domain is non-empty; in fact, by a construction described for $n = 1$ in [3], we can always assume preservation of injective maps.

If **C**, and hence $\mathbf{CoAlg}(\Sigma)$, has coproducts, then closedness under **E**-sinks is equivalent to closedness under quotients and coproducts, provided that every **E**-sink contains a small **E**-sink (this is the case in $\mathbf{Set}^n$). However, it is often just as easy to argue directly via **E**-sinks, e.g. in the proof of the following rather general sufficient condition for $\mathcal{M}$-coreflexivity (and, hence, existence of final coalgebras):

Lemma 9. *Let* $\Sigma : \mathbf{Set}^n \to \mathbf{Set}^n$. *If* **B** *is a subcategory of* $\mathbf{CoAlg}(\Sigma)$ *determined by formulas of the form*

$$\phi \equiv \forall s : S \bullet P(s),$$

where S *is one of the carriers of the coalgebra, and*

$$P(s) \implies P(h_S(s))$$

for each coalgebra homomorphism h *with* S-*component* h_S, *then* **B** *is closed under jointly surjective sinks in* $\mathbf{CoAlg}(\Sigma)$.

It is easy to see that this condition is satisfied for the modal logic formulas described above.

We are now ready to state the main existence result:

Theorem 10. *Let Sp be the specification*
    Sp_1 **then cofree** { Sp_2 }.
Call the sorts from Sp_1 *observable* sorts. *Let the specification* Sp_2 *consist of (no more than)*

- *declarations of* non-observable *sorts*
- *auxiliary* datatypes *that are* defined *over the other sorts using* **free** *with only equational axioms (or an equivalent construction)*
- *further operations called* observers *that each have exactly one non-observable argument and otherwise only observable arguments called* parameters, *and*
- *modal logic formulas, and*
- *(mandatory) further axioms (say, a redeclaration of the non-observable sorts as* **cotypes***) stating that domains of observers do not depend on parameters and form a disjoint cover of the respective non-observable sorts.*

Then Sp is conservative (model-extensive) over Sp_1.

Proof. From Proposition 1 together with monomorphicity of the auxiliary sorts, we know that the category of Sp_2-models over a given Sp_1-model is equivalent to a subcategory of Σ-coalgebras for some functor $\Sigma : \mathbf{Set}^n \to \mathbf{Set}^n$. It is easy to check that Σ is bounded, and hence admits a final coalgebra [26]; the proof is finished by appealing to Lemma 9 and Proposition 8.

Moreover, we have

Theorem 11. *If* DD *is a sequence of selector-based datatype definitions without subsorting, then*

    cofree { **cotypes** DD }   *and*   **cofree cotypes** DD

have the same semantics.

8 Tool Support

It is quite straightforward to extend the central analysis tool for CASL (the CASL tool set), to CoCASL. Of course, the more challenging task is to obtain good proof support. Here, we aim at extending the existing encoding of CASL in Isabelle/HOL [16] to CoCASL.

cogenerated types are no problem: coinduction is a second-order principle. The same holds for the infinite trees needed for **cofree cotypes**; actually, Isabelle/HOL already comes with such trees.

The greater difficulties come with specifications of the form **cofree** $\{SP\}$. If SP is flattenable and axiomatized within the modal logic, we can proceed similarly to the case of **cofree types**: the model of **cofree** SP is a subcoalgebra of the coalgebra of *all* behaviours (as specified by the corresponding **cofree type**), namely the largest subcoalgebra that satisfies the modal axioms.

More complex examples, such as nondeterministic automata or trees with unbounded branching, involve a free specification of output sorts of selectors (like lists or sets) within a **cofree** {...}. Here, in a first step, we proceed as above and encode the cofree type over the absolutely free type (only the branching may now be infinite, being determined by a datatype). Then the cofree type over the relatively free type is obtained as the quotient modulo the largest congruence [12]. In terms of tool support, this means the following:

- Equality of elements in the cofree datatype is obtained as before by coinductive reasoning (or via terminal sequence induction [21]), the difference with the absolutely free case being that the formulas in the free specification (e.g. associativity, commutativity, and idempotence in the case of finite sets) are now available for such proofs.
- Distinctness of elements is shown, again as before, by establishing that the behaviours are different. Here, the encoding of free specifications comes in: distinctness of two elements of a relatively free type is shown by separating the two elements by a congruence.

Remark 12. Above, we have seen two cases where free specifications within cofree specifications allow good technical handling:

- the output sorts of selectors for a cofree datatype may be given by a free specification, which is handled as described above;
- the modal formulas that restrict the elements of the cofree type may involve freely (or cofreely) specified predicates, which are dealt with in Isabelle by means of least and greatest fixed points.

Beyond these two cases, the situation remains somewhat unclear. E.g., the following specification is inconsistent:

spec FINALELEMENT = BOOL **then**
cofree {
   free type *Unit* ::= *1*
   op *el* : *Unit* → *Bool*   }

This seems to indicate that input sorts should not be restricted by equational axioms (the freeness constraint can be replaced by an equation here), or in fact by anything else except modal formulas; this is in agreement with suggestions made in [14]. On the other hand, observe that constraining output sorts is more or less mandatory: e.g., the specification of Moore automata (Figure 6) becomes meaningless if the freeness constraint (for the type of sets) is omitted – the model it describes is then just the singleton set (with a 'power set' consisting only of the empty set). In other words, enough of a handle must be provided to actually prove distinctness of observations.

Also, a proof principle for free specifications containing cofree specifications seems to be much harder to obtain. Here, we propose to avoid the outer free specification and use a generation axiom plus some characterization of equality by suitably chosen observers instead.

9   Conclusion and Related Work

We have introduced COCASL as a relatively easy extension of CASL. COCASL allows algebraic and coalgebraic specification to be mixed. We have shown that the well-known coalgebraic modal logic and even the modal μ-calculus can easily be expressed in COCASL, and give sufficient criteria for the existence of cofree models (also in the case of nested cofree and free specifications). Finally, we have shown how the existing coding of CASL into higher-order logic can be extended to COCASL.

COCASL is more expressive than other algebra-coalgebra combinations in the literature: [7] uses a simpler logic, while hidden algebra such as in BOBJ [24] and reachable-observable algebra such as in COL [4, 5] do not support cofree types (at least not at the level of basic specifications), which in particular means that corecursive definitions are not conservative. For example, a cogenerated specification of streams in COL with say, a *cons* operation, has also a model consisting of pairs (finite list, bit), where the finite list specifies the first part of the behaviour, and the bit specifies the (constant) behaviour afterwards. The corecursive definition of a flip stream (consisting of alternating zeros and ones) is then non-conservative.

By contrast, cofree types in COCASL support a style of specification separating the basic process type (with its data sorts, observers and other operations) from further, derived operations defined on top of this in a conservative way. Note that this is not a purely theoretical question: programming languages such as Charity [8] and Haskell [9] support infinite datastructures that correspond to the infinite trees in the behaviour algebras, and one should be able to specify that as many infinite trees as needed for all programs over some datastructure expressible in these languages are present in the models of a specification. The Haskell semantics for lazy datastructures (at least for the non-left-$\rightarrow$-recursive case) indeed comprises *all* infinite trees, i.e. is captured exactly by a behaviour algebra.

Unlike COL [4, 5], CoCasl does not simultaneously support a glass-box and a black-box view on a specification. However, we plan to develop a notion of behavioural refinement between CoCasl specifications. Then, the black-box/glass-box view of [4] could be expressed in CoCasl as a refinement of a black-box specification into a glass-box one, thus also providing a clear separation of concerns.

The *Coalgebraic Class Specification Language* CCSL [25], developed in close cooperation with the LOOP project [29], is based on the observation of [22] that coalgebras can give a semantics to classes of object–oriented languages. CCSL provides a notation for parameterized class specifications based on final coalgebras. Its semantic is based on a higher–order equational logic and it provides theorem proving support by compilers that translate CCSL into the higher–order logic of PVS and Isabelle. In its current version, CCSL does not support data type specifications with partial constructors, axioms or equations, i.e. it only supports free types in the sense of Casl. Recently CCSL has been extended by binary methods [28], which are supported in CoCasl via cogenerated constraints.

At the level of proof principles, recent research about circular coinduction [10] and terminal sequence induction [21] is expected to provide tactics for the encoding of CoCasl into Isabelle/HOL.

Acknowledgements

The authors wish to thank Christoph Lüth for useful discussions, Erwin R. Catesbeiana for providing the larger perspective, and the participants of an informal CoCasl and observability meeting, Hubert Baumeister, Michel Bidoit, Rolf Hennicker, Bernd Krieg-Brückner, Don Sannella, Andrzej Tarlecki, and Martin Wirsing, as well as Alexander Kurz, for intensive feedback to a draft version of this work.

References

1. J. Adámek, H. Herrlich, and G. E. Strecker, *Abstract and concrete categories*, Wiley Interscience, 1990.
2. M. Arbib and E. Manes, *Parametrized data types do not need highly constrained parameters*, Inform. Control **52** (1982), 139–158.
3. M. Barr, *Terminal coalgebras in well-founded set theory*, Theoret. Comput. Sci. **114** (1993), 299–315.
4. M. Bidoit and R. Hennicker, *On the integration of observability and reachability concepts*, Fossacs, LNCS, vol. 2303, Springer, 2002, pp. 21–36.
5. Michel Bidoit, Rolf Hennicker, and Alexander Kurz, *Observational logic, constructor-based logic, and their duality*, Theoret. Comput. Sci. **298** (2003), 471–510.
6. P. Burmeister, *Partial algebras – survey of a unifying approach towards a two-valued model theory for partial algebras*, Algebra Universalis **15** (1982), 306–358.

7. C. Cîrstea, *On specification logics for algebra-coalgebra structures: Reconciling reachability and observability*, LNCS **2303** (2002), 82–97.
8. R. Cockett and T. Fukushima, *About Charity*, Yellow Series Report 92/480/18, Univ. of Calgary, Dept. of Comp. Sci., 1992.
9. S. P. Jones et al., *Haskell 98: A non-strict, purely functional language*, (1999), http://www.haskell.org/onlinereport.
10. J. Goguen, K. Lin, and G. Rosu, *Conditional circular coinductive rewriting*, Automated Software Engineering, IEEE Press, 2000, pp. 123–131.
11. J. A. Goguen, *Hidden algebraic engineering*, Algebraic Engineering (Chrystopher Nehaniv and Masami Ito, eds.), World Scientific, 1999, pp. 17–36.
12. H. P. Gumm and T. Schröder, *Coalgebras of bounded type*, Math. Struct. Comput. Sci. **12** (2002), 565–578.
13. D. Kozen, *Results on the propositional mu -calculus*, Theoret. Comput. Sci. **27** (1983), 333–354.
14. A. Kurz, *Specifying coalgebras with modal logic*, Theoret. Comput. Sci. **260** (2001), 119–138.
15. _____, *Logics admitting final semantics*, Foundations of Software Science and Computation Structures, LNCS, vol. 2303, Springer, 2002, pp. 238–249.
16. T. Mossakowski, CASL: *From semantics to tools*, TACAS, LNCS, vol. 1785, Springer, 2000, pp. 93–108.
17. _____, *Relating* CASL *with other specification languages: the institution level*, Theoret. Comput. Sci. **286** (2002), 367–475.
18. T. Mossakowski, M. Roggenbach, and L. Schröder, CoCASL *at work – modelling process algebra*, CMCS 03, ENTCS, vol. 82(1), 2003.
19. T. Mossakowski, L. Schröder, M. Roggenbach, and H. Reichel, *Algebraic-coalgebraic specification in* CoCASL, Tech. report, University of Bremen, available at http://www.informatik.uni-bremen.de/~lschrode/papers/cocasl.ev.ps.
20. P. D. Mosses (ed.), CASL – *the common algebraic specification language. Reference manual*, Springer, to appear.
21. D. Pattinson, *Expressive logics for coalgebras via terminal sequence induction*, Tech. report, LMU München, 2002.
22. H. Reichel, *An approach to object semantics based on terminal co-algebras*, Math. Struct. Comput. Sci. **5** (1995), 129–152.
23. _____, *A uniform model theory for the specification of data and process types*, WADT, LNCS, vol. 1827, Springer, 2000, pp. 348–365.
24. G. Roşu, *Hidden logic*, Ph.D. thesis, Univ. of California at San Diego, 2000.
25. J. Rothe, H. Tews, and B. Jacobs, *The Coalgebraic Class Specification Language CCSL*, J. Universal Comput. Sci. **7** (2001), 175–193.
26. J. Rutten, *Universal coalgebra: A theory of systems*, Theoret. Comput. Sci. **249** (2000), 3–80.
27. A. Tarlecki, *On the existence of free models in abstract algebraic institutions*, Theoret. Comput. Sci. **37** (1985), 269–304.
28. H. Tews, *Coalgebraic methods for object–oriented languages*, Ph.D. thesis, Dresden Univ. of Technology, 2002.
29. J. van den Berg and B. Jacobs, *The LOOP compiler for Java and JML*, LNCS **2031** (2001), 299–312.

Translating Logics for Coalgebras

Dirk Pattinson

Institut für Informatik, LMU München

Abstract. This paper shows, that three different types of logics for coalgebras are institutions. The logics differ regarding the presentation of their syntax. In the first framework, abstract behavioural logic, one has a syntax-free representation of behavioural properties. We then turn to coalgebraic logic, the syntax of which is given as an initial algebra. The last framework, which we consider, is coalgebraic modal logic, the syntax of which is concretely given.

1 Introduction

This paper tries to contribute to the question, whether different types of logics, interpreted over coalgebras, carry the structure of an institution. Institutions, originally introduced by Goguen and Burstall, capture interplay between the transformation of systems and corresponding translations of logics. An institution therefore consists of two parts: A class of systems, and a class of logics, which can be used to describe properties of the systems under considerations. Both are linked by (semantical) transformation of systems and corresponding (syntactical) translation of the logics. If the systems, together with their logics, form an institution, we have the possibility to derive properties of a transformed system from properties of the original system, which makes the concept of institutions valuable in the stepwise process of building systems.

The class of systems we are dealing with in this paper, are coalgebras for an endofunctor on the category of sets. Coalgebras provide a uniform view on a large class of state-based systems, (see [20] for examples). In order to reason about coalgebraically modelled systems, modal logic has proven an appropriate tool ([10, 15, 8, 19, 7]).

Both the class of systems (coalgebras) and the corresponding class of (modal) logics are well understood – as long as we do not migrate between different types of systems (that is, between coalgebras for different functors) and leave the logics fixed. It is the purpose of this paper to add transformations between models and translation between the logics to the picture.

After recalling some basic terminology, we first address the question, whether *co*institutions are the appropriate framework in which one should consider logics for *co*algebras and their translations. We can rightfully say, that this is just a matter of taste: Every institution over a category Sig of signatures corresponds to a coinstitution over Sig^{op}, and vice versa (Proposition 1). In this light, we choose to work with coinstitution, which we feel are easier to work with in the

M. Wirsing, D. Pattinson, and R. Hennicker (Eds.): WADT 2002, LNCS 2755, pp. 393–408, 2003.
© Springer-Verlag Berlin Heidelberg 2003

context of coalgebras, mainly because we do not need to work with the dual category of signature morphisms.

Before dealing with translations on the logical side, we first study the transformation of models on semantical side. We argue that – working with coalgebras for endofunctors – natural transformations between functors provide us with a natural notion of signature morphism. This notion of signature morphism is then used to treat three different types of logics for coalgebras: abstract behavioural logic (the presentation of which is syntax free), coalgebraic logic (the syntax of which is abstract), and coalgebraic modal logic, where the syntax is concretely given.

We show, how to define translations between the logics for each of the three different types. It turns out that the institution property of abstract behavioural logic and coalgebraic modal logic is relatively easy to establish; in the case of coalgebraic logic, one needs a small extension of the syntax.

2   Preliminaries and Notation

In the whole paper, T denotes an endofunctor on the category Set of sets and functions.

2.1   Coalgebras

The definition of coalgebras (and their morphisms) dualises that of algebras for endofunctors:

Definition 1 (Coalgebras, Morphisms). *A T-coalgebra is a pair (C, γ) where C is a set and $\gamma : C \to TC$ is a function. A morphism between two T-coalgebras (C, γ) and (D, δ) is a function $f : C \to D$, which satisfies $Tf \circ \gamma = \delta \circ f$.*

Coalgebras, together with their morphisms, form a category, which we denote by CoAlg(T).

We think of coalgebras for an endofunctor as a general framework for state based systems, and we think of T as a *signature* for the T-coalgebras. Instantiating the framework with specific endofunctors (different signatures), we obtain different types of systems:

Example 1.   (i) Suppose $T_1 X = L \times X$ for a set L of labels. Then every state $c \in C$ of a T-coalgebra (C, γ) can be seen as producing an infinite trace of labels $l \in L$:

$$c = c_0 \xrightarrow{l_1} c_1 \xrightarrow{l_2} c_2 \longrightarrow \cdots$$

where $(l_k, c_k) = \gamma(c_{k-1})$ for $k > 0$.

(ii) For $T_2 X = (O \times X)^I$, the T_2-coalgebras are Mealy Automata: Given $(C, \gamma) \in$ CoAlg(T_2), a state $c \in C$ and an input $i \in I$, the transition function γ provides us with a new state $\pi_2 \circ \gamma(c)(i)$ and an output $o = \pi_1 \circ \gamma(c)(i) \in O$.

(iii) Suppose $TX = \mathcal{P}(X)^L$, where $\mathcal{P}$ is the covariant powerset functor. Then T-coalgebras are in 1-1 correspondence with labelled transition systems: Given $(C, \gamma) \in \mathsf{CoAlg}(T)$, put $c \xrightarrow{l} c'$ iff $c' \in \gamma(c)(l)$.

One of the appealing features of the general theory of coalgebras is, that T-coalgebras come with a meaningful built-in notion of behavioural equivalence:

Definition 2. *Suppose (C, γ) and $(D, \delta) \in \mathsf{CoAlg}(T)$. Then a pair of states $(c, d) \in C \times D$ is behaviourally equivalent, if there is $(E, \epsilon) \in \mathsf{CoAlg}(T)$ and a pair of morphisms $f : (C, \gamma) \rightarrow (E, \epsilon)$ and $g : (D, \delta) \rightarrow (E, \epsilon)$ such that $f(c) = g(d)$.*

This definition goes back to [11]; Rutten [20] has studied bisimulation, as defined by Aczel and Mendler [1] as fundamental notion of equivalence. Both notions agree if the signature functor preserves weak pullbacks; for functors not having this property, behavioural equivalence seems to be the more fundamental notion of equivalence (see [11] for discussion). In the examples, behavioural equivalence can be expressed as follows:

Example 2.   (i) Let $T_1 X = L \times X$ and suppose (C, γ) and $(D, \delta) \in \mathsf{CoAlg}(T_1)$. Then $(c, d) \in C \times D$ are behaviourally equivalent, if they produce the same trace of labels $l \in L$.

(ii) In the case $T_2 X (O \times X)^I$, every state $c \in C$ of a T_2-coalgebra (C, γ) defines a function $f_c : I^\omega \rightarrow O^\omega$ (given $i = (i_n)_{n \in \omega}$, let $c = c_0$ and $(o_n, c_{n+1}) = \gamma(c_n)(i_n)$. Put $f_c(i) = (o_n)_{n \in \omega}$). We obtain that two states are behaviourally equivalent, if the associated functions are equal.

(iii) For $T_3 X = \mathcal{P}(X)^L$ and (C, γ), $(D, \delta) \in \mathsf{CoAlg}(T_3)$, behavioural equivalence coincides with bisimulation, as used by Park [16] and Milner [14].

The significance of behavioural equivalence is that it identifies precisely those states, which cannot be distinguished from the outside. The logics, which we consider later, will all be invariant under behavioural equivalence.

2.2   Institutions

Institutions [21, 6] have been successfully used to describe the interplay between translation of logics and transformations of models along morphisms of signatures:

Definition 3. *Suppose Sig is a category (of signatures). An institution is a triple (Mod, Sen, Sig) where*

- *Mod : Sig $\rightarrow$ Cat^{op} associates categories of models to signatures*
- *Sen : Sig $\rightarrow$ Set associates a set of sentences (formulas) to every signature,* *and*
- *$\models$ is a family ($\models_S$) of relations $\models_S \subseteq \mathrm{Mod}(S) \times \mathrm{Sen}(S)$, indexed by the signatures $S \in$ Sig*

such that the satisfaction condition

$$\mathrm{Mod}(\sigma)(M) \models \phi \quad \Longleftrightarrow \quad M \models \mathrm{Sen}(\sigma)(\phi)$$

is satisfies for all $\sigma : S \rightarrow S'$, $M \in \mathrm{Mod}(S')$ and $\phi \in \mathrm{Sen}(S)$.

It is the purpose of the paper to establish the satisfaction condition for different types of logics, interpreted over coalgebras. For other examples of institutions, the reader is referred to the original paper by Goguen and Burstall [6]. Dually, we have

Definition 4. *A* coinstitution *over a category* Sig *of signatures consists of*

- *A functor* Mod : Sig → Cat
- *A functor* Sen : Sig → Set^{op}
- *A family* $\models_S$ *of relations* $\models_S \subseteq$ Mod$(S) \times$ Sen(S), *indexed by the objects S of* Sig,

such that the dual of the satisfaction condition

$$\text{Mod}(\sigma)(M) \models \phi \iff M \models \text{Sen}(\sigma)(\phi)$$

is satisfied for all $\sigma : S \to S'$, $M \in$ Mod(S) *and* $\phi \in$ Sen(S').

Note that, in a coinstitution, the translation Mod is covariant, whereas the translation Sen of sentences is contravariant. However, when dualising institutions, we do not obtain a new concept:

Proposition 1. *Suppose* Sig *is a category. Then there is a 1-1 correspondence between institutions over* Sig *and coinstitutions over* Sig^{op}.

Proof. Suppose (Mod, Set, Sig) is an institution over Sig. Then (Mod^{op}, Sen^{op}, $\models$) is a coinstitution over Sig^{op}; clearly this construction can be reversed.

In the light of this proposition, the concept of coinstitution is strictly speaking unnecessary. However, for the purposes of the present paper, we prefer to work with coinstitutions. This allows us to take a subcategory $\mathbb{S} \subseteq [T, T]$ of the category of endofunctors (instead of $\mathbb{S}^{op}$) as a category of signatures.

3   Translation of Models

One of the goals of this paper is to show that three different conceptions of modal logic for coalgebras give rise to an institution. All three logics will be interpreted over coalgebras for endofunctors on sets. Since we think of the underlying endofunctor T as a signature for the corresponding T-coalgebras, signature morphisms need to mediate between endofunctors on Set. The obvious notion for signature morphisms are therefore *natural transformations* (see [13]). Thus, our category Sig of signatures will have endofunctors as objects and natural transformations as morphisms, that is, we take Sig $\subseteq$ [Set, Set] as a (possibly non-full) subcategory of the functor category [Set, Set]. As far as signatures are concerned, this setup is common to all three types of logics, which we show to carry the structure of an institution. This section describes the model theoretic part, that is, the Mod functor, which translates models along signature morphisms. The translation between models described here is the same for all three conceptions of logics for coalgebras, which are later shown to carry the structure of an institution.

Before we start to study the translation of models (coalgebras) along signature morphisms, we first try to convince the reader that natural transformations are a indeed a natural choice for signature morphisms.

The key observation is the following:

Lemma 1. *Suppose* $T, S :$ Set $\rightarrow$ Set *and* $\sigma : S \rightarrow T$ *is a natural transformation. Then* $\sigma^\dagger :$ CoAlg$(S) \rightarrow$ CoAlg(T), *defined by* $\sigma^\dagger(C, \gamma) = (C, \sigma(C) \circ \gamma)$, *is functorial.*

Of course, this observation is not specific to the category of sets. We illustrate the use of natural transformations using the running examples introduced above.

Example 3. We consider natural transformations between the signatures discussed in Example 2.

(i) Every T_1-coalgebra (C, γ) can be viewed as Mealy automaton (that is, as T_2-coalgebra) if we simply ignore the input: put $\gamma'(c)(i) = \gamma(c)$ to obtain a transition function $\gamma' : C \rightarrow (C \times L)^I = T_2(C)$. On the level of natural transformations between the corresponding signature functors, this translation is accomplished by $\sigma : T_1 \rightarrow T_2$, with $\sigma(X) : L \times X \rightarrow (L \times X)^I$ defined by $\sigma(x)(i) = x$.

(ii) We can also view every Mealy automaton as a labelled transition system. Given a set I of inputs and O of outputs of the Mealy automaton, we put $L = O \times I$. Given $(C, \gamma) \in$ CoAlg(T_2), we obtain a labelled transition system (i.e. a T_3-coalgebra) by letting $\gamma'(c) = \{(i, o, c') \mid \gamma(c)(i) = (o, c')\}$. Using natural transformations, the situation is as follows: Consider $\sigma : T_2 \rightarrow T_3$, given by $\sigma(X) : (O \times X)^I \rightarrow \mathcal{P}(X)^{O \times I}$ where $\sigma_1(X)(f)(o, i) = \{x \in X \mid f(i) = (o, x)\}$. We obtain $(C, \gamma') = \sigma^\dagger(C, \gamma)$.

Note that we can also treat coalgebras for endofunctors, which depend on an additional parameter in our framework:

Example 4. Suppose $T : \mathbb{C} \times$ Set $\rightarrow$ Set, where $\mathbb{C}$ is an arbitrary category of parameters. In order to emphasise the fact that we think of the first component as parameter, we write $T_A(X)$ for $T(A, X)$. Given a morphism $f : A \rightarrow B \in \mathbb{C}$, we obtain a natural transformation $\sigma(X) = T(f, \mathrm{id}_X)$. Identifying $C \in \mathbb{C}$ with the endofunctor T_C, we can thus treat $\mathbb{C}$ as a category of signatures for coalgebras.

4 Abstract Behavioural Logic

This shows, that abstract behavioural logic can be endowed with the structure of an institution. Abstract behavioural logic was studied in [10, 11], where the term "logic" is understood in a very general sense:

Definition 5. *A* logic for coalgebras *is a set* $\mathcal{L}$ *(the* language *of the logic), together with a family* $\models$ *of relations, indexed by the* T-coalgebras, *such that* $\models_{(C,\gamma)} \subseteq C \times \mathcal{L}$.

We call a *logic* behavioural,

$$d \models_{(D,\delta)} \phi \iff c \models_{(C,\gamma)} \phi$$

for all formulas $\phi \in \mathcal{L}$ *and all behaviourally equivalent states* $(c, d) \in C \times D$*. As usual,* $(C, \gamma) \models \phi$ *iff* $\forall c \in C.c \models_{(C,\gamma)} \phi$*, and* $[\![\phi]\!]_{(C,\gamma)} = \{c \in C \mid c \models_{(C,\gamma)} \phi\}$*.*

The starting point of the investigations conducted in [10, 11] is the representation of formulas of a behavioural logic as subsets of the final T-coalgebra (assuming it exists). This representation can be formulated as follows:

Proposition 2. *Suppose* $(Z, \zeta) \in \mathsf{CoAlg}(T)$ *is final and* $\mathcal{L}$ *is a behavioural logic for* T*-coalgebras. Then*

$$[\![\phi]\!]_{(C,\gamma)} = !^{-1}([\![\phi]\!]_{(Z,\zeta)})$$

for all $(C, \gamma) \in \mathsf{CoAlg}(T)$ *and all* $\phi \in \mathcal{L}$*, where* $! : C \to Z$ *is the morphism given by finality.*

Proof. Immediate from the definition of behavioural equivalence.

Thus, every formula ϕ of a behavioural logic can be semantically represented as a subset of the final T-coalgebra. Thus, if (Z, ζ) is final in $\mathsf{CoAlg}(T)$, we can view $\mathcal{P}(Z)$ as behavioural logic with $c \models_{(C,\gamma)} \phi$ if $!(c) \in \phi$, where $\phi \in \mathcal{P}(Z)$ and $! : C \to Z$ is the final morphism:

Definition 6. *Suppose* (Z, ζ) *is final in* $\mathsf{CoAlg}(T)$*. The abstract behavioural logic* $\mathcal{A}_T = \mathcal{P}(Z)$ *has subsets of the final* T*-coalgebra as formulas. Satisfaction is given by* $c \models_{(C,\gamma)} \phi$ *if* $!(c) \in \phi$*.*

It is immediately obvious from the definition of behavioural logic, that abstract behavioural logic is indeed behavioural. We now add signature morphisms to the picture. So suppose $\sigma : S \to T$ is a natural transformation. If (Z_S, ζ_S) is final in $\mathsf{CoAlg}(S)$, then $\sigma^\dagger(Z_S, \zeta_S) \in \mathsf{CoAlg}(T)$, thus, assuming (Z_T, ζ_T) is final in $\mathsf{CoAlg}(T)$, we have a unique morphism $! : Z_S \to Z_T$, the inverse image of which induces a translation $\mathcal{A}_T \to \mathcal{A}_S$ between the abstract logics associated to T and S.

Proposition 3. *Suppose* $\sigma : S \to T$*, and* S, T *allow for final coalgebras* (Z_S, ζ_S) *and* (Z_T, ζ_T)*, respectively. Then*

$$(C, \gamma) \models \sigma^*(\phi) \iff \sigma^\dagger(C, \gamma) \models \phi$$

for all $(C, \gamma) \in \mathsf{CoAlg}(S)$ *and all* $\phi \in \mathcal{A}_T$*, where* $\sigma^* = !^{-1}$ *for the unique morphism* $! : \sigma^\dagger(Z_S, \zeta_S) \to (Z_T, \zeta_T)$*, given by finality.*

Proof. Suppose $(C, \gamma) \in \mathsf{CoAlg}(S)$ and consider the diagram where u is the morphism given by finality of (Z_S, ζ_S) and v is the morphism given by finality of (Z_T, ζ_T). Suppose $\phi \in \mathcal{A}_T$ and $c \in C$. Then $c \models_{(C,\gamma)} \sigma^*(\phi)$ iff $u(c) \in v^{-1}(\phi)$ iff $u \circ v(c) \in \phi$ iff $c \models_{\sigma^\dagger(C,\gamma)} \phi$, since $v \circ u : \sigma^\dagger(C,\gamma) \to (Z_T, \zeta_T)$ is equal to the unique morphism given by finality of (Z, ζ_T).

If we take some care in setting up our category of signatures Sig as to ensure that every endofunctor $T \in$ Sig admits a final coalgebra (otherwise abstract behavioural logic isn't meaningful), we obtain:

Theorem 1. *Suppose* Sig $\subseteq$ [Set, Set] *is a subcategory such that every* $T \in$ Sig *admits a final* T-*coalgebra. Let* $\mathrm{Sen}(T) = \mathcal{A}(T)$ *and* $\mathrm{Sen}(\sigma) = \sigma^\dagger$. *Then* (Sig, Mod, Sen, $\models$), *with* $\models$ *as in Definition 6, is a coinstitution.*

This theorem shows, that behavioural logics are an institution, if we replace the concrete syntax by a semantical abstraction. We now turn to coalgebraic logic, the language of which is given inductively as initial algebra.

5   Coalgebraic Logic

Coalgebraic Logic, due to Moss [15], is a modal logic, interpreted over coalgebras. The main feature of coalgebraic logic is the insight, that – on the level of T-coalgebras for an arbitrary endofunctor T – modal operators can be expressed using functor application. It turns out that coalgebraic logic, as originally defined by Moss [15] is *not* an institution: one cannot translate formulas along non-injective signature morphisms. However, adapting the definition slightly, we obtain a logic, which is an institution and into which coalgebraic logic can be conservatively embedded. In the original paper, the language of coalgebraic logic comprises a (in general proper) class of formulas, and is constructed by extending the endofunctor T to classes (assuming that T is standard and set-based). Here, we give an alternative (but equivalent) presentation of coalgebraic logic, which dispenses with the use of classes at the expense of assuming the existence of an inaccessible cardinal. Instead of assuming T to be standard and set-based, we assume that T is κ-accessible, for some inaccessible cardinal κ. In a nutshell, the accessibility condition assures that the image of T on a set is already determined by the image of T on sets of cardinality less than κ; this is a technical requirement wich ensures the existence of initial algebras, which constitute a part of the syntax of coalgebraic logic. We make this choice simply because we think that accessibility of an endofunctor is – for most readers – a more familiar concept than being standard and set based.

The second condition we have to require is, that T extends to an endofunctor $\hat{T}$ on the category Rel of sets and relations (we often write $A \leftrightarrow B$ for a relation $R \subseteq A \times B$). This extension is given by $\hat{T}X = TX$ for sets X and $\hat{T}R = T\pi_2 \circ (T\pi_1)^{-1}$, for a relation $R : A \leftrightarrow B$ with associated projections $\pi_1 : R \to A$ and $\pi_2 : R \to B$ (this is as in [15]). It is well known (the original reference is [4]), that functoriality of $\hat{T}$ is equivalent to T preserving weak pullbacks. We now introduce syntax and semantics of coalgebraic logic, where we assume throughout the section, that T is κ-accessible for some inaccessible κ and preserves weak pullbacks and denote the bounded powerset functor by $\mathcal{P}_\kappa$, that is, $\mathcal{P}_\kappa(X) = \{\mathfrak{x} \subseteq X \mid \mathsf{card}(\mathfrak{x}) < \kappa\}$.

Definition 7. *Let* $L_T = \mathcal{P}_\kappa + \mathcal{P}_\kappa \circ T$. *The syntax of coalgebraic logic is the carrier* $\mathcal{L}_T$ *of the initial* L_T-*algebra* $(\mathcal{L}_T, \iota_T)$.

If $(C, \gamma) \in \mathsf{CoAlg}(T)$, *put* $d_C : \mathcal{P}_\kappa \mathcal{P}(C) \to \mathcal{P}(C), d_C(\mathfrak{x}) = \cap \mathfrak{x}$ *and* $e_C : \mathcal{P}_\kappa T \mathcal{P}(C) \to \mathcal{P}(C)$, $e_C(x) = \{c \in C \mid \exists y \in x.(\gamma(c), y) \in \hat{T}(\epsilon_C)\}$, *where* $\epsilon_C \subseteq C \times \mathcal{P}(C)$ *is the membership relation.*

The semantics $\llbracket \cdot \rrbracket_{(C,\gamma)} : \mathcal{L}_T \to \mathcal{P}(C)$ *of* $\mathcal{L}_T$ *with respect to* (C, γ) *is the unique function with* $[d_C, e_C] \circ L_T(\llbracket \cdot \rrbracket_{(C,\gamma)}) = \llbracket \cdot \rrbracket_{(C,\gamma)} \circ \iota_T$. *If* $c \in \llbracket \phi \rrbracket_{(C,\gamma)}$, *we also write* $c \models_{(C,\gamma)} \phi$; *we drop the subscript* (C, γ) *whenever there is no danger of confusion; also* $(C, \gamma) \models \phi$ *iff* $c \models_{(C,\gamma)} \phi$ *for all* $c \in C$.

Note $\mathcal{L}$ contains $\mathsf{tt} = \bigwedge \emptyset$ and is closed under conjunctions of size $< \kappa$.

In the above definition, the auxiliary familiy of functions d_C is used to interpret conjunctions, and e takes care of the modalities. Note that the initial L_T-algebra $(\mathcal{L}_T, \iota_T)$ always exists since L_T is κ-accessible, see [2]. If $\mathsf{in}_1 : \mathcal{P}_\kappa \mathcal{L}_T \to \mathcal{P}_\kappa \mathcal{L}_T + \mathcal{P}_\kappa T \mathcal{L}_T$ and $\mathsf{in}_2 : T \mathcal{L}_T \to \mathcal{P}_\kappa \mathcal{L}_T + T \mathcal{L}_T$ denote the coproduct injections, we write $\bigwedge_T = \iota_T \circ \mathsf{in}_1$ and $\nabla_T = \iota \circ \mathsf{in}_2$. The language of coalgebraic logic can thus be described as the least set such that

$$\Phi \subseteq \mathcal{L}_T, \mathsf{card}(\Phi) < \kappa \implies \bigwedge_T \Phi \in \mathcal{L}_T$$

$$\phi \subseteq T \mathcal{L}_T \implies \nabla_T \phi \in \mathcal{L}_T$$

This presentation also highlights the (only) difference compared to Moss' original definition, where one does not take *subsets* of $T \mathcal{L}_T$ in the second clause, but *elements* of $T \mathcal{L}_T$.

If $(C, \gamma) \in \mathsf{CoAlg}(T)$, we then obtain

$$c \models \bigwedge_T \Phi \text{ iff } c \models \phi \text{ for all } \phi \in \Phi$$

$$c \models \nabla_T \Phi \text{ iff}(\gamma(c), \phi) \in \hat{T}(\models) \text{for some } \phi \in \Phi$$

for subsets $\Phi \subseteq \mathcal{L}_T$ of cardinality less than κ and $\phi \in T \mathcal{L}_T$.

We give a brief example of the nature of coalgebraic logic; for an in-depth discussion and more example see Moss' original article [15].

Example 5. Let $TX = L \times X$, where L is a set of labels; we drop the subscript "T" on $\mathcal{L}$ and ∇. As already mentioned, $\mathsf{tt} \in \mathcal{L}$ and obviously $\llbracket \mathsf{tt} \rrbracket = C$ for all $(C, \gamma) \in \mathsf{CoAlg}(T)$. If $l \in L$, we have $\{(l, \mathsf{tt})\} \subseteq T \mathcal{L}$, hence $\nabla \{(l, \mathsf{tt})\} \in \mathcal{L}$. Unravelling the definitions, one obtains $c \models \nabla \{(l, \mathsf{tt})\}$ if $\pi_1 \circ \gamma(c) = l$. In the same manner, one has $\nabla \{(m, \nabla \{(l, \mathsf{tt})\})\} \in \mathcal{L}$ for $m \in L$ with $c \models \nabla \{(m, \nabla \{(l, \mathsf{tt})\})\}$ iff the stream associated to c (cf. Example 1) begins with m and is followed by l.

Note that – if we restrict ourselves to singleton sets (as in the original paper [15]) as arguments of ∇, we cannot express the fact that a stream starts with l_0

or l_1 logically. This is the reason why coalgebraic logic, in its original formulation, fails to be a coinstitution: We cannot translate formulas along a signature morphism, which identifies two labels l_0 and l_1.

The generalisation of the original definition of coalgebraic logic does not allow us to distinguish bisimilar states. In other words, we have:

Proposition 4. $\mathcal{L}_T$ *is behavioural.*

Proof. It suffices to show that $f(c) \models_{(D,\delta)} \phi$ iff $c \models_{(C,\gamma)} \phi$ whenever $\phi \in \mathcal{L}_T$, $f : (C,\gamma) \to (D,\delta) \in \mathsf{CoAlg}(T)$ and $c \in C$. This follows from the fact that $f^{-1} : \mathcal{P}(D) \to \mathcal{P}(C)$ is a morphism of the L_T-algebras $(\mathcal{P}(D), [d_D, e_D])$ and $(\mathcal{P}(C), [d_C, e_C])$, where d_D, e_D, d_C and e_C are as in Definition 7.

To see that f^{-1} is a morphism of algebras, it suffices to show that

$$
\begin{array}{ccc}
\mathcal{P}_\kappa TP(D) & \xrightarrow{\;\mathcal{P}_\kappa T(f^{-1})\;} & \mathcal{P}_\kappa TP(C) \\
{\scriptstyle e_D}\downarrow & & \downarrow{\scriptstyle e_C} \\
\mathcal{P}(D) & \xrightarrow{\quad f^{-1} \quad} & \mathcal{P}(D)
\end{array}
$$

commutes. For $c \in C$ and $x \in \mathcal{P}_\kappa TPD$, we have

$$c \in e_C \circ \mathcal{P}_\kappa T(f^{-1})(x)$$
$$\text{iff } \exists y \in x.(Tf \circ \gamma(c), y) \in \hat{T}(\in_D)$$
$$\text{iff } \exists y \in x.(\delta \circ f(c), y) \in \hat{T}(\in_D)$$
$$\text{iff } c \in f^{-1} \circ e_D(x),$$

which shows the claim.

We now turn to show that coalgebraic logic forms an institution. Here, a little care is needed when setting up the category of signatures and the category of models: Recall that we have required T to be κ-accessible for some inaccessible κ. To show the satisfaction condition (and to define the appropriate translations), we need to restrict the cardinality of the models to $< \kappa$ and require that T restricts to the full subcategory of sets, which are of cardinality less than κ. Working with classes, this would be unnecessary – we would have to require the dual condition that T can be continuously extended to classes.

Definition 8. *A κ-accessible endofunctor is* below κ *if* $|TX| < \kappa$ *whenever* $|X| < \kappa$.

Most κ-accessible functors are indeed below κ. The prime example of a κ-accessible functor, which is not below κ is the constant functor with value κ. The following lemma gives a characterisation of functors below κ, which just depends on the value of the functor at 1.

Lemma 2. *Suppose T is κ-accessible. Then T is below κ if $|T1| < \kappa$.*

Proof. If $|X| < \kappa$, then the diagram $(\{x\} \hookrightarrow X \mid x \in X)$ is κ-filtered. The claim follows from κ being inaccessible and from the construction of κ-filtered colimits (see [3]).

In order establish the satisfaction condition, we additionally have to require that the natural transformation σ is compatible with the extensions $\hat{S}$ and $\hat{T}$ to relations. That is, we require that $G(\sigma) : \hat{S} \to \hat{T}$ is natural, where $G(\sigma)(X) = G(\sigma(X)) : \hat{S}X \nrightarrow \hat{T}X$ is defined as the graph of $\sigma(X)$, for X a set. In this case, we call σ *relational*.

Many natural transformations can be shown to be relational using the following criterion:

Lemma 3. *A natural transformation $\sigma : S \to T$ is relational, if every naturality square,*

$$
\begin{array}{ccc}
SA & \xrightarrow{\sigma(A)} & TA \\
{\scriptstyle Sf}\downarrow & & \downarrow{\scriptstyle Tf} \\
SB & \xrightarrow[\sigma(B)]{} & TB
\end{array}
$$

where $f : A \to B$, is a weak pullback.

Proof. Suppose A, B are sets and $R : A \nrightarrow B$ is a relation; we need to show that

$$
\begin{array}{ccc}
SA & \xrightarrow{G(\sigma(A))} & TA \\
{\scriptstyle \hat{S}(R)}\downarrow & & \downarrow{\scriptstyle \hat{T}(R)} \\
SB & \xrightarrow[G(\sigma(B))]{} & TB
\end{array}
$$

commutes in Rel.

First suppose that $(x, y) \in G(\sigma(B)) \circ \hat{S}(R)$. Thus there is some $x_1 \in \hat{S}(R)$ with $S\pi_1(x_1) = x$ and $\sigma(B) \circ S\pi_2(x_1) = y$. Put $y_1 = \sigma(R)(x_1)$. Then $T\pi_1(y_1) = \sigma(A)(x)$ and $T\pi_2(y_1) = y$, hence $(x, y) \in \hat{T}(R) \circ G(\sigma(A))$.

Now let $(x, y) \in \hat{T}(R) \circ G(\sigma(A))$. As above, there is $y_1 \in TR$ with $T\pi_1(y_1) = \sigma(A)(x)$ and $T\pi_2(y_1) = y$. Since

$$
\begin{array}{ccc}
SR & \xrightarrow{\sigma(R)} & TR \\
{\scriptstyle S\pi_1}\downarrow & & \downarrow{\scriptstyle T\pi_1} \\
SA & \xrightarrow[\sigma(A)]{} & TA
\end{array}
$$

is a weak pullback, there is $x_1 \in SR$ with $\sigma(R)(x_1) = y_1$ and $S\pi_1(x_1) = x$. Using naturality of σ, we obtain $S\pi_1(x_1) = x$ and $\sigma(B) \circ S\pi_2(x_1) = y$, so $(x, y) \in \sigma(B) \circ SR$.

Using the fact that products, coproducts, the powerset functor, identity functor and constant functors preserve weak pullbacks, we have the following criterion, which can be applied to a large class of signatures, obtained via parameterised functors (cf. Example 4).

Corollary 1. *Suppose T : Set $\times$ Set $\to$ Set is built using products, coproducts, the powerset functor, identity functor and constant functors only. Then, given $f : A \to B$, the natural transformation $T(f, \mathrm{id}) : T_A \to T_B$ is relational.*

Given a (not necessary relational) transformation $\sigma : S \to T$, we can define a translation $\sigma^* : \mathcal{L}_T \to \mathcal{L}_S$ as follows: Since $\mathcal{L}_T$ supports the structure ι_T of an initial T-algebra, every L_T-algebra structure $t : L_T\mathcal{L}_S \to \mathcal{L}_S$ defines a unique mapping $\sigma^* : \mathcal{L}_T \to \mathcal{L}_S$ with $\sigma^* \circ \iota_T = t \circ L_T(\sigma^*)$. So we have to find an appropriate L_T-algebra structure t on $\mathcal{L}_S$. We let $t = [t_1, t_2]$ where $t_1 : \mathcal{P}_\kappa \mathcal{L}_S \to \mathcal{L}_S$ is intersection and $t_2 : \mathcal{P}_\kappa \circ T\mathcal{L}_S \to \mathcal{L}_S$ is given as $t_2 = \nabla_T \circ \sigma(\mathcal{L}_S)^{-1}$. Note that $\sigma(\mathcal{L}_S)^{-1}$ maps $\mathcal{P}_\kappa(T\mathcal{L}_S) \to \mathcal{P}_\kappa(S\mathcal{L}_S)$.

Proposition 5. *Suppose $\sigma : S \to T$ is relational and $(C, \gamma) \in \mathsf{CoAlg}(S)$. Then*

$$(C, \gamma) \models \sigma^*(\phi) \quad\Longleftrightarrow\quad \sigma^\dagger(C, \gamma) \models \phi$$

for all $\phi \in \mathcal{L}_T$, provided $|C| < \kappa$.

Proof. Let $d = d_C$ and $e = e_C$ be as in Definition 7 and suppose $d^\dagger(\mathfrak{x}) = \bigcap \mathfrak{x}$ and $e^\dagger(\mathfrak{x}) = \{c \in C \mid \exists y \in x.(\sigma(C) \circ \gamma(c), y) \in \hat{T}(\in_C)\}$. Then, by definition of $[\![\cdot]\!]_T$, we have $[\![\cdot]\!]_T \circ [\bigwedge_T, \nabla_T] = [d^\dagger, e^\dagger] \circ \mathcal{P}_\kappa[\![\cdot]\!]_T + \mathcal{P}_\kappa T[\![\cdot]\!]_T$. Consider the following diagram:

$$
\begin{array}{ccc}
L_T\mathcal{L}_T \xrightarrow{L_T\sigma^*} L_T\mathcal{L}_S \xrightarrow{L_T[\![\cdot]\!]_S} L_T\mathcal{P}(C) \\
\Big\downarrow{\scriptstyle [\bigwedge_T, \nabla_T]} \quad \Big\downarrow{\scriptstyle \mathrm{id}+\sigma(\mathcal{L}_S)^{-1}} \quad \Big\downarrow{\scriptstyle \mathrm{id}+\sigma(\mathcal{P}C)^{-1}} \\
\quad L_S\mathcal{L}_S \xrightarrow{L_S[\![\cdot]\!]_S} L_S\mathcal{P}(C) \\
\quad \Big\downarrow{\scriptstyle [\bigwedge_S, \nabla_S]} \quad \Big\downarrow{\scriptstyle [d,e]} \\
\mathcal{L}_T \xrightarrow{\sigma^*} \mathcal{L}_S \xrightarrow{[\![\cdot]\!]_S} \mathcal{P}(C)
\end{array}
$$

The left hand square commutes by definition of σ^* and the lower right hand square by definition of $[\![\cdot]\!]_S$. We show that

(i) $[d, e] \circ (\mathrm{id} + \sigma(\mathcal{P}C)^{-1}) = [d^\dagger, e^\dagger]$
(ii) The top right corner commutes.

Both claims then entail the satisfaction condition as stated.

Ad 1: Since σ is relational, we have $\hat{T}(\in_C) \circ G(\sigma(C)) = G(\sigma(\mathcal{P}C)) \circ \hat{S}(\in_C)$. Now let $c \in C$ and $x \in \mathcal{P}_\kappa TP(C)$. We have

$$
\begin{aligned}
&c \in e^\dagger(x) \\
&\text{iff } \exists y \in x.(\gamma(c), y) \in G(\sigma(\mathcal{P}C)) \circ S(\in_C) \\
&\text{iff } \exists z \in \sigma(\mathcal{P}C)^{-1}(x).(\gamma(c), z) \in \hat{S}(\in_C) \\
&\text{iff } c \in e \circ \sigma(\mathcal{P}C)^{-1}(x).
\end{aligned}
$$

Ad 2: If R is a relation, we denote the opposite relation by R^{op}. Then, for a function f, we have $G(Tf)^{op} = \hat{T}((Gf)^{op})$, and similarly for S. Also note that $\hat{T}(G[\![\cdot]\!]_S^{op}) \circ G(\sigma(\mathcal{PC})) = G(\sigma(\mathcal{L}_S)) \circ \hat{S}(G[\![\cdot]\!]_S^{op})$ since σ is relational. Having said that, we obtain for $\phi \in \mathcal{P}_\kappa T\mathcal{L}_S$ and $c \in S\mathcal{PC}$:

$$c \in \mathcal{P}_\kappa S[\![\cdot]\!]_S \circ \sigma(\mathcal{L}_S)^{-1}(\phi)$$
$$\text{iff } \exists \psi \in \phi.(c,\psi) \in G(\sigma(\mathcal{L}_S)) \circ \hat{S}(G[\![\cdot]\!]_S^{op})$$
$$\text{iff } \exists \psi \in \phi.(c,\psi) \in \hat{T}(G[\![\cdot]\!]_S^{op}) \circ G(\sigma(\mathcal{PC}))$$
$$\text{iff } c \in \sigma(\mathcal{PC})^{-1} \circ \mathcal{P}_\kappa T[\![\cdot]\!]_S(\phi),$$

that is, the satisfaction condition holds.

Taking some care when choosing signatures and models, coalgebraic logic is a coinstitution ($\text{Mod}^\kappa(T)$ is the full subcategory of T-coalgebras with carrier $< \kappa$):

Theorem 2. *Suppose* $\text{Sig} \subseteq [\text{Set}, \text{Set}]$ *is a subcategory such that*

- *Each* $T \in \text{Sig}$ *is below* κ
- *Each* $\sigma : S \to T \in \text{Sig}$ *is relational.*

Then $(\text{Sig}, \text{Mod}^\kappa, \text{Sen}, \models)$, *with* $\text{Sen}(T) = \mathcal{L}_T$ *and* $\text{Sen}(\sigma) = \sigma^*$, *is a coinstitution.*

6   Coalgebraic Modal Logic

We have seen in the previous sections, that abstract behavioural logic and coalgebraic logic are coinstitutions. The formulation of abstract modal logic is completely syntax-free; the language of coalgebraic modal logic is abstract in that it is given as initial algebra. We now investigate coalgebraic modal logic, the language of which is concretely given as propositional logic, enriched with modal operators. Coalgebraic modal logic is based on the observation, that predicate liftings, which we now introduce, generalise modal operators from Kripke models to coalgebras for arbitrary signature functors.

Predicate liftings were first considered by Jacobs and Hermida [9] in the context of coinduction principles and later by Rößiger [18] and Jacobs [8] in the context of modal logic. There, as well as in the related paper [18], predicate liftings appear as syntactically defined entities, and naturality is a derived property. The notion of predicate lifting used in the present exposition is more general, and takes naturality as the defining property.

Definition 9. *A* predicate lifting *for* T *is a natural transformation* $\lambda : 2 \to 2 \circ T$, *where* $2 : \text{Set} \to \text{Set}^{op}$ *denotes the contravariant powerset functor.*

The next example shows, that predicate liftings do not only capture modal operators, but can also be used to interpret atomic propositions.

Example 6. Suppose $TX = \mathcal{P}(X) \times \mathcal{P}(A)$. Then every T-coalgebra (C, γ) defines a Kripke model $\mathbb{K}(C, \gamma) = (C, R, V)$ over the set A of atomic propositions: the accessibility relation is given by $(c, c') \in R$ iff $c' \in \pi_1 \circ \gamma(c)$ and for $a \in A$ we have $V(a) = \{c \in C \mid a \in \pi_2 \circ \gamma(c)\}$.

Now, for a set C, consider the operation $\lambda(C) : \mathcal{P}(C) \to \mathcal{P}(TC)$ given by $\lambda(C)(\mathfrak{c}) = \{(c', \mathfrak{a}) \in TC \mid c' \subseteq \mathfrak{c}\}$. It is easy to see that λ defines a predicate lifting for T. Now suppose $(C, \gamma) \in \mathsf{CoAlg}(T)$ and $\mathfrak{c} \subseteq C$, which we think of as the interpretation of a modal formula ϕ. Under the correspondence outlined above, we have $\gamma^{-1} \circ \lambda(C)(\mathfrak{c}) = \{c \in C \mid \forall c'.(c, c') \in R \implies c' \in \mathfrak{c}\}$, corresponding to the interpretation of the formula $\Box\phi$.

For the case of atomic propositions, consider the constant lifting, defined by $\alpha(C)(\mathfrak{c}) = \{(c', \mathfrak{a}) \in TC \mid a \in \mathfrak{a}\}$. Again, an easy calculation shows that α is a predicate lifting. Identifying T-coalgebras with Kripke models via the correspondence above, we obtain for $(C, \gamma) \in \mathsf{CoAlg}(T)$ and an arbitrary subset $\mathfrak{c} \subseteq C$ that $\gamma^{-1} \circ \alpha(\mathfrak{c}) = V(a)$, that is, the set of states which validate the proposition a.

This leads us to study propositional logic, enriched with predicate liftings, as a logic for coalgebras.

Definition 10. *Suppose $T :$ Set $\to$ Set and Λ is a set of predicate liftings for T. The language $\mathcal{L}(\Lambda_T)$ of coalgebraic modal logic associated with T and Λ is the least set according to the grammar*

$$\phi ::= \mathbf{ff} \mid \phi \to \psi \mid [\lambda]\phi \qquad (\lambda \in \Lambda).$$

Given $(C, \gamma) \in \mathsf{CoAlg}(T)$, the semantics $[\![\phi]\!]_{(C,\gamma)} = [\![\phi]\!]$ of formulas $\phi \in \mathcal{L}(\Lambda)$ is given by:

$$[\![\mathbf{ff}]\!] = \emptyset$$
$$[\![\phi \to \psi]\!] = (C \setminus [\![\phi]\!]) \cup [\![\psi]\!]$$
$$[\![[\lambda]\phi]\!] = \gamma^{-1} \circ \lambda(C)([\![\phi]\!]).$$

As usual, we write $c \models_{(C,\gamma)} \phi$ (and drop the subscript if there is no danger of confusion), if $c \in [\![\phi]\!]_{(C,\gamma)}$. As usual, we write $(C, \gamma) \models \phi$ if $c \models_{(C,\gamma)} \phi$ for all $c \in C$.

An easy induction on the structure of formulas shows, that coalgebraic modal logic cannot distinguish between states, which are behaviourally equivalent.

Lemma 4. *Coalgebraic modal logic is behavioural.*

Proof. By induction on the structure of formulas, one shows that $[\![\phi]\!]_{(C,\gamma)} = f^{-1}([\![\phi]\!]_{(D,\delta)})$ for $\phi \in \mathcal{L}(\Lambda)$ and a morphism of coalgebras $f : (C, \gamma) \to (D, \delta)$. The claim follows from the definition of behavioural equivalence.

We now investigate the effect of signature morphisms on formulas. The key observation is the following:

Lemma 5. *Suppose* $\sigma : S \to T$ *is natural and* λ *is a predicate lifting for* T. *Then* $\sigma^{-1} \circ \lambda$ *is a predicate lifting for* S.

The proof is a straightforward calculation, and therefore omitted. That is, a signature morphism $\sigma : S \to T$ translates the modal operators associated with T to modal operators for S. This defines an inductive translation between languages for S to languages for T:

Definition 11. *Suppose* $\sigma : S \to T$ *and suppose that* Λ_T, Λ_S *are sets of predicate liftings for* T *and* S, *respectively. Let* $\sigma^{-1}(\Lambda) = \{\sigma^{-1} \circ \lambda \mid \lambda \in \Lambda\}$. *If* $\Lambda_S \subseteq \sigma^{-1}(\Lambda_T)$, *we define* $\sigma^* : \mathcal{L}(\Lambda_S) \to \mathcal{L}(\sigma^{-1}(\Lambda_T))$ *by*

$$\sigma^*(\mathit{ff}) = \mathit{ff}$$
$$\sigma^*(\phi \wedge \psi) = \sigma^*(\phi) \wedge \sigma^*(\phi)$$
$$\sigma^*([\lambda]\phi) = [\sigma^{-1} \circ \lambda]\phi$$

Using this translation, we have the following property, which immediately entails the satisfaction condition:

Lemma 6. *Suppose* $\sigma : S \to T$ *and* Λ_T, Λ_S *are sets of predicate liftings for* T *and* S, *respectively, with* $\sigma^{-1}(\lambda_T) \subseteq \Lambda_S$. *Then*

$$[\![\sigma^*(\phi)]\!]_{(C,\gamma)} = [\![\phi]\!]_{\sigma^\dagger(C,\gamma)}$$

for all $(C, \gamma) \in \mathsf{CoAlg}(S)$ *and all* $\phi \in \mathcal{L}(\Lambda_T)$.

Proof. We proceed by induction on the structure of formulas and do the only interesting case $\phi = [\lambda]\psi$; by induction hypothesis we may assume that

$$[\![\sigma^*\psi]\!]_{(C,\gamma)} = [\![\psi]\!]_{\sigma^\dagger(C,\gamma)}.$$

We obtain

$$[\![\sigma^*([\lambda]\phi)]\!]_{(C,\gamma)} = \gamma^{-1} \circ \sigma^{-1} \circ \lambda(C)([\![\sigma^*(\psi)]\!]_{(C,\gamma)})$$
$$= (\gamma^\dagger)^{-1} \circ \lambda(C)([\![\psi]\!]_{\sigma^\dagger(C,\gamma)})$$
$$= [\![[\lambda]\psi]\!]_{\sigma^\dagger(C,\gamma)},$$

which finishes the proof.

Again, we have to pay some attention when setting up the category of signatures in order to obtain an institution.

Theorem 3. *Suppose* $\mathsf{Sig} \subseteq [\mathsf{Set}, \mathsf{Set}]$ *is a subcategory, and*

- Λ_T *is a set of predicate liftings for all* $T \in \mathsf{Sig}$, *and*
- $\sigma^{-1}(\Lambda_T) \subseteq \Lambda_S$ *for all* $\sigma : S \to T \in \mathsf{Sig}$.

Then $(\mathsf{Sig}, \mathsf{Mod}, \mathsf{Sen}, \models)$ *is an institution, where* $\mathsf{Sen}(T) = \mathcal{L}(\Lambda_T)$ *for* $T \in \mathsf{Sig}$ *and* $\mathsf{Sen}(\sigma) = \sigma^*$ *for* $\sigma : S \to T \in \mathsf{Sig}$.

7 Conclusions and Related Work

We have addressed the question whether logics for coalgebras can be translated along signature morphisms, as to form an institution. The answer was "in general yes, but one has to take a little care when setting up the framework."

It is well known, that algebras form institutions with respect to different kinds of logics. Therefore, one might be lead to expect that *co*algebras and their logics congregate in some kind of *co*institution. This is true to the same extent as coalgebras and their logics form an institution, since there is a one-to-one correspondence between institutions over a category $\mathbb{S}$ and coinstitutions over $\mathbb{S}^{op}$ (Proposition 1).

Hence, instead of showing that coalgebras and their logics form an institution, we can equivalently show that they are coinstitutions. We prefer the latter, since we feel more comfortable with a category $\mathbb{S} \subseteq [T, T]$ of signatures than with its opposite; but this is clearly just a matter of taste.

We then showed that the dual of the satisfaction condition holds for three different types of logics for coalgebras: Abstract Behavioural Logic, Coalgebraic Logic and Coalgebraic Modal Logic. The framework of abstract behavioural logic is based on the observation, that formulas of a behavioural logic can be represented as subsets of the final coalgebra, if the latter exists. This leads to a translation not of formulas, but of the associated representations, resulting in an institution (Theorem 1). For the second type of logic, Moss' coalgebraic logic, the syntax needed to be modified slightly to obtain an institution. We have showed that this modification does not increase the expressive power of the logic (Proposition 4) and gives rise to an institution (Theorem 2). The third framework which we have studied is coalgebraic modal logic, which – in contrast to the ones mentioned before – comes with a concrete syntax, given by a set of predicate liftings for the endofunctor under consideration. The key observation here is, that predicate liftings translate along signature morphisms (Lemma 5), thus giving rise to an inductively defined translation between logics for different signature functors. This translation is well-behaved, witnessed by the fact that coalgebraic modal logic also forms an institution (Theorem 3).

The question whether logics for coalgebras form institutions was also taken up in [5, 17]. In [5], the satisfaction condition was established for an inductively defined class of functors, so-called "Kripke polynomial functors", on a category of sorted sets. In contrast, our approach is purely semantical and can be seen to subsume the one-sorted case, treated in [5]. In [17], the satisfaction condition was only established for the case of coalgebraic modal logic. A purely semantical study about the relationship between categories of coalgebras for parameterised endofunctors was already carried out in [12].

References

1. P. Aczel and N. Mendler. A Final Coalgebra Theorem. In D. H. Pitt et al, editor, *Category Theory and Computer Science*, volume 389 of *Lect. Notes in Comp. Sci.*, pages 357–365. Springer, 1989.

2. J. Adámek. Free algebras and automata realizations in the language of categories. *Comment. Math. Univ. Carolinae*, 15:589–602, 1974.
3. F. Borceux. *Handbook of Categorical Algebra*, volume 2. Cambridge University Press, 1994.
4. A. Carboni, G. Kelly, and R. Wood. A 2-categorical approach to change of base and geometric morphisms I. *Cahiers de Topologie et Géometrié Différentielle Catégoriques*, 32(1):47–95, 1991.
5. Corina Cirstea. Institutionalizing coalgebraic modal logic. In L. Moss, editor, *Coalgebraic Methods in Computer Science (CMCS 2002)*, volume 65 of *Electr. Notes in Theoret. Comp. Sci.* Elsevier Science Publishers, 2002.
6. J. Goguen and R. Burstall. Institutions: Abstract Model Theory for Specification and Programming. *Journal of the Association for Computing Machinery*, 39(1), 1992.
7. R. Goldblatt. A calculus of terms for coalgebras of polynomial functors. In Marina Lenisa Andrea Corradini and Ugo Montanari, editors, *Coalgebraic Methods in Computer Science (CMCS'2001)*, volume 44 of *Electr. Notes in Theoret. Comp. Sci.*, 2001.
8. B. Jacobs. Many-sorted coalgebraic modal logic: a model-theoretic study. *Theoret. Informatics and Applications*, 35(1):31–59, 2001.
9. B. Jacobs and C. Hermida. Structural Induction and Coinduction in a Fibrational Setting. *Information and Computation*, 145:107–152, 1998.
10. A. Kurz. A Co-Variety-Theorem for Modal Logic. In *Proceedings of Advances in Modal Logic 2, Uppsala, 1998*. Center for the Study of Language and Information, Stanford University, 2000.
11. A. Kurz. *Logics for Coalgebras and Applications to Computer Science*. PhD thesis, Universität München, April 2000.
12. A. Kurz and D. Pattinson. Coalgebras and Modal Logics for Parameterised Endofunctors. Technical report, CWI, 2000.
13. S. MacLane. *Categories for the Working Mathematician*. Springer, 1971.
14. R. Milner. *Communication and Concurrency*. International series in computer science. Prentice Hall, 1989.
15. L. Moss. Coalgebraic Logic. *Annals of Pure and Applied Logic*, 96:277–317, 1999.
16. D. Park. Concurrency and Automata on Infinite Sequences. In P. Deussen, editor, *5th GI Conference*, volume 104 of *Lect. Notes in Comp. Sci.* Springer, 1981.
17. D. Pattinson. *Expressivity Results in the Modal Logic of Coalgebras*. PhD thesis, Universität München, June 2001.
18. M. Rößiger. Coalgebras and Modal Logic. In H. Reichel, editor, *Coalgebraic Methods in Computer Science (CMCS'2000)*, volume 33 of *Electr. Notes in Theoret. Comp. Sci.*, 2000.
19. M. Rößiger. From Modal Logic to Terminal Coalgebras. *Theor. Comp. Sci.*, 260:209–228, 2001.
20. J. Rutten. Universal Coalgebra: A theory of systems. *Theor. Comp. Sci.*, 249(1):3–80, 2000.
21. Andrzej Tarlecki. Institutions: An Abstract Framework for Formal Specifications. In E. Astesiano, H.-J. Kreowski, and B. Krieg-Brückner, editors, *Algebraic Fondations of System Specification*, volume 2. Springer, 1999.

Presenting and Combining Inference Systems
Presentations with Inference Rules*

Wiesław Pawłowski

Institute of Computer Science, Polish Academy of Sciences
ul. Abrahama 18, 81-825 Sopot, Poland
w.pawlowski@ipipan.gda.pl

Abstract. The paper discusses the problem of representing and combining inference systems for (abstract) *context institutions*, within the framework of *context presentations* [10]. As it turns out, thanks to the context information present in this setting, the inference rules for quantifier logics can be expressed and manipulated in a simple way, without referring to *binding operators* or *requirements* (cf. [12]).

1   Introduction

Many applications of logic in computing science concern composite domains. As a consequence, it is becoming widely accepted that *practically useful* logical formalisms should have a *compositional* nature, possibly reflecting the structure of the problem domain.

Since the development of the notion of *institution* [5], formalizing "an abstract model theory for specification and programming", the issue of combining logical systems started to attract interest of researchers. The notion of *parchment* [4] was originally invented as a tool for proving the *satisfaction condition* for institutions. In [6] parchments have been proposed as a framework for combining logics. The idea has been further refined and explored in a series of papers [7, 8, 3, 2].

Institutions and parchment-like structures describe the structural part of a logical system only. In particular, they do not take inference systems into account. The issue of combining inference systems has been investigated mostly in the context of *fibring logics* [11]. Some of the results from this field have been applied to parchment-like structures in [3] (for the case of propositional logics), and recently extended to quantifier logics in [2].

In what follows we shall discuss the problem of representing and combining inference systems for (abstract) *context institutions*, within the framework of *context presentations* [10]. As it turns out, thanks to the context information present in this setting, the inference rules for quantifier logics can be expressed and manipulated in a simple way, without referring to *binding operators* or *requirements* (cf. [12]).

* This research has been partially supported by the KBN grant No. 7 T11C 002 21.

M. Wirsing, D. Pattinson, and R. Hennicker (Eds.): WADT 2002, LNCS 2755, pp. 409–424, 2003.
© Springer-Verlag Berlin Heidelberg 2003

2   Context Presentations and Context Institutions

The aim of this section is mostly to recall some basic notions and facts from [10]. Let us start with *metasignatures* and *metastructures* which *presentations* use for describing the syntactic and semantic aspects of a logical system respectively.

2.1   Metalanguage Framework

Definition 1. *A* metasignature *is a six-tuple* $\Sigma = \langle S, \Omega, \Pi, V, C, Q \rangle$, *such that* $\langle S, \Omega, \Pi \rangle$ *is a relational signature,* $V \subseteq S$, C *is a family of sets indexed by natural numbers, and* Q *is a set.*

Informally speaking, metasignatures are just relational signatures enriched by symbols of *logical connectives*, (the family C), *quantifier symbols* (the set Q), and having a distinguished subset of sort names (the set V), for which we want to talk about *variables*. A metasignature morphism consist of a relational signature morphism preserving the *variable-sorts* i.e., elements of V, and two mappings relating the connectives and quantifier symbols (see [10] for details). The metasignatures and their morphisms (with an obvious definition of composition) constitute a category, which we shall denote by MSIG.

For every metasignature $L = \langle S, \Omega, \Pi, V, C, Q \rangle$, by $\mathsf{Syn}(L)$ we shall denote an algebraic signature $\langle S \uplus \{\star\}, \Omega^{\mathrm{Syn}} \cup \Pi^{\mathrm{Syn}} \cup C^{\mathrm{Syn}} \rangle$ such that:

- $\omega \in \Omega_{w,s}$ iff $\omega \in \Omega^{\mathrm{Syn}}_{T(w),T(s)}$
- $\pi \in \Pi_w$ iff $\pi \in \Pi^{\mathrm{Syn}}_{T(w),\mathbb{B}}$
- $c \in C_n$ iff $c \in C^{\mathrm{Syn}}_{\mathbb{B}\cdots\mathbb{B},\mathbb{B}}$

where $T(s)$ and $\mathbb{B}$ denote the injection of $s \in S$ and $\star$ into the disjoint union $S \uplus \{\star\}$ respectively.

The above construction in an obvious way extends to a *syntax functor*: Syn : MSIG $\rightarrow$ ALGSIG. In what follows, we shall also use two "sub-functors" of the functor Syn – Atm : MSIG $\rightarrow$ ALGSIG and Trm : MSIG $\rightarrow$ ALGSIG, such that:

- $\mathsf{Atm}(\langle S, \Omega, \Pi, V, C, Q \rangle) \mathrel{\hat{=}} \langle T(S) \cup \{\mathbb{B}\}, \Omega^{\mathrm{Syn}} \cup \Pi^{\mathrm{Syn}} \rangle$,
- $\mathsf{Trm}(\langle S, \Omega, \Pi, V, C, Q \rangle) \mathrel{\hat{=}} \langle T(S), \Omega^{\mathrm{Syn}} \rangle$

(the symbol $\hat{=}$ means "by definition").

The morphisms Atm_ℓ and Trm_ℓ are given by the respective components of the morphism Syn_ℓ. For every metasignature L, the signatures $\mathsf{Atm}(L)$ and $\mathsf{Trm}(L)$ are *sub-signatures* of $\mathsf{Syn}(L)$—hence the informal term "sub-functor".

Definition 2. *An* L*-metastructure* $\mathcal{A}$ *consists of:*

- *a* $\mathsf{Syn}(L)$*-algebra* A,
- *a* $T(V)$*-indexed set* $V_{\mathcal{A}}$, *such that for every* $s \in V$ $(V_{\mathcal{A}})_{T(s)} \subseteq |A|_{T(s)}$,
- *a set* $D_{\mathcal{A}}$, *such that* $D_{\mathcal{A}} \subseteq |A|_{\mathbb{B}}$,
- *for every symbol* $q \in Q$, *a partial function* $q_{\mathcal{A}} : \mathcal{P}(|A|_{\mathbb{B}}) \rightharpoonup |A|_{\mathbb{B}}$.

The set $|\mathcal{A}| \cong |A|$ is called the *carrier* of the metastructure $\mathcal{A}$, the subset $V_{\mathcal{A}}$—its set of *assignable values* (i.e., elements which can be "ranged-over" by variables). The set $|\mathcal{A}|_\mathbb{B}$, corresponding to the distinguished sort $\mathbb{B}$ of the "syntactic" signature $\mathsf{Syn}(L)$, plays the rôle of the set of *logical values*, and its subset $D_{\mathcal{A}}$ is the set of *designated elements*.

L-metastructures can be viewed as many-sorted algebras enriched by *generalized operations*—the partial functions corresponding to the symbols from Q. These operations are ment to represent the semantics of *quantifiers*. Metastructure morphisms are just algebraic homomorphisms preserving both the distinguished subsets (of *values* and *designated elements*) and the generalized operations (cf. [10]). The category of L-metastructures will be denoted by $\mathrm{MSTR}(L)$. For any metasignature morphism $\ell : L \to L'$, $\mathrm{MSTR}_\ell : \mathrm{MSTR}(L') \to \mathrm{MSTR}(L)$ denotes the obvious reduct functor.

2.2 Interpretation Structures and Presentations

Informally speaking, metasignatures can be used for describing the *grammar* of the language of a logical system while metastructures give the semantics of the language. This idea is formalized by notions of an *interpretation structure* and *presentation*.

Definition 3. *An* interpretation structure *is a triple $\langle L, \mathcal{M}, Int \rangle$, consisting of:*

- *a metalanguage signature $L \in |\mathrm{MSIG}|$*
- *a class of models $\mathcal{M} \in |\mathrm{CLASS}|$,*
- *an interpretation function (functor) $Int : \mathcal{M} \to \mathrm{MSTR}(L)$.*

Interpretation structures closely resemble *rooms* for *model-theoretic parchments* (cf. [8]). The main difference is that the latter are built over the notion of ordinary (many-sorted) algebra instead of metastructure.

Definition 4. *A triple $\langle \ell, m, int \rangle$ is an* interpretation structure morphism *from $\langle L, \mathcal{M}, Int \rangle$ to $\langle L', \mathcal{M}', Int' \rangle$ iff:*

- *$\ell : L \to L'$ is a metasignature morphism,*
- *$m : \mathcal{M}' \to \mathcal{M}$ is a function,*
- *$int : m ; Int \Rightarrow Int' ; \mathrm{MSTR}_\ell$ is a natural transformation.*

Since $\mathcal{M}'$ is a discrete category (a class) the natural transformation int is simply an $\mathcal{M}'$-indexed family of metastructure morphisms, such that for every $M' \in \mathcal{M}'$:

$$int_{M'} : Int(m(M')) \to \mathrm{MSTR}_\ell(Int'(M')).$$

The composition of $\langle \ell, m, int \rangle : IS_1 \to IS_2$ and $\langle \ell', m', int' \rangle : IS_2 \to IS_3$ is defined by:

$$\langle \ell, m, int \rangle ; \langle \ell', m', int' \rangle \cong \langle \ell ; \ell', m' ; m, (m'*int) ; (int'*\mathrm{MSTR}_\ell) \rangle.$$

Interpretation structures and their morphisms constitute a category which we shall denote by INTSTR.

Definition 5. *A* context presentation *is an arbitrary functor into the category of interpretation structures:*

$$\mathcal{P} : \text{SIG}^{\mathcal{P}} \to \text{INTSTR}.$$

We shall call $\text{SIG}^{\mathcal{P}}$ the *category of signatures* of the presentation $\mathcal{P}$.

Let $\mathcal{P} : \text{SIG}^{\mathcal{P}} \to \text{INTSTR}$ be a context presentation. To simplify notation, in the sequel we shall omit the superscript from the name of the category of signatures for $\mathcal{P}$.

For any signature $\Sigma \in |\text{SIG}|$ the presentation $\mathcal{P}$ assigns an interpretation structure $\mathcal{P}(\Sigma)$:

$$\Sigma \mapsto \mathcal{P}(\Sigma) = \langle\, L_{\Sigma}, \mathcal{M}_{\Sigma}, Int_{\Sigma} : \mathcal{M}_{\Sigma} \to \text{MSTR}(L_{\Sigma}) \,\rangle.$$

By "projecting" $\mathcal{P}$ to the first and second component respectively we obtain the *metalanguage functor* $\text{Lan}^{\mathcal{P}} : \text{SIG} \to \text{MSIG}$ and the *model functor* $\text{Mod}^{\mathcal{P}} : \text{SIG}^{op} \to \text{CLASS}$ for $\mathcal{P}$. Relationships between these two functors can be depicted as follows:

Definition 6. *Let* $\mathcal{P}_1 : \text{SIG}^{\mathcal{P}_1} \to \text{INTSTR}$ *and* $\mathcal{P}_2 : \text{SIG}^{\mathcal{P}_2} \to \text{INTSTR}$ *be presentations. A morphism from* $\mathcal{P}_1$ *to* $\mathcal{P}_2$ *is a pair* $\langle\, \Phi, \mu \,\rangle : \mathcal{P}_1 \to \mathcal{P}_2$ *such that:*

- $\Phi : \text{SIG}^{\mathcal{P}_1} \to \text{SIG}^{\mathcal{P}_2}$ *is a functor,*
- $\mu : \Phi\, ; \mathcal{P}_2 \Rightarrow \mathcal{P}_1$ *is a natural transformation.*

Presentations and presentation morphisms constitute a category which we shall denote by PRES.

We shall now give two simple examples and describe a presentation $\mathcal{EQL}$ for *many-sorted equational logic* and presentation $\mathcal{FOL}$ for *many-sorted first-order logic* (without equality). Other examples of presentations, as well as examples of their morphisms, can be found in [9, 10].

Example 1 (Many-sorted equational logic). The category of signatures for $\mathcal{EQL}$ is the category of algebraic signatures ALGSIG. The metalanguage functor $\text{Lan}^{\mathcal{EQL}}$: ALGSIG → MSIG is defined as follows:

- $\text{Lan}^{\mathcal{EQL}}(\langle\, S, \Omega \,\rangle) \,\hat{=}\, \langle\, S, \Omega, \Pi, S, C, \emptyset \,\rangle$, where $\Pi_{ss} \,\hat{=}\, \{\equiv\}$ for $s \in S$, and all other elements of Π, and all C_n for $n \geq 0$ are empty.
- for every $\sigma : \Sigma_1 \to \Sigma_2$ in ALGSIG, the morphism $\text{Lan}^{\mathcal{EQL}}_{\sigma} : \text{Lan}^{\mathcal{EQL}}(\Sigma_1) \to \text{Lan}^{\mathcal{EQL}}(\Sigma_2)$ is defined as σ for the symbols coming from Σ, and as the identity for the *equations* from Π_{ss}, for $s \in S$.

The model functor $\text{Mod}^{\mathcal{EQL}}$ for every signature Σ in ALGSIG returns the class of all Σ-algebras $\text{ALG}(\Sigma)$. For every algebraic signature morphism σ, the function $\text{Mod}^{\mathcal{EQL}}_{\sigma}$ is the usual algebraic σ-reduct operation.

Let $\Sigma = \langle S, \Omega \rangle$ be an algebraic signature. For every Σ-model (algebra) M, we shall now define the metastructure $Int_\Sigma(M)$. Let A_M be a $\mathsf{Syn}^{\mathcal{EQL}}(\Sigma)$-algebra such that:

- for every sort $s \in S$, let $|A_M|_{T(s)} \mathrel{\hat{=}} |M|_s$,
- $|A_M|_{\mathbb{B}} \mathrel{\hat{=}} \{\mathtt{tt}, \mathtt{ff}\}$,
- for every $\omega : s_1 \ldots s_n \to s_0$, $\omega_{A_M} \mathrel{\hat{=}} \omega_M$,
- for every s in S and a_1, a_2 in $|A_M|_{T(s)}$, $\equiv_{A_M} (a_1, a_2)$ equals $\mathtt{tt}$ if and only if a_1 and a_2 are identical.

For every $s \in S$, as the *set of values* $|V^{Int_\Sigma(M)}|_{T(s)}$ for the sort $T(s)$, let us take the whole[1] set $|A_M|_{T(s)}$, and as the set of *designated elements* $D^{Int_\Sigma(M)}$—the singleton set $\{\mathtt{tt}\}$.

For every algebraic signature morphism $\sigma : \Sigma_1 \to \Sigma_2$ and every algebra $A \in \mathsf{Mod}^{\mathcal{EQL}}_{\Sigma_2}$, let int^σ_A be the identity morphism.

Example 2 (Many-sorted first-order logic). The category of signatures for $\mathcal{FOL}$ is the category of *relational signatures* RELSIG, whose objects are triples $\langle S, \Omega, \Pi \rangle$, where S is the set of *sort names*, Ω is an $S^* \times S$-indexed set of *operation symbols* and Π is an S^*-indexed set of *relational symbols*. Morphisms in RELSIG are defined in the usual way. The metalanguage functor $\mathsf{Lan}^{\mathcal{FOL}} : \mathrm{RELSIG} \to \mathrm{MSIG}$ is defined as follows:

- $\mathsf{Lan}^{\mathcal{FOL}}(\langle S, \Omega, \Pi \rangle) \mathrel{\hat{=}} \langle S, \Omega, \Pi, S, C, \{\forall\} \rangle$, where $C_1 \mathrel{\hat{=}} \{\neg\}$, $C_2 \mathrel{\hat{=}} \{\wedge\}$, and the sets: C_0 and C_n for $n \geq 3$ are empty,
- for every relational signature morphism $\sigma : \Sigma_1 \to \Sigma_2$, the morphism $\mathsf{Lan}^{\mathcal{FOL}}_\sigma : \mathsf{Lan}^{\mathcal{FOL}}(\Sigma_1) \to \mathsf{Lan}^{\mathcal{FOL}}(\Sigma_2)$ is given by σ, for symbols coming from Σ, and is defined as the identity for *connectives* and the *quantifier*, i.e. for the symbols: $\neg$, $\wedge$ and $\forall$.

The functor $\mathsf{Mod}^{\mathcal{FOL}} : \mathrm{RELSIG}^{op} \to \mathrm{CLASS}$, for every relational signature Σ, returns the class of all *relational Σ-structures* (defined in a standard way). For every $\sigma : \Sigma_1 \to \Sigma_2$ in RELSIG, the function $\mathsf{Mod}^{\mathcal{FOL}}_\sigma$ is the corresponding σ-reduct operation.

Let us now define the interpretation function for $\Sigma = \langle S, \Omega, \Pi \rangle$. Let M be an arbitrary relational Σ-structure. We have to define the metastructure $Int_\Sigma(M)$. Let A_M be a $\mathsf{Syn}^{\mathcal{FOL}}(\Sigma)$-algebra, such that:

- the *carrier* $|A_M|$ and the interpretation of operation symbols from Ω are defined as in the case of $\mathcal{EQL}$,
- for every relational symbol $\pi : s_1 \ldots s_n$ and a_i in $|A_M|_{T(s_i)}$, $i = 1 \ldots n$ the value of $\pi_{A_M}(a_1, \ldots, a_n)$ equals $\mathtt{tt}$ if and only if the tuple $\langle a_1, \ldots, a_n \rangle$ belongs to π_M,
- $\neg_{A_M}$ and $\wedge_{A_M}$ are the usual negation and conjunction.

---

[1] In general, the set of values could be a proper subset of the carrier, e.g. as in the case of *partial* first-order logic (cf. [10]).

The *set of values* and the *set of designated elements* are defined in the same way as for $\mathcal{EQL}$. The generalized operation $\forall_{Int_{\Sigma}(M)}$, interpreting the $\forall$ quantifier is defined as follows:

$$\forall_{Int_{\Sigma}(M)}(B) \,\hat{=}\, \begin{cases} \mathtt{ff} & \text{if } \mathtt{ff} \in B \\ \mathtt{tt} & \text{otherwise.} \end{cases}$$

where $B \subseteq |A_M|_{\mathbb{B}}$ is an arbitrary subset.

For every signature morphism $\sigma : \Sigma_1 \to \Sigma_2$ in RELSIG, and every model $M_2 \in \mathsf{Mod}_{\Sigma_2}^{\mathcal{FOL}}$, the morphism $int_{M_2}^{\sigma}$ is the identity.

2.3   Towards Syntax – Metaformulae

As we have mentioned, we want to treat metasignatures as *grammars* for the languages of terms and formulae. This idea is made precise through the construction of a *metaformula functor*

$$\mathsf{MFrm} : \mathrm{MSIG} \to \mathrm{sDGM}(\mathrm{SET})$$

where sDGM(SET) is the category of *small diagrams* in SET[2]. Below, we shall briefly sketch this construction, since (a slight modification of) it will also be used in the next section for defining *schematic metaformulae*.

Let $\mathcal{V}$ be an arbitrary (denumerable) vocabulary of *variable symbols* with a fixed choice function $choice : (\mathcal{P}(\mathcal{V})\backslash\emptyset) \to \mathcal{V}$, i.e. a function such that $choice(V) \in V$ for all nonempty $V \subseteq \mathcal{V}$.

For every metasignature $L = \langle\, S, \Omega, \Pi, V, C, Q \,\rangle$ let MCTXT$_L$ be the full subcategory of the category of substitutions $\mathcal{T}_{\mathsf{Trm}(L)}$, whose objects are $T(S)$-sorted sets X of elements of $\mathcal{V}$ such that $X_{T(s)} = \emptyset$ for $s \notin V$. This category will be called the category of L-*metacontexts*.

For any L-metacontext X, using (almost) the usual first-order syntax approach, we define the set of metaformulae $\mathsf{MFrm}_L(X)$. The construction preceeds by induction over the set of all L-metacontexts. Let $\{\, \mathsf{MFrm}_L(X) \mid X \in |\mathrm{MCTXT}_L| \,\}$ be the smallest family of sets satisfying the following conditions:

$$|T_{\mathsf{Atm}(L)}(X)|_{\mathbb{B}} \subseteq \mathsf{MFrm}_L(X)$$

$$\frac{c \in C_n \qquad \varphi_1 \ldots \varphi_n \in \mathsf{MFrm}_L(X)}{c(\varphi_1, \ldots, \varphi_n) \in \mathsf{MFrm}_L(X)}$$

$$\frac{\varphi \in \mathsf{MFrm}_L(X \cup \{v\}_s) \qquad v = fresh(X) \qquad q \in Q \qquad s \in V}{q\,v^s.\varphi \in \mathsf{MFrm}_L(X)}$$

where $fresh(X) \approx choice(\mathcal{V}\setminus|X|)$. The first condition expresses the fact that every *atomic metaformula* is a metaformula. The remaining two requirements say

---

[2] The objects in sDGM(SET) are pairs $\langle\, \mathsf{A}, \mathsf{F} \,\rangle$, where A is a *small* category and $\mathsf{F} : \mathsf{A} \to \mathrm{SET}$ is a functor. A morphism from $\langle\, \mathsf{A}, \mathsf{F} \,\rangle$ to $\langle\, \mathsf{B}, \mathsf{G} \,\rangle$ is a pair $\langle\, \mathsf{H}, \alpha \,\rangle$, such that $\mathsf{H} : \mathsf{A} \to \mathsf{B}$ is a functor and $\alpha : \mathsf{F} \Rightarrow \mathsf{H}; \mathsf{G}$ is a natural transformation. Composition of morphisms $\langle\, \mathsf{H}, \alpha \,\rangle : \langle\, \mathsf{A}_1, \mathsf{F}_1 \,\rangle \to \langle\, \mathsf{A}_2, \mathsf{F}_2 \,\rangle$ and $\langle\, \mathsf{H}', \beta \,\rangle : \langle\, \mathsf{A}_2, \mathsf{F}_2 \,\rangle \to \langle\, \mathsf{A}_3, \mathsf{F}_3 \,\rangle$ is defined as $\langle\, \mathsf{H}; \mathsf{H}',\ \alpha\,; (\mathsf{H} * \beta) \,\rangle$.

that the set $\mathsf{MFrm}_L(X)$ is closed under applications of *connectives* and *quanti-fiers*.

The requirement that $v = \mathit{fresh}(X)$ guarantees that the resulting metafor-mulea are *normalized* wrt. a suitably defined notion of syntactic substitution (in the sense of [13, 9]). Hence, we can avoid the complexity of working with equiva-lence classes (wrt. α-conversion) of metaformulae. The syntax translation along metasignature morphisms is defined in a standard way. It is not difficult to show that the construction indeed defines a functor from MSIG to sDGM(SET).

By the *base* of an arbitrary functor $\mathsf{F} : \mathsf{A} \rightarrow$ sDGM(SET) we shall mean a functor $\mathsf{base}(\mathsf{F}) : \mathsf{A} \rightarrow$ sCAT, such that $\mathsf{base}(\mathsf{F})(A) \cong \mathsf{C}_A$ and $\mathsf{base}(a : A_1 \rightarrow A_2) \cong \mathsf{C}_a$, where $\mathsf{F}(A) = \mathsf{C}_A : \mathsf{C}_A \rightarrow$ SET and $\mathsf{F}(a) = \langle \mathsf{C}_a : \mathsf{C}_{A_1} \rightarrow \mathsf{C}_{A_2}, \alpha \rangle$. For example, the base for the metaformula fuctor MFrm is the *metacontext functor* MCtxt : MSIG $\rightarrow$ sCAT, which to every metasignature L assigns the category of L-metacontexts MCTXT$_L$ and for every metasignature morphism gives the appropriate metacontext translation.

2.4 From Presentation to Context Institution

In [9, 10] universal constructions in the category PRES of presentations have been proposed as a framework for systematic construction of logical systems. Among the objects of PRES the "really interesting ones" are those for which we can directly construct a corresponding *context institution*.

Definition 7. *A context institution consists of:*

- *a category* SIG *of signatures,*
- *a formula functor* Frm : SIG $\rightarrow$ sDGM(SET)
- *a model functor* Mod : SIG^{op} $\rightarrow$ CLASS,
- *a valuation functor* Val : ELTS(Mod) $\rightarrow$ sDGM(SET),

such that $\mathsf{base}(\mathsf{Val}) ; (\_)^{op} = \mathsf{base}(\pi ; \mathsf{Frm})$, *where* ELTS(Mod) *denotes the cate-gory of elements[3] for* Mod *and* $\pi :$ ELTS(Mod) $\rightarrow$ SIG *is the obvious projection functor, plus for every: signature* Σ, *model* M *in* Mod$_\Sigma$, *and context* $\Gamma \in |\mathsf{Ctxt}_\Sigma|$ *(where* Ctxt $\cong \mathsf{base}(\mathsf{Frm})$*),*

- *a binary* satisfaction relation:

$$M[\_] \models_{\Sigma, \Gamma} \_ \ \subseteq \ \mathsf{Val}_{\Sigma, M}(\Gamma) \times \mathsf{Frm}_\Sigma(\Gamma)$$

such that suitably defined Satisfaction *and* Substitution *conditions hold (see [10] for details).*

Due to the lack of space we shall not give any examples of context institutions here. The interested Reader may consult [10, 9] for more information, motivation

---

[3] The objects in ELTS(Mod) are pairs $\langle \Sigma, M \rangle$ s.t. $\Sigma \in |\mathsf{SIG}|$ and $M \in |\mathsf{Mod}_\Sigma|$. A morphism from $\langle \Sigma_1, M_1 \rangle$ to $\langle \Sigma_2, M_2 \rangle$ is a signature mprphism $\sigma : \Sigma_1 \rightarrow \Sigma_2$ s.t. $\mathsf{Mod}_\sigma(M_2) = M_1$ (for more on categories of elements see [1]).

as well as the definition of *context institution morphism*. In what follows the category of context institutions will be denoted by CONINS.

The condition which enables us to construct a context institution $\mathcal{I}(\mathcal{P})$ out of a context presentation $\mathcal{P}$ is called *logicality*. In what follows $\mathcal{I}(\mathcal{P})$ will be called the context institution *generated by* $\mathcal{P}$.

Definition 8. *A presentation* $\mathcal{P}$: SIG $\rightarrow$ INTSTR *is called* logical *iff:*

– *for every signature Σ in* SIG, *and every model M in* Mod$_\Sigma$, *all the generalized operations in the metastructure* $Int_\Sigma(M)$ *are total,*
– *for every signature morphism $\sigma : \Sigma \rightarrow \Sigma'$ and every model M' in* Mod$_{\Sigma'}$ *the morphism* $int^\sigma_{M'} : Int_\Sigma(\mathrm{Mod}_\sigma(M')) \rightarrow \mathrm{MSTR}_{\mathrm{Lan}_\sigma}(Int_{\Sigma'}(M'))$:
  - reflects *the designated elements,*
  - is surjective on *the set of assignable values.*

The notion of logicality extends to presentation morphisms yielding a category of *logical presentations* denoted by LOGPRES.

For an arbitrary logical presentation $\mathcal{P}$ the context institution $\mathcal{I}(\mathcal{P})$ *generated by* $\mathcal{P}$, has the same category of signatures as $\mathcal{P}$. Also the model functor for $\mathcal{I}(\mathcal{P})$ is the model functor for $\mathcal{P}$ (see Sect. 2.2). The *formula functor* is defined as the composition Lan$^\mathcal{P}$; MFrm. Valuations are (suitably indexed) functions between contexts and *carriers* of the metastructures interpreting models (given by *Int*). For every signature Σ, every Σ-model M and every Σ-context X the satisfaction relation is given by:

$$M[v] \models^{\mathcal{I}(\mathcal{P})}_{\Sigma,X} \phi \quad \text{iff} \quad [\![\phi]\!]_v \in D_{Int_\Sigma(M)}.$$

where $[\![\_]\!]_v$ denotes a *semantic interpretation function* for formulae.

In other words, a formula ϕ is *satisfied* by a valuation v of the context X in the model M, if and only if, the value of its semantic interpretation corresponding to v belongs to the set of *designated elements* of the metastructure $Int_\Sigma(M)$ (see [10, 9] for details). The construction of $\mathcal{I}(\mathcal{P})$ extends to a functor $\mathcal{I}(\_)$: LOGPRES $\rightarrow$ CONINS (cf. [9], Theorem 23 and Proposition 24).

In an obvious way both presentations described above—$\mathcal{EQL}$ and $\mathcal{FOL}$—are logical. Many more examples of logical presentations can be found in [9, 10].

3   Schematic Metaformulae

To proceed towards our goal of introducing *inference rules* for context presentations we need to define expressions which, when appropriately "instantiated", would denote formulae of the object logic. We shall call such expressions *schematic metaformulae*. They constitute a functor

$$\mathrm{SMFrm} : \mathrm{MSIG} \rightarrow \mathrm{sDGM}(\mathrm{SET}) \tag{1}$$

which we define below. In what follows we shall frequently use the term "s-metaformula" instead of "schematic metaformula".

In addition to the "ordinary" variables (elements of $\mathcal{V}$) s-metaformulae can contain variables denoting formulae of the logic in question. Those *formula variables* have to constitute a part of the extended notion of *context* for s-metaformulae. Defining *schematic metacontexts* we have to be careful to avoid "incompatibilities" between different occurrences of the same formula variable in a given s-metaformula. To clarify it let us consider the expression

$$qx^\tau.\phi \;\Rightarrow\; qx^\sigma.\phi \tag{2}$$

containing two occurences of a formula variable ϕ.

Assume that we want the whole expression to be interpreted in the set of metaformulae over a given metacontext X. Then, because of the way the family of sets $\mathsf{MFrm}_L(X)$ has been defined, the first occurrence of ϕ should be evaluated within the set of metaformulae over $X \cup \{x\}_\tau$. At the same time, for the same reason the second occurrence has to be evaluated in the set of metaformulae over $X \cup \{x\}_\sigma$[4]. In general these two sets would be different. Since in a given instance of the quasi s-metaformula (2) we want both occurences of ϕ to denote the same metaformula we shall put the information about the *binding context* of a formula variable into the schematic metacontext.

Let $\mathcal{FV}$ be the dictionary (set) of *formula variables*. We shall assume that $\mathcal{V}$ and $\mathcal{FV}$ are disjoint. For any metasignature $L = \langle\, S, \Omega, \Pi, V, C, Q \,\rangle$ by SMCTXT_L we shall denote a category whose object are triples $\langle\, \bar{s}, X, \Phi \,\rangle$ such that $\bar{s} \in V^*$, X is an L-metacontext and Φ is an V^*-sorted set of elements from $\mathcal{FV}$. We shall also say, that a formula variable $\phi \in \Phi_{s_1 \dots s_n}$ has a *binding context* $s_1 \dots s_n$. A morphism in SMCTXT_L from $\langle\, \bar{s}, X, \Phi \,\rangle$ to $\langle\, \bar{s}, Y, \Psi \,\rangle$, where $\Phi \subseteq \Psi$ is an arbitrary metacontext morphism $t : X \to Y$.

In what follows, for any metacontext X and any $s \in V$ by $s(X)$ we shall denote the metacontext $X \cup \{fresh(X)\}_s$. This notation extends to any sequence $\bar{s} = s_1 \dots s_n$ by putting $\bar{s}(X) = s_1(\dots s_n(X)\dots)$.

Let $\{\, \mathsf{SMF}_L\langle\, \bar{s}, X, \Phi \,\rangle \mid \langle\, \bar{s}, X, \Phi \,\rangle \in |\mathrm{SMCTXT}_L| \,\}$ denotes the least family of sets such that $\mathsf{SMF}_L\langle\, \bar{s}, X, \Phi \,\rangle \subseteq \mathsf{SMF}_L\langle\, \bar{s}, X, \Psi \,\rangle$ for $\Phi \subseteq \Psi$ and satisfying the following additional closure conditions:

$$\frac{\bar{s} \in V^* \qquad \varphi \in |T_{\mathsf{Atm}(L)}(\bar{s}(X))|_\mathbb{B}}{\varphi \in \mathsf{SMF}_L\langle\, \bar{s}, X, \emptyset \,\rangle} \qquad\qquad \frac{\phi \in \mathcal{FV} \qquad \bar{s} \in V^*}{\phi \in \mathsf{SMF}_L\langle\, \bar{s}, X, \{\phi\}_{\bar{s}} \,\rangle}$$

$$\frac{c \in C_n \qquad \varphi_1 \dots \varphi_n \in \mathsf{SMF}_L\langle\, \bar{s}, X, \Phi \,\rangle}{c(\varphi_1, \dots, \varphi_n) \in \mathsf{SMF}_L\langle\, \bar{s}, X, \Phi \,\rangle}$$

$$\frac{\varphi \in \mathsf{SMF}_L\langle\, s{::}\bar{s}, X, \Phi \,\rangle \qquad v = fresh(\bar{s}(X)) \qquad q \in Q \qquad s \in V}{q\,v^s.\varphi \in \mathsf{SMF}_L\langle\, \bar{s}, X, \Phi \,\rangle}$$

$$\frac{\varphi \in \mathsf{SMF}_L\langle\, \bar{s}, X, \Phi \,\rangle \qquad s \in V}{[\varphi]_s \in \mathsf{SMF}_L\langle\, s{::}\bar{s}, X, \Phi \,\rangle}$$

---

[4] More formally, instead of the variable x we should have used $fresh(X)$ (with appropriate sort "decoration").

The last clause represents an "inclusion" $[\_]_s$ between the sets $\mathsf{SMF}_L\langle\,\bar{s}, X, \varPhi\,\rangle$ and $\mathsf{SMF}_L\langle\,s{::}\bar{s}, X, \varPhi\,\rangle^5$.

As in the case of metaformulae, because of the "freshness" requirement for the variable v above, each of the sets $\mathsf{SMF}_L\langle\,\bar{s}, X, \varPhi\,\rangle$ contains only expressions which are *normalized* wrt. syntactic substitution of terms for variables, i.e. expressions invariant under the identity substitution.

For any metasignature L the *schematic metaformula functor* SMFrm_L to every schematic metacontext $\langle\,\bar{s}, X, \varPhi\,\rangle$ assigns the set $\mathsf{SMF}_L\langle\,\bar{s}, X, \varPhi\,\rangle$. For any s-metacontext morphism $t : \langle\,\bar{s}, X, \varPhi\,\rangle \to \langle\,\bar{s}, Y, \varPsi\,\rangle$, its image under SMFrm_L is a suitably defined substitution operation. Due to the lack of space we shall not give its almost obvious definition here. For a metasignature morphism $\ell : L_1 \to L_2$ its image under SMFrm is a family of functions translating for every s-metacontext $\langle\,\bar{s}, X, \varPhi\,\rangle$ s-metaformulae from $\mathsf{SMF}_{L_1}\langle\,\bar{s}, X, \varPhi\,\rangle$ to the set $\mathsf{SMF}_{L_2}\langle\,\ell^*(\bar{s}), \mathsf{MCtxt}_\ell(X), \mathsf{SSet}_{\ell^*}(\varPhi)\,\rangle$ [6].

Note, that there is a "natural inclusion" $\alpha : \mathsf{MFrm} \Rightarrow \mathsf{SMFrm}$. For every metasignature L the functor $\alpha_L^{ctx} : \mathrm{MCTXT}_L \to \mathrm{SMCTXT}_L$ sends each metacontext X to the s-metacontext $\langle\,\epsilon, X, \emptyset\,\rangle$ and every morphism $f : X \to Y$ to itself. For any metacontext X the function $\alpha_L^{frm}(X) : \mathsf{MFrm}_L(X) \to \mathsf{SMFrm}_L\langle\,\epsilon, X, \emptyset\,\rangle$ is the identity.

4   Inference Rules and Extended Presentations

Let L be an arbitrary metasignature. Taking schematic metaformulae as the basis, we shall now define the notion of an L-*inference-rule*.

Definition 9. *An L-inference-rule is a triple $r = \langle\,\langle\,\bar{s}, X, \varPhi\,\rangle, prem(r), conc(r)\,\rangle$ such that:*

- $\langle\,\bar{s}, X, \varPhi\,\rangle$ *is a schematic metacontext,*
- $prem(r) \in \mathcal{P}(\mathsf{SMFrm}_L\langle\,\bar{s}, X, \varPhi\,\rangle)$,
- $conc(r) \in \mathsf{SMFrm}_L\langle\,\bar{s}, X, \varPhi\,\rangle$

Using translations of s-metacontexts and s-metaformulae provided by the functor SMFrm we can easily see that the inference rules actually define a functor

$$\mathsf{Rules} : \mathrm{MSIG} \to \mathrm{SET}$$

which for every metasignature L assigns the set of all L-rules.

Definition 10. *An extended interpretation structure is a pair $\langle\,IS, R\,\rangle$ consisting of:*

- *an interpretation structure $IS = \langle\,L, \mathcal{M}, Int\,\rangle$,*
- *a set of L-rules $R \subseteq \mathsf{Rules}(L)$.*

---

[5] Technically speaking $[\_]_s$ is an obvious extension (by taking *formula variables* into account) of the operation of performing an "inslusion substitution" $\iota : X \to s(X)$. In particular, $[\_]_s$ "re-normalizes" its argument wrt. $s(X)$.

[6] By ℓ^* we denote the obvious extension of the *sort-component* of ℓ to sequences. SSet is the functor "creating" categories of *sorted sets* (cf. [10]).

A morphism from $\langle IS_1, R_1 \rangle$ *to* $\langle IS_2, R_2 \rangle$ *is an arbitrary morphims* $\langle \ell, m, int \rangle$: $IS_1 \to IS_2$ *in* INTSTR *such that* $\mathsf{Rules}_\ell(R_1) \subseteq R_2$.

The category of *extended interpretation structures* and their morphisms will be denoted by EINTSTR.

Proposition 1. *The category* EINTSTR *is cocomplete.*

Proof. The category INTSTR is cocomplete ([9], Theorem 13). Similarily cocomplete is the category of metasignatures ([9], Proposition 18). Using these two facts it is easy to verify that the colimits in EINTSTR are computed via colimits in INTSTR and a suitable category of "signed" sets of rules.

Definition 11. *An* extended context presentation *is an arbitrary functor into the category of extended interpretation structures:*

$$\mathcal{P} \colon \mathrm{SIG}^{\mathcal{P}} \to \text{EINTSTR}.$$

Defining the morphisms similarly as in the case of (ordinary) presentations we obtain a category of *extended presentations* EPRES.

Corollary 1. *The category* EPRES *of extended presentations is complete.*

Proof. Completeness is a consequence of the fact that EPRES is a category of functors "into" a cocomplete category.

Example 3 (Many-sorted equational logic cntd.). Let us take the presentation $\mathcal{EQL}$ for many-sorted equational logic, defined in Example 1. We shall turn it into an extended presentation $\mathcal{EQL}$: ALGSIG $\to$ EINTSTR by augmenting for every algebraic signature $\Sigma = \langle S, \Omega \rangle$ the original interpretation structure $\mathcal{EQL}(\Sigma)$ by the set of rules R_Σ, consisting of the *reflexivity, symmetry, transitivity* and *congruence* rules. For example for all $s \in S$ the transitivity rule looks as follows:

$$\frac{x^s \equiv y^s \qquad y^s \equiv z^s}{x^s \equiv z^s} \quad \langle \epsilon, \{x, y, z\}_s, \emptyset \rangle$$

Please note, that there is no *substitution rule* among the rules in R_Σ. It may seem strange at first, but as we shall learn in the next section, there is a good reason for this "omission".

Example 4 (Many-sorted first-order logic cntd.). To obtain an extended presentation for the first-order logic (without equality) it is enough to augment each interpretation structure for the presentation $\mathcal{FOL}$ by inference rules corresponding to the axiom schemata of the classical propositional calculus, the rule of *modus ponens* and the following rules:

$$\frac{}{[\forall x^s . \phi]_s \Rightarrow \phi} \langle s, \emptyset, \{\phi\}_s \rangle \qquad \frac{[\phi]_s \Rightarrow \psi}{[\phi]_s \Rightarrow [\forall x^s . \psi]_s} \langle s, \emptyset, \{\{\phi\}_\epsilon, \{\psi\}_s\} \rangle$$

for all $s \in S$.

4.1   From Extended Presentations to Entailment Systems

The notion of *logicality* introduced for ordinary presentations can without any change be used for the extended case—we do not require anything extra about the inference rules of *extended logical presentations*. We shall denote the resulting category by eLogPres.

Let $\mathcal{P}$: Sig $\rightarrow$ eIntStr be a logical extended presentation. In this section we shall define the *entailment system* generated by the rules of $\mathcal{P}$ and discuss its semantic interpretation within the context institution $\mathcal{I}(\mathcal{P})$. Let us start with the notion of a *valuation* of a *schematic metacontext*.

Definition 12. *A* valuation *of a schematic metacontext* $\langle \bar{s}, X, \Phi \rangle$ *in an L-metacontext Y is an arbitrary pair* $\langle t, \rho \rangle$ *such that* $t : X \rightarrow Y$ *is a metacontext morphism and* ρ *is an V^*-indexed function from Φ to* $\langle \mathsf{MFrm}_L(\bar{s}(Y)) \mid \bar{s} \in V^* \rangle$.

Every $\langle t, \rho \rangle$ defines an *instantiation function* $\{ \_ \}_{\langle t, \rho \rangle}$: $\mathsf{SMFrm}_L \langle \bar{s}, X, \Phi \rangle \rightarrow \mathsf{MFrm}_L(\bar{s}(Y))$. Due to the lack of space we are not able to present it here[7]. Using the above concept we are ready to define the notion of an *instantiation* for inference rules. Let $r = \langle \langle \bar{s}, X, \Phi \rangle, prem(r), conc(r) \rangle$ be an *L-inference-rule*.

Definition 13. *An* instantiation *of r in an L-metacontext Y via* $\langle t, \rho \rangle$ *is a pair*

$$\langle \{ \{\varphi\}_{\langle t, \rho \rangle} \mid \varphi \in prem(r) \}, \{conc(r)\}_{\langle t, \rho \rangle} \rangle.$$

In such a case we shall call the metaformula $\{conc(r)\}_{\langle t, \rho \rangle}$ *an* immediate consequence *of* $\{ \{\varphi\}_{\langle t, \rho \rangle} \mid \varphi \in prem(r) \}$ *via r in Y.*

For any (extended) presentation $\mathcal{P}$ we can define a *sentence functor* $\mathsf{Sen}^{\mathcal{P}}$: Sig $\rightarrow$ Set such that

$$\mathsf{Sen}^{\mathcal{P}}(\Sigma) \cong \bigcup_{X \in |\mathsf{Ctxt}^{\mathcal{P}}_{\Sigma}|} \{X\} \times \mathsf{Frm}^{\mathcal{P}}_{\Sigma}(X)$$

for any signature Σ (with an obvious action on morphisms). In plain words, Σ-*sentences* are just "Σ-formulae with context".

The inference rules of $\mathcal{P}$ define an *entailment system*, i.e. a family $Ent(\mathcal{P}) \cong \{ \vdash^{\mathcal{P}}_{\Sigma} \mid \Sigma \in |\mathsf{Sig}| \}$ such that

- each $\vdash^{\mathcal{P}}_{\Sigma} \subseteq \mathcal{P}(\mathsf{Sen}^{\mathcal{P}}_{\Sigma}) \times \mathsf{Sen}^{\mathcal{P}}_{\Sigma}$ and
- whenever $\Gamma \vdash^{\mathcal{P}}_{\Sigma_1} \phi$ and $\sigma : \Sigma_1 \rightarrow \Sigma_2$ then $\mathsf{Sen}^{\mathcal{P}}_{\sigma}(\Gamma) \vdash^{\mathcal{P}}_{\Sigma_2} \mathsf{Sen}^{\mathcal{P}}_{\sigma}(\phi)$.

The entailment system $Ent(\mathcal{P})$ is given as follows: $\Gamma \vdash^{\mathcal{P}}_{\Sigma} \phi$ iff there exists a finite sequence $\phi_1, \ldots, \phi_n$ of Σ-sentences such that

---

[7] The definition is quite obvious, with the only "delicate" point being the normalization requirement for the result.

- $\phi_n = \phi$
- for all $i \leq n$ either $\phi_i \in \Gamma$ or ϕ_i is an immediate consequence of some of the sentences from $\{\phi_1, \ldots, \phi_{i-1}\}$ via one of the rules from R_Σ[8] or ϕ_i is a *substitution instance* of one of the sentences from $\{\phi_1, \ldots, \phi_{i-1}\}$ [9].

It should be clear now why we have "omitted" the *substitution rule* from the Example 3. It is because the substitution rule is a built-in "metarule", i.e. it is a part of our notion of *consequence* in the entailment system for $\mathcal{P}$.

For any signature Σ in SIG let us introduce the following *validity* satisfaction relation between Σ-models of $\mathcal{P}$ and sentences from $\mathsf{Sen}^{\mathcal{P}}$:

$$M \Vdash_\Sigma \langle X, \phi \rangle \quad \text{iff} \quad \text{for all } v \in \mathsf{Val}_{\Sigma, M}^{\mathcal{I}(\mathcal{P})}(X) \quad M[v] \models_{\Sigma, X}^{\mathcal{I}(\mathcal{P})} \phi$$

Using the *semantic consequence* corresponding to $\Vdash_\Sigma$ we can define the notion of *soundness* for rules. Let r be a Σ-rule in $\mathcal{P}$, i.e. $r \in R_\Sigma$ where $\mathcal{P}(\Sigma) = \langle \mathsf{Lan}_\Sigma^{\mathcal{P}}, \mathsf{Mod}_\Sigma^{\mathcal{P}}, \mathit{Int}_\Sigma, R_\Sigma \rangle$.

Definition 14. *The rule r is sound iff for every Σ-context X and every $\Xi \subseteq \mathsf{Frm}_\Sigma^{\mathcal{P}}(X)$, $\varphi \in \mathsf{Frm}_\Sigma^{\mathcal{P}}(X)$, if φ is an immediate consequence of Ξ via r in X, then $\{\langle X, \xi \rangle \mid \xi \in \Xi\} \Vdash_\Sigma \langle X, \varphi \rangle$.*

Using the validity satisfaction we can also easily define the *semantic soundness* and *semantic completeness* of the entailment system for $\mathcal{P}$ given above.

Definition 15. *The entailment system $\mathit{Ent}(\mathcal{P})$ is semantically sound if from $\Gamma \vdash_\Sigma^{\mathcal{P}} \phi$ it follows that $\Gamma \Vdash_\Sigma \phi$ for every signature Σ, $\Gamma \subseteq \mathsf{Sen}_\Sigma^{\mathcal{P}}$, and $\phi \in \mathsf{Sen}_\Sigma^{\mathcal{P}}$. It is semantically complete if the implication in the other direction holds.*

The following proposition verifies that our notion of entailment for $\mathcal{P}$ is intuitively "correct".

Proposition 2. *The entailment system $\mathit{Ent}(\mathcal{P})$ is semantically sound, if and only if, all its rules are sound.*

4.2 Putting Extended Presentations Together

The category of *extended logical presentations* ELOGPRES is not complete (for the same reasons as LOGPRES is not complete [10]). Therefore if we want to combine extended logical presentations, we will have to use a broader category EPRES for this purpose[10].

---

[8] Strictly speaking, the metaformula being the second component of ϕ has to be an immediate consequence of some of the metaformulae corresponding to the sentences $\{\phi_1, \ldots, \phi_{i-1}\}$. This requires of course these premises and ϕ to have a common metacontext.

[9] It means that there is an index $k < i$ and a context morphism $t : X \to Y$ in $\mathsf{Ctxt}_\Sigma^{\mathcal{P}}$, where $\mathsf{Ctxt}^{\mathcal{P}} = \mathsf{base}(\mathsf{Frm}^{\mathcal{P}})$, such that $\mathsf{Frm}_\Sigma^{\mathcal{P}}(t)(\varphi_k) = \varphi$ and $\phi_k = \langle X, \varphi_k \rangle$ and $\phi = \langle Y, \varphi \rangle$.

[10] Limits in EPRES are just limits in PRES extended by the combined inference rules. Therefore all the observations about combining ordinary presentations remain true fro the case of EPRES (cf. [10]).

The interesting thing is, how the inference rules behave in the combination process. Let us take the following example of a pullback square in EPRES:

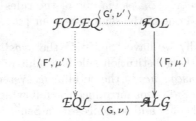

As it turns out, the presentation $\mathcal{FOLEQ}$ is logical and the resulting entailment system is *semantically sound*. However, no matter how complete both $Ent(\mathcal{EQL})$ and $Ent(\mathcal{FOL})$ are, their combination is *not complete*. For example let

$$\phi = \langle X,\, t_1 \equiv t_2 \Rightarrow (P(t_1) \Rightarrow P(t_2)) \rangle$$

where P is a unary predicate symbol and t_i's are terms. Then $\emptyset \Vdash \phi$, but obviously ϕ cannot be deduced from the empty set in $Ent(\mathcal{FOLEQ})$. The problem is that the nature of the interaction between the equality and other formulae is not described by any of the rules from the component presentations. So we cannot hope for a general "completeness preservation" result here. Fortunately, as one can show, the situation with soundness is better.

Proposition 3. *If the limit in* EPRES *of a diagram* D, *consisting of extended logical presentations only, is logical, then it preserves the soundness of the component inference systems.*

5   Concluding Remarks

We have extended the notion of *context presentation* from [9, 10] by augmenting it by *inference rules*. The resulting framework seems to cover many interesting examples, although—due to the lack of space—only two simple ones were briefly sketched here.

For any *extended presentation* $\mathcal{P}$ we have defined an entailment system $Ent(\mathcal{P})$ generated by the rules of $\mathcal{P}$ and the corresponding notions of *soundness* and *completeness* wrt. the context institution $\mathcal{I}(\mathcal{P})$.

Thanks to the completeness of the category EPRES of *extended presentations* we can use the standard categorical "machinery" as a tool for building logical systems with inference rules in a compositional way. Of course the process is rather far from being "automatic" (except for the simple cases). Assuming that both the components and the result are *logical* we can assure *soundness preservation*

The framework as presented, although quite powerful already, does not allow rules with *schematic substitution* such as the induction principle for example. We believe that adding them is possible and moreover, since the notion of *substitution* is a first-order citizen in context presentations, the extended framework should enjoy analogous meta-properties as the ones discussed above.

A precise relationship between our approach and the "fibred parchment approach" presented in [2] is yet to be established. One difference is of course the treatment of variables and quantifiers, where our approach seems to be more uniform and general (e.g. it admits *open formulae* and *substitutions* as first-class citizens). The two approaches also differ in the assumed structure of the *algebra of truth values*, where we follow the *model-theoretic pachments* approach [8], whereas [2] uses a Tarskian *closure operation*.

Another important difference concerns the problem of *completeness preservation*. As we have seen in Section 4, in our approach completeness might be lost in the process of combination, whereas [2] contains a suitable completeness preservation result. The reason for the seeming discrepancy is the notion of *fullness* adopted in [2], which allows to keep the completeness of the resulting logic. In a sense, it comes for the price of obtaining a "non-standard" result however. For example, even at the semantical level, the combination of (full versions of) propositional modal logic and first-order logic as described in [2] does not coincide with the usual first-order modal logic. The solution we have adopted follows a more "traditional" interpretation, where although the completeness might be lost, the logic obtained as the result of the combination seems to (at least) semantically coincide with the "intended one".

References

1. F. Borceux. *Handbook of Categorical Algebra 1, Basic Category Theory*, volume 50 of *Encyclopedia of Mathematics and its Applications*. Cambridge University Press, 1994.

2. C. Caleiro, P. Gouveia, and J. Ramos. Completeness results for fibered parchments. This volume.

3. C. Caleiro, P. Mateus, J. Ramos, and A. Sernadas. Combining logics: Parchments revisited. In M. Cerioli and G. Reggio, editors, *Recent Trends in Algebraic Development Techniques. 15th International Workshop WADT 2001 Joint with the CoFI WG Meeting. Genova, Italy, April 2001. Selected papers*, volume 2267 of *Lecture Notes in Computer Science*, pages 48–70. Springer-Verlag, 2001.

4. J. A. Goguen and R. M. Burstall. A study in the foundations of programming methodology: Specifications, institutions, charters and parchments. In D. Pitt, S. Abramsky, A. Poigné, and D. Rydeheard, editors, *Proc. Conference on Category Theory and Computer Programming*, volume 240 of *Lecture Notes in Computer Science*, pages 313–333. Springer-Verlag, 1986.

5. J. A. Goguen and R. M. Burstall. Institutions: Abstract model theory for specification and programming. *Journal of the Association for Computing Machinery*, 39:95–146, 1992.

6. T. Mossakowski. Using limits of parchments to systematically construct institutions of partial algebras. In M. Haveraaen and O.-J. Dahl, editors, *Recent Trends in Data Type Specifications. 11th Workshop on Specification of Abstract Data Types joint with the 8th General COMPASS Workshop. Oslo, Norway, September 1995. Selected Papers*, volume 1130 of *Lecture Notes in Computer Science*, pages 379–393. Springer-Verlag, 1996.

7. T. Mossakowski, A. Tarlecki, and W. Pawłowski. Combining and representing logical systems. In E. Moggi, editor, *Category Theory and Computer Science, CTCS'97, Santa Margherita Ligure, Italy, 1997, Proceedings*, volume 1290 of *Lecture Notes in Computer Science*, pages 177–198. Springer-Verlag, 1997.

8. T. Mossakowski, A. Tarlecki, and W. Pawłowski. Combining and representing logical systems using model-theoretic parchments. In F. Parisi Presicce, editor, *Recent Trends in Algebraic Development Techniques. 12th International Workshop WADT'97. Tarquinia, Italy, June 1997. Selected papers*, volume 1376 of *Lecture Notes in Computer Science*, pages 349–364. Springer-Verlag, 1998.

9. W. Pawłowski. *Contextual Logical Systems for the Foundations of Software Specification and Development (in Polish)*. PhD thesis, Institute of Computer Science, Polish Academy of Sciences, Warsaw, 2000.

10. W. Pawłowski. Presentations for abstract context institutions. In M. Cerioli and G. Reggio, editors, *Recent Trends in Algebraic Development Techniques. 15th International Workshop WADT 2001 Joint with the CoFI WG Meeting. Genova, Italy, April 2001. Selected papers*, volume 2267 of *Lecture Notes in Computer Science*, pages 256–279. Springer-Verlag, 2001.

11. A. Sernadas, C. Sernadas, and C. Caleiro. Fibring of logics as a categorial construction. *Journal of Logic and Computation*, 9(2):149–179, 1999.

12. A. Sernadas, C. Sernadas, C. Caleiro, and T. Mossakowski. Categorical fibring of logics with terms and binding operators. In D. Gabbay and M. de Rijke, editors, *Frontiers of Combining Systems 2*, pages 295–316. Research Studies Press, 2000.

13. A. Stoughton. Substitution revisited. *Theoretical Computer Science*, 59:317–325, 1988.

Monad-Independent Dynamic Logic in HasCasl

Lutz Schröder and Till Mossakowski

BISS, Department of Computer Science, University of Bremen

Abstract. Monads have been recognized by Moggi as an elegant device for dealing with stateful computation in functional programming languages. In previous work, we have introduced a Hoare calculus for partial correctness of monadic programs. All this has been done in an entirely monad-independent way. Here, we extend this to a monad-independent dynamic logic (assuming a moderate amount of additional infrastructure for the monad). Dynamic logic is more expressive than the Hoare calculus; in particular, it allows reasoning about termination and total correctness. As the background formalism for these concepts, we use the logic of HasCasl, a higher-order language for functional specification and programming.

Introduction

One of the central concepts of modern functional programming is the encapsulation of side effects via monads following the seminal paper [8]. In particular, state monads are used to emulate an imperative programming style in the functional programming language Haskell [20]. Monads can be used to abstract from a particular notion of computation, since they model a wide range of computational effects: e.g., stateful computations, non-determinism, partiality, exceptions, input, and output can all be viewed as monadic computations, and so can various combinations of these concepts such as non-deterministic stateful computations.

Moggi [8] has suggested a Hoare calculus for a state monad with state interpreted as global store. We have generalized this in [17] to a monad-independent Hoare calculus. However, Hoare logic in general is concerned only with partial correctness and does not allow reasoning about termination or total correctness. The right framework for studying the latter is dynamic logic as introduced in [13]. Here, we examine the infrastructure that is needed in order to develop dynamic logic in a monad-independent way, and show that this does indeed make sense when instantiated to the usual monads mentioned above.

The formalism is embedded into the logic of HasCasl, a higher order language for functional specification and programming based on the first order algebraic specification language Casl [1,3]. This allows expressing programs *and* their expected properties within one and the same language, so that we obtain a unified interpretation of dynamic logic, where formulas consist of programs and logical expressions.

Related work includes evaluation logic, with its two rather different semantics as defined by Pitts [10] and Moggi [9], respectively. The approach of [10] is *local*,

M. Wirsing, D. Pattinson, and R. Hennicker (Eds.): WADT 2002, LNCS 2755, pp. 425–441, 2003.

but depends additionally on a hyperdoctrine over the monad. The semantics of [9] is given entirely in terms of the monad, but has a *global* nature. We use the term 'dynamic logic' to emphasize that our approach is local; however, unlike in [10], we require no structure beyond the monad, thus reconciling the approaches of [10] and [9].

1   HASCASL

The language HASCASL has been introduced in [16] as a higher order extension of CASL, based on the partial λ-calculus. We give a brief summary of the language.

Any HASCASL specification determines a *signature* consisting of classes, types, and operations, and associated *axioms* that the operations are required to satisfy. Basic types are introduced by means of the keyword **type**. Types may be parameterized by type arguments; e.g., we may write

> **var** $a : Type$
> **type** *List a*

and obtain a unary type constructor *List*. There are built-in type constructors (with fixed interpretations) $\_ \times \_$ for product types, $\_ \to ? \_$ and $\_ \to \_$ for partial and total function types, respectively, *Pred* $\_$ for predicate types, and a unit type *Unit*.

Next, an *operator* is a constant of some type, declared by

> **op** $f : t$

where t is a type. Since types may contain type variables, operators can be polymorphic in the style of ML.

From the given operators, we may form higher order terms: a term is either a variable, an application, a tuple, or a λ-abstraction. Such terms may be used in *axioms* formulated, to begin, in what we shall call the *external logic*. This logic offers the usual logical connectives (conjunction, negation etc.) as well as universal and existential quantifiers, where the outermost universal quantifications may also be over type variables, strong and existential equality denoted by $=$ and $\stackrel{e}{=}$, respectively, and definedness assertions *def* α (the latter feature and the distinction between the various equalities are related to partial functions; cf. [1] for a detailed discussion).

The semantics of a HASCASL specification is the class of its (set-theoretic) *intensional Henkin models*: a function type need not contain all set-theoretic functions, and two functions that yield the same value on every input need not be equal; see [16] for a discussion of the rationale behind this. If desired, extensionality of models may be forced by means of an axiom expressible within the language.

A consequence of the intensional semantics is the presence of an intuitionistic *internal logic* that lives within λ-terms. One can specify an *internal equality* (for which the symbol $=$ is built-in syntactical sugar) to be used within λ-terms, which then allows *specifying* the full set of logical operations and quantifiers of intuitionistic logic; this is carried out in detail in [16]. There is built-in syntactical sugar for the internal logic, invoked by means of the keyword **internal** which signifies that formulas in the following block are to be understood as formulas of the internal logic.

By means of the internal logic, one can then specify a class of complete partial orders and fixed point recursion in the style of HOLCF [14]. On top of this, syntactical sugar is provided that allows recursive function definitions in the style used in functional programming, indicated by the keyword **program**.

HasCasl supports *type classes*. These are declared in the form

<class C

and are to be understood as subsets of the syntactical universe of all types. Types as well as type variables can be restricted to belong to an assigned class, e.g. by writing

type $t : C$

In particular, axioms and operators may be polymorphic over classes. Classes may be subclasses of each other, and they may have generic instances. By attaching polymorphic operators and axioms to a class, one achieves a similar effect as with Haskell's type classes.

In a similar vein, one can add *constructor classes* to HasCasl. They can be interpreted as predicates on the syntactical universe of abstracted type expressions (also called *pseudotypes*), such as $\lambda a : Type \bullet a \to? List\ a$. As for type classes, there are constructor subclasses; types, operators, axioms may be polymorphic over constructor classes; and this polymorphism is semantically coded by collections of instances. A typical example of a constructor class is the class of monads (see Figure 1).

In summary, HasCasl is a language that allows both property-oriented specification and functional programming; executable HasCasl specifications may easily be translated into Haskell programs.

2 Monads for Computations

On the basis of the seminal paper [8], monads are being used for encapsulating side effects in modern functional programming languages; in particular, this idea is one of the central concepts of Haskell [5]. Intuitively, a monad associates to each type A a type TA of computations of type A; a function with side effects that takes inputs of type A and returns values of type B is, then, just a function of type $A \to TB$. This approach abstracts away from particular notions of computation such as store, non-determinism, non-termination etc.; a surprisingly large amount of reasoning can in fact be carried out independently of the choice of such a notion.

More formally, a monad on a given category $\mathbf{C}$ can be defined as a *Kleisli triple* $\mathbb{T} = (T, \eta, \_^*)$, where $T : \mathrm{Ob}\,\mathbf{C} \to \mathrm{Ob}\,\mathbf{C}$ is a function, the *unit* η is a family of morphisms $\eta_A : A \to TA$, and $\_^*$ assigns to each morphism $f : A \to TB$ a morphism $f^* : TA \to TB$ such that

$$\eta_A^* = id_{TA}, \quad f^*\eta_A = f, \quad \text{and} \quad g^*f^* = (g^*f)^*.$$

This description is equivalent to the more familiar one via an endofunctor with unit and multiplication [7]. 'Functions with side effects' are then modeled in the *Kleisli category* of $\mathbb{T}$, which has the same objects as $\mathbf{C}$ and $\mathbf{C}$-morphisms $A \to TB$ as morphisms from A to B.

In order to support a language with finitary operations and multi-variable contexts, one needs a further technical requirement: a monad is called *strong* if it is equipped with a natural transformation $t_{A,B} : A \times TB \to T(A \times B)$ called *tensorial strength*, subject to certain coherence conditions (see e.g. [8]); this is equivalent to enrichment of the monad over $\mathbf{C}$ [8].

Example 1 ([8]). Computationally relevant monads on **Set** include

- stateful computations with possible non-termination: $TA = (S \to? (A \times S))$, where S is a fixed set of states and $\_ \to? \_$ denotes the partial function type;
- (finite) non-determinism: $TA = \mathcal{P}_{fin}(A)$, where $\mathcal{P}_{fin}$ is the finite power set;
- exceptions: $TA = A + E$, where E is a fixed set of exceptions;
- interactive input: $TA = \mu\gamma.A + (U \to \gamma)$, where U is a set of input values;
- non-deterministic stateful computations: $TA = (S \to \mathcal{P}_{fin}(A \times S))$;
- continuations: $TA = (A \to R) \to R$, where R is a type of *results*.

Other typical examples of monads are the list monad, where TA is the type of lists over A, and the free Abelian group monad, where TA consists of expressions of the form $\sum_{i=1}^{m} n_i a_i$, with $n_i \in \mathbb{Z}$ and $a_i \in A$ for $i = 1, \dots, m$.

Figure 1 shows a specification of monads in HASCASL. As an example of an instance for this type class, a specification of the state monad is shown in Figure 2. Since the operations of the monad are functions in the model, the monads thus specified are automatically strong, strength being equivalent to enrichment. The notation is (almost) identical to the one used in Haskell, i.e. the unit is denoted by *ret*, and the operator $\_ \gg \_$ denotes, in the above notation, the function $(x, f) \mapsto f^*(x)$. This specification is the basis for a built-in sugaring in the form of a Haskell-style do-notation: for monadic expressions e_1 and e_2,

$$\text{do } x \leftarrow e_1; \ e_2$$

abbreviates $e_1 \gg \lambda x \bullet e_2$. Further details will be discussed below.

A slight complication concerning the axiomatization arises from the fact that partial functions are involved. Note that the second unit law $f^*\eta = f$ has been replaced by two axioms, one stating that the said equation holds on the domain of f, and another one stating that $f^*\eta$, while possibly having a larger domain of definition than f, behaves like f under binding. This ensures that standard monads such as the state monad with its usual definition (under which f^* is always a total function; cf. Figure 2 and the recent discussion on [4]) are actually subsumed, while leaving the essence of the proposed calculus untouched. Moreover, for the sake of simplicity of the further treatment, we have included the mono requirement (stating that *ret* is injective) in the specification.

Reasoning about a category equipped with a strong monad is greatly facilitated by the fact that proofs can be conducted in a *meta-language* introduced in [8], which we here adapt to deal with partial functions. Although we do not a priori work in a category (this would require working out the details of how

```
spec  MONAD = INTERNALLOGIC then
     class  Monad : Type → Type {
     vars   T : Monad;  A, B, C, D : Type
     ops    _ ≫ _ : T A → (A →? T B) →? T B;
            ret : A → T A                        }
     internal {
     forall x, y : A;  y : T A;  f : A →? T B;  g : B →? T C;  a : A →? B
        • def (f x) ⇒ ((ret x) ≫ f) = f x
        • y ≫ (λx : A • ret (a x) ≫ f) = y ≫ (λx : A • f (a x))
        • (y ≫ ret) = y
        • ((y ≫ f) ≫ g) = (y ≫ (λx : A • f x ≫ g))
        • ret x = ret y ⇒ x = y
            }
```

Fig. 1. The constructor class of monads

```
spec  STATE = MONAD then
     type instance ST : Monad
     vars  A, B : Type
     internal {
     types S;
            ST A := S →? (A × S)
     forall x : A;  y : ST;  f : A →? ST B
        • ret x = λs : S • (x, s)
        • (y ≫ f) = λs1 : S • let (s2, z) = y s1 in f z s2      }
```

Fig. 2. Specification of the state monad

the specification of monads induces a monad on the classifying category as constructed in [15]), the meta-language is still applicable here, as its logic can be obtained from the axioms in Figure 1. The crucial features of this language are

- A type operator T; terms of type TA are called (A-valued) *programs* or *computations*;
- an polymorphic operator ret : $A \to TA$ corresponding to the unit;
- a binding construct, which we here denote in Haskell's do style:

$$\text{do } x \leftarrow p;\ q$$

is interpreted by means of the tensorial strength and Kleisli composition [8]; this is equivalent the do-notation introduced above. Intuitively, do $x \leftarrow p;\ q$ computes p and passes the results on to q. Binding satisfies an associative law and three unit laws corresponding to the axioms in Figure 1. We denote nested do expressions like do $x \leftarrow p$; do $y \leftarrow q$; ... by do $x \leftarrow p; y \leftarrow q$; .... Repeated nestings such as do $x_1 \leftarrow p_1, \dots, x_n \leftarrow p_n$; q are denoted in the form do $\bar{x} \leftarrow \bar{p}$; q. Term fragments of the form $\bar{x} \leftarrow \bar{p}$ are called *program sequences*. Variables x_i that do not appear later on may be omitted from the notation.

On top of a monad, one can generically define control structures such as a while loop. Such definitions require general recursion, which is realized in HASCASL by means of fixed point recursion on cpos. Thus, one has to restrict to monads that allow lifting a cpo structure on A to a cpo structure on the type TA of computations in such a way that the monad operations become continuous. This is laid out in detail in [17].

3   The Generic Approach to Side Effects and State

We now discuss monad-independent notions of program properties such as side-effect freeness and determinism, as well as logical aspects of boolean-valued programs with effects. These concepts will be used to obtain a monad-independent notion of state, which forms the foundation for the interpretation of the box and diamond operators of dynamic logic introduced in the next section. The notions of (deterministic) side-effect freeness and validity of formulas with effects have been introduced in [17]; the concepts of termination, state, and state discloser are new. We will phrase all definitions in terms of the meta-language for monads, using the fixed notation $\mathbb{T}$ for the monad, T for the associated type constructor etc. throughout.

Definition 2. A program p is called *side-effect free* if

$$(\text{do } y \leftarrow p;\ \text{ret} *) = \text{ret} * \qquad (\text{shorthand: } sef(p)),$$

where $*$ is the unique element of the unit type.

Side-effect free programs have the expected properties:

Lemma 3. *If p is side-effect free, then*

$$(\text{do } p;\ q) = q$$

for each program q, provided that q is defined. Moreover, if $p, q : T1$, then

$$(\text{do } q;\ p) = q.$$

Example 4. A program p is side-effect free

- in the state monad iff p terminates and does not change the state;
- in the non-determinism monad iff p always has at least one possible outcome;
- in the exception monad iff p terminates normally;
- in the interactive input monad iff p never reads any input;
- in the non-deterministic state monad iff p does not change the state and always has at least one possible outcome (i.e. never gets stuck).

Remark 5. A program p is called *stateless* if it factors through ret, i.e. if it is just a value inserted into the monad. For example, in the state monad, stateless-ness means that the program neither changes nor reads the state (p is stateless iff p *exists* in the sense of [8]). Stateless programs are side-effect free, but not vice versa.

We will want to regard programs that return truth values as formulas with side effects in a modal logic setting. A basic notion we need for such formulas is that of global validity, which we denote explicitly by a 'global box' $\boxdot$:

Definition 6. Given a term ϕ of type $T\Omega$, where Ω denotes the type of internal truth values, $\boxdot\phi$ abbreviates

$$\phi = \text{do } \phi; \text{ ret } \top.$$

If ϕ is side-effect free, then $\boxdot\phi$ simplifies to $\phi = \text{ret } \top$; otherwise, the formula above ensures that the right hand side has the same side-effect as ϕ.

Remark 7. Note that the equality symbol $=$ in the definition of the formula $\boxdot\phi$ above is strong equality. In particular, in the classical case $\boxdot\phi$ is true if ϕ is undefined.

Example 8. In the monads of Example 1, satisfaction of $\boxdot\phi$ amounts to the following:

- in the state monad: successful execution of ϕ from any initial state yields $\top$;
- in the non-determinism monad: ϕ yields at most the value $\top$ (or none at all)
- in the exception monad: ϕ yields $\top$ whenever it terminates normally.
- in the interactive input monad: the value eventually produced by ϕ after some combination of inputs is always $\top$;
- in the non-deterministic state monad: execution of ϕ from any initial state yields at most the value $\top$;

In order to perform proofs about the logic introduced below, we require an auxiliary calculus for judgements of the form $[\bar{x} \leftarrow \bar{p}]_G \phi$, which intuitively state that ϕ holds after $\bar{x} \leftarrow \bar{p}$, where ϕ is an actual formula of the internal logic (i.e. $\phi : \Omega$). The idea is to work with formulas that have all state-dependence shoved to the outside, so that the usual logical rules apply to the remaining part. Formally, $[\bar{x} \leftarrow \bar{p}]_G \phi$ abbreviates (slightly deviating from [17])

$$(\text{do } \bar{x} \leftarrow \bar{p}; \text{ ret}(\bar{x}, \phi)) = \text{do } \bar{x} \leftarrow \bar{p}; \text{ ret}(\bar{x}, \top).$$

The degenerate case $[\,]_G \phi$ is (by the mono requirement) equivalent to ϕ. Note that $[\bar{x} \leftarrow \bar{p}; \bar{y} \leftarrow \bar{q}]_G \phi$ is properly weaker than $[\bar{x} \leftarrow \bar{p}]_G [\bar{y} \leftarrow \bar{q}]_G \phi$.

The intuition behind this definition is the same as for $\boxdot$. Indeed, in many cases, $[\bar{x} \leftarrow \bar{p}]_G \phi$ is semantically equivalent to

$$\boxdot \text{ do } \bar{x} \leftarrow \bar{p}; \text{ ret } \phi;$$

monads for which this equivalence holds will be called *simple*. A monad with rank over **Set**, presented by a signature and equations, is simple if, in each equation, the two sides contain the same variables; this covers e.g. the exception monad, the non-determinism monad, and the list monad. Moreover, the usual state monads are simple, although their known equational presentation [11] does not satisfy the variable requirement. In general, however, $\boxdot$ do $\bar{x} \leftarrow \bar{p}$; ret ϕ

$$(\wedge I) \quad \frac{[\bar{x} \leftarrow \bar{p}]_G \, \phi \quad [\bar{x} \leftarrow \bar{p}]_G \, \psi}{[\bar{x} \leftarrow \bar{p}]_G \, \phi \wedge \psi} \qquad (mp) \quad \frac{\phi \Rightarrow \psi \quad [\bar{x} \leftarrow \bar{p}]_G \, \phi}{[\bar{x} \leftarrow \bar{p}]_G \, \psi} \qquad (eq) \quad \frac{[\bar{x} \leftarrow \bar{p}]_G \, q_1 = q_2 \quad [\bar{x} \leftarrow \bar{p}; y \leftarrow q_1; \bar{z} \leftarrow \bar{r}]_G \, \phi}{[\bar{x} \leftarrow \bar{p}; y \leftarrow q_2; \bar{z} \leftarrow \bar{r}]_G \, \phi}$$

$$(app) \quad \frac{[\bar{x} \leftarrow \bar{p}]_G \, \phi \quad y \notin FV(\phi)}{[\bar{x} \leftarrow \bar{p}; y \leftarrow q]_G \, \phi} \qquad (pre) \quad \frac{[\bar{y} \leftarrow \bar{q}]_G \, \phi}{[\bar{x} \leftarrow p; \bar{y} \leftarrow \bar{q}]_G \, \phi} \qquad (\eta) \quad \frac{}{[x \leftarrow \mathrm{ret}\, a]_G \, x = a}$$

$$(ctr) \quad \frac{[\ldots; x \leftarrow p; y \leftarrow q; \bar{z} \leftarrow \bar{r}]_G \, \phi}{[\ldots; y \leftarrow (\mathrm{do}\ x \leftarrow p;\ q); \bar{z} \leftarrow \bar{r}]_G \, \phi} \quad (x \notin FV(\phi) \cup FV(\bar{r}))$$

Fig. 3. The auxiliary calculus

is properly weaker than $[\bar{x} \leftarrow \bar{p}]_G \, \phi$. E.g., in the free Abelian group monad, one has $\boxed{G}$ do $x \leftarrow (a - b)$; ret $\perp$, but not $[x \leftarrow (a - b)]_G \perp$. Similarly, the continuation monad fails to be simple. For purposes of presentation, we shall henceforth *assume simplicity* of the monad at hand; this can, however, be avoided at the price of having slightly more complicated definitions.

In Figure 3, double lines indicate that a rule works in both directions. The set of free variables of p is denoted by $FV(p)$. Rule (pre) is subject to the usual variable condition on x (i.e. x does not occur freely in undischarged assumptions). The calculus is sound:

Theorem 9. *If $[\bar{y} \leftarrow \bar{q}]_G \, \psi$ is deducible from $[\bar{x} \leftarrow \bar{p}]_G \, \phi$ by the rules of Figure 3, then $([\bar{x} \leftarrow \bar{p}]_G \, \phi) \Rightarrow ([\bar{y} \leftarrow \bar{q}]_G \, \psi)$ holds in the internal logic.*

An important derived rule is

$$(sef) \quad \frac{[\bar{x} \leftarrow \bar{p}; q; \bar{z} \leftarrow \bar{r}]_G \, \phi}{[\bar{x} \leftarrow \bar{p}; \bar{z} \leftarrow \bar{r}]_G \, \phi} \quad (sef(q)).$$

For side-effect free programs, we can now express determinacy:

Definition 10. A side-effect free program p is *deterministically side-effect free (dsef)* if

$$[x \leftarrow p; y \leftarrow p]_G \, x = y \qquad \text{(shorthand: } dsef(p) \text{)}.$$

Stateless programs are dsef. In most of the running examples, all side-effect free programs are dsef, with the unsurprising exception of the monads where non-determinism is involved. In these cases, a side-effect free program is dsef iff it is deterministic. *The subtype of TA formed by the deterministically side-effect free computations will be denoted by DA throughout.*

Deterministically side-effect free subterms of programs can be handled notationally in a more relaxed way. The basis for this is the following:

Lemma 11. *Let r be a program, and let p and q be dsef. Then*

(i) $(\text{do } x \leftarrow p; y \leftarrow p;\ r) = \text{do } x \leftarrow p;\ r[y/x]$, *and*
(ii) $(\text{do } x \leftarrow p; y \leftarrow q;\ r) = \text{do } y \leftarrow q; x \leftarrow p;\ r$.

Corollary 12 (Structural rules). *Let $\phi : \Omega$ be a formula, and let p and q be dsef. Then*

(i) $([\ldots; x \leftarrow p; y \leftarrow p; \bar{z} \leftarrow \bar{r}]_G\, \phi) \Leftrightarrow [\ldots; x \leftarrow p; \bar{z} \leftarrow \bar{r}[y/x]]_G\, \phi[y/x]$;
(ii) $([\ldots; x \leftarrow p; y \leftarrow q; \ldots]_G\, \phi) \Leftrightarrow [\ldots; y \leftarrow q; x \leftarrow p; \ldots]_G\, \phi$;

Since side-effect freeness amounts to 'context weakening', we can thus safely allow terms and formulas in which terms of type DA occur in places where a term of type A is expected. More precisely, if $\bar{x} = (x_1, \ldots, x_n)$ is a list of variables of types $A_1, \ldots, A_n$, then we admit terms $q[\bar{x}/\bar{p}]$ where the x_i are substituted by terms $p_i : DA_i$; such a term is just an abbreviation for do $\bar{x} \leftarrow \bar{p};\ q$, with well-definedness guaranteed by Lemma 11. Similarly, $[\bar{y} \leftarrow \bar{q}]_G\, \phi[\bar{x}/\bar{p}]$ abbreviates $[\bar{y} \leftarrow \bar{q}; \bar{x} \leftarrow \bar{p}]_G\, \phi$, with well-definedness guaranteed by Corollary 12.

A further abstraction concerns the termination of programs:

Definition 13. A program p *terminates* if

$$[\bar{x} \leftarrow \bar{q}; p]_G\, \phi \quad \text{implies} \quad [\bar{x} \leftarrow \bar{q}]_G\, \phi$$

for each program sequence $\bar{x} \leftarrow \bar{q}$ and each ϕ.

E.g., in the non-determinism monad, p terminates iff $p \neq \emptyset$. In the state monad, $p : S \to? (A \times S)$ terminates iff p is total. All side-effect free programs terminate.

We can now give a monad-independent definition of state:

Definition 14. A *state* is an element $s : T1$ such that s terminates and such that there exists, for each dsef program $p : DA$, a (necessarily unique) element $a : A$ such that

$$(\text{do } s;\ p) = \text{do } s;\ \text{ret } a.$$

A state s is called *forcible* if, for each terminating program p,

$$s = \text{do } p;\ s.$$

Example 15. In our running examples, the notions of state and forcible state explicate as follows:

- the states of the state monad are the constant state transformers $\lambda s : S \bullet (*, t)$, where $t \in S$. All of these states are forcible. Of course, the set of constant state transformers is isomorphic to the set S, i.e. the definition does indeed capture the original states. The situation is essentially the same in the non-deterministic state monad, where the states are the constant deterministic state transformers $\lambda s : S \bullet \{(*, t)\}$.
- Both the exception monad and the non-determinism monad have only one state ('running'), namely the unique terminating element of $T1$ — i.e. $*$ in the exception monad, and $\{*\}$ in the non-determinism monad. In both cases, this state is forcible.

- The states of the interactive input monad are the elements of $T1$, i.e. the U-branching trees with trivially labelled leaves. None of these states are forcible, in accordance with the intuition that one cannot unread input.

We will henceforth denote the type of forcible states by S. (S can be thought of as being defined as a subtype of $T1$.)

The main result of this section is a structure theorem concerning the type DA of deterministically side-effect free programs. This theorem will be applied to the case of dsef formulas in Section 4. It relies on the existence of a program that 'gives away' the present state:

Definition 16. A dsef program $d : DS$ is called a *state discloser* if

$$\text{do } x \leftarrow d; \ x$$

is side-effect free (i.e. equal to ret $*$, thus even stateless).

Example 17. In monads with only one forcible state s, such as the non-determinism monad or the exception monad, there is, trivially, a state discloser in the shape of the term ret s. In the state monad, we may identify the set of forcible states with the originally given set S of states (cf. Example 15); then, the element

$$\lambda s : S \bullet (s, s)$$

of DS is a state discloser. This works analogously for the non-deterministic state monad. In this sense, the notion of state discloser axiomatizes the lookup-operator mentioned in [9]. The interactive input monad does *not* have a state discloser, since here S and, hence, TS are empty.

A final prerequisite is the introduction (only for purposes of meta-reasoning) of a unique description operator into the internal language: the term $\iota a : A \bullet \phi$ will denote the element a that satisfies ϕ if a unique such element exists, and will otherwise be undefined. By results of [18], this is a definitional language extension and thus unproblematic.

The announced structure theorem confirms the intuition that deterministically side-effect free computations are essentially state-dependent values:

Theorem 18. *If $\mathbb{T}$ has a state discloser d, then for each type A,*

$$DA \cong (S \to A),$$

where the isomorphism maps $y : DA$ to

$$\lambda s : S \bullet \iota a : A \bullet (\text{do } s; \ y = \text{do } s; \ \text{ret } a)$$

and its inverse maps $f : S \to A$ to

$$\text{do } x \leftarrow d; \ \text{ret } f(x).$$

By Example 17, this theorem applies to most of our running example monads, except the interactive input monad which does not have a state discloser (and for which indeed the structure theorem fails to hold, since there are no forcible states), and the continuation monad.

Remark 19. By the above theorem, the state discloser is unique if it exists, being the element of DS that is mapped to the identity on S under the isomorphism $DS \cong (S \to S)$.

4 Interpreting Dynamic Logic

The approach to Hoare logic pursued in [17] was partly driven by the concept of interpreting formulas and programs within one and the same framework, that is, in the HASCASL internal logic, respectively in the meta-language over an arbitrary monad. This required giving a semantics to Hoare triples $\{\phi\}$ p $\{\psi\}$; for this purpose, a notion of global validity was sufficient. By contrast, formulas of dynamic logic allow a nesting of modal operators of the nature 'after execution of p' and the usual connectives of first order logic. This means informally that the state is changed according to the effect of p within the scope of the modal operator, but is 'restored' outside that scope. E.g., in a dynamic logic formula such as

$$[p]\,\phi \implies [q]\,\psi,$$

the subformulas ϕ and ψ are evaluated in modified states, but $[p]\,\phi$ and $[q]\,\psi$ are evaluated in the *same* state.

This means that the semantics of $[p]\,\phi$ must be side-effect free, although p may have side-effects that affect ϕ. Moreover, one will expect that a formula evaluates to a deterministic truth value (although p may be non-deterministic). Thus, it is reasonable to require that *formulas are interpreted as deterministically side effect-free Ω-valued programs*. We state explicitly

Definition 20. A *formula* (of dynamic logic) is a term $\phi : D\Omega$.

(Recall here the the notation for dsef subterms introduced after Corollary 12.)

The question is now if $D\Omega$ has enough structure to allow the interpretation of the diamond and box operators $\langle p \rangle$ and $[p]$ of dynamic logic. The interpretation can be introduced *axiomatically* in a rather straightforward manner. To begin, observe that $D\Omega$ is made into a partial order by putting

$$\phi \leq \psi \iff (\phi \Rightarrow \psi)$$

for $\phi, \psi : D\Omega$. The crucial requirement for dynamic logic is, then, the existence of lower and upper deterministically side-effect free approximations for Ω-valued program sequences:

Definition 21. We say that $\mathbb{T}$ *admits (propositional) dynamic logic* if there exist, for each $\phi : T\Omega$, a formula $\Box\phi : D\Omega$ such that

$$([\bar{x} \leftarrow \bar{p}]_G\, x_i \Rightarrow \Box\phi) \iff ([\bar{x} \leftarrow \bar{p}; a \leftarrow \phi]_G\, x_i \Rightarrow a)$$

for each program sequence $\bar{x} \leftarrow \bar{p}$ containing $x_i : \Omega$ and, dually, a formula $\Diamond \phi : D\Omega$ such that

$$([\bar{x} \leftarrow \bar{p}]_G \Diamond\phi \Rightarrow x_i) \iff ([\bar{x} \leftarrow \bar{p}; a \leftarrow \phi]_G a \Rightarrow x_i).$$

$\mathbb{T}$ *admits quantified dynamic logic* if, additionally, $D\Omega$ is a complete lattice. In this case, joins and meets in $D\Omega$ are denoted by $\bigvee$ and $\bigwedge$, respectively.

(N.B.: since the internal logic is in general intuitionistic, one cannot define $\langle \bar{y} \leftarrow \bar{q} \rangle$ as $\neg[\bar{y} \leftarrow \bar{q}]\neg$.) The formulas $\Box\phi$ and, dually, $\Diamond\phi$ are uniquely determined (in contrast to the modalities in [10], which are part of the interpretation structure):

Lemma 22. *If* $\mathbb{T}$ *admits dynamic logic, then* $\Box\phi$ *is the greatest formula (w.r.t. the ordering introduced above)* $\psi : D\Omega$ *such that*

$$[a \leftarrow \psi; b \leftarrow \phi]_G a \Rightarrow b.$$

If Theorem 18 is applicable, then $D\Omega$ is order-isomorphic to $S \to \Omega$ equipped with the pointwise order, hence a complete lattice with joins and meets formed pointwise. It follows that there exist, for $\phi : T\Omega$, (unique) formulas $\Box\phi$ and $\Diamond\phi$ that satisfy the properties in Lemma 22. Under two weak additional assumptions (hoped to be dispensed with in the future), we can prove that in this case, $\mathbb{T}$ admits dynamic logic.

Definition 23. A state $s : S$ is called *logically splitting* if, for all program sequences $\bar{x} \leftarrow \bar{p}$ and $\bar{y} \leftarrow \bar{q}$ and for each formula ϕ,

$$[\bar{x} \leftarrow \bar{p}; s; \bar{y} \leftarrow \bar{q}]_G \phi \quad \text{implies} \quad [\bar{x} \leftarrow \bar{p}]_G [s; \bar{y} \leftarrow \bar{q}]_G \phi.$$

(The converse implication holds universally.) In the running examples, all forcible states are logically splitting; we conjecture that this is in fact always the case.

Definition 24. A monad is *logically regular* if, for each $\phi : T\Omega$,

$$[a \leftarrow \phi]_G (([b \leftarrow \phi]_G b) \Rightarrow a).$$

Classically, *all monads are logically regular*. We conjecture that this holds also in the intuitionistic case.

Theorem 25. *If* $\mathbb{T}$ *is logically regular and has a state discloser, and if all forcible states are logically splitting, then* $\mathbb{T}$ *admits quantified dynamic logic.*

Example 26. In those of our running example monads that have only one forcible state, $\Box\phi$ and $\Diamond\phi$ are just truth values. E.g., in the exception monad, $\Box\phi$ is true iff ϕ either throws an exception or returns $\top$, while $\Diamond\phi$ is true iff ϕ returns $\top$. In the nondeterminism monad, $\Box\phi$ holds iff $\phi \subset \{\top\}$, and $\Diamond\phi$ holds iff $\top \in \phi$.

In the various state monads, $\Box\phi$ and $\Diamond\phi$ depend on the state s. E.g., in the non-deterministic state monad, $\Box\phi$ holds in a state s iff $\phi(s) \subset \{\top\} \times S$, and $\Diamond\phi$ holds in s iff there exists s' such that $(\top, s') \in \phi(s)$.

Remark 27. The interactive input monad does admit quantified dynamic logic, although it does not have a state discloser. In fact, it is unclear whether there are monads that fail to admit dynamic logic; a good candidate for such a counterexample is the continuation monad. This is subject of further research.

If $\mathbb{T}$ admits dynamic logic, then one has the usual syntax of dynamic logic with the modal operators given by

$$[\bar{x} \leftarrow \bar{p}] \, \phi := \Box \, \text{do} \ \bar{x} \leftarrow \bar{p}; \ \phi \quad \text{and} \quad \langle \bar{x} \leftarrow \bar{p} \rangle \phi := \Diamond \, \text{do} \ \bar{x} \leftarrow \bar{p}; \ \phi,$$

and the logical connectives defined by substitution of dsef terms as explained after Corollary 12. Actual truth values, i.e. terms of type Ω, appearing in dynamic logic formulas are implicitly cast to $D\Omega$ via ret; it is easy to see that this is compatible with the interpretation of the logical connectives and hence does not lead to confusion. If $\mathbb{T}$ admits quantifies dynamic logic, then one can also interpret universal and existential quantification by putting

$$\forall x : A \bullet \phi(x) = \bigwedge_{x:A} \phi(x) \quad \text{and} \quad \exists x : A \bullet \phi(x) = \bigvee_{x:A} \phi(x).$$

Note that, as in the monad-independent Hoare logic of [17], we admit program *sequences* inside the modal operators in order to accommodate reasoning about intermediate results. A side benefit of this (e.g. in comparison to [10]) is the possibility of expressing natural axioms concerning composite sequences (Axioms (seq$\Box$) and (seq$\Diamond$) in Figure 4).

In quantified dynamic logic, the modal operators are interrelated in a way analogous to quantifiers in intuitionistic logic:

Theorem 28. *If $\mathbb{T}$ admits quantified dynamic logic, then $\Diamond\phi$ is expressible as*

$$\forall a : \Omega \bullet \Box(\phi \Rightarrow a) \Rightarrow a,$$

Validity of formulas is, as usual in modal logics, defined via the global box:

Definition 29. A formula $\phi : D\Omega$ is *valid* if $\boxdot \phi$ holds, i.e. if $\phi = \text{ret} \top$.

Figure 4 shows a generic Hilbert style proof calculus for dynamic logic (we shall not be concerned with deduction in quantified dynamic logic here). The necessitation rule is subject to the usual constraints on the variables $\bar{x}$. The first five axioms are the standard axioms of the K-fragment of intuitionistic modal logic as given e.g. in [12, 19], with a slight variation of the third axiom — the usual form $\neg\langle p \rangle \bot$ is a special case of the second of the two dual forms given. All intuitionistic propositional tautologies are implicitly included here. Moreover, there are four axioms concerning composition of program sequences. The last three axioms concern programs that enjoy particular properties as defined above; these properties are side conditions that are expected to be discharged outside the calculus. However, the axioms are applicable to stateless programs ret a without further ado, so that the two generic program constructors (sequential composition and return) are covered.

As in the case of the generic Hoare rules of [17], axioms that deal with monad-specific program constructs are necessarily missing here; typical examples include the nondeterministic choice and iteration constructs of standard dynamic logic. Such specific formulas come in via the specifications of particular monads; in fact, they often turn out to be a good choice for actual axioms of the specification.

Rules:

$$(\text{nec}) \ \frac{\phi}{[\bar{x} \leftarrow \bar{p}]\,\phi} \qquad (\text{mp}) \ \frac{\phi \Rightarrow \psi; \quad \phi}{\psi}$$

Axioms:

(K1) $[\bar{x} \leftarrow \bar{p}]\,(\phi \Rightarrow \psi) \Rightarrow [\bar{x} \leftarrow \bar{p}]\,\phi \Rightarrow [\bar{x} \leftarrow \bar{p}]\,\psi$

(K2) $[\bar{x} \leftarrow \bar{p}]\,(\phi \Rightarrow \psi) \Rightarrow \langle\bar{x} \leftarrow \bar{p}\rangle\phi \Rightarrow \langle\bar{x} \leftarrow \bar{p}\rangle\psi$

(K3□) $\text{ret}\,\phi \Rightarrow [p]\,\text{ret}\,\phi$

(K3◇) $\langle p\rangle\,\text{ret}\,\phi \Rightarrow \text{ret}\,\phi$

(K4) $\langle\bar{x} \leftarrow \bar{p}\rangle(\phi \vee \psi) \Rightarrow (\langle\bar{x} \leftarrow \bar{p}\rangle\phi \vee \langle\bar{x} \leftarrow \bar{p}\rangle\psi)$

(K5) $(\langle\bar{x} \leftarrow \bar{p}\rangle\phi \Rightarrow [\bar{x} \leftarrow \bar{p}]\,\psi) \Rightarrow [\bar{x} \leftarrow \bar{p}]\,(\phi \Rightarrow \psi)$

(seq□) $[\bar{x} \leftarrow \bar{p}; y \leftarrow q]\,\phi \iff [\bar{x} \leftarrow \bar{p}]\,[y \leftarrow q]\,\phi$

(seq◇) $\langle\bar{x} \leftarrow \bar{p}; y \leftarrow q\rangle\phi \iff \langle\bar{x} \leftarrow \bar{p}\rangle\,\langle y \leftarrow q\rangle\phi$

(ctr□) $[x \leftarrow p; y \leftarrow q]\,\phi \implies [y \leftarrow (\text{do } x \leftarrow p;\ q)]\,\phi \ (x \notin FV(\phi))$

(ctr◇) $\langle x \leftarrow p; y \leftarrow q\rangle\phi \impliedby \langle y \leftarrow (\text{do } x \leftarrow p;\ q)\rangle\phi \ (x \notin FV(\phi))$

(dsef□) $[x \leftarrow p]\,\phi \iff \phi[x/p]$ ((if p is dsef)

(dsef◇) $\langle x \leftarrow p\rangle\phi \iff \phi[x/p]$ (if p is dsef)

(tm) $\langle p\rangle\top$ (if p terminates)

(sef□) $[p]\,\phi \implies \phi$ (if p is *sef*)

(sef◇) $\langle p\rangle\phi \impliedby \phi$ (if p is *sef*)

(ssef) $[p]\,\phi \iff \phi \iff \langle p\rangle\phi$ (if p is strongly *sef*)

Fig. 4. The generic proof calculus for propositional dynamic logic

---

The proof calculus is sound:

Theorem 30. *Let $\mathbb{T}$ admit dynamic logic. If a formula ψ can be deduced from formulas $\phi_1, \ldots, \phi_n$ by means of the rules of Figure 4, then validity of the ϕ_i implies validity of ψ, i.e. $\bigwedge \boxed{a}\phi_i \implies \boxed{a}\psi$.*

The proof relies on the auxiliary calculus of Figure 3. Completeness of the calculus is the subject of further research.

Remark 31. If $\mathbb{T}$ is simple, then the converses of Axioms (ctr□), (ctr◇) hold.

5   From Partial to Total Correctness

Classically as well is in the monad-independent setting, dynamic logic subsumes Hoare logic: The monad-independent Hoare triples introduced in [17] can be expressed in dynamic logic as

$$\{\phi\}\ \bar{x} \leftarrow \bar{p}\ \{\psi\} : \Longleftrightarrow (\phi \Rightarrow [\bar{x} \leftarrow \bar{p}]\ \psi).$$

An obvious expressive advantage of dynamic logic over Hoare logic is its local nature — i.e. formulas hold locally, in any given state. By contrast, Hoare triples only provide global axiomatizations, where each formula is separately quantified over all states. Beyond this, a crucial feature of dynamic logic is its ability to express termination.

In the Hoare calculus, *non-termination* can be expressed by stating that the postcondition $\perp$ holds. However, Hoare logic is too weak to express termination, while the formula $\langle p\rangle\top$ of dynamic logic indeed states that p terminates in the monad-independent sense defined above (which has the expected meaning in concrete monads). We now can give a meaning to Hoare triples for total correctness by interpreting them as partial correctness plus termination:

$$[\phi]\ \bar{x} \leftarrow \bar{p}\ [\psi] : \Longleftrightarrow (\phi \Rightarrow (\langle\bar{x} \leftarrow \bar{p}\rangle\top \wedge [\bar{x} \leftarrow \bar{p}]\ \psi)).$$

The iteration rule from [17] (*iter b p e* iterates $p : A \to TA$, starting from the initial value $e : A$, while $b : A \to T(bool)$ is satisfied) can be adjusted to cope with total correctness, along the lines of [6]:

$$x : A \text{ in } \Gamma$$
$$t : A \to TB$$
$$\_ < \_ : B \times B \to \Omega \text{ is well-founded}$$

(iter-total) $\dfrac{\Gamma \rhd [\phi \wedge b\ x \wedge \text{ret}(*(t\ x) = z)]\ y \leftarrow p\ x\ [\phi[x/y] \wedge \text{ret}(*(tx) < z)]}{\Gamma \rhd [\phi[x/e]]\ y \leftarrow iter\ b\ p\ e\ [\phi[x/y] \wedge \neg(b\ y)]}$

while the other rules can be carried over from the Hoare calculus for partial correctness.

Using the monad-independent Hoare calculus, we have, for instance, axiomatized the join operator in the non-determinism monad by

$$\{\phi\}\ x \leftarrow p\ \{\chi_1\} \wedge \{\phi\}\ x \leftarrow q\ \{\chi_2\} \implies \{\phi\}\ x \leftarrow p[\!]q\ \{\chi_1 \vee \chi_2\}.$$

With dynamic logic, a more fine-grained specification of non-determinism is possible:

$$\langle p[\!]q\rangle\phi \Longleftrightarrow (\langle p\rangle\phi \vee \langle q\rangle\phi)$$
$$[p[\!]q]\ \phi \Longleftrightarrow ([p]\ \phi \wedge [q]\ \phi)$$

From the second axiom, we easily get the Hoare rule above by taking ψ to be $\chi_1 \vee \chi_2$. But we get more than that: the first axiom implies that $p[\!]q$ terminates *if and only if* p_1 or p_2 terminates. This is not expressible in Hoare logic. Moreover, it is now easy, using the rule (iter-total) above, to prove termination of the non-deterministic variant of Euclid's algorithm for computing the greatest common divisor (the partial correctness of which has been proved as an example in [17]).

6   Conclusion

Building on results on monad-independent reasoning about program properties developed in [17], we have designed a monad-independent dynamic logic and a representation of this logic in the internal logic of HASCASL, i.e. essentially in intuitionistic partial higher order logic. The main problem here was to provide a monad-independent semantics of the dynamic modal operators. As a solution, we have introduced an axiomatic method which imposes additional constraints on the monad. This method is complemented by results that, under suitable conditions, allow the extraction of abstract states from a given monad, giving rise to a structure theorem for 'dynamic truth values' which guarantees the interpretability of the modal operators. The structure theorem is both of independent interest and the basis for further research into the question of whether a monad can generally be decomposed into aspects of input, output, and state.

Given the semantics of the modal operators, we have introduced a generic proof system for dynamic logic over an arbitrary monad, where the rules and axioms are proved as lemmas about the encoding. We have thus ended up with a logic that allows dynamic reasoning about computations with side effects, leaving the actual nature of the side effects open. In practice, one will aim at performing a large amount of verification in this generic setting, and switch to instantiations of the calculus for particular monads only in the more detailed analysis. As an example, we have laid out how operations in the non-determinism monad can be axiomatized by means of dynamic logic formulas. A library of monad definitions in HASCASL will make extensive use of this axiomatization principle; the compositionality of such axiomatizations w.r.t. monad combination is subject of further research.

One of the crucial features of dynamic logic is its ability to express termination of programs. As an example application, we have shown how to obtain a termination rule for a generic iteration construct; i.e. not only partial correctness, but also total correctness lends itself to monad-independent reasoning. This corroborates Moggi's claim [8] that the logic of monads is the right setting for reasoning about computations with effects.

Acknowledgements

This work forms part of the DFG-funded project HasCASL (KR 1191/7-1). The authors wish to thank Christoph Lüth for useful comments and discussions.

References

1. E. Astesiano, M. Bidoit, H. Kirchner, B. Krieg-Brückner, P. D. Mosses, D. Sannella, and A. Tarlecki. CASL: the Common Algebraic Specification Language. *Theoret. Comput. Sci.*, 286:153–196, 2002.
2. CoFI. The Common Framework Initiative for algebraic specification and development, electronic archives. http://www.brics.dk/Projects/CoFI.
3. CoFI Language Design Task Group. CASL – The CoFI Algebraic Specification Language – Summary, version 1.0. Documents/CASL/Summary, in [2], July 1999.
4. The Haskell mailing list. http://www.haskell.org/mailinglist.html, 2002.

5. S. P. Jones, J. Hughes, L. Augustsson, D. Barton, B. Boutel, W. Burton, J. Fasel, K. Hammond, R. Hinze, P. Hudak, T. Johnsson, M. Jones, J. Launchbury, E. Meijer, J. Peterson, A. Reid, C. Runciman, and P. Wadler. Haskell 98: A non-strict, purely functional language. 1999. http://www.haskell.org/onlinereport.
6. J. Loeckx and K. Sieber. *The Foundations of Program Verification.* Wiley, 1987.
7. S. Mac Lane. *Categories for the Working Mathematician.* Springer, 1997.
8. E. Moggi. Notions of computation and monads. *Inform. and Comput.*, 93:55–92, 1991.
9. E. Moggi. A semantics for evaluation logic. *Fund. Inform.*, 22:117–152, 1995.
10. A. Pitts. Evaluation logic. In *Higher Order Workshop*, Workshops in Computing, pages 162–189. Springer, 1991.
11. G. Plotkin and J. Power. Notions of computation determine monads. In *Foundations of Software Science and Computation Structures*, volume 2303 of *LNCS*, pages 342–356. Springer, 2002.
12. G. Plotkin and C. Stirling. A framework for intuitionistic modal logic. In *Theoretical Aspects of Reasoning about Knowledge*. Morgan Kaufmann, 1986.
13. V. Pratt. Semantical considerations on Floyd-Hoare logic. In *Foundations of Conputer Science*, pages 109–121. IEEE, 1976.
14. F. Regensburger. HOLCF: Higher order logic of computable functions. In *Theorem Proving in Higher Order Logics*, volume 971 of *LNCS*, pages 293–307, 1995.
15. L. Schröder. Classifying categories for partial equational logic. In *Category Theory and Computer Science*, volume 69 of *ENTCS*, 2002.
16. L. Schröder and T. Mossakowski. HASCASL: Towards integrated specification and development of functional programs. In *Algebraic Methodology and Software Technology*, volume 2422 of *LNCS*, pages 99–116. Springer, 2002.
17. L. Schröder and T. Mossakowski. Monad-independent Hoare logic in HASCASL. In M. Pezze, editor, *Fundamental Aspects of Software Engineering*, volume 2621 of *LNCS*, pages 261–277, 2003.
18. D. S. Scott. Relating theories of the λ-calculus. In *To H.B. Curry: Essays in Combinatory Logic, Lambda Calculus and Formalisms*, pages 403–450. Academic Press, 1980.
19. A. K. Simpson. *The Proof Theory and Semantics of Intuitionistic Modal Logic.* PhD thesis, University of Edinburgh, 1994.
20. Philip Wadler. How to declare an imperative. *ACM Computing Surveys*, 29:240–263, 1997.

Preserving Properties in System Redesign: Rule-Based Approach[*]

Milan Urbášek

Technical University Berlin, Germany
Institute for Software Technology and Theoretical Computer Science
urbasek@cs.tu-berlin.de

Abstract. This paper deals with stepwise development of systems based on rule-based approach. Modeling using this approach usually starts with a rough model of a system which is refined in further steps. This approach is based on rules and transformations as known from theory of HLR systems. Preservation of certain system properties during this process is of importance. In the context of Petri nets the developed property preserving rules and transformations restrict the variety of modeling possibilities to refinement of the system by additional details. The concept for conceptual change and redesign of a part of the modeled system is of interest. Up to now the request for a large change of structure of the modeled system forced the redevelopment of the whole system from origin. Otherwise the property preserving rules could not be employed and the tedious investigation of system properties had to be done for the final system. In this paper we describe the possibility of building new property preserving rules from other ones which are suitable for redesign of system's parts.

1 Introduction

The incremental approach to development of system models has been investigated in many papers. Well accepted technique of the incremental development is the stepwise refinement of modeled system. This technique refines a coarse model of a system until the necessary level of detail is reached. The important question of a stepwise design is how to preserve certain system properties. Then, they need not to be checked repeatedly during the design of the system. This gives rise to the concept of property preserving development and redesign. The approach in this paper is based on transformations in High-Level Replacement Systems (HLR Systems) as introduced in [EHKP91a]. The Q-theory introduced

---

[*] This work is part of the joint research project "DFG- Forschergruppe PETRINETZ-TECHNOLOGIE" supported by the German Research Council (DFG). This work has also been done within the project CEZ:J22/98:262200012 "Research in Information and Control Systems" and supported by the Grant Agency of the Czech Republic under the grant No.102/01/1485 "Environment for Development, Modelling, and Application of Heterogeneous Systems".

M. Wirsing, D. Pattinson, and R. Hennicker (Eds.): WADT 2002, LNCS 2755, pp. 442–456, 2003.
© Springer-Verlag Berlin Heidelberg 2003

in [Pad96] extends the theory of transformations by property preserving refinement morphisms. The necessary notions from the theory of HLR systems are summarized in Section 2.

The idea of a transformation $N_1 \Rightarrow N_2$ in HLR systems uses the so-called double pushout approach as known from graph transformations. If such a transformation is extended by a property preserving morphism q assuring that certain system property ϕ is preserved then such a transformation is called Q-transformation. The general results concerning compatibility of Q-transformations with categorical structuring techniques like union and fusion and other results including Church Rosser and Parallelism Theorems are available.

Disadvantage of this approach is that it is not possible to completely redesign a part of your system and preserve certain system properties via transformations if such a change is necessary from practical or other reasons. A stepwise refinement usually does not allow the change of the design concept because the refinement of the existing structure is allowed only.

The property preserving system redesign is based on the following. Having two Q-transformations preserving certain properties, say the transformations $N_0 \overset{q_1}{\Rightarrow} N_1$ and $N_0 \overset{q_2}{\Rightarrow} N_2$, there can be a new direct transformation $N_1 \Rightarrow N_2$ constructed.

If a system property ϕ is respected and preserved by transformations $N_0 \Rightarrow N_1$ and $N_0 \Rightarrow N_2$ then as a logical conclusion, the property ϕ is preserved by the transformation $N_1 \Rightarrow N_2$. The newly derived transformation is called *PB-induced transformation*[1] as its construction is based on pullbacks. The PB-induced transformations are formally introduced in Section 3.

The major advantage of this approach is that it is possible to build up new property preserving rules and transformations according to old ones without the explicit definition of a property preserving refinement morphism. This allows switching from one design concept to another without restarting design of the system from the origin.

An example of an application is liveness preserving transformations of Petri nets. We will discuss the application of this approach on the design and redesign of simple Producers-Consumers system in Section 4.

2   Formal Framework of Net Model Transformation

The idea of net transformation systems is one of the possible instantiations of the more general idea of high-level transformation systems (see [EHKP91b]). In this section we summarize the necessary definitions and results. To present all the theoretical results in detail is beyond the scope of the paper. They can be found in the literature cited here.

The next definition introduces rules, transformations and net transformation systems formally for a given category **NET** of low- or high-level nets. More about the underlying theory can be found in [PER95] and in [Pad99]. We will assume

---

[1] This is a different notion than pullback transformation of open graphs.

to have a suitable category **NET** of nets and net morphisms, i.e. a category satisfying so called HLR conditions stated in Appendix A.

Definition 1 (Rules, Transformations, Net Transformation Systems).

1. A rule $p = (L \xleftarrow{l} K \xrightarrow{r} R)$ in a category **NET** consists of the nets L, K and R, called left-hand side, interface, and right-hand side, respectively, and two morphisms $K \xrightarrow{l} L$ and $K \xrightarrow{r} R$ with both morphisms $l, r \in \mathcal{M}$, a suitable class of injective morphisms in **NET**.

2. Given a rule $p = (L \xleftarrow{l} K \xrightarrow{r} R)$, a direct transformation $G \xRightarrow{p} H$ from a net G to a net H is given by two pushout diagrams **(1)** and **(2)** in the category **NET** as shown below.

$$
\begin{array}{ccccc}
L & \xleftarrow{l} & K & \xrightarrow{r} & R \\
{\scriptstyle m}\downarrow & (1) & {\scriptstyle k}\downarrow & (2) & \downarrow{\scriptstyle n} \\
G & \xleftarrow{g} & C & \xrightarrow{h} & H
\end{array}
$$

The morphisms $L \xrightarrow{m} G$ and $R \xrightarrow{n} H$ are called occurrences of L in G and R in H, respectively. By an occurrence of rule $r = (L \xleftarrow{l} K \xrightarrow{r} R)$ in a net G we mean an occurrence of the left-hand side L in G.

In fact, the occurrence morphism m has to satisfy a specific condition, called gluing condition, in order to be able to apply the rule p to the net G.

3. Given a category of nets **NET** together with a suitable class of injective morphisms $\mathcal{M}$, a net transformation system $H = (\mathcal{S}, \mathcal{P})$ in $(\mathbf{NET}, \mathcal{M})$ is given by a start net $\mathcal{S} \in |\mathbf{NET}|$, and a set of rules $\mathcal{P}$.

Informally, a rule $r = (L \xleftarrow{l} K \xrightarrow{r} R)$ is given by three nets L, K, and R. Moreover, K is a subnet of both L and R expressed by the morphisms l and r. Application of a rule to the net G is a net model transformation of G. The transformation means replacing a subnet specified by the left-hand side of the rule with the net specified by the right-hand side. More precisely, we first identify the subnet L in G. Then we delete those parts of the subnet L which are not subnets of the interface net K. This results in an intermediate net C, where in a further step we add the difference of R and K to the preserved subnet C to obtain the transformed net H. In case the left-hand side is empty, we simply add the right-hand side to the first net.

Although the net transformation framework is a suitable concept for stepwise development of systems, very often there is a need to consider more general morphisms for refinement or abstraction in addition. The main idea is to enlarge the category of nets by $\mathcal{Q}$-morphisms in the sense of [Pad96] in order to formulate refinement/abstraction morphisms.

More precisely, another category of nets **QNET** with a distinguished class of morphisms $\mathcal{Q}$, called $\mathcal{Q}$-morphisms, is employed. The category **QNET** enriches the net transformation system defined in $(\mathbf{NET}, \mathcal{M})$ and yields the notion of $\mathcal{Q}$-morphisms and $\mathcal{Q}$-transformations. The class of $\mathcal{Q}$-morphisms has to satisfy additional requirements called $\mathcal{Q}$-conditions (see [Pad96]) to be adequate for refinement or abstraction. The formal definitions are given below.

Definition 2 (Class of Q-morphisms). *Let* **QNET** *be a category, so that* **NET** *is a subcategory* **NET** $\subseteq$ **QNET** *and Q a class of morphisms in* **QNET**.

1. *The morphisms in Q are called Q-morphisms, or refinement/abstraction morphisms.*
2. *The Q-morphisms must satisfy so called Q-conditions, namely closedness of Q, preservation of pushouts, and inheritance of Q-morphisms under pushouts and coproducts.*

Moreover, let $\mathcal{O} = (\mathcal{O}^q)_{q\in Q}$ be an indexed class of adequate occurrence morphisms in **QNET** with respect to a refinement morphism $q \in Q$. These occurrence morphisms restrict applicability of rules as defined in Theorem 1 below.

Definition 3 (Q-Rules and Q-Transformations).

1. *A (preserving) Q-rule (p, q) is given by a rule $p = (L \xleftarrow{l} K \xrightarrow{r} R)$ in* **NET** *and a Q-morphism $q : L \to R$, so that $q \circ l = r$ in* **QNET**.
2. *A respecting Q-rule (p, q) is given by a rule $p = (L \xleftarrow{l} K \xrightarrow{r} R)$ in* **NET** *and a Q-morphism $q : R \to L$, so that $q \circ r = l$ in* **QNET**.

The next theorem states that Q-morphisms are preserved by transformations.

Theorem 1 (Induced Transformations and Pushouts in QNET). *Let* **QNET** *be a supercategory of* **NET** *according to Definition 2.*

1. *Given a preserving Q-rule (p, q) and a transformation $G \xRightarrow{p} H$ in* **NET** *with an occurrence $m \in \mathcal{O}^q$ defined by the pushouts (1) and (2), there is a unique $q' \in Q$, such that $q' \circ g = h$ and $q' \circ m = n \circ q$ in* **QNET**. *The transformation $(G \xRightarrow{p} H, q' : G \to H)$, or $G \xRightarrow{(p,q')} H$ for short, is called Q-transformation. Moreover, $R \xrightarrow{n} H \xleftarrow{q'} G$ is pushout of $G \xleftarrow{m} L \xrightarrow{q} R$ in* **QNET**.
If morphisms in Q preserve some property ϕ then we have the following:

$$G \models \phi \quad \text{implies} \quad H \models \phi.$$

2. *Given a respecting Q-rule (p, q) and a transformation $G \xRightarrow{p} H$ with an occurrence $n \in \mathcal{O}^q$ in* **NET** *defined by the pushouts (1) and (2), there is a unique $q' \in Q$, such that $q' \circ h = g$ and $q' \circ n = m \circ q$ in* **QNET**. *The transformation $(G \xRightarrow{p} H, q' : H \to G)$, or $G \xRightarrow{(p,q')} H$ for short, is called Q-R-transformation. Moreover, $L \xrightarrow{m} G \xleftarrow{q'} H$ is pushout of $H \xleftarrow{n} R \xrightarrow{q} L$ in* **QNET**.
If morphisms in Q respect some property ϕ then we have the following:

$$G \models \phi \quad \text{implies} \quad H \models \phi.$$

$$
\begin{array}{ccccc}
& & q & & \\
& & \overset{\frown}{\cdots} & & \\
L & \xleftarrow{\ \ l\ \ } & K & \xrightarrow{\ \ r\ \ } & R \\
{\scriptstyle m}\downarrow & (1) & {\scriptstyle k}\downarrow & (2) & \downarrow{\scriptstyle n} \\
G & \xleftarrow{\ \ g\ \ } & C & \xrightarrow{\ \ h\ \ } & H \\
& & \underset{\smile}{\cdots} & & \\
& & q' & &
\end{array}
$$

The proof can be found in [Pad96].

Preserving and respecting transformations differ in the direction of Q-morphisms q, q'. The pushout squares **(1)** and **(2)** represent a transformation in the category **NET** (as in Definition 1). The Q-morphism q is a refinement/abstraction morphism in **QNET**. The morphism q' is the induced morphism (according to Theorem 1) which belongs to the class of Q-morphisms in **QNET** as well.

There are many other interesting results available in the general theory of HLR systems. The most important are Church-Rosser theorem, parallelism theorem, and compatibility with horizontal structuring techniques (union and fusion). These results can be found in [EHKP91b,Pad96].

3   PB-Induced Transformations and System Redesign

The Q-theory as discussed in previous section is very useful for enhancement of rule-based refinement by refinement/abstraction morphisms. But these refinement/abstraction morphisms may become quite complicated morphisms as in case of collapsing and abstracting morphisms published in [UP02,Urb02]. For many applications even the generalization of simple morphisms is of interest as in the case of generalized collapsing morphisms (see [Urb02,Urb03]). However, generalization of quite sophisticated morphisms may be tiring. Therefore, we try to develop new rule according to other rules such that some system properties will be preserved without explicit definition of Q-morphisms.

In next paragraphs we will develop the concept of PB-induced rules which are related to D-concurrent rules published in [EHKP91b,EHKPP90]. Our PB-induced rules are a special case of D-concurrent rules when we use the trick that all rules in double pushout approach are symmetric.

We will adopt the pullback construction to define induced rules and transformations. Further we will discuss property preservation under these rules and transformations.

Our goal is to extend the existing commutative diagram of two property preserving Q-transformations below by another transformation $N_1 \Rightarrow N_2$ such that system properties will be kept unchanged and there is no explicit property preserving morphism from N_1 to N_2 or vice versa.

We will consider property respecting Q-rules in this section. Property preserving Q-rules are dealt with similarly.

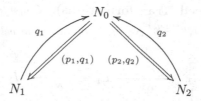

For our approach we will consider HLR category satisfying HLR2 conditions. HLR2 conditions (see Definition 6 in Appendix A) are more restrictive than HLR conditions (see Definition 5 in Appendix A) but for practical cases they are satisfied by many categories of nets.

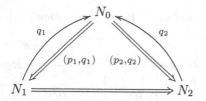

We will construct new rule (and new transformation $N_1 \Rightarrow N_2$) from two rules with the same left-hand side net as in definition below.

Definition 4 (PB-Induced rules). *Given two Q-rules* $r_1 = (p_1 = (L_1 \leftarrow I_1 \rightarrow R_1), q_1)$ *and* $r_2 = (p_2 = (L_1 \leftarrow I_2 \rightarrow R_2), q_2)$ *as below.*

$$R_1 \overset{q_1}{\underset{\mathcal{M}}{\leftarrow}} I_1 \overset{}{\underset{\mathcal{M}}{\rightarrow}} L_1 \overset{q_2}{\underset{\mathcal{M}}{\leftarrow}} I_2 \overset{}{\underset{\mathcal{M}}{\rightarrow}} R_2$$

We define a PB-induced rule $pbind(r_1, r_2)$ *as a rule*

$$R_1 \overset{\mathcal{M}}{\leftarrow} L_0 \overset{\mathcal{M}}{\rightarrow} R_2$$

where net (object) L_0 *is obtained by calculation of the pullback as below, and morphisms* $L_0 \rightarrow R_1$, $L_0 \rightarrow R_2$ *are obtained by composition of morphisms* $L_0 \rightarrow I_1 \rightarrow R_1$ *and* $L_0 \rightarrow I_2 \rightarrow R_2$ *as in the diagram below.*

$$
\begin{array}{ccc}
L_0 & \overset{\mathcal{M}}{\longrightarrow} I_2 \overset{\mathcal{M}}{\longrightarrow} R_2 \\
{\scriptstyle \mathcal{M}}\downarrow \quad (PB) \quad \downarrow{\scriptstyle \mathcal{M}} \\
R_1 \overset{}{\underset{\mathcal{M}}{\leftarrow}} I_1 & \overset{}{\underset{\mathcal{M}}{\longrightarrow}} L_1
\end{array}
$$

Note that morphisms $L_0 \rightarrow I_1$ *and* $L_0 \rightarrow I_2$ *are* $\mathcal{M}$*-morphisms due to the inheritance of* $\mathcal{M}$ *morphisms under pullback (see Appendix A, Definition 5 part 3) and that a composition of two* $\mathcal{M}$ *morphisms is an* $\mathcal{M}$ *morphism when HLR2 categories are considered.*

Lemma 1 (PB-Induced Transformations). *Consider two Q-transformations $N' \overset{(r_1, q_1)}{\Longrightarrow} N_1$ and $N' \overset{(r_2, q_2)}{\Longrightarrow} N_2$ as in the diagram below.*

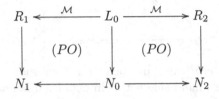

Then there exists an induced PB-transformation $N_1 \overset{r_3}{\Longrightarrow} N_2$ according to the rule $r_3 = pbind(r_1, r_2)$:

$$
\begin{array}{ccccc}
R_1 & \xleftarrow{\;\mathcal{M}\;} & L_0 & \xrightarrow{\;\mathcal{M}\;} & R_2 \\
\downarrow & & \downarrow & & \downarrow \\
 & (PO) & & (PO) & \\
N_1 & \longleftarrow & N_0 & \longrightarrow & N_2
\end{array}
$$

where $N_0 \to N_1$ and $N_0 \to N_2$ are defined via the pullback of $C_2 \to N'$ and $C_1 \to N'$ (see diagram below).

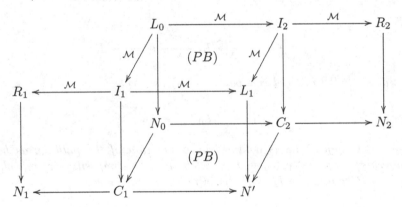

Proof. The proof comprises following steps.

1. L_0, N_0 and corresponding morphisms are defined as pullbacks in the diagram above. It has to be shown the existence of the morphism $L_0 \to N_0$ and the commutativity of the whole diagram.
   Due to universal property of the pullback $N_0 \to C_1$ and $N_0 \to C_2$ and existence of composition morphisms $L_0 \to C_2$ and $L_0 \to C_1$ there exists a unique morphism $L_0 \to N_0$ such that the diagram commutes.

2. The composition morphisms $L_0 \to R_1$ and $L_0 \to R_2$ are $\mathcal{M}$-morphisms as HLR2 conditions guarantee that a class of $\mathcal{M}$-morphisms is closed under composition.

3. Morphisms $N_0 \to C_1$, $N_0 \to C_2$, $C_1 \to N'$, and $C_2 \to N'$ are in $\mathcal{M}$ due to HLR 1 condition 3 - Inheritance of $\mathcal{M}$ under pushouts (see Definition 5 in Appendix A).

4. Due to condition 3 (Cube lemma) of HLR2 conditions (see Appendix A) the diagrams

$$
\begin{array}{ccc}
I_1 & \longleftarrow L_0 \longrightarrow & I_2 \\
\downarrow & \downarrow & \downarrow \\
C_1 & \longleftarrow N_0 \longrightarrow & C_2
\end{array}
$$

are pushouts.

5. Due to closedness of a pushout construction under composition, the diagrams

$$
\begin{array}{ccc}
R_1 & \longleftarrow L_0 \longrightarrow & R_2 \\
\downarrow & \downarrow & \downarrow \\
N_1 & \longleftarrow N_0 \longrightarrow & N_2
\end{array}
$$

are pushouts. Therefore they form the requested transformation.

The presented PB-induced rules and transformations can be constructed in HLR systems satisfying HLR2 conditions.

Next we present an example of calculation of a PB-induced rule from two rules. The calculation of a PB-induced transformation from two given transformations is analogous.

Consider two rules in the Figure 1(a),(b). The construction of a PB-induced rule starts with a calculation of a pullback as in the Figure 1(c). The final PB-induced rule according to the definition is obtained by composition of morphisms and is shown in the Figure 1(d).

A special case of PB-induced rules and transformations is valid for HLR systems satisfying HLR conditions only. In this case not only left-hand side nets are the same but also both interface nets as in the diagram below. The constructed rule $p = (R_1 \leftarrow I_1 = I_2 \to R_2)$ and transformation $N_1 \overset{p}{\Longrightarrow} N_2$ is a simple case of more general construction of PB-induced rules.

$$
\begin{array}{ccc}
R_1 \xleftarrow{\;\mathcal{M}\;} I_1 = I_2 \xrightarrow{\;\mathcal{M}\;} R_2 \\
\downarrow \qquad \downarrow \searrow \qquad \downarrow \\
\qquad\qquad L_1 \\
N_1 \longleftarrow C_1 = C_2 \;\rule{0pt}{0pt}\;\longrightarrow N_2 \\
\qquad\qquad \searrow\; \downarrow \\
\qquad\qquad N'
\end{array}
$$

Theorem 2 (Preservation of Properties via PB-Transformations). Consider two $\mathcal{Q}$-transformations $N_0 \overset{(p_1,q_1)}{\Longrightarrow} N_1$ and $N_0 \overset{(p_2,q_2)}{\Longrightarrow} N_2$ and the PB-induced transformation $N_1 \Longrightarrow N_2$ as in the commuting diagram below.

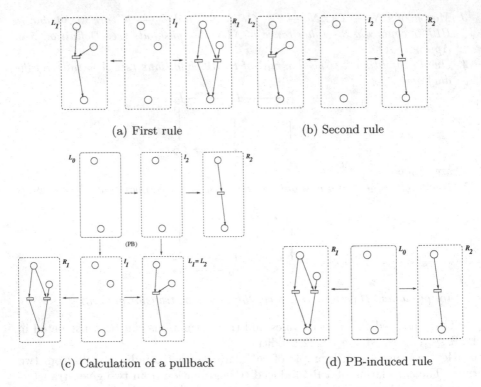

(a) First rule   (b) Second rule

(c) Calculation of a pullback   (d) PB-induced rule

Fig. 1. Example of a calculation of a PB-induced rule

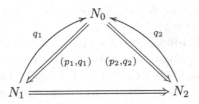

If both Q transformations $N_0 \overset{(r_1,q_1)}{\Longrightarrow} N_1$ and $N_0 \overset{(r_2,q_2)}{\Longrightarrow} N_2$ preserve and respect certain system property ϕ then the PB-induced transformation $N_1 \Longrightarrow N_2$
preserves and respects the system property ϕ, too.

Proof. Both Q transformations preserve and respect the system property ϕ, i.e.

$$N_1 \models \phi \qquad \Leftrightarrow \qquad N_0 \models \phi$$

$$N_2 \models \phi \qquad \Leftrightarrow \qquad N_0 \models \phi$$

As a logical conclusion of these two premises we get

$$N_1 \models \phi \qquad \Leftrightarrow \qquad N_2 \models \phi$$

This corollary is important when new property preserving rules and transformations are built up from existing ones.

The major advantage of PB-induced rules is that it is possible to build up new property preserving rules and transformations according to old ones without the explicit definition of a property preserving refinement morphism. This allows one to switch from one design concept to another without restarting design of the system from the origin and therefore to overcome the limitations of simple stepwise refinement.

In general, more complex redesign rules than PB-induced rules could be constructed. Construction of PB-induced rules may assure preservation of system properties. If other redesign rules are used, then preservation of system properties has to be proven by other techniques.

4   Example: Design and Redesign of Producers-Consumers System

We will illustrate our concept on an example of a *Producers-Consumers system*. Many variations of such systems can be found in producing lines or manufacturing processes. The Producers-Consumers system is a typical example of a system which is not bounded. The main analysis questions of the system concentrate on liveness. Analysis of liveness may reveal possible deadlocks in the behavior of the modeled system. A simple version of the *Producers-Consumers system* might look like the one in the Figure 2(a). There are two producing lines involved in the process of producing two parts needed for the final products. After producing these parts are assembled and delivered to the customer for being consumed. After the first abstract view of the system model designers usually focus on details of the system and try to refine it. During this refinement the techniques of hierarchical decomposition are employed. The focus is on methods preserving main properties of the modeled system.

Here we are concerned with the liveness preservation of the refinement and we focus on the special case of transition refinement, one of the basic refinement techniques. In [GPU01,Urb02] it has been shown that a special kind of transition refinement preserves and respects liveness of the net. We will not present formal definitions and theorems due to lack of space. They can be found in cited papers. The liveness preserving transition refinement of nets is based on collapsing morphisms. These rules replace one transition by a certain kind of a live net called *live in-out cycle*.

Figure 2(c) shows a transformation rule which refines the boldface transition on the left-hand side of the rule and replaces it by a more complex expression of the behavior of the system on the right-hand side. The boldface part of the right-hand side of the rule is a live in-out cycle.

The idea of a transformation is that a left-hand side subnet (in our case *Delivering*) is mapped to an existing Petri net. Then, the image of this subnet is deleted except for the interface (the middle part of the rule, called also gluing

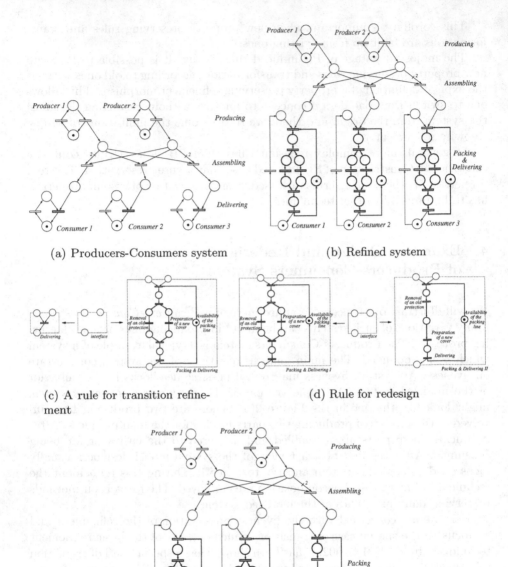

(a) Producers-Consumers system

(b) Refined system

(c) A rule for transition refinement

(d) Rule for redesign

(e) Final system

Fig. 2. Design and redesign of producers-consumers system

object) in the Petri net. The right-hand side (*Packing & Delivering*) of the rule is then glued to the interface which remains unchanged.

In our example the refining step corresponds to the refinement of the product delivering phase into two phases. The first one is a packing phase, the other phase is proper delivering of a product. Packing involves two parallel subprocesses. At one particular moment exactly one product can be in the packing line. This condition is modeled by the marked place – *Availability of the packing line.*

If we refine the boldface transitions in the Figure 2(a), we obtain the system in the Figure 2(b). There were three transitions in the system replaced according to the rule for transition refinement. This approach is often used in hierarchical decomposition modeling.

Remember that one rule can be applied several times (in our case three times). This fact is one of the main advantages of the rule-based approach. The transformation is described as fully local and the application therefore does not depend on the other parts of the net.

Liveness preserving transformations are limited to the refinement of transitions by live in-out cycles. Live in-out cycles are quite restricted live nets. This approach has been generalized in [Urb02]. In this paper so-called live boxes are admitted for transition refinement. In both cases it is possible to replace one transition by a subnet only. It is impossible to replace subnet by subnet. Using results from Section 3, replacement of one live in-out cycle (or live box) by another is possible. Even such a change guarantees preservation of liveness. For more details see [Urb02,Urb03]. Thus, the rule in Figure 2(d) is liveness preserving rule as boldface parts on left- and right-hand side are live in-out cycles. This rule allows changing parallel process of packing and delivering into sequential process. The redesign of the consumer 3 according to this rule yields the final net in Figure 2(e).

All nets in Figures 2(a),(b),(e) are live in the sense that every transition can become enabled from any reachable marking of the net. Liveness of the abstract place/transition system *Producers-Consumers system* in Figure 2(a) has to be proven by standard techniques. Liveness of the designed system in Figure 2(b) can be deduced easily by employing results from [GPU01,UP02]. The liveness of the net in Figure 2(e) can be induced from results in Section 3 and in [Urb02,Urb03].

The use of liveness preserving transformations guarantees the refined system to be live if the original system was.

5   Conclusion

The application of rule-based approach to stepwise development of systems has been presented in this paper. We introduced so-called PB-induced rules and transformations which are built as a combination of two other rules and transformations. We have shown that PB-induced transformations may be liveness preserving. We have demonstrated usability of this approach on the design and redesign of a simple live producers-consumers system.

The PB-induced transformation can replace one live in-out cycle by another while preserving liveness. This is a great advantage for system designers. They need not to refine the system stepwise only. They can choose certain part of a system and redesign it when necessary. The behavioral properties (liveness in this example) remain unchanged. This avoids the tedious total redesign of the system, understanding the old parts of the code, unnecessary costs of new development, etc. The presented concept can be applied to HLR systems in general, not only to Petri net based HLR systems as presented here.

Moreover, this approach can be applied to other system models having certain property, even if the models were not build up in a rule-based way.

References

[EHKP91a] H. Ehrig, A. Habel, H.-J. Kreowski, and F. Parisi-Presicce. From graph grammars to high level replacement systems. In *4th Int. Workshop on Graph Grammars and their Application to Computer Science, LNCS 532*, pages 269–291. Springer Verlag, 1991.

[EHKP91b] H. Ehrig, A. Habel, H.-J. Kreowski, and F. Parisi-Presicce. Parallelism and concurrency in High Level Replacement Systems. *Math. Struc. in Comp. Science*, 1:361–404, 1991.

[EHKPP90] H. Ehrig, A. Habel, H.-J. Kreowski, and F. Parisi-Presicce. Parallelism and concurrency in high level replacement systems. Technical Report 90-35, Technical University of Berlin, 1990.

[ERRW02] H. Ehrig, W. Reisig, G. Rozenberg, and H. Weber, editors. *Advances in Petri Nets: Petri Net Technology for Communication Based Systems.* LNCS 2472. Springer, 2003, to appear.

[GPU01] M. Gajewsky, J. Padberg, and M. Urbášek. Rule-Based Refinement for Place/Transition Systems: Preserving Liveness-Properties. Technical Report 2001-8, Technical University of Berlin, 2001.

[Pad96] J. Padberg. *Abstract Petri Nets: A Uniform Approach and Rule-Based Refinement*. PhD thesis, Technical University Berlin, 1996. Shaker Verlag.

[Pad99] J. Padberg. Categorical Approach to Horizontal Structuring and Refinement of High-Level Replacement Systems. *Applied Categorical Structures*, 7(4):371–403, December 1999.

[PER95] J. Padberg, H. Ehrig, and L. Ribeiro. Algebraic high-level net transformation systems. *MSCS*, 2:217–256, 1995.

[UP02] M. Urbášek and J. Padberg. Preserving liveness with rule-based refinement of place/transition systems. In Society for Design and Process Science (SDPS), editors, *Proc. IDPT 2002: Sixth World Conference on Integrated Design and Process Technology, CD-ROM*, page 10, 2002.

[Urb02] M. Urbášek. New Safety Property and Liveness Preserving Morphisms of P/T Systems. Technical Report 2002-14, Technical University Berlin, 2002.

[Urb03] M. Urbášek. Categorical Net Transformations for Petri Net Technology. PhD Thesis, Technical University Berlin, 2003, to appear.

A   HLR Conditions

In this appendix the formal definitions of HLR and HLR2 conditions are given.

Definition 5 (HLR Conditions). *Given a category* **NET** *(of nets) and a distinguished class $\mathcal{M}$ of morphisms in* **NET** *the following conditions 1 – 7 are called HLR-conditions:*

1. *Existence of $\mathcal{M}$-pushouts*
   For objects A,B,C and morphisms $C \leftarrow A \rightarrow B$, where at least one is in $\mathcal{M}$ there exists a pushout $C \rightarrow D \leftarrow B$.
2. *Existence of $\mathcal{M}$-pullbacks*
   For objects B,C,D and morphisms $B \rightarrow D$, $C \rightarrow D$ as in diagram (1) below, where both morphisms are in $\mathcal{M}$ there exists a pullback $C \leftarrow A \rightarrow B$.
3. *Inheritance of $\mathcal{M}$*
   (a) For each pushout diagram (1) as below the morphism $A \rightarrow B \in \mathcal{M}$ implies $C \rightarrow D \in \mathcal{M}$.
   (b) For each pullback diagram (1) as below the morphism $B \rightarrow D \in \mathcal{M}$ and $C \rightarrow D \in \mathcal{M}$ implies $A \rightarrow B \in \mathcal{M}$ and $A \rightarrow C \in \mathcal{M}$.
4. *Existence of binary coproducts and compatibility with $\mathcal{M}$*
   (a) For each pair of objects A, B there is a coproduct A+B with the universal morphisms $A \rightarrow A + B$ and $B \rightarrow A + B$.
   (b) For each pair of morphisms $A \xrightarrow{f} A'$ and $B \xrightarrow{g} B'$ in $\mathcal{M}$ the coproduct morphism $A + B \xrightarrow{f+g} A' + B'$ is also in $\mathcal{M}$.
5. *Existence of coequalizers*
   NET *has coequalizers.*
6. *$\mathcal{M}$-pushouts are pullbacks*
   Pushouts of $\mathcal{M}$-morphisms are pullbacks.
7. *$\mathcal{M}$-pushout-pullback-decomposition*
   For each diagram below, we have: If (1+2) is a pushout , (2) is a pullback and $A \rightarrow C$, $B \rightarrow D$, $E \rightarrow F$, $B \rightarrow E$ and $D \rightarrow F$ are $\mathcal{M}$-morphisms, then also (1) is a pushout.

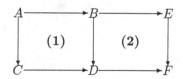

Definition 6 (HLR2 Conditions). *Given a category* **NET** *(of nets) and a distinguished class $\mathcal{M}$ of morphisms in* **NET** *the following conditions 1 – 3 are called HLR2 conditions:*

1. *HLR-Conditions (see Def. 5).*
2. *$\mathcal{M}$-Morphisms are closed under composition and isomorphisms*
   (a) Given morphisms $A \rightarrow B$ and $B \rightarrow C$ in $\mathcal{M}$ then $A \rightarrow B \rightarrow C$ is in $\mathcal{M}$.
   (b) Each **NET***-isomorphism is in $\mathcal{M}$.*

3. *Cube-pushout-pullback lemma*
   Given a commutative cube:

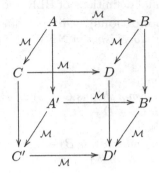

where all morphisms in the top- and bottom diagrams are in $\mathcal{M}$, the top diagram is a pullback, and front and right diagrams are pushouts then we have:

The bottom diagram is a pullback if and only if the back and left diagrams are pushouts.

Author Index

Lecture Notes in Computer Science

For information about Vols. 1–2820
please contact your bookseller or Springer-Verlag

Vol. 2854: J. Hoffmann, Utilizing Problem Structure in Planning. XIII, 251 pages. 2003. (Subseries LNAI)

Vol. 2855: R. Alur, I. Lee (Eds.), Embedded Software. Proceedings, 2003. X, 373 pages. 2003.

Vol. 2856: M. Smirnov, E. Biersack, C. Blondia, O. Bonaventure, O. Casals, G. Karlsson, George Pavlou, B. Quoitin, J. Roberts, I. Stavrakakis, B. Stiller, P. Trimintzios, P. Van Mieghem (Eds.), Quality of Future Internet Services. IX, 293 pages. 2003.

Vol. 2857: M.A. Nascimento, E.S. de Moura, A.L. Oliveira (Eds.), String Processing and Information Retrieval. Proceedings, 2003. XI, 379 pages. 2003.

Vol. 2858: A. Veidenbaum, K. Joe, H. Amano, H. Aiso (Eds.), High Performance Computing. Proceedings, 2003. XV, 566 pages. 2003.

Vol. 2859: B. Apolloni, M. Marinaro, R. Tagliaferri (Eds.), Neural Nets. Proceedings, 2003. X, 376 pages. 2003.

Vol. 2860: D. Geist, E. Tronci (Eds.), Correct Hardware Design and Verification Methods. Proceedings, 2003. XII, 426 pages. 2003.

Vol. 2861: C. Bliek, C. Jermann, A. Neumaier (Eds.), Global Optimization and Constraint Satisfaction. Proceedings, 2002. XII, 239 pages. 2003.

Vol. 2862: D. Feitelson, L. Rudolph, U. Schwiegelshohn (Eds.), Job Scheduling Strategies for Parallel Processing. Proceedings, 2003. VII, 269 pages. 2003.

Vol. 2863: P. Stevens, J. Whittle, G. Booch (Eds.), «UML» 2003 – The Unified Modeling Language. Proceedings, 2003. XIV, 415 pages. 2003.

Vol. 2864: A.K. Dey, A. Schmidt, J.F. McCarthy (Eds.), UbiComp 2003: Ubiquitous Computing. Proceedings, 2003. XVII, 368 pages. 2003.

Vol. 2865: S. Pierre, M. Barbeau, E. Kranakis (Eds.), Ad-Hoc, Mobile, and Wireless Networks. Proceedings, 2003. X, 293 pages. 2003.

Vol. 2867: M. Brunner, A. Keller (Eds.), Self-Managing Distributed Systems. Proceedings, 2003. XIII, 274 pages. 2003.

Vol. 2868: P. Perner, R. Brause, H.-G. Holzhütter (Eds.), Medical Data Analysis. Proceedings, 2003. VIII, 127 pages. 2003.

Vol. 2869: A. Yazici, C. Şener (Eds.), Computer and Information Sciences – ISCIS 2003. Proceedings, 2003. XIX, 1110 pages. 2003.

Vol. 2870: D. Fensel, K. Sycara, J. Mylopoulos (Eds.), The Semantic Web - ISWC 2003. Proceedings, 2003. XV, 931 pages. 2003.

Vol. 2871: N. Zhong, Z.W. Raś, S. Tsumoto, E. Suzuki (Eds.), Foundations of Intelligent Systems. Proceedings, 2003. XV, 697 pages. 2003. (Subseries LNAI)

Vol. 2873: J. Lawry, J. Shanahan, A. Ralescu (Eds.), Modelling with Words. XIII, 229 pages. 2003. (Subseries LNAI)

Vol. 2874: C. Priami (Ed.), Global Computing. Proceedings, 2003. XIX, 255 pages. 2003.

Vol. 2875: E. Aarts, R. Collier, E. van Loenen, B. de Ruyter (Eds.), Ambient Intelligence. Proceedings, 2003. XI, 432 pages. 2003.

Vol. 2876: M. Schroeder, G. Wagner (Eds.), Rules and Rule Markup Languages for the Semantic Web. Proceedings, 2003. VII, 173 pages. 2003.

Vol. 2877: T. Böhme, G. Heyer, H. Unger (Eds.), Innovative Internet Community Systems. Proceedings, 2003. VIII, 263 pages. 2003.

Vol. 2878: R.E. Ellis, T.M. Peters (Eds.), Medical Image Computing and Computer-Assisted Intervention - MICCAI 2003. Part I. Proceedings, 2003. XXXIII, 819 pages. 2003.

Vol. 2879: R.E. Ellis, T.M. Peters (Eds.), Medical Image Computing and Computer-Assisted Intervention - MICCAI 2003. Part II. Proceedings, 2003. XXXIV, 1003 pages. 2003.

Vol. 2880: H.L. Bodlaender (Ed.), Graph-Theoretic Concepts in Computer Science. Proceedings, 2003. XI, 386 pages. 2003.

Vol. 2881: E. Horlait, T. Magedanz, R.H. Glitho (Eds.), Mobile Agents for Telecommunication Applications. Proceedings, 2003. IX, 297 pages. 2003.

Vol. 2883: J. Schaeffer, M. Müller, Y. Björnsson (Eds.), Computers and Games. Proceedings, 2002. XI, 431 pages. 2003.

Vol. 2884: E. Najm, U. Nestmann, P. Stevens (Eds.), Formal Methods for Open Object-Based Distributed Systems. Proceedings, 2003. X, 293 pages. 2003.

Vol. 2885: J.S. Dong, J. Woodcock (Eds.), Formal Methods and Software Engineering. Proceedings, 2003. XI, 683 pages. 2003.

Vol. 2886: I. Nyström, G. Sanniti di Baja, S. Svensson (Eds.), Discrete Geometry for Computer Imagery. Proceedings, 2003. XII, 556 pages. 2003.

Vol. 2887: T. Johansson (Ed.), Fast Software Encryption. Proceedings, 2003. IX, 397 pages. 2003.

Vol. 2888: R. Meersman, Zahir Tari, D.C. Schmidt et al. (Eds.), On The Move to Meaningful Internet Systems 2003: CoopIS, DOA, and ODBASE. Proceedings, 2003. XXI, 1546 pages. 2003.

Vol. 2889: Robert Meersman, Zahir Tari et al. (Eds.), On The Move to Meaningful Internet Systems 2003: OTM 2003 Workshops. Proceedings, 2003. XXI, 1096 pages. 2003.

Vol. 2891: J. Lee, M. Barley (Eds.), Intelligent Agents and Multi-Agent Systems. Proceedings, 2003. X, 215 pages. 2003. (Subseries LNAI)

Vol. 2893: J.-B. Stefani, I. Demeure, D. Hagimont (Eds.), Distributed Applications and Interoperable Systems. Proceedings, 2003. XIII, 311 pages. 2003.

Vol. 2895: A. Ohori (Ed.), Programming Languages and Systems. Proceedings, 2003. XIII, 427 pages. 2003.

Vol. 2897: O. Balet, G. Subsol, P. Torguet (Eds.), Virtual Storytelling. Proceedings, 2003. XI, 240 pages. 2003.

Vol. 2899: G. Ventre, R. Canonico (Eds.), Interactive Multimedia on Next Generation Networks. Proceedings, 2003. XIV, 420 pages. 2003.

Vol. 2901: F. Bry, N. Henze, J. Małuszyński (Eds.), Principles and Practice of Semantic Web Reasoning. Proceedings, 2003. X, 209 pages. 2003.

Vol. 2902: F. Moura Pires, S. Abreu (Eds.), Progress in Artificial Intelligence. Proceedings, 2003. XV, 504 pages. 2003. (Subseries LNAI).

Vol. 2905: A. Sanfeliu, J. Ruiz-Shulcloper (Eds.), Progress in Pattern Recognition, Speech and Image Analysis. Proceedings, 2003. XVII, 693 pages. 2003.